Heat and Mass Transfer - Advances in Science and Technology Applications

Edited by Alfredo Iranzo

Published in London, United Kingdom

IntechOpen

Supporting open minds since 2005

Heat and Mass Transfer - Advances in Science and Technology Applications
http://dx.doi.org/10.5772/intechopen.77466
Edited by Alfredo Iranzo

Part of IntechOpen Book Series: Heat and Mass Transfer, Volume 2
Book Series Editor: Yong Ren

Contributors
Amina.Lyria Deghal Cheridi, Ahcene Loubar, Abla Chaker, Huaizhi Han, Yaning Zhang, Bingxi Li, Ruitian Yu, Quan Zhu, Zhengying Wei, Jun Du, Vitor Rosa, Deovaldo Moraes Júnior, Cemil Koyunoğlu, Ildebrando Pérez-Reyes, Cornelio Alvarez-Herrera, Jonathan Rodriguez-Campos, Jorge Flores-Velazquez, Dmitry G Luchinsky, Ekaterina Ponizovskaya-Devine, Vasil Hafiychuk, Michae Khasin, Michael Foygel, Barbara Brown, Jared Sass, Antonio Montes, Enrique Martinez De La Ossa, Clara Pereyra, Muhammad Wakil Shahzad, Kim Choon Ng, Muhammad Burhan, Wei Wu, Jibin Hu, Shihua Yuan, Panagiotis Sismanis, Luis Antonio Dávalos-Orozco, Nencho Stanev Deliiski, Natalia Yordanova Tumbarkova, Armando Gallegos-Muñoz, Martín Picón-Núñez, Fabian Luna-Cabrera, Francisco Elizalde-Blancas, Juan Manuel Belman-Flores, Margarita Zaeva, Anatoly Tsirlin, Fadi Alnaimat, Yasir Rashid, Martin Ortiz, Alfredo Iranzo

First published in London, United Kingdom, 2019 by IntechOpen
IntechOpen is the global imprint of INTECHOPEN LIMITED, registered in England and Wales, registration number: 11086078, The Shard, 25th floor, 32 London Bridge Street
London, SE19SG - United Kingdom
Printed in Croatia

British Library Cataloguing-in-Publication Data
A catalogue record for this book is available from the British Library

Additional hard and PDF copies can be obtained from orders@intechopen.com

Heat and Mass Transfer - Advances in Science and Technology Applications
Edited by Alfredo Iranzo
p. cm.
Print ISBN 978-1-78984-446-7
Online ISBN 978-1-78984-447-4
eBook (PDF) ISBN 978-1-78985-416-9
ISSN 2631-6196

For EU product safety concerns:
IN TECH d.o.o., Prolaz Marije Krucifikse Kozulić 3, 51000 Rijeka, Croatia, info@intechopen.com or visit our website at intechopen.com.

IntechOpen Book Series

Heat and Mass Transfer

Volume 2

Alfredo Iranzo obtained a Chemical Engineering degree from the University of Zaragoza (2000), where he collaborated with LITEC/CSIC (Laboratory of Research in Fluid Dynamics and Combustion Technologies). He then joined ANSYS Germany as a computational fluid dynamics (CFD) engineer in Otterfing, Germany. He obtained an MSc in Thermal Energy Systems in 2010 and a PhD from the Universidad de Sevilla (2011). He currently works as a researcher at the Energy Engineering Department at the University of Sevilla. His research activity is focused on proton-exchange membrane fuel cells and CFD applications. He has published more than 30 scientific publications and two book chapters. Alfredo Iranzo is a member of the editorial board of *Engineering Applications of Computational Fluid Mechanics* and subject editor of the *International Journal of Hydrogen Energy*.

Editor of Volume 2: Dr. Alfredo Iranzo
Energy Engineering Department, School of Engineering, University of Sevilla, Sevilla, Spain

Book Series Editor: Yong Ren
University of Nottingham Ningbo, China

Scope of the Series

Heat and mass transfer are ubiquitous transport phenomena in manifold fields from the natural environment and living organisms to engineering processes.
This book series will cover the latest advances in applying fundamental heat and mass transfer theory and newly developed technologies and materials for solving practical industrial engineering, biomedical and health care problems, while tackling challenges arising from limitations of existing theory and methods that may be associated with simplified assumptions, inaccuracy of fabrication process and/or measurement instruments. This book series aims to promote developments in basic research on heat and mass transfer phenomena and serve as a platform for the exchanging of new ideas to boost further development.

Topics may include:

- New analytical modelling and numerical methods of engineering thermodynamics
- New technology or methods for material thermal property measurement and correlation
- Heat and mass transfer in flow flow and chemical reactions
- Environmental applications of heat and mass transfer
- Biomedical applications of heat and mass transfer
- Heat and mass transfer in development of functional material

Contents

Preface

This book aims at providing researchers and practitioners with a valuable compendium of the latest significant advances in the field of heat and mass transfer. The book covers a set of topics of particular interest ranging from heat transfer in power generation to heat and mass transfer in new and emerging technologies. The book comprises 19 chapters arranged in eight sections, which include: Introduction, Heat Transfer in Industrial Processes, Heat Transfer in Power Generation, Heat and Mass Transfer in New and Emerging Technologies, Multi-phase Flows and Phase Change, Thermodynamics and Thermophysical Properties, Mass Transfer Processes and Theoretical Approaches for Heat Transfer Problems. Several chapters within the book include theoretical and numerical modeling and simulation developments, which are increasingly contributing to the current state of the art in heat and mass transfer.

This book has been possible thanks to the outstanding contributions of many scientists, researchers, and engineers in the field of heat and mass transfer, and also the exceptional work of the professionals at IntechOpen. All contributions are sincerely acknowledged and truly appreciated..

I sincerely believe that the content of the book will be of interest to researchers, practitioners, and graduate students within the many disciplines and applications involved

Dr. Alfredo Iranzo
Energy Engineering Department,
School of Engineering,
University of Sevilla,
Sevilla, Spain

Section 1

Introduction

Chapter 1

Introductory Chapter: Heat and Mass Transfer - Advances in Science and Technology Applications

Alfredo Iranzo

1. Introduction

Heat and mass transfer is the core science for many industrial processes as well as technical and scientific devices. Automotive, aerospace, power generation (both by conventional fossil-based technologies and also by renewable energies), industrial equipment and rotating machinery, materials and chemical processing, and many other industries are involving heat and mass transfer processes.Consequently, both heat transfer and mass transfer are disciplines of the highest significance in technology and industry and obviously also in research and academia. As an example, over 10,000 journal publications are published every year with the latest scientific developments.

This book, titled *Heat and Mass Transfer: Advances in Science and Technology Applications*, intends to provide researchers and practitioners with a valuable compendium of significant advances in the field of heat and mass transfer, focusing both on the scientific and research fields and also on applications within industrial scenarios. There is a clear rationality in presenting both heat and mass transfer within a single book, as it must be considered that both processes are governed by very similar mathematical equations, particularly in the case of diffusion and convection (however, heat radiation is not having a similar process in mass transfer).

The book includes 19 chapters, arranged in a set of 8 sections according to the logic and significance of its contents. The book comprises sections focused on industrial applications, such as heat transfer in industrial processes or heat transfer in power generation. An extensive collection of chapters dealing with heat and mass transfer process relevant to industry is included, such as steam boilers, heat exchanger optimization, cryogenics, concentrated solar power (CSP), additive manufacturing processes for metals, supercritical CO_2 precipitation, falling film evaporators, or non-Newtonian fluids in mixing vessels. Other sections such as thermodynamics or theoretical approaches for heat transfer problems are involving chapters more focused on scientific aspects, with topics covering bio-convective linear stability, diffusion of Fe_2B layers on steels, Rayleigh convection in viscoelastic fluids, irreversible thermodynamics, or calculation of thermal properties, among others. The book also contains sections covering heat and mass transfer in new and emerging technologies, multiphase flows and phase change, and mass transfer processes.

Experimental setups and results are covered within the book, but as modeling and simulation techniques are increasingly contributing to the development of

IntechOpen

heat and mass transfer research and applications, several chapters are involving theoretical and numerical modeling and simulation developments and examples, also including computational fluid dynamic (CFD) applications. The application of optimization methods such as response surface methodology (RSM) and genetic algorithm (GA) is also covered within the book. At the end of each chapter, a rich number of references are included for ensuring the continuity and further reading on each particular topic.

Some of the classic and well-known references covering both introductory and advanced topics are the books from Incropera et al. [1] or Rohsenow et al. [2]. A recognized handbook also covering many aspects in mass transfer is written by Cheremisinof and Adams [3] as well as the book by Bird, Stewart, and Lightfoot on transport phenomena [4]. For heat and mass transfer topics in chemical engineering, the handbook from Perry and Green [5] is also recommended.

I sincerely believe that the content of the book will be of interest to researchers, practitioners, and graduate students within the many disciplines and applications involved.

Author details

Alfredo Iranzo
University of Sevilla, Sevilla, Spain

*Address all correspondence to: airanzo@us.es

References

[1] Incropera F, DeWitt D, Bergman T, Lavine A. Fundamentals of Heat and Mass Transfer. 6th ed. New York: John Wiley & Sons; 2006. p. 997. ISBN: 978-0-471-45728-2

[2] Rohsenow W, Hartnett J, Cho Y. Handbook of Heat Transfer. 3rd ed. New York: McGraw-Hill; 1998. ISBN: 9780070535558

[3] Cheremisinof N, Adams R. Handbook of Heat and Mass Transfer. 2nd ed. Houston: Gulf; 1986. ISBN: 9780872014114

[4] Bird B, Stewart W, Lightfoot E, Edwin N. Transport Phenomena. 2nd ed. New York: John Wiley; 2002. p. 895. ISBN: 0-471-41077-2

[5] Perry R, Green D. Perry's Chemical Engineers' Handbook. 8th ed. New York: McGraw-Hill; 2008. ISBN: 9780071542098

Section 2

Heat Transfer in Industrial Processes

Chapter 2

Design of Industrial Falling Film Evaporators

Muhammad Wakil Shahzad, Muhammad Burhan and Kim Choon Ng

Abstract

The high performance evaporators are important for process industries such as food, desalination and refineries. The falling film evaporators have many advantages over flooded and vertical tubes that make them best candidate for processes industries application. The heat transfer area is the key parameter in designing of an evaporator and many correlations are available to estimate the size of tube bundle. Unfortunately, most of the correlation is available only for pure water and above 322 K saturation temperatures. Out of these conditions, the areas are designed by the extrapolation of existing correlations. We demonstrated that the actual heat transfer values are 2–3-fold higher at lower temperature and hence simple extrapolated estimation leads to inefficient and high capital cost design. We proposed an accurate heat transfer correlation for falling film evaporators that can capture both, low temperature evaporation and salt concentration effectively. It is also embedded with unique bubble-assisted evaporation parameter that can be only observed at low temperature and it enhances the heat transfer. The proposed correlation is applicable from 280 to 305 K saturation temperatures and feed water concentration ranges from 35,000 to 95,000 ppm. The uncertainty of measured data is less than 5% and RMS of regressed data is 3.5%. In this chapter, first part summarized the all available correlations and their limitations. In second part, falling film evaporation heat transfer coefficient (FFHTC) is proposed and model is developed. In the last part, experimentation is conducted and FFHTC developed and compared with conventional correlations.

Keywords: heat transfer, falling film, evaporator design, seawater evaporator

1. Introduction

The falling film evaporators currently leading in the processes industries because they have many advantages over submerged tubes evaporators. In the past, the vertical tubes evaporators were considered as most efficient but now they have been replaced by falling film evaporators due to its distinctive nature of operation. The submerged and vertical tubes evaporators are normally not fast responsive to many operational parameters. On the other hand, the falling film evaporators respond fast to feed quality and heat source supplied. These properties make them very efficient to operate across small temperature differences so that they can be arranged in cascading manners for maximum efficiency. In addition, falling film evaporators have many other advantages such as:

1.1 Advantages of falling film evaporators

The falling film evaporators have following advantages over flooded evaporators:

1. Compact design due to improved heat transfer.
2. Improved wettability provides uniform heat transfer properties across the tubes.
3. Less charge requirement, two to three times the evaporation rate.
4. Fast operation and short contact time for working fluid, favourable for food industry.
5. Falling film washed away the deposition on tubes that minimize the chances of fouling on tube surfaces.

The falling film evaporators also have many advantages over vertical tubes evaporators such as:

1. Smaller size as compared to vertical tubes for same capacity due to high heat transfer coefficient.
2. Falling film evaporator tubes are available with different corrugations to enhance the heat transfer rates as compared to vertical tube evaporators.
3. Multi pass tube bundle design for required operation as compared to single pass vertical tubes evaporators with limited operations.
4. Larger length to diameter ratio evaporator design is possible as compared to vertical tubes that help to enhance wettability and minimize the chances of dry-outs and flooding.
5. The compact design reduces the overall piping work as compared to single pass large size vertical design.
6. Due to vertical stacking arrangement, the overall footprint can be small in large industries by using falling film evaporators.

It can be seen clearly that falling film evaporators have many advantages over submerged and vertical tubes evaporators but there is lack of heat transfer data at sub-atmospheric temperature. Particularly, below 323 K, there is no data available in the literature. This was one of the main motivations for this work.

2. Heat transfer review for falling film evaporators

An efficient design of falling film evaporators is important especially for food and desalination industries. Conventionally, the empirical and theoretical correlations available in the literature are employed for the heat transfer area estimation. Most of correlations are based on different refrigerants and at near atmospheric temperature. Only few correlations are available for pure water for 322 K and above saturation temperatures. **Table 1** provides the detail of many researchers' work related to heat transfer correlations. It also highlighted the studies of different operational parameters impact on heat transfer.

Reference	Investigators	Detail
Basic correlation development		
[1]	Uche et al.	Investigation of heat transfer coefficient for different heat source temperature and different flow velocities for vertical and horizontal tubes evaporators
[2]	Ribatski and Jacobi	Heat transfer coefficient values development for water and other refrigerants for horizontal evaporators fitted with single and multi tubes
[3]	Adib et al.	Experimentation on vertical tubes evaporators for heat transfer coefficient investigation. They found good agreement with published data [4–8]
[9]	Parken and Fletcher	Correlation development for non-boiling conditions
[10]	Han and Fletcher	Heat transfer coefficient correlation development for falling film evaporators for boiling conditions above 322 K saturation temperatures
[11–13]	Fujita et. al	Analytical model development for falling film evaporators with R-11 refrigerant. The measured accuracy was ± 20%
Operational parameters impact investigation		
[11, 14–17]	Liu et al., Fujita et al., Yang et al., Parken et al., Ribatski et al.	Film Reynold number investigation on heat transfer coefficient and main conclusions are: a. Heat transfer increase with Reynold number b. Heat transfer decrease to its minimum value and then increase c. Heat transfer increase to maximum and then drop
[18]	Lorenz and Yung	Investigation of single tube and multi tubes arrangement on heat transfer. They also found that the critical Reynold number is below 300 and single tube have good heat transfer as compared to array of tubes
[19]	Thome et al.	Heat transfer study on different tube geometry such as Gewa-B, plain surface, turbo-BII HP and high heat flux tubes. They found a stark difference in heat transfer values
[11]	Fujita et al.	Tubes array and feed header impact on heat transfer was studied and showed that top row has low heat transfer due to direct exposure of feed
[14]	Liu et al.	Tubes geometry impact was investigated and concluded that roll worked tubes heat transfer is 3–4 fold higher them smooth tubes
[20]	Aly et al.	Film thickness impact was studied and found that thickness has negative impact on heat transfer
[21–23]	Moeykens et al. and Chang et al	Impact of different refrigerants such as R-141b, R22, R123 and R-134a were studied and found enhanced heat transfer by additional distribution plats
[24]	Bourouni et al.	The characteristic dimensions effect was investigated and found that heat transfer enhanced significantly with increase in evaporator size

Table 1.
Literature summary on heat transfer coefficient investigation and operational parameters impact.

References	Correlation
Xu et al. [25]	$h_{evaporation} = 05.169 \times 10^{-11} \cdot \left[\frac{h_{fg} \cdot g \cdot \rho_l^2 \cdot D^2}{\Delta t^2 \cdot \mu_l} \right]^{-0.333} \left(\frac{\bar{\delta}}{D} \right)^{-0.422 \Delta t^{0.503}} \cdot \left(1 + \frac{\delta_{max} - \delta_{min}}{\bar{\delta}} \right)^{5.708}$ Heat source 50°C, deionized liquid, evaporator with copper tubes horizontally arranged
Fujita et al. [11]	First tube: $Nu = \left((Re_f)^{-23} + 0.008 (Re_f)^{0.3} (Pr)^{0.25} \right)^{1/2}$ Second to last tubes: $Nu = \left((Re_f)^{-23} + 0.01 (Re_f)^{0.3} (Pr)^{0.25} \right)^{1/2}$ Refrigerant Freon R-11, copper tubes with electrical heaters, diameter 25 mm
Han and Fletcher [10]	$h_{evaporation} = 0.0028 \cdot \left[\frac{\mu_l^2}{g \cdot \rho_l^2 \cdot k_l^3} \right]^{-0.333} (Re_\Gamma)^{0.5} (Pr)^{0.85}$ Pure water, 49–127°C, electrically heated single horizontal tube, OD—50.8 mm, thickness—1.7 mm, length—254 mm
Bourouni et al. [24]	$h_f = 2.2 \cdot \left[\frac{\nu_f^2}{g \cdot k_l^3} \right]^{-0.333} \cdot \left(\frac{H}{OD} \right)^{0.1} \cdot (Re_f)^{-0.333}$ Pure water, 60 and 90°C, polypropylene horizontal tubes aero-evaporator, OD—25.4 mm

DOI: http://dx.doi.org/10.5772/intechopen.84230

References	Correlation
Chun and Seban [26]	$h_{\text{film}} = 0.821.\left[\frac{\mu_l^2}{g.\rho_l^2.k_l^3}\right]^{-0.333}(\text{Re}_\Gamma)^{-0.22}$ Pure water, 46–118°C, vertical single tube evaporator with electrical heater, tube 28 mm diameter and 292 mm long
Alhusseini et al. [27]	Laminar regime: $h_{\text{laminar}} = 2.65.(\text{Re})^{-0.158}(\text{Ka})^{0.0563}$ Mixed regime: $h = \left(h^{5_{\text{laminar}}} + h^{5_{\text{turbulent}}}\right)^{1/5}$ Pure water and Propylene glycol
Shmerler et al. [28]	$h_E = 0.0038.(\text{Re}_\Gamma)^{0.35}(\text{Pr})^{0.95}$ Vertical tube evaporator with electrical heat, water as a working fluid, tubes 25.4 mm diameter and 781 mm long
Chien et al.[29]	$\text{Nu}_{\text{cv}} = 0.0386.(\text{Re}_\Gamma)^{0.09}.(\text{Re}_\Gamma)^{0.986}$ Horizontal tubes evaporator with R245fa refrigerants, operational temperature 5 and 20°C

Table 2.
Review of heat transfer coefficient correlations for different evaporator design and operation conditions.

The heat transfer correlations available in the literature are based on different parameters and they have some limitations. **Table 2** showed most famous and widely accepted heat transfer coefficient correlations and their limitations.

The most commonly used correlation is proposed by Han and Fletcher for horizontal tube falling film evaporators. They developed this correlation for pure water at saturation temperature ranges from 322 to 393 K. It can be noticed that there are two major gaps in available literature; firstly, no data is available for evaporation heat transfer for below 322 K and secondly, there is lack of data for salt solution as boiling point elevation changes with salt concentration. These two factors are important for processes industries falling film evaporators design as most of processes are performed below 322 K such as in food and desalination industries [30–42]. This was the main motivation of this study, to provide detailed parameters for falling film evaporators design for process industries. We developed falling film heat transfer coefficient (FFHTC) correlation for saline water evaporation from 280 to 305 K saturation temperatures. We also demonstrated the effect of salt concentration on heat transfer and LMTD. This will help to design efficient falling film evaporators for processes industries.

3. Falling film heat transfer coefficient development

The idea was to modify the famous and well accepted Han and Fletcher's correlation to incorporate the different salt concentration effect and expanded to low range temperature evaporation. This will help to fill two major gaps as mentioned earlier in processes industries evaporators design.

3.1 Theoretical model

The dimensionless terms such as Nusselt, Reynolds and Prandtl numbers in the Han and Fletcher's correlation are adequate to incorporate the liquid film thermal effect in heat transfer. As per steam properties table, the specific volume of steam is rapidly changes at low temperature and it might have significant effect on heat transfer. At low temperature, the generation of microbubbles at tubes surfaces rapidly detach due to low density and it agitate the thermal barrier formed by liquid film. The conventional heat transfer correlations are not able to capture this effect. The heat transfer enhancement due to micro-bubble generation and detaching is an important phenomenon at low temperature and need to be captured in heat transfer correlation for efficient evaporator design.

The basic form of Han and Fletcher's correlation is shown in Eq. (1).

$$\frac{h_{\text{evap}}\left(\frac{\mu_l^2}{g\rho_l^2}\right)^{1/3}}{k_l} = \text{Nu} = 0.0028(\text{Re}_\Gamma)^{0.5}(\text{Pr})^{0.85} \tag{1}$$

The constants and indices can be found from the boundary conditions of falling film evaporators. The heat supplied to the evaporator can be calculated by the energy balance of hot water circulation through the tubes as presented in Eq. (2).

$$Q_{\text{in}} = m_{\text{ch},w}\text{Cp}_{\text{ch},w}\left(T^o_{\text{ch},w} - T^i_{\text{ch},w}\right) \tag{2}$$

The overall heat transfer coefficient (U_{overall}) can be calculated by using the saturation temperatures of evaporator and log mean temperature difference (LMTD) parameters as shown in Eq. (3).

$$UA_{overall} = \frac{m_{ch,w} Cp_{ch,w} \left(T_{ch,w}^{out} - T_{ch,w}^{in} \right)}{\left\{ \frac{\left(T_{ch,w}^{out} - T_{sat} \right) - \left(T_{ch,w}^{in} - T_{sat} \right)}{\ln \frac{\left(T_{ch,w}^{out} - T_{sat} \right)}{\left(T_{ch,w}^{in} - T_{sat} \right)}} \right\}} \tag{3}$$

The Dittus-Boelter correlation can be applied to investigate the local heat transfer coefficient for falling film evaporators as shown in Eq. (4).

$$Nu = 0.023 Re^{0.25} Pr^{n} \tag{4}$$

Now, falling film heat transfer coefficient for evaporation can be calculated by applying Eqs. (1)–(4). The material resistance is neglected due to very thin tube wall (less than 0.7 mm). Eq. (5) presents the calculation process for falling film heat transfer coefficient.

$$\frac{1}{UA} = \left(\frac{1}{hA} \right)_{tubeside} + R_{wall} + \left(\frac{1}{hA} \right)_{outside} \tag{5}$$

The unknown parameters in Eq. (5) are calculated by the planned experiments as discussed in the following sections.

3.2 Experimental apparatus

The pilot facility of adsorption desalination (AD) cycle in Mechanical Engineering (ME) Department of NUS is utilized to investigate the unknown parameters for FFHTC correlation development. The AD pilot facility is shown in **Figure 1**.

The AD cycle has four major components such as (a) reactor beds packed with adsorbent, (b) evaporator, (c) condenser and (d) circulation pumps. In addition, there is also a conditioning facility and pre-treatment facility to perform test at an accurate conditions. The flow schematic of AD cycle is shown in **Figure 2**.

To investigate the falling film heat transfer coefficient, evaporator is designed with horizontal tubes arranged in staggered manner. There are four rows of tubes and each row has 12 tubes installed in four pass arrangements. The tubes are

Figure 1.
Adsorption cycle pilot installed at NUS, Singapore (published with PI permission [43, 44]).

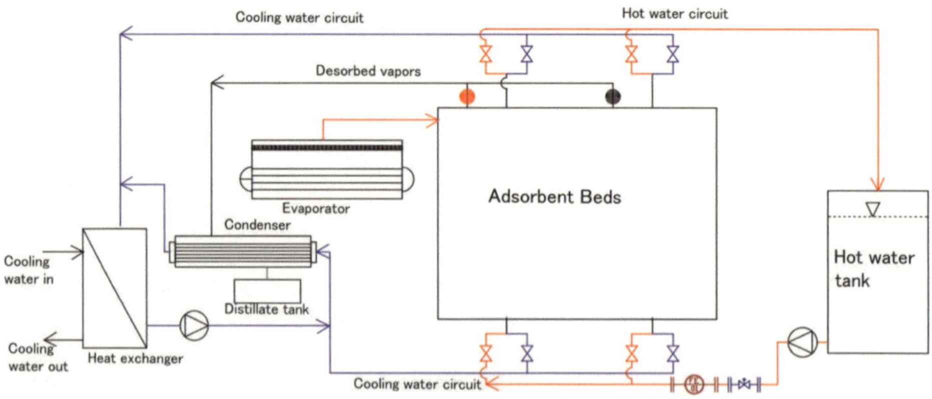

Figure 2.
Adsorption cycle flow schematic with detailed components (published with author's permission [43, 44]).

Parameters	Values	Units
Number of tubes	52	
Length of each tube	2000	mm
Tube outer diameter	25	mm
Tube thickness	1.0	mm
No of passes	2	
Shell diameter	600	mm
Shell length	1800	mm

Table 3.
Adsorption cycle evaporator design parameters.

fabricated with special outside and inside profile to enhance heat transfer. The design parameters of evaporator are given in **Table 3**.

3.2.1 Experimental procedure

There are three liquid circuits in the system those are important to control and maintain for a successful experiment. Firstly, the chilled water circulation through the tubes of evaporator to maintain required saturation temperature. An accurate thyristor controlled heater is installed to control chilled water temperature within $\pm$ 0.15 K. A vacuum rated feed pump help to spray water from pool of evaporator below tubes bundle to the tubes surface. To maintain the liquid level in the evaporator, the evaporated quantity refluxed back from condenser as a close loop.

Secondly, the cold water supply to the adsorption bed to remove the heat of adsorption. The adsorber bed directly communicates to evaporator to adsorb the vapors and release the heat of adsorption. This heat must be removed to maintain the vapor uptake otherwise it can be drooped to very low quantity. The cooled water flow through the cooling tower on the rooftop to reject heat to the ambient.

Lastly, the heat source to the desorber bed to regenerate the adsorbent. Once the adsorber bed fully saturated, it cannot take more vapor and it has to be regenerated for next adsorption process. The hot water is circulated through the tubes of the bed to supply heat of desorption to the adsorbent. The hot water temperature is maintained either by heater or solar thermal collectors.

Since whole system is operating at sub-atmospheric pressure so it is required to remove the non-condensable gases. A vacuum pump is connected to all the major components to remove non-condensable in case on any leakage. **Table 4** shows the operation parameters of AD cycle.

The system is instrumented with highly accurate sensors to extract real time data. For example, for pressure measurements, Yokogawa pressure transducers are installed. These sensors can measure 0–60 kPa (abs) with accuracy of ±0.25%. Similarly, liquid flow is measured by KROHNE flow meters (accuracy ± 0.5%) and temperatures are recorded by OMEGA 10 kΩ thermistors (accuracy ± 0.15 K). All sensors are connected to Agilent system for data logging.

Parameters	Values	Units
Chilled water flow rate	50	LPM
Sea water flow rate (Γ)	1.8	LPM/m of tube length
Feed water salinity range	35,000–95,000	ppm

Table 4.
Experimental operational parameters of adsorption pilot.

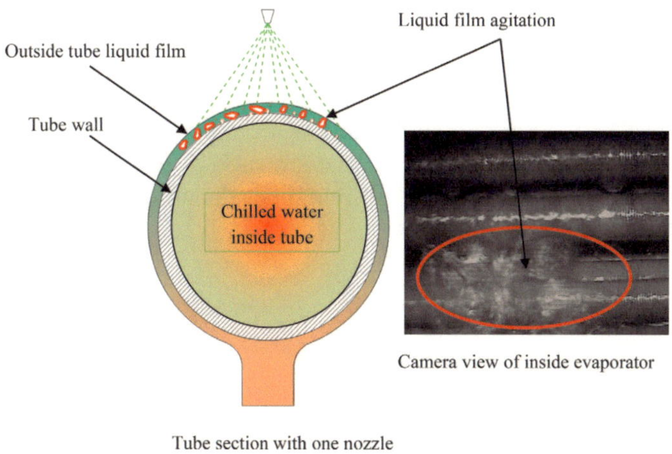

Figure 3.
Micro-bubbles agitation of liquid film on evaporator tube surfaces captured by camera (published with author's permission [43, 44]).

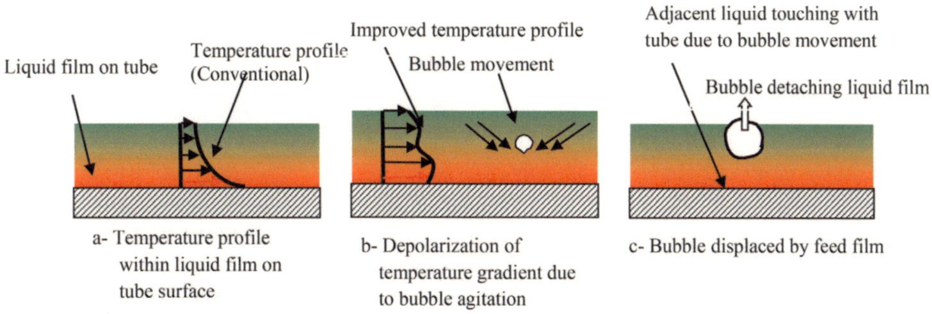

Figure 4.
Conventional thermal barrier braking phenomenon due to bubble agitation (published with author's permission [43, 44]).

To capture the event of micro-bubble formation at low pressure, a high speed camera was installed on evaporator. The camera successfully captured the agitation of liquid film on tube surface due to formation and detaching of micro-bubbles as shown in **Figure 3**. The phenomenon of breaking the liquid thermal barrier due to film agitation is presented in **Figure 4** step by step. The natural temperature gradient within liquid film on tubes surface is the major bottle neck in heat transfer. The micro-bubble generation at low temperature agitates this barrier due to low density and produce turbulence as also captured by camera. The micro-bubble, firstly agitate the liquid film and break thermal barrier that enhance heat transfer. Secondly, when it moves up due to low density, it draw heat and provide space to adjacent liquid to have direct contact with tube surface that helps faster heat transfer rates.

4. Results and discussion

The overall heat transfer coefficient (U) was calculated at assorted heat source and salt concentrations. The evaporator chilled water temperature was varies from 10 to 40°C and salt concentration from 35,000 to 95,000 ppm. The typical trend is presented in **Figure 5** at 90,000 ppm salt concentration. The similar trend was observed at other concentration values.

The two important results can be concluded, firstly, the U values drop over 25% due to salt concentration at lower temperature but this impact is not very significant at higher temperature. This might be due to propertied change at higher temperature. Secondly, The U values are higher at lower temperature and this is due to micro-bubble generation and detaching phenomenon as described earlier. The same trend of U values at all concentrations strengthens the argument of micro-bubble enhanced heat transfer phenomenon.

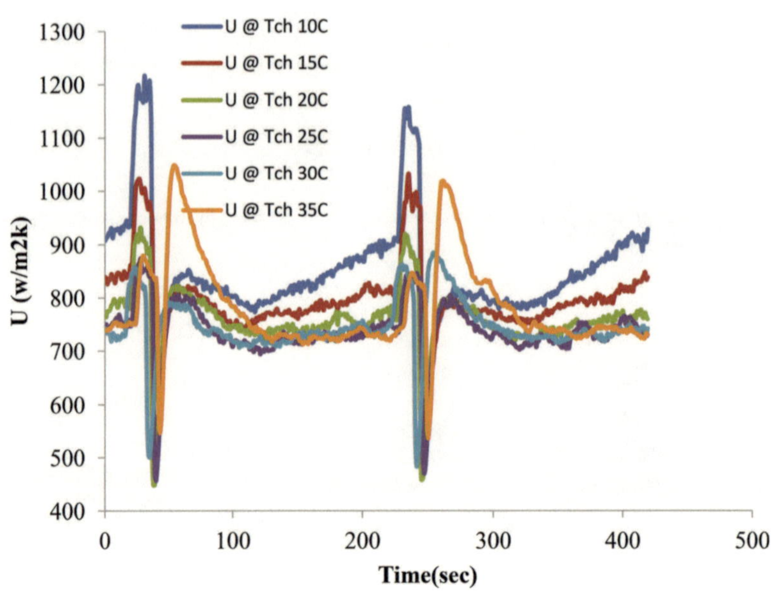

Figure 5.
Overall heat transfer coefficient profiles at 90,000 ppm salt concentration and different chilled water temperatures (with author's permission [43, 44]).

The falling film heat transfer coefficient (FFHTC) values are then calculated by using the methodology presented in the earlier section and presented in **Figure 6**. It can be noticed that FFHTC follows the same trend as U values at assorted heat source and salt concentrations.

The noticeable point in the plot is the heat transfer coefficient values drop initially with drop in evaporator vapor space temperature and achieve minimum values at 300 K. Once the vapor space temperature dropped further down, the heat transfer values start increasing. The increasing trend is even sharper below 295 K vapor space temperature and this is because of rapid change in vapor specific volume. The vapor specific volume change can divide the evaporation processes into three categories; namely, film surface evaporation, transition and micro-bubble assisted evaporation. The sharp change in specific volume below 295 K help to generate micro-bubble that detach from tube surface immediately due to low density and agitate the thermal barrier resulting increase in heat transfer rates. This phenomenon is observed and captured for the first time and named as "micro-bubble assisted film evaporation".

It can be clearly noticed that micro-bubbles play an important role at low temperature to enhance the heat transfer. The traditional heat transfer coefficient correlations are not able to capture this unique phenomenon. All correlations available in the literature can only work in film surface evaporation zone. Their extrapolation to capture transition and micro-bubble assisted zone also cannot predict an accurate value and heat exchanger designed based on these values cannot perform up to the level. Hence there is an urgent need for the development of an accurate heat transfer coefficient correlation to capture these two zones for efficient heat exchanger design.

A new correlation is proposed for falling film heat transfer coefficient that can efficiently capture transition and micro-bubble assisted evaporation at assorted salt concentration. The proposed model was written in FORTRAN and fitted with experimental data conditions. All important parameters such as heat flux, flow velocity and vapor properties were also included. Most importantly, the salt concentration and vapor specific volume parameters those were missing in conventional correlations are also embedded in the proposed correlation as shown in Eq. (6) [48, 49].

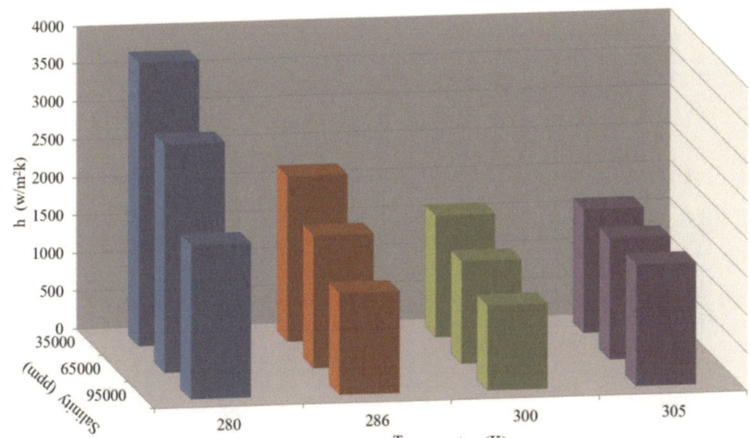

Figure 6.
Experimental film evaporation heat transfer coefficient profiles at different saturation temperature and different salt concentrations (with author's permission [43, 44]).

$$h_{fallingfilm} = \left\{0.279\left(\frac{\mu_l{}^2}{g.\rho_l{}^2K_l{}^3}\right)^{-0.333}(R_e)^{-2.18}(P_r)^{4.0}\left(2.\exp\left(\frac{S}{30000}\right)-1\right)^{-0.45}\right.$$
$$\left.\cdot\left(\frac{T_{evap}}{322}\right)\right\} + \left\{0.875\left(\frac{q}{DT}\right)\cdot\left(\frac{V_{evap}}{52.65}\right)\right\} \tag{6}$$

The proposed correlation is applicable from 280 to 305 K saturation temperatures. It also captures the feed water concentration ranges from 35,000 to 95,000 ppm. The film Reynolds number (ReΓ) ranges from 45 to 90 and Prandtl number (Pr) from 5 to 10. In proposed correlation, the first term control the thermally driven evaporation and second terms capture bubble assisted evaporation phenomenon that is missing in the conventional correlations. The proposed model results are presented in **Figure 7**. It can be noticed that model has good agreement with experimental results. The uncertainty of measured data is less than 5% and RMS of regressed data is 3.5%.

Conventionally, the Han and Fletcher correlation is applied in the industry for low temperature rages with its extrapolated results. The comparison of actual heat transfer values calculated by the experiments is compared with extrapolated Han and Fletcher values and it can be observed from **Figure 8** that there is huge difference. The conventional Han and Fletcher correlation can only capture film evaporation zone accurately but bubble assisted evaporation is totally out of range. The unique feature of "bubble assisted evaporation" can only be captured by the proposed falling film heat transfer coefficient correlation that boost heat transfer 2–3 fold. As a result, for process industries where the saturation temperature is below 295 K, the evaporator can be compact and low cost as compared to current design. The proposed correlation is timely and important for efficient design of falling film evaporator for process industries.

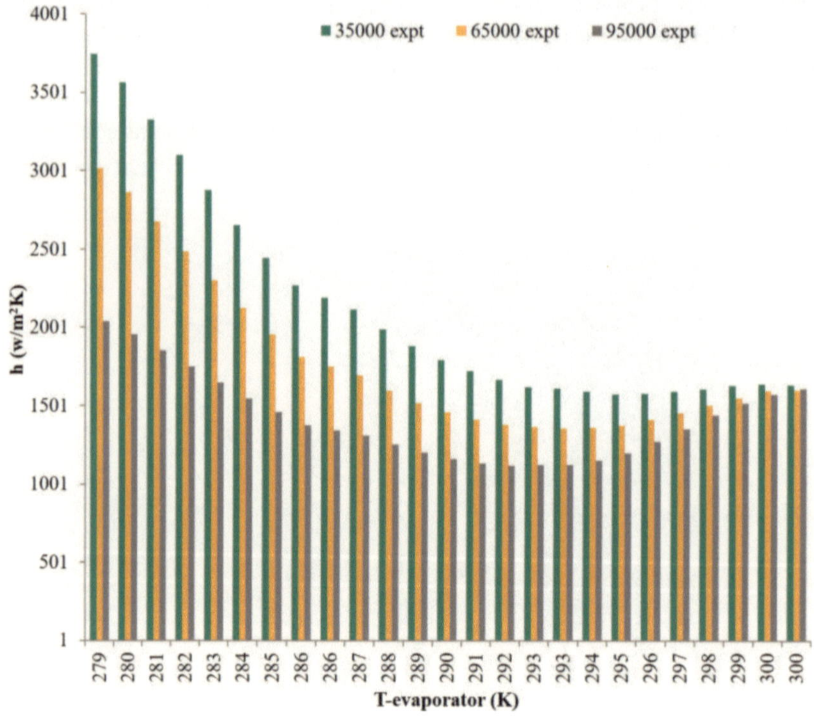

Figure 7.
The proposed falling film heat transfer coefficient correlation with experimental results (with author's permission [43, 44]).

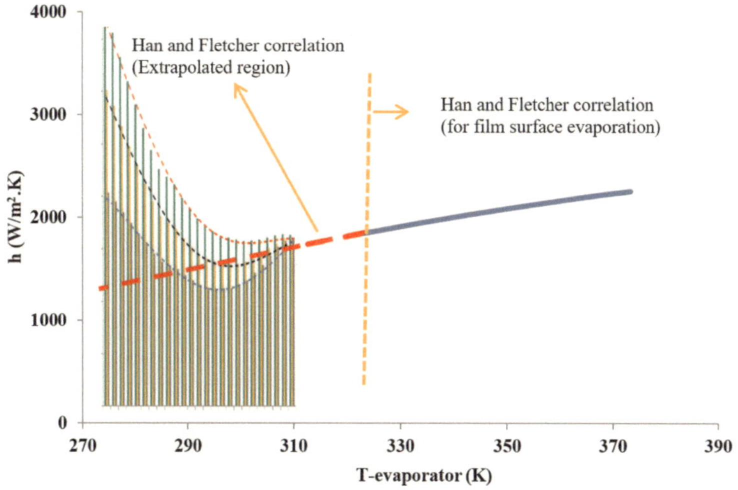

Figure 8.
Proposed falling film heat transfer coefficient compared with conventional Han & Fletcher's extrapolated values (with author's permission [43, 44]).

5. Summary

The horizontal falling film evaporators have many advantages over submerged and vertical tubes evaporators. Currently, there is no heat transfer correlation that can capture evaporation at low temperature especially below 295 K with different salt concentration. This is very important for efficient design of process evaporators. A horizontal tube falling film heat transfer coefficient correlation is proposed to capture effect, low temperature and salt concentration. It is demonstrated that the actual heat transfer values at low temperature can be 2–3 fold higher than the estimated values due to unique bubble-assisted evaporation phenomenon. The proposed correlation is applicable from 280 to 305 K saturation temperatures and feed water concentration ranges from 35,000 to 95,000 ppm. The uncertainty of measured data is less than 5% and RMS of regressed data is 3.5%.

Acknowledgements

Authors would like to thanks to KAUT and NUS for this study. The data is reproduced by the author's permission [44].

Nomenclature

μ_l	liquid viscosity (kg/m-s)
k_l	liquid conductivity (W/m K)
Pr	Prandtl number
q	input heat flux (W/m^2)
T_{evap}	evaporator saturation temperature (K)
$T_{saturation}$	evaporator saturation temperature (K)
$T_{ch,\ in}$	chilled water inlet temperature (K)
v_g	vapor specific volume (m^3/kg)
ΔT	$T_{ch,out} - T_{evap}$

ρ_l	liquid density (kg/m^3)
Re_Γ	Film Reynolds number
S	feed water salinity (ppm)

Abbreviations

EHTC	evaporation heat transfer coefficient
FFEHTC	falling film evaporation heat transfer coefficient
MED	multi-effect desalination
MSF	multi stage flash evaporation
AD	adsorption desalination
LMTD	log mean temperature difference
ppm	part per million

Author details

Muhammad Wakil Shahzad*, Muhammad Burhan and Kim Choon Ng
Water Desalination and Reuse Center (WDRC), King Abdullah University of Science and Technology, Thuwal, Saudi Arabia

*Address all correspondence to: muhammad.shahzad@kaust.edu.sa

References

[1] Uche J, Artal J, Serra L. Comparison of heat transfer coefficient correlations for thermal desalination units. Desalination. 2002;**152**:195-200

[2] Ribatski G, Jacobi AM. Falling film evaporation on horizontal tubes—A critical review. International Journal of Refrigeration. 2005;**28**:635-653

[3] Adib TA, Heyd B, Vasseur J. Experimental results and modeling of boiling heat transfer coefficients in falling film evaporator usable for evaporator design. Chemical Engineering and Processing. 2009;**48**: 961-968

[4] Chun KR, Seban RA. Heat transfer to evaporating liquid films. Transactions of the the American Society of mechanical Engineers: Journal of Heat Transfer. 1971;**93**(C):391-396

[5] Prost JS, Gonzaalez MT, Urbicain MJ. Determination and correlation of heat transfer coefficients in a falling film evaporator. Journal of Food Engineering. 2006;**4**(73):320-326

[6] Ahmed SY, Kaparthi R. Heat transfer studies of falling film heat exchangers. Indian Journal of Technology. 1963;**1**: 377-381

[7] McAdams WH, Drew TB, Bays GS. Heat transfer to falling—Water films. Transactions of American Society of mechanical Engineers. 1940;**62**:627

[8] Herbert LS, Stern UJ. An experimental investigation of heat transfer to water in film flow. Canadian Journal of Chemical Engineering. 1968; **46**:401-407

[9] Parken WH, Fletcher LS. An experimental and analytical investigation of heat transfer to thin water films on horizontal tubes. Report UVA-526078-MAE. University of Virginia; 1977. pp. 77-101

[10] Han JC, Fletcher LS. Falling film evaporation and boiling in circumferential and axial grooves on horizontal tubes. Industrial and Engineering Chemistry Process Design and Development. 1985;**24**:570-575

[11] Fujita Y, Tsutsui M. Experimental investigation of falling film evaporation on horizontal tubes. Heat Transfer-Japanese Research. 1998;**27**:609-618

[12] Fujita Y, Tsutsui M, Zhou Z-Z. Evaporation heat transfer of falling films on horizontal tube—Part 1, analytical study. Heat Transfer-Japanese Research. 1995;**24**:1-16

[13] Fujita Y, Tsutsui M, Zhou Z-Z. Evaporation heat transfer of falling films on horizontal tube—Part 2, experimental study. Heat Transfer-Japanese Research. 1995;**24**:17-31

[14] Liu ZH, Yi J. Falling film evaporation heat transfer of water/salt mixtures from roll-worked enhanced tubes and tube bundle. Applied Thermal Engineering. 2002;**22**(1):83-95

[15] Yang LP, Shen SQ. Experimental study of falling film evaporation heat transfer outside horizontal tubes. In: Conference on Desalination and the Environment; Halkidiki, Greece. 2007

[16] Parken WH et al. Heat-transfer through falling film evaporation and boiling on horizontal tubes. Journal of Heat Transfer—Transfer The American Society of mechanical Engineers. 1990; **112**(3):744-750

[17] Ribatski G, Thome JR. Experimental study on the onset of local dryout in an evaporating falling film on horizontal plain tubes. Experimental Thermal and Fluid Science. 2007;**31**(6):483-493

[18] Lorenz JJ, Yung D. Film breakdown and bundle-depth effects in horizontal tube, falling-film evaporators. Journal of Heat Transfer—Transfer The American Society of mechanical Engineers. 1982; **104**(3):569-571

[19] Roques JF, Thome JR. Falling films on arrays of horizontal tubes with R-134a. Part II: Flow visualization, onset of dry out, and heat transfer predictions. Heat Transfer Engineering. 2007;**28**(5): 415-434

[20] Aly G, Al-Hadda A, Abdel-Jawad M. Parametric study on falling film seawater desalination. Desalination. 1897;**65**:43-55

[21] Moeykens S, Pate MB. Spray evaporation heat transfer performance of R-134a on plain tubes. ASHRAE Transactions. 1994;**100**(2):173-184

[22] Moeykens S, Kelly JE, Pate MB. Spray evaporation heat transfer performance of R-123 in tube bundles. ASHRAE Transactions. 1996;**102**(2): 259-272

[23] Chang TB, Chiou JS. Spray evaporation heat transfer of R-141b on a horizontal tube bundle. International Journal of Heat and Mass Transfer. 1998;**42**:1467-1478

[24] Bourouni K, Martin R, Tadrist L, Tadrist H. Modelling of heat and mass transfer in a horizontal tube falling film evaporators for water desalination. Desalination. 1998;**116**:165-184

[25] Xu L, Ge M, Wang S, Wang Y. Heat transfer film coefficients of falling film horizontal tube evaporators. Desalination. 2004;**166**:223-230

[26] Chun KR, Seban RA. Heat transfer to evaporating liquid films. The American Society of mechanical Engineers Journal of heat transfer. 1971; **11**:391-396

[27] Alhusseini AA, Tuzla K, Chen JC. Falling film evaporation of single component liquids. International Journal of Heat and Mass Transfer. 1998;**41**(12):1623-1632

[28] Shmerler JA, Mudawwar I. Local evaporative heat transfer coefficient in turbulent free-falling liquid films. International Journal of Heat and Mass Transfer. 1988;**31**(4):731-742

[29] Chien LH, Tsai YL. An experimental study of pool boiling and falling film evaporation on horizontal tubes in R-245fa. Applied Thermal Engineering. 2011;**31**:4044-4054

[30] Shahzad MW, Burhan M, Ng KC. The fallacy of energy efficiency for comparing desalination plants: A standard primary energy approach. Nature Clean Water. DOI: 10.1038/s41545-018-0028-4

[31] Shahzad MW, Burhan M, Ghaffour N, Ng KC. A multi evaporator desalination system operated with thermocline energy for future sustainability. Desalination. 2018;**435**: 268-277

[32] Shahzad MW, Burhan M, Son HS, Seung Jin O, Ng KC. Desalination processes evaluation at common platform: A universal performance ratio (UPR) method. Applied Thermal Engineering. 2018;**134**:62-67

[33] Ng KC, Shahzad MW. Sustainable desalination using ocean thermocline energy. Renewable and Sustainable Energy Reviews. 2018;**82**:240-246

[34] Shahzad MW, Burhan M, Ng KC. Pushing desalination recovery to the maximum limit: Membrane and thermal processes integration. Desalination. 2017;**416**:54-64. DOI: 10.1016/j.desal.2017.04.024

[35] Ng KC, Shahzad MW, Son HS, Hamed OA. An exergy approach to

efficiency evaluation of desalination. Applied Physics Letters. 2017;**110**: 184101. DOI: 10.1063/1.4982628

[36] Shahzad MW, Burhan M, Li A, Ng KC. Energy-water-environment nexus underpinning future desalination susta inability. Desalination. 2017;**413**:52-64

[37] Shahzad MW, Ng KC. An improved multi-evaporator adsorption desalination cycle for GCC countries. Energy Technology. 2017. DOI: 10.1002/ente.201700061

[38] Shahzad MW, Ng KC, Thu K. Future sustainable desalination using waste heat: Kudos to thermodynamic synergy. Environmental Science: Water Research & Technology. 2016;**2**:206-212

[39] Thu K, Kim Y-D, Shahzad MW, Saththasivam J, Ng KC. Performance investigation of an advanced multi-effec t adsorption desalination (MEAD) cycle. Applied Energy. 2015;**159**:469-477

[40] Shahzad MW, Thu K, Kim Y-d, Ng KC. An experimental investigation on MEDAD hybrid desalination cycle. Applied Energy. 2015;**148**:273-281

[41] Ng KC, Thu K, Seung Jin O, Li A, Shahzad MW, Ismail AB. Recent developments in thermally-driven seawater desalination: Energy efficiency improvement by hybridization of the MED and AD cycles. Desalination. 2015;**356**:255-270

[42] Shahzad MW, Ng KC, Thu K, Saha BB, Chun WG. Multi effect desalination and adsorption desalination (MEDAD): A hybrid desalination method. Applied Thermal Engineering. 2014;**72**:289-297

[43] Shahzad MW, Myat A, Gee CW, Ng KC. Bubble-assisted film evaporation correlation for saline water at sub-atmospheric pressures in horizontal-tube evaporator. Applied Thermal Engineering. 2013;**50**:670-676

[44] Shahzad MW. The hybrid multi-effect desalination (MED) and the adsorption (AD) cycle for desalination [doctoral thesis]. National University of Singapore; 2013

Chapter 3

Review Heat Transfer of Non-Newtonian Fluids in Agitated Tanks

Vitor da Silva Rosa and Deovaldo de Moraes Júnior

Abstract

The heating and cooling of non-Newtonian liquids in tanks with mechanical impellers are operations commonly employed as chemical reactors, heat exchangers, distillers, extractors, thinners and decanters. In particular, the design of heat exchangers (jackets, helical coils, spiral coils and vertical tubular baffles) in tanks requires the prior knowledge of the rheology of the liquid for the calculation of the convection coefficients and the Reynolds number, in order to obtain the area thermal exchange. This chapter aimed to present the basic concepts of tanks with agitation, non-Newtonian liquids, hydrodynamics, heat transfer and, finally, with a practical design example for engineers and undergraduate students.

Keywords: non-Newtonian liquid, tank agitation, heat transfer, rheology

1. Introduction

The agitation and mixing of liquids are an operation commonly used in chemical, petrochemical, food and pharmaceutical processes. In general, the aforementioned operation is performed in tanks (usually cylindrical shape) with mechanical impellers [1]. Generally, agitation refers to forcing a fluid by mechanical means to flow in circulatory or similar pattern inside a vessel [2]. Mixing is the random distribution into and through one another, of two or more initially separate phases [3]. Mixing is achieved by moving material from one region to another [4].

The quality of the mixture is the main parameter that is associated with the efficiency of the heat transfer and mass transfer operations occurring in tanks with agitation and mixing.

Most of the liquids stirred and mixed in tanks, are polymer solutions and paint mixtures, as well as mineral pulps and food pastes, which exhibit a non-Newtonian rheology [5].

Thus, this article will address a chapter with the main concepts on the hydrodynamics of agitation, and, subsequently, the basic parameters of heat transfer in the agitation of non-Newtonian liquids independent of time are detailed, because they represent about 85% of all liquids occurring in industrial processes. At the end of the article, an example of the design of a tank with heating and agitation of a pseudoplastic liquid will be presented, applying the equations that will be presented in the course of this text.

2. Non-Newtonian liquids

Non-Newtonian liquids have a non-linear relationship between shear stress and shear rate, so that the apparent viscosity of this liquid is a function of the rate applied by the source promoter of the amount of movement.

Rheology classifies non-Newtonian liquids into three large groups: (a) time-independent, (b) time-dependent and (c) viscoelastic [6].

Liquids independent of time have their apparent viscosity varying depending on the shear rate and the temperature (**Figure 1**).

The rheological model of Ostwald de Waele [7] or model of the law of the powers, as presented in Eq. (1), has a good adjustment to the data obtained experimentally in rheometers and viscometer, in the shear rate range between 10 and 1000 s^{-1}.

$$\tau = k\left(\nabla\vec{v}\right)^n \quad (1)$$

The constant k (consistency factor) and the exponent N (consistency index) are obtained experimentally in viscometers and rheometers. When index n is less than 1, the liquid is pseudoplastic with decrease in apparent viscosity according to the variation in shear rate, such as the food liquids and polymer solutions. With index n equals 1 the liquid is Newtonian as water and hydrocarbons. In case of index n greater than 1, the liquid is dilating with increase in apparent viscosity as the increase in shear rate, such as the starch suspensions and some mineral sludge.

Viscoplastic liquids, which require a yield stress, are adjusted satisfactorily by the Herschel-Bulkley model (Eq. (2)).

$$\tau - \tau_0 = k\left(\nabla\vec{v}\right)^n \quad (2)$$

The yield stress (τ_0) is obtained experimentally being a constant parameter of each type of liquid.

The apparent viscosity can be calculated from Eqs. (1) and (2) replacing the shear stress term by the following definition (Eq. (3)):

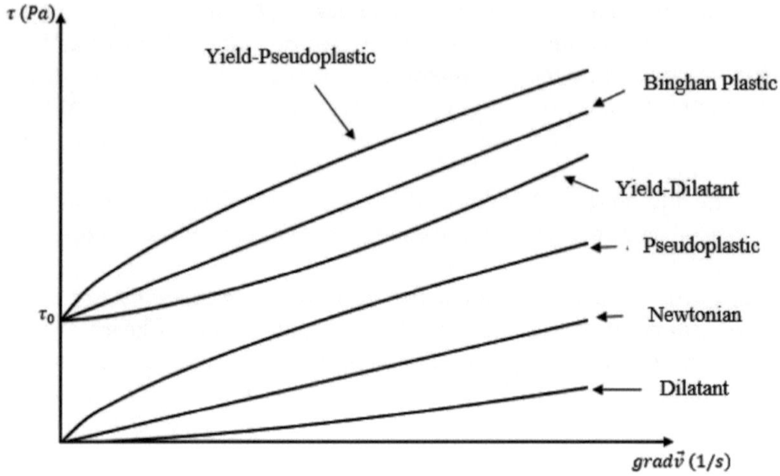

Figure 1.
Classification of time-independent liquids.

$$\eta = \tau / \nabla\left(\vec{v}\right)^{n} \qquad (3)$$

There are other models (Casson, Ellis, and Carreau) that have a more accurate adjustment of the apparent viscosity variation with the shear rate, when compared with the model of the law of the powers. However, these models have several constants that must be experimentally determined.

Time-dependent liquids (**Figure 2**) are classified as thixotropics with decreased apparent viscosity with time and reoptical with increased apparent viscosity over time [8].

The thixotropic fluid has an ascending curve similar to pseudoplastics, however, due to hysteresis (phenomenon caused by the variation of apparent viscosity with time), the relaxation of the liquid occurs by a downward curve that differs from the curve up. In the case of the reopetic liquid, the inverse occurs, and the ascending curve is similar to the dilating fluid.

The influence of time on apparent viscosity variation is difficult to achieve and with little accuracy by theoretical models, however, liquids such as food pastes have a great variation in hysteresis, which should be considered in the design of a tank with agitation. The most practical form is from tests in a rheometer in the so-called round-trip test. The sample is subjected to countless cycles of ascending and descending in the shear rate, in order to raise curves similar to those shown in **Figure 2**.

In the design of tanks with mechanical impellers and even other unitary operations (such as pumping and heat exchangers), the variation of hysteresis is considered almost negligible in these liquids, so that the model of the law of the powers satisfies design requirements perfectly.

Finally, the last class of non-Newtonian liquids is called viscoelastic. These liquids have at low shear rates a behavior tending to the solid state, and parameters such as modulus of elasticity are obtained experimentally in vibratory rheometers. In the case of submission to high shear rates, these liquids have similar behavior to

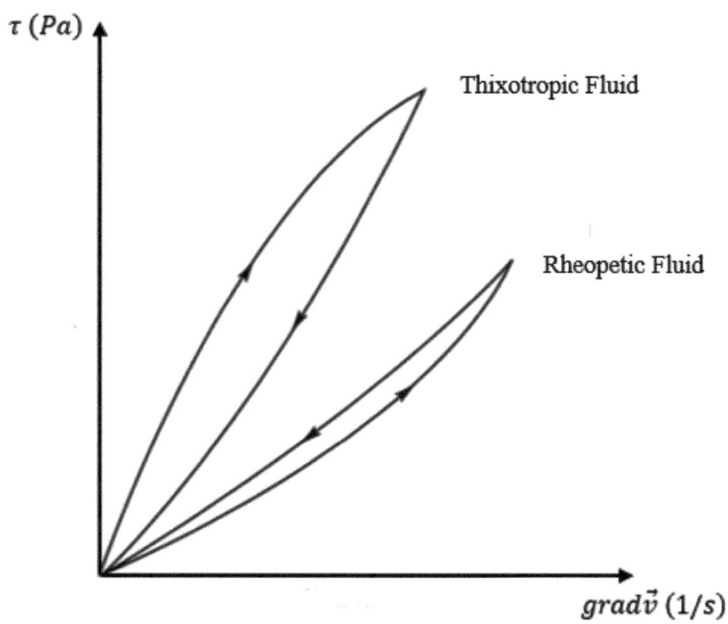

Figure 2.
Time-dependent liquids.

the liquid phase as the anionic polyacrylamide, which is used as an agent for flocculant water treatment plants.

A curious effect of viscoelastic liquids during a stirring and blending operation is the so-called effect of Weissenberg, which is characterized by the liquid being fixed in the impeller during its rotation and climbing through the shaft. Therefore, when this type of liquid is agitated, the rotations employed must be low (lower than 50 rpm) and it is recommended to use a helical impeller of double tape.

The heat transfer in the agitation of Newtonian liquids and non-Newtonian liquids is intimately linked to hydrodynamics, because the flow occurring in the fluid and in the peripheries of the tank influences significantly how the heat is transmitted between the promoter source for all the liquid that will be warmed or cooled.

3. Hydrodynamics of agitation and mixing of non-Newtonian liquids

The flow in stirring and mixing units is a function of the type of mechanical impeller used, which can be of axial type and radial type.

The axial type impeller promotes a flow predominantly parallel to the impeller shaft, directing the liquid to the base of the tank. The most common axial impeller used in industrial processes is the pitched blade turbine (PBT), which can have from two to four paddles and angles between 30 and 60°, the conventional one, a PBT with four paddles inclined to 45° (**Figure 3A**). They are recommended for mixing suspended solids and in processes requiring low to moderate turbulence (with Reynolds number between 4000 and 40,000). There are other types of axial impellers such as the helical tape (for liquids with high viscosities), the naval propeller (for mixtures of immiscible liquids with moderate turbulence) and the propellants with teeth (for high shear in the mixture of pigments) [9, 10].

Radial type impellers promote fluid flow, in the output of the same and in the direction of the tank wall, and therefore, the power consumption by the electric motor is higher when compared to an axial impeller. The most common radial impeller used in industrial processes is the Rushton turbine (RT) with six flat blades or flat Six Blade Impeller (**Figure 3B**). There are turbines of four paddles to eight paddles, but they are poorly used and, in the case of liquids with high viscosity, it is recommended to use radial impeller type anchor [11].

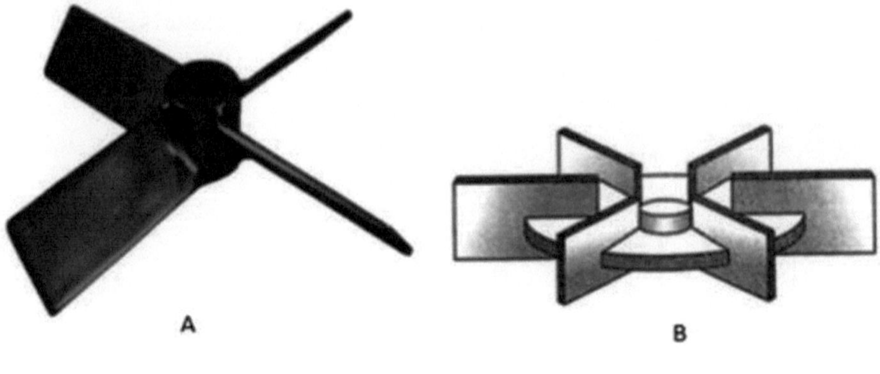

Figure 3.
(A) Pitched blade turbine and (B) Rushton turbine.

The radial impeller is used for mixtures of mixed and immiscible liquids and gases with liquids, usually in situations that require major turbulence, aiming at rapid mixing times (in the range of 9–14 s).

A big problem with the processing and blending in tanks was the standardization of the dimensions that the tank must possess in relation to its peripherals (impeller, impeller position in the tank, baffles and draft tubes). Rushton et al. [12] presented several relationships based on experimental results, in which if followed, the quality of the mixture is potentiated, the amount of plates of the tank is minimized (diameter equal to the height to tank with lid) and the power consumption is close to the smallest possible, which in terms of engineering is excellent.

Figure 4 presents a scheme of the standard dimensions proposed by the aforementioned researchers. When a tank is designed out of these specifications, this tank is called non-standard.

The main design parameters of a tank with stirring and mixing (regardless of the operation in which the equipment is intended) are the power consumption, the mixing time, the thermal efficiency and the efficiency of the mass transfer.

Non-Newtonian liquids have high apparent viscosities, usually in the range of 500–2000 cP so that the mechanical impeller when stirring this liquid will cause a viscous energy dissipation, which is directly proportional to the consumption of useful power by the impeller. In the project, the introduction of energy into the system and calculation of its contribution in the source of heating or cooling in the tank should be taken into account.

The viscous dissipation function for a stirring system is written according to Eq. (4) [14].

$$\phi_V = \left(2\tau_{rr}{}^2 + 2\tau_{\theta\theta}{}^2 + 2\tau_{ZZ}{}^2 + \tau_{rZ}{}^2 + \tau_{r\theta}{}^2 + \tau_{Z\theta}{}^2\right)/\eta^2 \tag{4}$$

Eq. (4) is considered the components of normal stresses in the three directions in cylindrical coordinates and the shear components in the radial direction with the axial and tangential planes and in the axial direction with the tangential plane. The shear stress components are suitable for a rheological model appropriate to the type of liquid that will be agitated and heated or cooled.

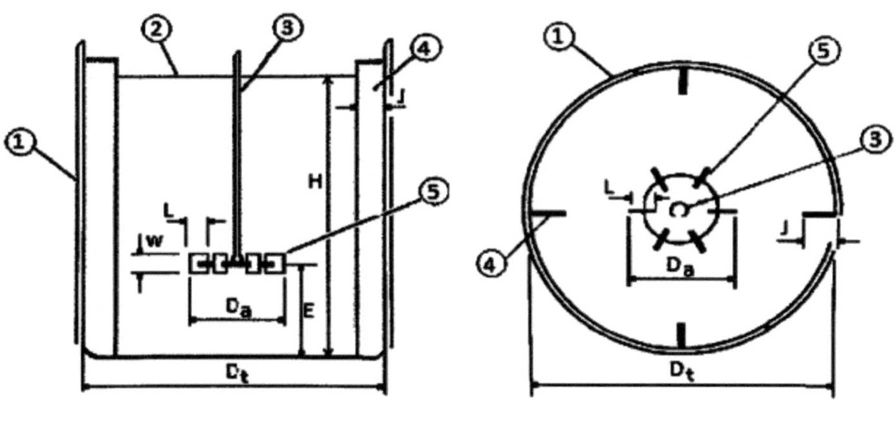

Figure 4.
Geometric relations fixed by Rushton, Costich and Everett 7: 1) tank wall; 2) height of liquid level; 3) mechanical impeller shaft; 4) baffle; 5) mechanical impeller. With the following relationships: $S_1 = D_t/D_a = 3; S_2 = E/D_a = 1; S_3 = L/D_a = 1/4; S_4 = W/D_a = 1/5; S_5 = J/D_t = 0.1$ *and* $S_6 = H/D_t = 1$ *Moraes Júnior e Moraes [13].*

Considering a pseudoplastic liquid, shear stresses can be calculated by the model of the Power Law (Eq. (1)). Finally, the power consumed by the mechanical impeller is given by Eq. (5).

$$P = \int \eta \phi_V r dr d\theta dz \tag{5}$$

The analytical solution of Eq. (5) is not feasible due to the complexity of the viscous dissipation function by involving nonhomogeneous partial derivatives of high order.

Another alternative is the solution by semi-empirical models obtained by Buckingham's Pi theory, which relates the response variable (in this case the power consumption) with the independent variables through the fundamental quantities.

In Eq. (6), the empirical model is presented for calculating power consumption as a function of adimensional numbers.

$$\frac{P}{\rho N^3 {D_a}^5} = K' \left(\frac{N{D_a}^2\rho}{\eta}\right)^{a'} \left(\frac{D_a N^2}{g}\right)^{b'} (geometry)^{c'} \tag{6}$$

The first term to the left in Eq. (6) is the number of power (N_P); on the right limb evaluating from left to right has itself the Reynolds number, the number of Froude, and the geometric relations. It is noteworthy that the constant K′ and the exponents a ′, b′ and c′ are obtained experimentally and are functions of the type of impeller, the type of tank, the presence of baffles and the rheology of the fluid in agitation.

In most agitation systems, the same contains baffles, so the number of Froude becomes negligible in relation to the Reynolds number. Eq. (6) can be rewritten for a given geometry simplified in Eq. (7).

$$N_p = K'(Re)^{a'} \tag{7}$$

It can be noted that the Reynolds number (flow parameter) has a significant effect on power consumption, as well as in heat transfer, as discussed in topic 3.

In the agitation and mixing of Newtonian liquids, the calculation of the Reynolds number is simple, because the apparent viscosity reduces the dynamic viscosity, which is tabulated as a function of the temperature for several liquids. In the case of non-Newtonian liquids the calculation is no longer simple and becomes a problem of closure, because a rheological model needs to be chosen and its parameters determined experimentally.

Considering that the liquid follows the model of the power law (valid for pseudoplastics and dilatants), the apparent viscosity is calculated by Eq. (8).

$$\eta = k\left(\nabla\vec{v}\right)^{n-1} \tag{8}$$

Thus, the Reynolds number is written as (Eq. (9)):

$$Re = \frac{N{D_a}^2\rho}{k\left(\nabla\vec{v}\right)^{n-1}} \tag{9}$$

The shear rate required for calculating the Reynolds number can be calculated by solving the equation of the amount of motion applied to a control volume (in this case the tank with agitation and mixing), as shown in Eq. (10).

$$\rho Dv/D\theta = -\nabla p + \nabla(\eta \nabla v) + \rho g \tag{10}$$

As in Eq. (5), the solution of Eq. (10) is unfeasible, since the velocity components in the three directions and the temporal component are relevant in agitation and mixing systems. Besides that, the internal geometry of the tank (through the presence of the mechanical impeller and the baffles) makes the analytical solution impracticable. An alternative to numerical solution is the determination of some function that describes the variation of shear rate in relation to rotation of the mechanical impeller, thus, it is not necessary to directly determine the shear rate, which would enable the calculation of the Reynolds number.

Metzner and Otto [15] performed a pioneering work in the agitation and mixing of non-Newtonian liquids in the search for this function that relates the shear rate and the rotation of the mechanical impeller. The researchers worked with tanks in the range of 6–22 inches stirring carboxymethylcellulose (CMC) solutions, Carbopol and the Attasol. All these substances are polymers widely used in industrial processes and it was found that in aqueous solutions, these polymers follow the model of the law of the powers.

By following the model of the law of the powers, these polymers in aqueous solution have their apparent viscosity as a function only of temperature and variation of the shear rate, which decreases the complexity of the problem.

The researchers presented a method to relate, in an experimental way, the shear rate with the rotation of the mechanical impeller, based on the following assumptions: (a) the consistency index (n) of the power law was adopted as constant, despite there is a slight variation of this parameter with the shear rate, but in terms of design, this variation is negligible; (b) the flow of non-Newtonian liquids occurs preferentially in the laminar regime so that there is no detachment of the boundary layer which is the surface of the mechanical impeller and (c) the variation of the shear rate occurs exclusively due to the rotation of the mechanical impeller and not in relation to the rheology of the liquid.

With the last premise, it was assumed that there is a mean shear rate that varies only in function of rotation so that this parameter represents all the fluctuations of shear rate occurring during the agitation of the liquid. In Eq. (11), the definition of this mean parameter is presented:

$$\overline{\nabla \vec{v}} = k_s N \tag{11}$$

The constant k_s is determined experimentally according to the type of mechanical impeller and its geometry and the tank under analysis. **Table 1** shows the values of the constant k_s for some types of mechanical impellers. It is noteworthy that Eq. (11) is valid only for non-Newtonian liquids independent of the pseudoplastics-type time (n < 1).

Impeller	Number of baffles	D_a (m)	D_t/D_a	$k_s(n < 1)$
Radial with 6 flat blades	0	0.051–0.20	1.3–5.5	11.5 ± 1.5
Radial with 6 flat blades	4	0.051–0.20	1.3–5.5	11.5 ± 1.5
Axial with 4 paddles	4	0.12	2.13	10.0
Anchor	0	0.28	1.02	11.0 ± 1.5

Table 1.
Values of k_s *for various types of mechanical impellers [5].*

In Eq. (12), the Reynolds of Metzner and Otto is presented, and the use of it depends on the knowledge of the rheological parameters of the model of the power law (k and n) and the constant k_s.

$$Re_{MO} = \frac{N^{2-n}.D_a{}^2\rho}{k.k_s{}^{n-1}} \tag{12}$$

Calderbank and Moo-Young [16] disagreed with Metzner and Otto [15] in determining the function that relates the shear rate with the rotation only in one of the premises; for them the variation of the shear rate besides relying on the rotation of the mechanical impeller, is also a function of the rheology of the liquid.

Thus, the researchers used a similarity approach with the Reynolds of Metzner and Reed [17] (Eq. (13)) modified for a stirring system.

$$Re_{MR} = \frac{D_a{}^2N\rho(8N)^{1-n}}{k}\left(\frac{4n}{3n+1}\right)^n \tag{13}$$

The constant 8 that appears in Eq. (13) is obtained from the original definition of Reynolds by Metzner and Reed, valid for the flow of non-Newtonian liquids inside pipes. Thus, Eq. (13) was generalized replacing the constant 8 with a constant B, which is the function of rheology and the shear rate of the fluid, as presented in Eq. (14) (Reynolds of Calderbank and Moo-Young [16]):

$$Re_{CM} = \frac{D_a{}^2N\rho(BN)^{1-n}}{k}\left(\frac{4n}{3n+1}\right)^n \tag{14}$$

The value of parameter B is calculated using Eq. (15) for pseudoplastic liquids, with an error of approximately 10%:

$$B = (D_t/D_a)^2/\left[(D_t/D_a)^2 - 1\right] \tag{15}$$

If the tank follows the standard geometry model proposed by Rushton, Costich and Everett [10], the value of parameter B is 11.

Tanguy et al. [18] presented a more reticent view as to the observation of Calderbank and Moo-Young [16] in relation to the dependence of the shear rate with the rheology of the liquid.

The researchers investigated with various types of impellers such as anchor-type and observed that variations in the behavior index (n) in the range 0.3–0.95 did not provide a significant variation of the constant k_s and parameter B. Thus, they concluded that the Metzner and Otto model [15] for being simpler should be applied.

However, the concept presented by the researchers can be extended to the axial impellers with four paddles and radial turbine type. For example, the value of the constant k_s for the radial impeller turbine is 11.5, while the value of B is 11, presenting a deviation of 4.3%, which in terms of engineering design is little significant.

Soon it is recommended in tank projects with agitation, mixing and heat transfer of non-Newtonian liquids, the use of Reynolds by Metzner and Otto [15] is recommended for being simpler to calculate and with excellent accuracy.

The researchers previously mentioned found a relation to the shear rate with the rotation of the mechanical impeller through an empirical model based on data obtained in an experimental way. However, in the last decades, the numerical solution by computational fluid dynamics (CFD) has been employed to solve the equation of the amount of movement (Eq. (10)) for complex geometries, which enables the determination of the field of speeds and thus the shear stresses at any point in the domain.

Ameur and Bouzit [14] studied the power consumed by a two-blade radial mechanical impeller in a tank without baffles in the agitation of a pseudoplastic liquid. The authors used the model of the law of the powers (Eq. (8)) for the prediction of the apparent viscosity and varied the index of behavior of the liquid between 0.7 and 1.0.

With the results obtained in the simulation, a theoretical model was determined to predict the number of power (Eq. (16)) according to a generalized Reynolds number (Eq. (17)), the height of the liquid level (H), diameter impeller (D_a), tank diameter (D_t) and rheology of the liquid:

$$N_p = \frac{0.000209}{Re_o}\left(114.1 + 0.56e^{\frac{n}{3}}\right)\left[170 - 166e^{\left(-\frac{H}{1.2D_t}\right)}\right]\left[6.2e^{\left(\frac{D_a}{0.23D_t}\right)}\right] \tag{16}$$

$$Re_0 = \frac{\rho N^{2-n} {D_a}^2}{k} \tag{17}$$

Eq. (16) has validity for agitation of pseudoplastic liquids in tanks without chicane with Reynolds between 0.1 and 10. The results obtained in the simulation were validated with the experimental work of Bertrand and Couderc [19] so that there was an excellent adherence of the experimental data with Eq. (16).

In terms of design, this work contemplate very low Reynolds number. In most cases found industrially, the values for the Reynolds are in the range of 200 to 800, so that the results obtained in the simulation should be uses with caution.

A phenomenon of difficult prediction in the agitation of Newtonian and non-Newtonian liquids is turbulence, which can be defined in a few words such as a transient, rotational event and fluctuations in velocity components. These variables make the analysis of the equation of the movement very complex, originating terms, such as the Reynolds tensor, which requires experimental parameters to give a closure to the equation.

Wu [20] studied the application of six turbulence models, such as the "Reynolds stress model", in the agitation of non-Newtonian fluid in a tank with mechanical impeller. The tank under study is an anaerobic reactor and non-Newtonian fluid is a mud characterized as pseudoplastic following the model of the law of powers with consistency index (n) ranging from 0367 to 1000.

Turbulence models are applied with the agitation from an axial impeller with four flat blades at 45° (PBT) and a modification of the same with curved blades. The author presents with detail the mathematical treatment of the constitutive equations and the difficulty of obtaining the results in the simulation by CFD, because in this case, he considered the transient process (which exponentially increases computational time). Finally, the results were validated experimentally and there was a distancing of 30% of the results.

Taking into account the complexity from the analysis of turbulence in tanks with agitation, the variation of rheological properties and errors of the numerical solution, the deviation presented is acceptable and the study becomes very relevant for the theme in question.

Sossa-Echeverria and Taghipour [21] evaluated the use of axial-type impellers placed on the lateral side of the tank aiming to stir pseudoplastic liquids. The CFD tool was used for the analysis of the vector field formed by the impellers, and the study was conducted in the laminar region, with Reynolds between 10 and 200.

The results were experimentally validated with the agitation of Carbopol solutions and the field of velocities was obtained by the particle image velocimetry (PIV) technique. In this technique, spherical particles of a reflective material are placed in the tank and a laser positioned orthogonal to the tank wall, which emits radiation and the same is reflected in the moving particles, thus enabling in real time the determination of vectors. There was an excellent adherence between the experimental data and those predicted by the simulation, so the authors provided a diagram of the number of power according to the Reynolds number.

In general, in terms of design, CFD simulation has been quite useful in estimating parameters during agitation without the need to perform a large amount of experiments, however, it should not be forgotten that the experimental results are validation of the models obtained in the simulations.

4. Heat transfer

4.1 Calculation of the thermal exchange area

Heat exchangers are chemical and biochemical reactors and in tanks with mechanical impellers are the jackets, the helical coils, the spiral coils and the vertical tube baffles [22]. The vertical tube baffles in addition to exchanging heat with the stirring fluid avoid the formation of vortices that can be harmful to the quality efficiency of the mixture [23].

Optimally, the heat exchange system of a tank is projected on a permanent basis through the classical design equation (Eq. (18)):

$$A = Q/U.LMTD \tag{18}$$

The coefficient U usually called the overall heat transfer coefficient or the global coefficient of performance of the heat exchanger is the parameter that relates the mechanisms of thermal exchange (conduction, convection and radiation), the flow, the properties of the liquid and the geometry of the control volume with the thermal exchange area.

In Eq. (18), the coefficient U is assumed to be constant independent of the temperature variation and liquid viscosity, besides being invariant with the process time. In the case of liquids with high viscosities (such as heavy oils), the coefficient U may have variations during the process due to the viscosity gradient that occurs as a function of temperature, in which case this problem is eliminated with the introduction of caloric temperature (valid for Newtonian liquids).

However, in the case of non-Newtonian liquids, the approach of the caloric temperature is not adequate due to the variation of the apparent viscosity with the shear rates caused by the rotation of the mechanical impeller. In any case, the transient regime is very prominent in the heating or cooling of non-Newtonian liquids, which makes it impossible to use Eq. (18) for calculating the thermal exchange area [24].

Rosa and Moraes [1] presented a deduction from the first law of thermodynamics applied to a tank in Newtonian fluid agitation for transient regime operations; however, at that moment the effect of viscous dissipation in the process as described in Item 2 of this text was not considered, which for non-Newtonian liquids is indispensable.

In order to contemplate the effect of viscous dissipation, consider a tank perfectly insulated in which a non-Newtonian liquid is placed and a mechanical impeller. This liquid will be heated by another liquid through a heat exchanger (which can be a jacket or serpentine or vertical tubular chicane). The operation will occur in batch. Eq. (19) illustrates the first law of thermodynamics in terms of rate:

$$Q_{vc} - W_{vc} + w_e\left(h_e + v_e^2/2 + gz_e\right) - w_s\left(h_s + v_s^2/2 + gz_s\right) = dE_{vc}/d\theta \tag{19}$$

Consider the following hypotheses for design: (1) As the tank is perfectly insulated, there are no liquid flows of heat between the tank and the external environment, (2) the kinetic and potential energy variations are negligible in relation to the enthalpies and (3) perfectly agitated tank. The term regarding to the work is related to the contribution of the viscous dissipation generated by the rotation of the impeller, which will be considered.

The derivative present in Eq. (19) is referent to the total energy variation of the control volume. With the hypotheses mentioned above, this derivative summarizes the variation of internal energy so that it is equal to the variation of enthalpy for incompressible liquids. Finally, the enthalpy can be written in function of the specific heat as the constant pressure and the derivative is summarized as the "bulk" temperature variation (perfectly agitated tank condition).

Eq. (20) is a presented Eq. (18) after the hypotheses mentioned:

$$w_h c_{ph}(T_1 - T) - W_{vc} = Mc_{pc}\frac{dt_b}{d\theta} \tag{20}$$

The left-hand side of Eq. (18) is a source of thermal exchange, so that it can be written according to coefficient U (Eq. (21)).

$$w_h c_{ph}(T_1 - T) = UALMTD \tag{21}$$

Replacing Eq. (21) in Eq. (20), you have:

$$UA\{(T_1 - T)/[ln(T_1 - t_b/T - t_b)]\} - W_{vc} = Mc_{pc}\frac{dt_b}{d\theta} \tag{22}$$

The integration of Eq. (22) depends on two analyses: (1) The coefficient U must be constant during the process; otherwise, it is necessary to know its variation according to the apparent viscosity and time. (2) It is necessary to determine an expression for the work (viscous dissipation).

The work provided by the impeller to the liquid is equivalent to the power consumption, so the same can be calculated by the power number, according to Eq. (23).

$$W_{vc} = N_p \rho N^3 D_a^{\,5} \tag{23}$$

According Rosa [24], the heating of non-Newtonian liquids of the pseudoplastic type occurs with the constant U coefficient. Therefore, replacing Eq. (23) in Eq. (22), considering the constant U coefficient and integrating, you have:

$$ln\left(\frac{K_1 w_h c_{ph}(T_1 - T_b) - W_{vc}}{K_1 w_h c_{ph}(T_1 - T_{b_0}) - W_{vc}}\right) = -\frac{K_1 w_h c_{ph}}{Mc_{pc}}\theta \tag{24}$$

Eq. (24) should be solved by trial and error for the constant K_1 until the convergence between the right and left limbs occurs. The thermal exchange area is calculated from Eq. (25).

$$A = \frac{w_h c_{ph}}{U} ln\left(\frac{1}{1 - K_1}\right) \tag{25}$$

The solution of Eq. (25) implies the determination of coefficient U, which will be detailed in Section 4.2.

4.2 Overall heat transfer coefficient

Mathematically, the coefficient U is written according to the thermal exchange mechanisms however, in the heating and cooling of Newtonian and non-Newtonian liquids in tanks, convection is the predominant mechanism, so that the conduction and radiation can be despising, according to Eq. (26).

$$\frac{1}{U} = \frac{1}{h_i} + \frac{1}{h_o} \tag{26}$$

The internal n coefficient convection (h_i) is referring to the liquid traversing the interior of the heat exchanger, in this case the jacket or the serpentine or the vertical tubular baffle. Usually these liquids are Newtonians, so that the expressions for the calculation of the internal coefficient of convection are known in the literature.

Rosa and Moraes [1] discussed in detail the main equations used for the coefficient h_i in tanks with agitation, so in this topic, the main discussion will be in relation to the external convection coefficient (h_o).

The coefficient h_o represents the mechanism of forced convection occurring in the agitation of the non-Newtonian liquid, which depends on the geometry of the tank and the mechanical impeller, the flow and the physical properties of the liquid. The convection coefficient is calculated on the external surface of the heat exchanger based on the condition of equal thermal flow between the liquid traversing the interior of the surface and the liquid being agitated (Eq. (27)).

$$h_o = -\frac{k'(\partial T/\partial r)_{r=R}}{(T_s - T_{00})} \tag{27}$$

The solution of Eq. (27) depends on the determination of the temperature profile of the non-Newtonian liquid in agitation. The temperature profile can be calculated from the application of the energy equation (Eq. (28)) to the control volume (tank with agitation).

$$\rho c_p DT/D\theta = \mathrm{k}'\nabla^2 T + \eta\Phi' + \beta' TDp/D\theta + q''' \tag{28}$$

It is not possible to calculate the convection coefficient in an analytical way that the solution of Eq. (28) depends on the previous solution of the equation of the movement quantity (Eq. (10)) and as previously seen is not feasible due to the complex geometry of the tank.

As an alternative, as in power consumption, the convection coefficient is obtained by an empirical model, obtained through the Buckingham's Pi theory (Eq. (29)):

$$\frac{h_o D_t}{k'} = K\left(\frac{N D_a{}^2 \rho}{\eta}\right)^a \left(\frac{C_p \eta}{k'}\right)^b \left(\frac{\eta}{\eta_w}\right)^c \tag{29}$$

In Eq. (29), the coefficient h_o is calculated from the number of the Nusselt, being a function of the number of Reynolds, Prandtl number and relation between the apparent viscosities calculated at the "bulk" temperature of the liquid in agitation and the viscosity calculated at the temperature of the external surface of the heat exchanger. The constant and exponents of Eq. (29) are experimentally determined according to the type of tank, the type of mechanical impeller, the type of heat exchanger and whether the tank contains baffles.

However, the use of Eq. (29) depends on two parameters: (1) The choice of a rheological model suitable for non-Newtonian fluid in agitation and (2) The model that will relate the shear rate occurring with the rotation of the mechanical impeller.

As quoted, most of the non-Newtonian liquids are pseudoplastics and follow the model of the law of the powers, so that the shear rates and the rotation of the mechanical impeller are mainly related by the models of Metzner and Otto [15] and Calderbank and Moo-Young [16]. But as if seen ahead, several researchers have used other ways to relate the shear rate with the rotation of the mechanical impeller.

Carreau et al. [25] conducted a study in the heating and cooling of pseudoplastic fluids (aqueous solutions of carboxymethylcellulose and Carbopol 934) in a tank with an internal diameter of 0.76 m equipped with an axial impeller with four paddles inclined to 45° and a simple jacket. The study was carried out in transient regimen and the authors neglected the effect of viscous dissipation caused by mechanical impeller to agitation fluid.

The authors used the method of Wilson similar to using by Chilton et al. [26] to obtain the parameters of Eq. (29); however, they noticed a convergence problem in the method in relation to obtaining the exponent of the number of Prandtl, due to the variation of the apparent viscosity with the rotation of the mechanical impeller.

In this way, the authors introduced a concept called differential viscosity (μ_d) that is nothing more than the relationship between the derivative of shear stress by the shear rate; however, for values of shear rates above 500 s^{-1}, the differential viscosity remains constant, as shown in Eq. (30). This is a reasonable consideration because at high rotations of the mechanical impeller, the apparent viscosity of pseudoplastic fluids tends to remain constant:

$$\mu_d = \left(d\tau/d\nabla\vec{v}\right)_{\nabla\vec{v}\to 00} \tag{30}$$

The authors presented a variation of the Reynolds number based on the concept of Metzner and Otto [15] and Calderbank and Moo-Young [16] as shown in Eq. (31).

$$Re_{Carreau} = \frac{N^{2-n}{D_a}^2\rho}{(k/8)\left(\frac{6n+2}{n}\right)^n} \tag{31}$$

The authors proposed two correlations, one for heating and one for cooling, according to Eqs. (32) and (33), respectively:

$$Nu = 3.41(Re_{Carreau})^{0.67}(Pr_{Carreau})^{0.33} \tag{32}$$

$$Nu = 1.43(Re_{Carreau})^{0.67}(Pr_{Carreau})^{0.33} \tag{33}$$

In Eqs. (32) and (33), the absence of the relative term of viscosities in temperature "*Bulk*" by the temperature of the wall is noted, which caused a great difference between the proportionality constants, 3.41 for the heating and 1.43 for the cooling, although the inclinations are the same, based on the equality of the exponents of the Reynolds and Prandtl numbers. Eqs. (32) and (33) have an error of 11.8 and 14.0%, respectively.

In order to engage the heating and cooling phenomena in a single equation, the authors used the differential viscosity as a function of a modified exponent, as shown in Eq. (34).

$$Nu = 1.474(Re_{Carreau})^{0.70}(Pr_{Carreau})^{0.33}\left(\frac{\mu}{\mu_{dw}}\right)^{0.24/n} \tag{34}$$

The error of Eq. (34) is 19.3%, slightly higher than the error of Eqs. (30) and (31). In terms of engineering design, the Nusselt equation, if possible, should be able to predict heating and cooling in the same system of agitation, aiming at ease for the engineer, but without losing the predictability of the phenomenon. Eq. (34) has validity for Reynolds between 100 and 5000, Prandtl in the range of 100–800, and consistency index between 0.343 and 0.633.

Hagedorn and Salamone [27] carried out a study of heating pseudoplastic fluids in a tank operating in a batch with a jacket, aiming to obtain an expression that would allow calculating the convection coefficient with axial, radial and anchor impellers. The authors applied the constitutive equations of continuity, amount of movement and energy in cylindrical coordinates and resolved them with dimensional analysis, as presented in Eq. (35). In the dimensional analysis, the authors used the concept of Metzner and Otto [15] for the average shear rate:

$$Nu = K{Re_{MO}}^{\left[\frac{a}{n+1}+b\right]}{Pr_{MO}}^{c}(Vi_{MO})^{d}\left(\frac{D_t}{D_a}\right)^{e}\left(\frac{W}{D_a}\right)^{f}n^{g} \tag{35}$$

Eq. (35) is valid for Reynolds between 35 and 680,000, Prandtl in the range of 2–23,600 and consistency index between 0.36 and 1.0. A large variation of the Reynolds number is observed, which is explained using water (low viscosity) and high rotations. **Table 2** shows the exponents of Eq. (35) for the impellers used in the work.

Sandall and Patel [28] analyzed the heating of pseudoplastic fluids in tank equipped with Jacket, using two types of mechanical impellers, radial turbine with six flat blades and an anchor-type impeller. The fluids used were aqueous solutions of Carbopol and also two Newtonian fluids (water and glycerin), aiming to increase the number of Reynolds as it was done in the work of Hagedorn and Salamone [27].

The tank has an internal diameter of 0.18 m and baffles were used with the radial impeller and, with the anchor-type impeller, the tank was without baffles. The average shear rate required for calculating the adimensionals of Eq. (29) was based on the concept of Calderbank and Moo-Young [16]. Eqs. (36) and (37) present the Nusselt expressions for the radial impeller and anchor-type, respectively:

$$Nu = 0.315(Re_{CM})^{0.67}(Pr_{CM})^{0.33}(Vi_{CM})^{0.12} \tag{36}$$

$$Nu = 0.482(Re_{CM})^{0.67}(Pr_{CM})^{0.33}(Vi_{CM})^{0.12} \tag{37}$$

Eq. (36) has validity for Reynolds between 80 and 93,000, Prandtl in the range of 2.1–644 and consistency index between 0.35 and 1.0. Similarly, Eq. (37) is valid for Reynolds in the range 320–89,600, Prandtl between 2.1 and 644 and consistency index in the range from 0.35 to 1.0.

Comparing the error of the Nusselt equation to the turbine-type impeller (Eq. (36)), around 18%, with the error of the equation proposed by Hagedorn and

Impeller	K	a	b	c	d	e	f	g
Anchor	0.56	1.43	0	0.30	0.34	—	—	0.54
Paddle	2.51	0.96	0.15	0.26	0.31	−0.46	0.46	0.56
Axial	0.55	1.28	0	0.30	0.32	−0.40	—	1.32
Radial	3.57	1.25	0	0.24	0.30	—	0	0.78

Table 2.
Exponents of Eq. (35). Pseudoplastic fluids in tank with jacket.

Salamone [27], about 26.8%, it is noted that the model of Sandall and Patel [28] has a better fit to the phenomenon observed.

In 1971, Martone and Sandall [29] did a study of heat transfer in the agitation of sludge composed of calcium carbonate in volumetric concentrations of 12%, 15%, 23% and 26% and also water and glycerin to obtain a wide range of the Reynolds number. Rheology of these slurries follows the rheological model of Binghan (Eq. (38)), which has a yield stress.

$$\tau = \tau_0 + \mu_B \nabla \vec{v} \tag{38}$$

In this way, the number of Reynolds and, Prandtl and viscous ratio of Eq. (29) should be modified according to the initial voltage. Calcium carbonate slurries are heterogeneous suspensions in such a way that the influence of the volumetric fraction of the solid in the Nusselt number should be accounted for. Therefore, Eq. (29) is rewritten as presented in Eq. (39).

$$\frac{Nu}{(\varphi/\varphi - 1)^a} = K\left(\frac{D_a{}^2 N\rho}{\mu_b + (\tau_0/BN)}\right)^a \left(\frac{\mu_{ba} + (\tau_0/BN)}{k}\right)^b \left(\frac{\mu_b + (\tau_0/BN)}{(\mu_a + (\tau_0/BN))_w}\right)^c \tag{39}$$

Table 3 presents the constant and exponents of Eq. (39) for the radial impeller turbine and anchor used in the study, as well as the range of validity of the adimensional variables.

Heinlein and Sandall [30] used the heating of pseudoplastic fluids (aqueous solutions of Carbopol) and Binghan fluids (watery sludges) in a tank with an internal diameter of 0.18 m and anchor-type impeller, as shown in Eq. (40). The average shear rate used was based on the concept of Metzner and Otto [15].

$$Nu = C_1 (Re_{MO})^{0.50} (Pr_{MO})^{0.33} (Vi_{MO})^{0.18} \tag{40}$$

The values of the constant C_1 of Eq. (40) for some tank diameter ratios by impeller diameter is provided in **Table 4**.

Mitsuishi and Miyairi [31] considered that the rheological model described by the law of powers is not comprehensive enough to represent the entire amplitude of shear stress variation with deformation rates. Thus, they proposed to conduct an experimental study in the investigation of the heat transfer of non-Newtonian fluids in agitated tanks based on the Ellis rheological model, as presented in Eq. (41).

Impulsor	K	a	b	c	d	Reynolds	Prandtl	φ (%vol)
Radial	0.536	0.66	0.33	0.14	0.065	196–79,800	2.2–702	0–0.0113
Âncora	0.315	0.66	0.33	0.20	0.072	336–94,800	2.2–621	0–0.0116

Table 3.
Parameters of Eq. (39).

D_t/D_a	C_1	Error (%)
1.027	0.691	17.4
1.100	0.531	29.7
1.220	0.546	29.7

Table 4.
Constant values C_1 of Eq. (40).

$$\tau = -\left\{\frac{1}{\eta_0}\left[1+\left(\frac{\sqrt{\frac{1}{2}(\tau:\tau)}}{\tau_{\frac{1}{2}}}\right)^{\alpha'-1}\right]\right\}\nabla\vec{v} \tag{41}$$

The rheological model of Ellis can estimate shear stresses for shear rate values from 0.01 to 1000, due to the model having three adjustment constants; η_0, the apparent viscosity at the limit of the shear rate tending to zero, $\tau_{1/2}$; the shear stress corresponding to 50% of the variation of the deformation rate; and α, a constant of adjustment of the experimental data.

The tank under study has a diameter of 0.4 m and two radial impellers with two paddles with two diameters were used, one of 0.20 m and the other of 0.322 m. The tank doesn't contain baffles. The heating of the aqueous solutions of carboxymethylcellulose (CMC) and polyethylene (PO) was given by a simple jacket. Due to the complexity of Eq. (41), the authors arrived to a conclusion that the predominant flow with the radial impeller is of the tangential type. By similarity analysis, they deduced that the flow between two coaxial cylinders (as in a Searle-type-viscometer), as shown in **Figure 5**, was the same thing that happened during the agitation.

In **Figure 5**, an inconvenience is observed, referring to the hypothesis adopted by the authors regarding the similarity of agitation with the flow between the coaxial cylinders. As the smallest impeller used has 50% of the tank diameter and the other around 80.5%, a large part of the flow present in the agitation will be without phenomenological explanation, which in terms of design, is difficult to apply. The authors presented an equation for the apparent tangential viscosity, as shown in Eq. (42).

In the case of non-Newtonian fluids with high viscosities, the torque generated by the mechanical impeller when moving the fluid inside the tank should be considered; thus the apparent viscosity is defined in relation to the voltage generated in the tank wall, considering the hypothesis of not slipping the fluid near the wall. Thus, this apparent viscosity is also called apparent viscosity of potency (Eq. (43)):

$$\eta_{tan} = \frac{\eta_0}{2}\left\{\frac{[1-(D_a/D_t)2]/(D_a/D_t)2}{\frac{1}{2}\left(\frac{1}{(D_a/D_t)2}-1\right)+\frac{(T_{or})^{\alpha-1}}{2\alpha}\left(\frac{1}{(D_a/D_t)^{2\alpha}}-1\right)}\right\} \tag{42}$$

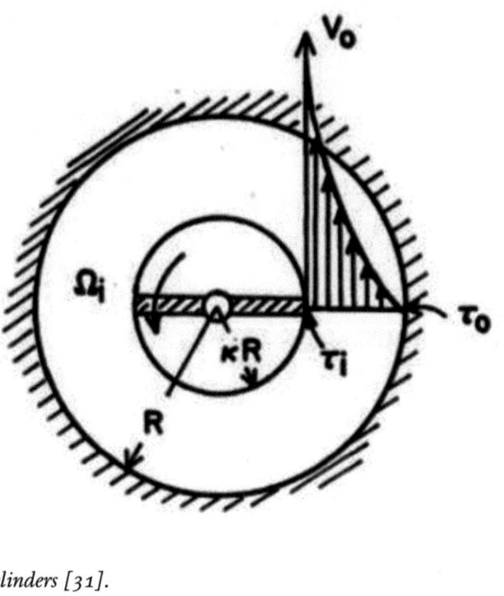

Figure 5.
Flow between two coaxial cylinders [31].

$$\eta_{pot} = \eta_0 \Big/ \left\{ 1 + \frac{4}{\alpha+3} \left[T_{orw} \left(\alpha 8N \frac{\eta_0}{\tau_{\frac{1}{2}}} \right) \right]^{\alpha-1} \right\} \tag{43}$$

With the two apparent viscosities defined by Eqs. (42) and (43), the expressions for the number of Nusselt tangential and on the wall, are obtained by Eqs. (44) and (45), respectively.

$$Nu_{tan} = K_{tan} (Re_{tan})^{a''} (Pr_{tan})^{0.33} (Vi_{tan})^{0.14} \tag{44}$$

$$Nu_{pot} = K_{pot} (Re_{pot})^{a'''} (Pr_{pot})^{0.33} (Vi_{pot})^{0.14} \tag{45}$$

The values of Nusselt tangential and Nusselt on the wall (power) are obtained graphically through a dimensionless variable defined as j_{Tan} and j_{Pot} respectively. You can observe in **Figure 6** the graphic ratio for the impeller with 50% of the tank diameter. The study is conducted by Mitsuishi and Miyairi [31] and also analysis of the heat transmission with a helical impeller wrapped in a *draft tube*.

Shamloo and Edwards [32] studied the heat transfer in the agitation of Newtonian and non-Newtonian fluids with high viscosities in a tank with a diameter of 0.15 m and another tank of 0.40 m. The mechanical impeller chosen was the helical type, and the heating jacket was the spiral type. The fluids used were chocolate, aqueous solutions of carboxymethylcellulose, glycerin, lubricating oils, silicon, sucrose solution, and solutions of Carbopol 940.

The authors proposed a model based on Eq. (29) for the Nusselt number adding two terms referring to the helical impeller: (1) Term referring to the number of impeller blades and (2) term referring to the distance between the impeller tip to the tank wall, as shown in the Eq. (46).

$$Nu = 0.568 (Re_o)^{0.23} (Pr_o)^{0.23} (n_b)^{0.23} (C'/D_t)^{-0.54} \tag{46}$$

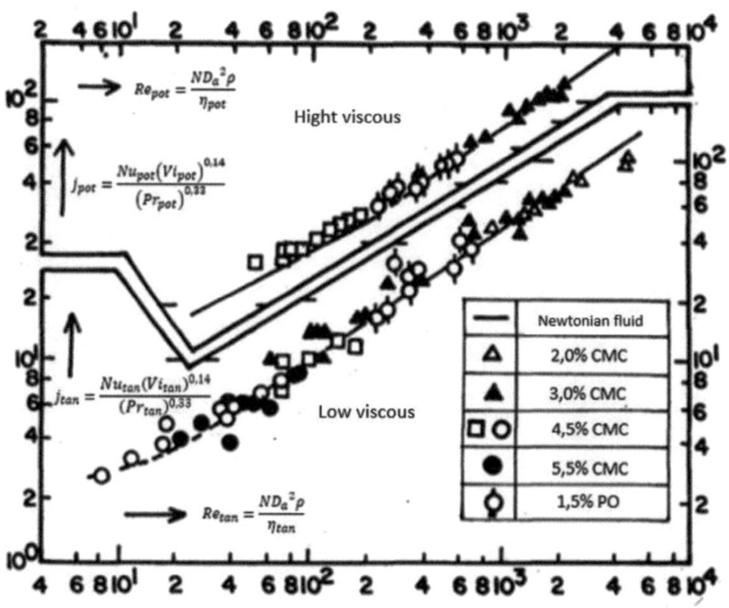

Figure 6.
Graphical relations for Eqs. (44) and (45) [31].

The Reynolds number and the number of Prandtl are given by Eqs. (47) and (48), respectively. It is noteworthy that these equations are proposed modified by Shamloo and Edwards [32], in which the average shear rate is initially based on Metzner and Otto [15] and adapted for helical impeller as presented in Eq. (49).

$$Re_o = ND_a{}^2\rho/\{[k34 - 144(c'/D_t)]N\}^{n-1} \tag{47}$$

$$Pr_o = c_p\{[k34 - 144(c'/D_t)]N\}^{n-1} \tag{48}$$

$$\nabla\vec{v} = [34 - 144(C'/D_t)]N \tag{49}$$

It is observed in Eq. (46) that the term for apparent viscosity variation was not included due to the narrow range of Reynolds number used that was between 0.01 and 10, which characterizes a fully laminar flow so that in this type of flow, the variation between the viscosities at the temperature *bulk* and the wall is negligible.

Suryanarayanan et al. [33] conducted a study of heating and cooling of pseudoplastic fluids (carboxymethylcellulose solutions) in a tank with baffles containing a jacket and a tank containing a helical serpentine. The impeller used in the study was a turbine type with four flat blades. The authors varied the impeller diameter in 78, 118 and 152 mm, the liquid level in the tank between 90 and 214 mm, the diameter of the helical serpentine in the range of 169–278 mm and the serpentine tube in the range of 15.9–22.2 mm. The average shear rate used in this work was based on the concept of Calderbank and Moo-Young [16].

In this study, the authors aimed to determine the effect of the variation of the level height, the diameter of the serpentine and the heat exchange itself, in addition to the traditional analysis of the number of Reynolds and Prandtl in the number of Nusselt. Eq. (50) presents the model obtained for the radial impeller with the use of the helical serpentine, with an experimental error of 7.1%, and in an analogous way and Eq. (51) provides to the model for the jacket with a 7.8% error:

$$Nu = 0.21(Re_{CM})^{0.66}(Pr_{CM})^{0.33}\left(\frac{D_a}{D_t}\right)^{0.17}\left(\frac{H}{D_t}\right)^{0.13}\left(\frac{D_c}{D_t}\right)^{-0.29}\left(\frac{D_i}{D_t}\right)^{-0.45} \tag{50}$$

$$Nu = 0.22(Re_{CM})^{0.63}(Pr_{CM})^{0.33}\left(\frac{D_a}{D_t}\right)^{0.14}\left(\frac{H}{D_t}\right)^{0.09}\left(\frac{D_c}{D_t}\right)^{-0.21}\left(\frac{D_i}{D_t}\right)^{-0.35} \tag{51}$$

It is observed in Eqs. (50) and (51) that the authors neglected the effects of the apparent viscosity ratio at the temperature *bulk* apparent viscosity at the wall temperature, since they added to the model proposed in Eq. (29) four terms referring to the effect of the stirring system geometry. The relationship of these geometric terms generates very small numbers, although the exponents are significant, in such a way that more than 90% of the response of the Nusselt number in these equations is given only by the variation of the number of Reynolds and Prandtl. The validity range of the 62 equations for Reynolds between is 200 and 21,700, Prandtl in the range of 49–1220, and consistency index between 0.47 and 1.0.

Regarding the range of application of Eqs. (50) and (51), it should be satisfied for the Reynolds and Prandtl numbers and the employment of turbine type impeller with four flat blades.

Kai and Shengyao [34] did a study of heating and cooling of non-Newtonian fluids in agitated tanks with a different approach from the authors. Eq. (29) was modified in the following respects: (1) The Reynolds number contemplates the power consumed by the mechanical impeller as well as the Prandtl number, (2) a term relative to the influence of the impeller diameter ratio by the tank diameter, and (3) a term relative to the Impeller blades, quantity of impellers on the shaft and blade angulation. In Eq. (52) these modifications are in a condensed manner:

$$Nu = K(Re')^a (Pr')^b (Vi')^c \left(\frac{D_a}{D_t}\right)^d \left(\frac{n_b n_n b sin\omega}{H}\right)^e \tag{52}$$

The Reynolds number is calculated from Eq. (53) based on the concepts provided by Kai and Shengyao [34] presented in Eqs. (54), (55) and (56) and similarly, the Prandtl number in Eq. (57).

$$Re' = \frac{N D_a{}^2 \rho}{k} 0.4^{(1-n)} N^{[2-f(2-n)](1-n)/n} \tag{53}$$

$$f' = \exp(-m Re_m) \tag{54}$$

$$\nabla\vec{v} = k''^{1/n} N^{[2-f'(2-n)]/n} \tag{55}$$

$$k'' = K_p D_a{}^5 \rho / 2\pi k_1 V k \left(D_a{}^2 \rho / k\right)^{f'} \tag{56}$$

$$Pr' = \frac{c_p k}{k'} 0.4^{(1-n)} N^{[2-f(2-n)](1-n)/n} \tag{57}$$

The parameters m and k depend on the type of flow, the viscosity of the fluid and the type of impeller, however, the authors found the values of 0.00705 and 0.4 for m and k, respectively, based on the best conditions for adjusting the mathematical models.

As an example of Eq. (52), for cooling with vertical tube baffles and radial impeller, with a range of Reynolds between 26 and 6310, according to Eq. (58).

$$Nu = 1.19(Re')^{0.67} (Pr')^{0.33} (Vi')^{0.17} \left(\frac{n_p n_n b sin\omega}{H}\right)^{0.74} \tag{58}$$

Hai Devotta and Rao [35] studied the heat transfer with Newtonian and non-Newtonian fluids in a stirring system with the use of helical impeller and heating jacket. Two tanks were used, one with 75 mm and the other with 80 mm diameter, the height of the liquid level and the number of impeller blades ranged. The average shear rate was calculated from Eq. (59).

In Eq. (59), the expression for the calculation of the Nusselt number is displayed, which is valid for Reynolds in the range of 10–1000 and the consistency index between 0.45 and 1.0.

$$Nu = 0.55(Re)^{0.48} (Pr)^{0.33} (Vi)^{0.14} (H/D_t)^{-0.44} \tag{59}$$

Triveni et al. [5] conducted a study in the heating and cooling of castor oil, liquid soap, carboxymethylcellulose solutions and calcium carbonate solutions. The heat transmission was promoted in a tank with a diameter of 0.29 m, by a single-type helical serpentine immersed in the tank, and the impellers used were an anchor-type and a radial turbine. The average shear rate was based on the concept of Calderbank and Moo-Young [16].

In this study, the authors incorporated Eq. (29), the term concerning the formation of vortices (number of Froude), the influence of natural convection (number of Grashof) and, finally, the influence of aeration on heat transmission, as presented in Eq. (60).

$$Nu = K(Re_{CM})^a (Pr_{CM})^b (Vi_{CM})^c (Fr_0)^d (Gr_{CM})^e \left(\frac{v_g}{ND_a}\right)^f \tag{60}$$

The influence of aeration, described by the term $\frac{v_g}{ND_a}$ is irrelevant in increasing heat transmission by both the axial impeller and the radial impeller. However, the effects of natural convection are considerable in the anchor-type impeller due to its low rotation of operation, favoring the formation of natural convection currents in the system. However, in relation to the turbine-type impeller, this effect is negligible due to the great turbulence achieved by this impeller. The Froude number presents significance only in systems without baffles or with low rotations, as in the case of the anchor-type impeller.

Eq. (61) presents the prediction of the number of Nusselt with the turbine-type impeller, despising the effects of natural convection and aeration:

$$Nu = 0.514 {Re_{CM}}^{0.598} {Pr_{CM}}^{0.335} {Vi_{CM}}^{0.112} {Fr_0}^{-0.179} \tag{61}$$

Several works present countless forms of the Nusselt equation for each type of non-Newtonian fluid in agitation in the tanks, for example, Pimenta and Campos [36], who studied the effects of viscoelasticity on the heat transmission of Non-Newtonian solutions in tanks equipped with helical serpentine and laminar flow. Viscoelasticity can be represented by the Weissenberg number (Eq. (62)), which represents the relaxation time of the fluid after the application of a shear rate.

The authors also incorporated the Nusselt number prediction model with the effects of secondary flows occurring around the surface of the helical serpentines, characterized by the number of Dean (Eq. (63)). The proposed Nusselt equation (Eq. (64)) is valid for the heating of carboxymethylcellulose and xanthan gum, with Prandtl between 17 and 203, with a 30% error in the calculation of the Nusselt number. The average shear rate was based on the concept of Metzner and Otto [15]:

$$Wi = \lambda'' v / D_{ic} \tag{62}$$

$$De = Re_{MO} (D_{ic}/D_c)^{1/2} \tag{63}$$

$$Nu = 0.486 \left[\left(\frac{3n+1}{4n}\right)^{0.275n} \left(0.717 + 0.993 \frac{D_{ic}}{D_c}\right) (Re_{MO} Pr_{MO})^{0.275} \right] \left(1 + 0.728\, De^{0.225}\right) (Wi + 1)^{0.011(n-1)} \tag{64}$$

5. Example of project

One 1 m^3 internal diameter tank and containing four baffles will be used for the heating of 794 kg of an aqueous solution of carboxymethylcellulose (CMC) concentration of 1.0% (w/w) from 20 to 40°C. The heating will be carried out through a simple jacket, in which you will go through hot water with inlet temperature of 60°C with a flow rate of 2000 kg/h. The mechanical impeller employed is of radial turbine type with six flat blades, which provides a rotation of 100 rpm. Calculate the thermal exchange area (design) that the jacket must possess to make the solution warm up in 40 min. The tank is insulated with rigid polyurethane foam, so that heat loss to the external environment can be considered negligible. The dimensions of the tank and its internals follow the proposals by Rushton et al. [12]—See **Figure 4**. The solution of CMC is pseudoplastic type, following the model of the law of the powers for variation of apparent viscosity. **Table 5** shows the physical and transport properties of cold and hot liquids in the average heating and cooling temperatures.

The process described occurs in transient and batch regime, so the thermal change area of the jacket is calculated from Eqs. (65) and (66) previously presented in Eqs. (24) and (25) (Item 4.1).

$$ln\left(\frac{K_1 w_h c_{ph}(T_1 - T_b) - W_{vc}}{K_1 w_h c_{ph}(T_1 - T_{b_0}) - W_{vc}}\right) = -\frac{K_1 w_h c_{ph}}{Mc_{pc}}\theta \tag{65}$$

$$A = \frac{w_h c_{ph}}{U} ln\left(\frac{1}{1 - K_1}\right) \tag{66}$$

The non-Newtonian solution for being pseudoplastic was treated rheologically by the model of the power law and the relationship between the rotation of the mechanical impeller and the shear rate, described by the model of Metzner and Otto [15].

Initially, the Reynolds number of the solution at the average temperature of the non-Newtonian liquid should be calculated, in this case 30°C, according to Eq. (67).

$$Re_{MO} = \frac{N^{2-n}.{D_a}^2\rho}{k.{k_s}^{n-1}} = 340.83 \tag{67}$$

With N of (100/60) rps, n of 0.66, D_a of 0.33m (1/3 de D_t), ρ of 1010 kg/m³, k of 1.468 and k_s of 11.5 (**Table 1**). Usually, the number of power is obtained in graphs according to the type of impeller and the presence or not of baffles in the tank. In this example, the impeller is of radial type with six flat blades, so that the work of Metzner and Otto [15] shows the power number curve with the Reynolds number in the laminar flow.

Therefore, the power number read in the graph is presented in Eq. (68) and the work provided by the impeller the solution in Eq. (69).

$$N_p = 3.8 \tag{68}$$

$$W_{vc} = N_p \rho N^3 {D_a}^5 = 69.5W \tag{69}$$

With ρ of 1010 kg/m³, N of (100/60) rps e D_a of 0.33 m

Replacing the provided data and the calculated work in Eq. (69), by trial and error, you get the constant K_1 with w_h of 0.556 kg/s, C_{ph} of 4180 J/kg°C (in T, de 60°C), T_1 of 60°C, T_b of 40°C, T_{bO} de 20°C, C_{pc} of 4580 J/kg°C e M of 794 kg.

Property	Water (45°C)	CMC 1% (30°C)
ρ (kg/m³)	985.62	1010
C_p JKG°C)	4189.9	4580
K′ (W/M°C)	0.686	0.624
μPa.S	0.000609	—
K (Pa. S^n)	—	1.468 (30°C); 0.95 in (45°C)
n	—	0.66 (30°C); 0.69 in (45°C)
β (1/s)	0.00013	—

Table 5.
Physical and transport properties [22].

$$K_1 = 0.47 \tag{70}$$

Therefore, Eq. (70) can be rewritten as:

$$A = \frac{1475.51}{U} \tag{71}$$

The coefficient U is obtained from the internal and external convection coefficients, as presented in Eq. (72) (previously presented in Eq. (26)).

$$\frac{1}{U} = \frac{1}{h_i} + \frac{1}{h_o} \tag{72}$$

The internal coefficient convection (h_i) is referring to hot water traversing the jacket. According to Silveira [37], the coefficient h_i can be calculated for a simple jacket with the heating fluid seeping upward according to Eq. (73).

$$h_i = k' 0.15 \left(\frac{c_p \mu}{k'}\right)^{\frac{1}{3}} \left[\frac{\rho^2 g \beta' LMTD}{\mu^2}\right]^{\frac{1}{3}} = 836.9 W/m^{2\circ}C \tag{73}$$

With properties on the T_m of 45°C, β' of $1.3 \times 10^{-4} {}^\circ C^{-1}$ and LMTD obtained by $\frac{60-30}{\ln\left[\frac{60-40}{30-20}\right]} = 43.3°C$.

The external coefficient convection will be determined by the model provided by Hagedorn and Salamone [27] for the turbine-type radial impeller (Eq. (74)). Noting that Eq. (74) already has the exponents referring to the radial impeller, obtained previously in **Table 2**.

$$\frac{h_o D_t}{k'} = 3.57 Re_{MO}{}^{0.75} Pr_{MO}{}^{0.24} (Vi_{MO})^{0.30} n^{0.78} \tag{74}$$

The number of Prandtl and the relationship between viscosities are calculated in Eqs. (75) and (76), respectively.

$$Pr_{MO} = c_p k (k_s N)^{n-1} / k' = 3926.16 \tag{75}$$

$$Vi = \eta / \eta_w = \left[k (k_s N)^{n-1}\right]_{t_b} / \left[k (k_s N)^{n-1}\right]_{Tw} = 1.42 \tag{76}$$

with n of 0.66 for T_b of 30°C and n of 0.69 for T_w 45°C.

In Eq. (76), the "bulk" temperature was the average heating of the non-Newtonian solution and the temperature in the wall (T_w) was considered the average temperature of the heating fluid (hot water at 45°C). The jacket or any other heat exchanger in tanks is made of copper or some other metal that has high thermal conductivity, in order to resist the heat transmission between the jacket wall and the tank being negligible.

Replacing Eqs. (67), (75) and (76) in Eq. (74) with K from 0.624 W/m°C and D_t from 1 m, you have:

$$h_o = 1052.86 \ W/m^{2\circ}C \tag{77}$$

Soon, the coefficient U by Eq. 72:

$$U = 466.27 \ W/m^{2\circ}C \tag{78}$$

Finally, the thermal exchange area is obtained by replacing the coefficient U in Eq. (71):

$$A = 3.16\, m^2 \tag{79}$$

The total height of the tank is calculated by the sum of the useful height (referring to the liquid level) and a safety margin to avoid transshipments. The height of the tank has the same value of the inner diameter, in this case 1 meter. Therefore, in addition to 15%, it is

$$H_{total} = H + 0.15H = 1.15m \tag{80}$$

The tank has a total lateral area of 3.61 m^2, while the jacket had its design area in 3.16 m^2, which shows the coherence of the calculation obtained.

6. Conclusion

In order to present a compact and practical way for undergraduate students and professionals in the field, the basic concepts of agitation with non-Newtonian liquids in tanks with mechanical impellers were presented in the text. The text also included a brief review of the design equations for tanks with heat exchange for non-Newtonian liquids. The work was completed with an example of a heating design with a pseudoplastic liquid jacket.

Acknowledgements

The authors acknowledge the Universidade Santa Cecília and Rodrinox Ind. Com. Ltda for all the research support and execution.

Nomenclature

A	thermal exchange area (m^2)
Cp	specific heat (J/kg°C)
Cp_c	specific heat of the cold fluid (J/kg°C)
Cp_h	hot fluid specific heat (J/kg°C)
D_a	diameter of mechanical impeller (m)
D_t	internal diameter of tank (m)
E	distance from impeller to bottom of tank (m)
E_{VC}	total energy in control volume (J)
g	gravitational acceleration (m/s ^{2})
h_e	specific enthalpy of the input mass flows in the control volume (J/kg)
h_s	specific enthalpy of output mass flows in control volume (J/kg)
h_i	internal convection coefficient (W/m^2/°C)
h_o	external convection coefficient (W/m^2/°C)
H	height of liquid level (m)
J	width of baffles (m)
k	consistency factor of the law model of the Powers (Pa.s^n)
k'	thermal conductivity (W/m°C)
L	blade length of mechanical impeller (m)

LMTD	logarithmic average of temperature differences (°C)
M	fluid mass in tank (kg)
n_b	number of mechanical impeller blades
n	powers Law consistency index
N	rotation of mechanical impeller (rpm)
p	pressure (Pa)
P	power consumed by mechanical impeller (W)
Q	heat generation rate (W/m^2)
Q	heat transfer rate (W)
Q_{VC}	heat transfer rate between control volume (W)
T	hot fluid outlet temperature (°C)
T_1	hot fluid inlet temperature (°C)
T_m	average hot fluid temperature (°C)
t_b	"bulk" temperature (°C)
U	overall heat transfer coefficient ($W/m^{2o}C$)
v	velocity (m/s)
V	tank volume (m^3)
$\dot{V}$	volumetric flow rate (m^3/s)
w_h	hot fluid mass flow (kg/s)
W	blade width of mechanical impeller (m)
W_{Vc}	work provided to control volume by impeller rotation (W)

Greek letters

ρ	specific mass (kg/m^3)
μ	dynamic Viscosity (Pa.S)
μ_b	Binghan model Viscosity (Pa.S)
μ_d	differential viscosity ($Pa.s^n$)
μ_{Dw}	differential viscosity at wall temperature ($Pa.s^n$)
η	apparent viscosity (Pa.S)
η_w	apparent viscosity at wall temperature (Pa.S)
η_0	apparent viscosity with shear rate at zero (Pa.S)
θ	time (min)
β	coefficient of thermal expansion
Φ	viscous dissipation (J/kg)
∇	operator Nabla
τ	shear stress (Pa)
τ_0	initial shear stress (Pa)
ω	angle of the blades of the mechanical impeller
λ	fluid relaxation time (s^{-1})

Adimensionals Numbers

Fr_0	Froude number, $Fr = N^2 D_a / g$
Gr	number of Grashof, $Gr = \beta' g \Delta T {D_a}^3 \rho^2 / \mu^2$
N_P	power number, $N_P = P / \rho N^3 {D_a}^5$
Naked	Nusselt number for stirring system, $Nu_o = h_o D_t / k'$
Pr_{Mo}	Metzner and Otto's Prandtl number, $Pr_{MO} = c_p k (k_s N)^{n-1} / k'$
Pr_{Cm}	Prandtl number of Calderbank and Moo-Young, $Pr_{CM} = c_p k \left(BN \left(\frac{4n}{3n+1} \right)^{n/(1-n)} \right)^{n-1} / k'$

Pr_o	Prandtl number of Shamloo and Edwards $Pr_o = c_p\{[k34 - 144(c'/D_t)]N\}^{n-1}$
Re	Reynolds number for agitation, $Re = ND_a{}^2\rho/\mu$
Re_{Mo}	Reynolds number of Metzner and Otto, $Re_{MO} = N^{2-n}D_a{}^2\rho/k(k_s)^{n-1}$
Re_{MR}	Reynolds number of Metzner and Reed, $Re_{MR} = \frac{\rho v^{2-n}D_i{}^n}{8^{n-1}k}\left(\frac{4n}{3n+1}\right)^n$
Re_{Cm}	Reynolds number of Calderbank and Moo-Young, $Re_{CM} = ND_a{}^2\rho/k\left(BN\left(\frac{4n}{3n+1}\right)^{n/(1-n)}\right)^{n-1}$
Re_m	modified Reynolds number, $Re_m = N^{2-n}D_a{}^2\rho/k$
Re_{Carr}	Reynolds Number of Carreau, $Re_{Carreau} = \frac{N^{2-n}D_a{}^2\rho}{(k/8)\left(\frac{6n+2}{n}\right)^n}$
Re_o	Reynolds Number of Shamloo and Edwards $Re = ND_a{}^2\rho/\{[k34 - 144(c'/D_t)]N\}^{n-1}$
Vi	relationship between the apparent viscosity of the fluid in the "bulk" temperature by the viscosity of the fluid in the wall temperature, $Vi = \eta/\eta_w$

Author details

Vitor da Silva Rosa* and Deovaldo de Moraes Júnior
Chemical Engineering Department, Santa Cecília University, São Paulo, Brazil

*Address all correspondence to: victor@unisanta.br

References

[1] Rosa VS, Moraes Júnior D. In: Murshed SMS, Lopes MM, editors. Design of Heat Transfer Surfaces in Agitated Vessels, Heat Exchangers. Rijeka: IntechOpen; 2017. DOI: 10.5772/66729

[2] Geankoplis CJ, Hersel AA, Lepek DH. Transport Processes and Separation Process Principles. 5ª ed. Boston: Prentice Hall; 2018

[3] McCabe WL, Smith JC, Harriot P. Unit Operations of Chemical Engineering. 7ª ed. Boston: McGraw-Hill; 2005

[4] Coulson JM, Richardson JF, Backhurst JR, Harker JH. Chemical Engineering. 5ª ed. Oxford: Butterworth Heinemann; 1996

[5] Triveni B, Vishwanadham B, Madhavi T, Venkateshwar S. Mixing studies of non-Newtonian fluids in an anchor agitated vessel. Chemical Engineering Research and Design. 2008; **88**(7):809-818

[6] Tanner RI. Engineering Rheology. 2ª ed. New York: Oxford University Press; 2002

[7] Ostwald VW. Ueber die geschwindigkeitsfunktion der viskosität disperses systeme. Colloid & Polymer Science. 1925;**36**:157-167

[8] Chhabra RP, Richardson JF. Non-newtonian Flow and Applied Rheology. Engineering Applications. 2ª ed. Oxford: Butterworth-Heinemann; 2011

[9] Green DW, Southard MA. Perry's Chemical Engineers' Handbook. 9ª ed. New York: Mc-Graw Hill; 2018

[10] Albright LF. Albright's Chemical Engineering Handbook. Boca Raton: CRC Press; 2008

[11] Paul EL, Atiemo-Ubeng SM. Handbook of Mixing. 1ª ed. New Jersey: Wiley-Interscience; 2004

[12] Rushton JH, Costich EW, Everett HJ. Power characteristics of mixing impellers. Chemical Engineering Progress. 1950;**46**(8):395-404

[13] Moraes Júnior D, Moraes MS. Laboratório de operações unitárias I. 1ª ed. Brasil: Edição do autor; 2011

[14] Ameur H, Bouzit M. Power consumption for stirring shear-thinning fluids by two-blade impeller. Energy. 2013;**50**:326-332

[15] Metzner AB, Otto RE. Agitation of non-Newtonian fluids. AICHE Journal. 1957;**3**(1):3-10

[16] Calderbank PH, Moo-Young MB. The power characteristics of agitators for the mixing of Newtonian and non-Newtonian fluids. Transactions. Institute of Chemical Engineers. 1961; **39**:337-347

[17] Metzner AB, Reed JC. Flow of non-Newtonian fluids – Correlation of the laminar transition, and turbulent-flow regions. AICHE Journal. 1955;(4): 434-440

[18] Tanguy PA, Thibault F, De La Fuente EB. A new investigation of the Metzner-Otto concept for anchor mixing impellers. Canadian Journal of Chemical Engineering. 1996;**74**: 222-228

[19] Bertrand J, Couderc IP. Evaluation of the power consumption in agitation of viscous Newtonian or pseudoplastic liquids by two-bladed, anchor or gate agitators. Chemical Engineering Research and Design. 1985;**63**:259 e63

[20] WU B. CFD investigation of turbulence models for mechanical

agitation of non-Newtonian fluids in anaerobic digesters. Water Research. 2011;**45**:2082-2094

[21] Sossa-Echeverria J, Taghipour F. Computational simulation of mixing flow of shear thinning non-Newtonian fluids with various impellers in a stirred tank. Chemical Engineering and Processing. 2015;**93**:66-78

[22] Rosa VS, Taqueda MES, Paiva JL, Moraes MS, Moraes Júnior D. Nusselt's correlations in agitated tanks using the spiral coil with Rushton turbine and PBT 45° impeller. Comparison with tanks containing vertical tube baffles. Applied Thermal Engineering. 2017;**110**: 1331-1342

[23] Rosa VS, Souza Pinto TC, Santos AR, Moino CAA, Tamião K, Lia LRB, et al. External coefficient of heat transfer by convection in mixed vessels using vertical tube baffles. Industrial & Engineering Chemistry Research. 2013; **52**(6):2434-2438

[24] Rosa VS. Heat transfer and scale-up in tanks with mechanical impellers in operation with non-Newtonian fluids (doctoral thesis). Brazil: University of São Paulo; 2017

[25] Carreau P, Charest G, Corneille JL. Heat transfer to agitated non-Newtonian fluids. The Canadian Journal of Chemical Engineering. 1966;**44**(1):3-8

[26] Chilton TH, Drew TB, Jebens RH. Heat transfer coefficients in agitated vessels. Industrial and Engineering Chemistry. 1944;**36**(6):510-516

[27] Hagedorn D, Salamone JJ. Batch heat transfer coefficients for pseudoplastic fluids in agitated vessels. Industrial and Engineering Chemistry Process Design and Development. 1967; **6**(4):469-475

[28] Sandall OC, Patel KG. Heat transfer to non-Newtonian pseudoplastic fluids in agitated vessels. Industrial and Engineering Chemistry Process Design and Development. 1970;**9**(1):139-144

[29] Martone JA, Sandall OC. Heat transfer to Bingham plastic slurries in agitated vessels. Industrial and Engineering Chemistry Process Design and Development. 1971;**10**(1):86-91

[30] Heinlein HW, Sandall OC. Low Reynolds number heat transfer to non-Newtonian fluids in anchor-agitated vessels. Industrial Engineering Chemistry Process Development. 1972; **11**:490-495

[31] Mitsuishi N, Miyairi Y. Heat transfer to non-Newtonian fluids in an agitated vessel. Journal of Chemical Engineering of Japan. 1973;**6**(5):415-420

[32] Shamlou PA, Edwards MF. Heat transfer to viscous Newtonian and non-Newtonian fluids for helical ribbon mixers. Chemical Engineering Science. 1986;**41**:1957-1967

[33] Yanarayanan S, Mujawar BA, Raja Rao M. Heat transfer to pseudoplastic fluids in an agitated vessel. Industrial and Engineering Chemistry Process Design and Development. 1976;**15**(4): 564-569

[34] Kai W, Shengyao Y. Heat transfer and power consumption of non-Newtonian fluids in agitated vessels. Chemical Engineering Science. 1989;**44**: 33-40

[35] Hai CL, Devotta I, Rao PG. Heat transfer to viscous Newtonian and non-Newtonian fluids using helical ribbon agitator. Chemical Engineering Journal. 2000;**79**:73-77

[36] Pimenta TA, Campos JBLM. Heat transfer coefficients from Newtonian and non-Newtonian fluids flowing in laminar regime in a helical coil. International Journal of Heat and Mass Transfer. 2013;**58**(1):676-690

[37] Silveira RG. Análise térmica de Sistema de refrigeração direta em fermentadores cilindrocônicos. Dissertação de mestrado – Universidade Estadual Paulista. 116 p. 2009

Chapter 4

Heat and Mass Transfer in Outward Convex Corrugated Tube Heat Exchangers

Huaizhi Han, Bingxi Li, Yaning Zhang, Quan Zhu and Ruitian Yu

Abstract

Heat and mass transfer in outward convex corrugated tube heat exchangers is of significant importance for the optimization, fabrication, and application of outward convex corrugated tube heat exchangers. This chapter gives a deep investigation of the heat and mass transfer in outward convex corrugated tube heat exchangers. Based on the experimental setup developed, the performances of a novel outward convex corrugated tube heat exchanger are presented. Simulation methods are then used to detail the heat and mass transfer at tube side and shell side of the outward convex corrugated tube heat exchanger, and these include the flow structure, temperature distribution, and turbulence kinetic energy. Heat and mass transfer enhancements of the outward convex corrugated tube heat exchanger are also studied, and they are from tube side, shell side, and overall system aspects. Finally, multi-objective optimization of the outward convex corrugated tube heat exchanger is conducted to obtain the optimal performances through using Response Surface Methodology (RSM) and Non-dominated Sorting Genetic Algorithm (NSGA-II). Main conclusions and future outlook are then briefly stated and summarized. We firmly believe that the contents presented in this chapter can not only enrich the knowledge of heat exchangers but also develop methods for studying heat exchangers.

Keywords: heat and mass transfer, heat exchanger, outward convex corrugated tube, enhancement, optimization

1. Introduction

The outward convex corrugated tube heat exchanger (CTHE) is a novel kind of shell and tube heat exchangers, which can be applied in many applications. Designing this kind of heat exchangers is considerable flexibility because the geometrical structure can be varied easily by altering the tube diameter, length, and arrangement [1, 2]. The exchanger can be designed for suffering high pressure condition. The exchangers are applied primarily for single phase and phase change heat transfer application. They could also be used for heat transfer applications with high operating temperature and/or pressure.

Figure 1 shows a bundle of outward convex corrugated tubes (CT) fabricated in the tubesheets, which is the most significant components in the CTHE. Two kinds of

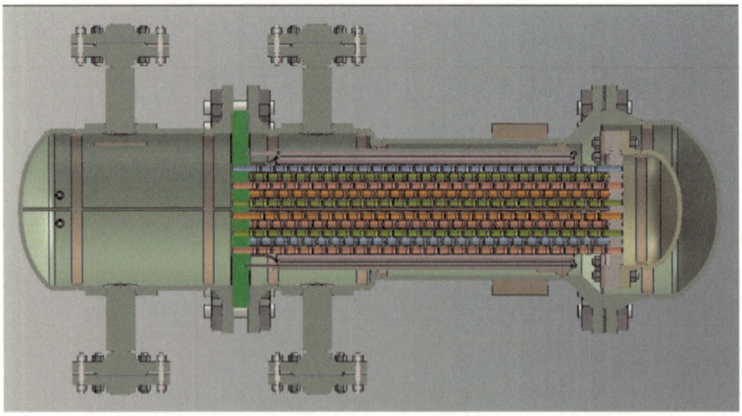

Figure 1.
The schematic of out outward convex corrugated tube heat exchangers.

fluids flow inside and outside of CT, respectively. Except the tube bundles, the major components of this exchanger also include shell, front-end head, and rear-end head.

The exchangers could be widely used in industry for the following reasons. (a) Wide capacity and operating conditions, such as from high vacuum to ultrahigh pressure (over 100 MPa) and from cryogenics to high temperatures (about 1100°C). (b) Special operating conditions: vibration, heavy fouling, highly viscous fluids, erosion, corrosion, toxicity, radioactivity, multicomponent mixtures, and so on. (c) The most versatile exchangers, made from a variety of metal and nonmetal materials (such as graphite, glass, and Teflon) and range in size from small to supergiant surface area. (d) Extensively applications: petroleum-refining and chemical industries; as steam generators, condensers, boiler feedwater heaters, and oil cooler in power plants; as condensers and evaporators in some air-conditioning and refrigeration applications; in waste heat recovery applications; and in environmental control [3–5].

The main difference between the CTHE and traditional heat exchangers is the adopted tube type. Traditionally, the inward intermittent or continuous type corrugated tubes are employed, as an example for both helically corrugated and transverse corrugated tubes, owing to their ease of realization. However, in engineering devices, it is necessary to adopt CT, which could be conveniently and periodically inspected with complete accessibility [6].

A schematic view of the CT configuration currently investigated is shown in **Figure 2**. The structure parameters of the outward corrugated tube include inner diameter (D), tube length (L), corrugation height (H), corrugation pitch (p), corrugation crest radius (R), and corrugation trough radius (r).

The design and improvement of the CT are considered a significant aspect of researches in terms of heat and mass transfer. Almost all of the heat transfer augmentation techniques have been introduced to improve the overall thermo-hydraulic performance. Thus, these techniques achieved reductions in the size and cost of heat exchangers.

2. Performances of outward convex corrugated tube heat exchangers

2.1 Manufacture of corrugated tube heat exchangers

Manufacture consideration could be divided into manufacturing equipment, processing, and other qualitative criteria. The equipment considerations determine

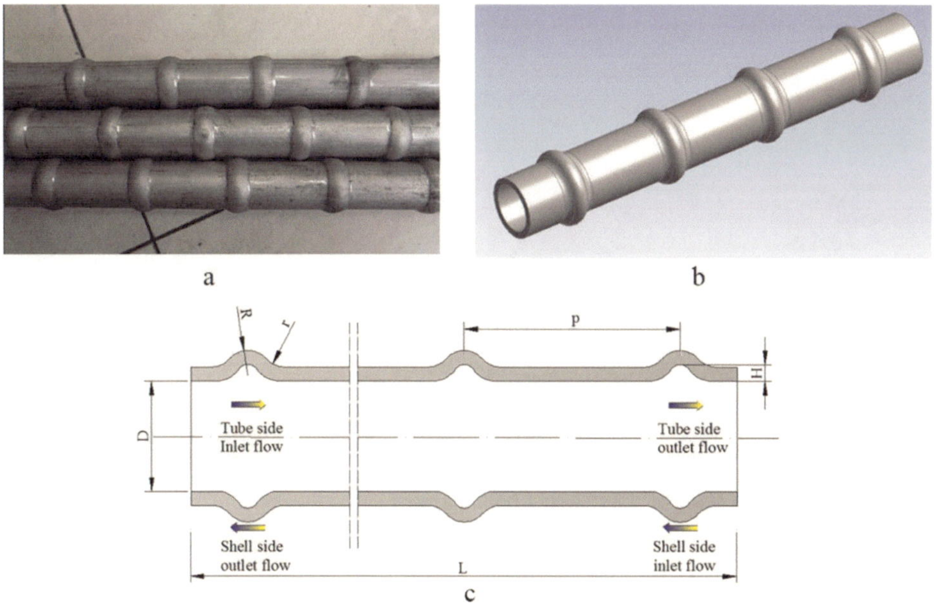

Figure 2.
The real and schematic view of the outward convex corrugated tube.

which design could be selected, which include existing and new tooling, availability and limitations of equipment, offline production, and investment funding. Processing considerations make sure how individual parts and components of a heat exchanger are manufactured and assembled, which including manufacture of individual parts, stacking of a heat exchanger core and eventual brazing, mounting of pipes, washing/cleaning of the exchanger, and leak testing in the system. When a heat exchanger is designed, the manufacturing equipment and the complete processing considerations must be evaluated previously, particularly for an extended surface heat exchanger [11, 12].

In the novel tube and shell heat exchanger, the structure of the outward convex corrugated tube is special, composed of alternating corrugated segment and straight pipe section. The main difference from traditional heat exchanger is the adopted structure, so the manufacture processing for the novel tube type is highlighted in this section. The working conditions of the heat exchanger are mainly for high temperature and pressure operation condition. To ensure the safe operation of heat exchanger, a thick-walled stainless steel tube with strong pressure resistance is selected as the base tube. For example, the mechanical properties of stainless steel tube material are as follows: yield strength is 390 MPa, material hardening index is 0.148, material strength coefficient is 764 MPa, material anisotropy coefficient is 0.83, material modulus of elasticity is 207GPa, and Poisson ratio is 0.28.

The outward convex corrugated tube is manufactured according to high pressure hydraulic bulking based on the smooth stainless steel tube. The hydraulic bulking equipment is 10,000 KN. As shown in **Figure 3(a)**, the equipment is assembled with 400 MPa internal high-pressure forming system, which is mainly composed of the supercharger, two horizontal push cylinder hydraulic servo system, and computer control system. The manufacturing process needs to be supplemented with the corresponding mold, installed on the hydraulic bulking equipment. The mold consists of three parts, which includes upper module as shown in **Figure 3(b)**, lower module as shown in **Figure 3(c)** and sealing punch. The inner mosaic block with corrugation shape is inserted in the mold as shown in

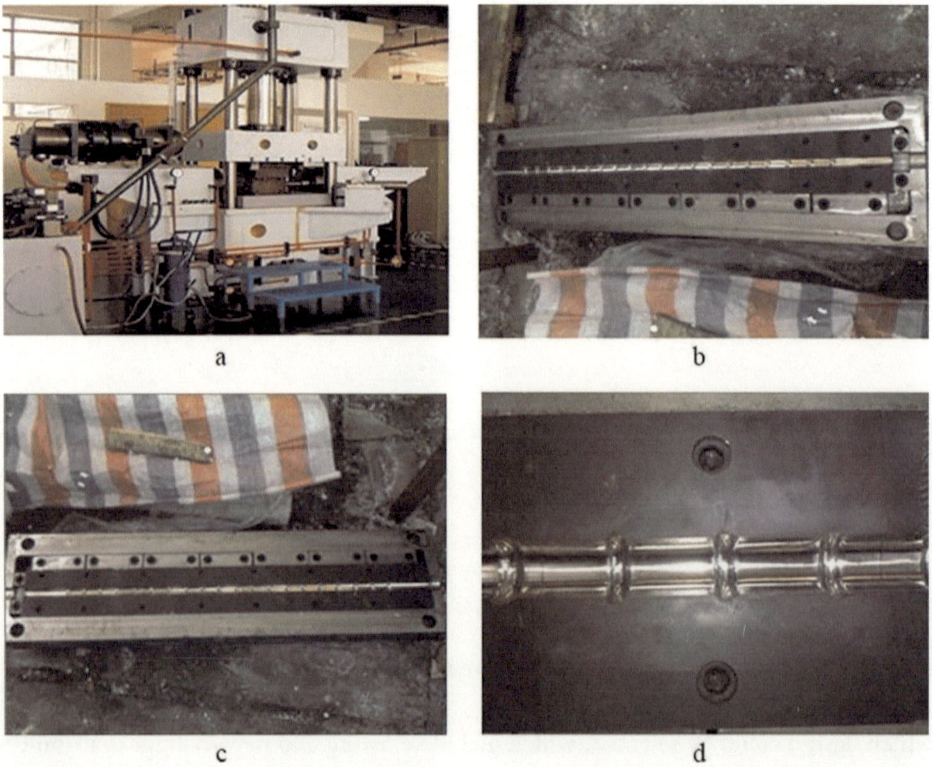

Figure 3.
The hydraulic bulging machine and mold. Based on the modified order as the above sticky. (a) Hydraulic bulking equipment; (b) upper module; (c) lower module; (d) inner mosaic block.

Figure 3(d). High pressure liquid (water or oil) is provided inside the smooth stainless steel tube and finally hydroforms the outward convex corrugated tube.

2.2 Layout of corrugated tube heat exchangers

In order to test the heat transfer and resistance performance of the corrugated tube heat exchangers, experimental study on the corrugated tube heat exchanger must be performed. We adopted steady-state techniques to establish the relationship between *Nu* and *Re*. Different data acquisition and reduction methods are used, depending on whether the test fluid is primarily a gas (air) or a liquid. A gas to gas heat exchange will be conducted in our experimental test.

The schematic of the experimental apparatus for outward corrugated tube is depicted in **Figure 4**. The system comprises a screw air compressor (the highest discharging pressure is 1.3 MPa, and the air displacement is fixed at 1.81 m^3/min), two pressure-regulating valves (0.3 MPa on the hot circuit and 0.9 MPa on the cold circuit), a heater (the temperature range is 50–500°C), a test section (operating with two groups of switching valves), a measuring system (two critical Venturi flowmeters, two pressure transducers, and two temperature transducers), a data acquisition system (DAS), and a pipe system (304 stainless steel tube).

The experimental medium was air, which was compressed by the helical-lobe compressor to a pressure of 1.25 MPa. The system is made of stainless steel devices and consists of the hot circuit and cold circuit. The pressure-regulating valves adjust the air pressure to 0.3 MPa on the hot circuit and 0.9 MPa on the cold circuit with an accuracy of ±2%. The critical Venturi flowmeters control the mass flow rate in the

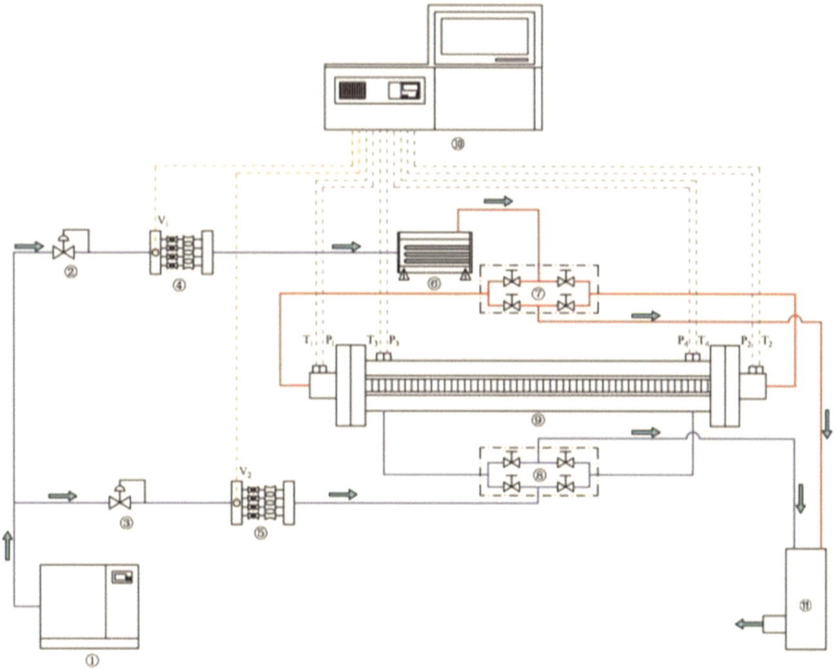

Figure 4.
System drawing of test bed. 1. Screw air compressor, 2. Pressure-regulating valve in hot circuit, 3. Pressure-regulating valve in cold circuit, 4. Critical Venturi flowmeter in hot circuit, 5. Critical flow meters in cold circuit, 6. Air heater, 7. Switching valves in hot circuit, 8. Switching valves in cold circuit, 9. Test section, 10. Data acquisition system, 11. Muffler.

hot and cold circuits. The air in the hot circuit is heated by the heater exchanger and then flows into the tube side of the test section, whereas the air in the cold circuit directly flows into the shell side. The section has a detachable structure, which enables convenient changes in various tube components. Moreover, the valve group in the vicinity of the test section makes the air flow into the tube, through either inlet of the tube side or the shell side, thus creating a uniform-current flow and a counter-current flow for each respective flow direction. Finally, the hot air and the cold air complete the heat exchange in the annular tubes of the test section, and then noise of them will be reduced through the muffler.

In the measuring system, the mass flow rates can be measured with two critical Venturi flowmeters on both circuits, with an accuracy of ±0.2%. The flow meter in the hot circuit was installed before the air heater because hot air may damage the flow meter or reduce the measurement accuracy (precision). After the heater, a temperature transducer was installed to monitor the air temperature. The DAS obtained the flow rate signal, which was transferred to a programmable logic controller (PLC) in the industrial personal computer (IPC), and the accuracy of the transformation module was ±0.05%. The pressure and temperature transducers were installed at the inlet and outlet of the section to measure the pressure and temperature of the air on both sides. All thermocouples were calibrated with an accuracy of ±0.1% of the test data. The pressure drop of the test section was measured with pressure transducers, which have an accuracy of ±0.2% and a measuring range of 0–5 kPa. The values were collected and displayed on the IPC and were automatically recorded.

The uncertainty is estimated with the method suggested by Kline and Moffat. As mentioned above, the measurement uncertainties of tube length and tube diameters are about 0.05 and 0.1%, respectively. In addition, the measurement accuracy of temperature is 0.14%, the measurement error of the differential pressure meter is 2.06%, and the critical Venturi flowmeter has a precision of 3.11%. According to the uncertainty propagation equation, the uncertainties in the values of experimental parameters like the Reynolds number, Nusselt number, and friction factor are 3.89, 4.41, and 4.87%, respectively.

2.3 Data acquisition

The main purpose of our experimental study is to construct the relationship among the heat transfer rate q, heat transfer surface area A, heat capacity rate c of each fluid, overall heat transfer coefficient U, and fluid terminal temperatures [10]. To conduct the heat transfer analysis of an exchanger, the basic relationships that are applied for this purpose are the energy balance based on the first law of thermodynamics, as outlined in Eq. (1).

$$Q = \dot{m}\ (i_2\text{-}i_1) \tag{1}$$

where $\dot{m}$ is the rate of mass flow, i_1 and i_2 represent the inlet and outlet enthalpies of the fluid, and Q is the heat transfer rate between hot fluid and cold fluid.

As shown in **Figure 5**, a two-fluid counterflow exchanger is considered to present variables relating to its thermal performance. Although flow arrangement may be different for different exchangers, the basic concept of modeling remains the same. The following analysis is intended to introduce important variables for heat exchanger.

If the fluids do not undergo a phase change and have constant specific heats with $di = c_p \cdot dT$, heat transfer rate released from the hot fluid (Q_h) and absorbed by the cold air (Q_c) can be expressed as

$$Q_h = \dot{m}_h C_{p,h}(T_{h_1} - T_{h_2}) \tag{2}$$

and

$$Q_c = \dot{m}_c c_{p,c}(T_{c_2} - T_{c_1}) \tag{3}$$

The subscripts h and c refer to the hot and cold fluids, and the numbers 1 and 2 designate the fluid inlet and outlet conditions, respectively.

Thus, the average value of the heat transfer rate is calculated as

$$Q_m = \frac{Q_h + Q_c}{2} \tag{4}$$

Eq. (5) reflects a convection-conduction heat transfer phenomenon in a two-fluid heat exchanger. The temperature difference between the hot and cold fluids ($\Delta T = T_h - T_c$) constantly changes along with heat exchanger. Therefore, in order to conveniently analyze the heat transfer performance of heat exchanger, it is

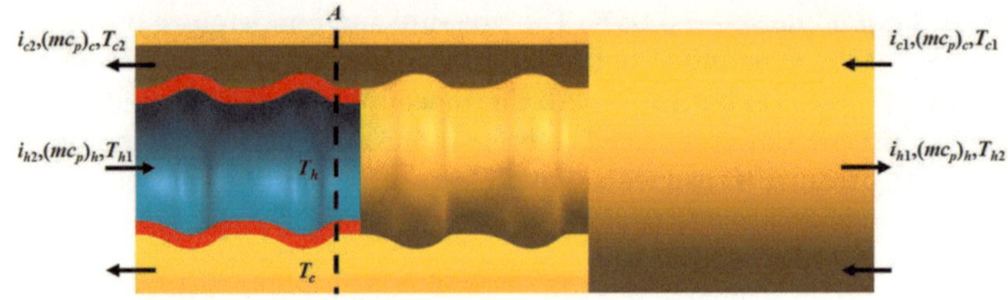

Figure 5.
The energy balance for the hot and cold fluids of a two-fluid heat exchanger.

important to establish an appropriate mean value of the temperature difference between the hot and cold fluids such that the total heat transfer rate Q between the fluids can be determined from

$$Q = UA\Delta T_{\mathrm{m}} \tag{5}$$

The heat transfer rate Q is proportional to the heat transfer area A, the average overall heat transfer coefficient based on the area U, and mean temperature difference $\Delta T_{\max}$ ΔT_{m} between the two fluids. This means that temperature difference is a log-mean temperature difference (for counterflow and parallel-flow exchangers).

$$\Delta T_{\mathrm{m}} = \frac{\Delta T_{\max} - \Delta T_{\min}}{\ln \frac{\Delta T_{\max}}{\Delta T_{\min}}} \tag{6}$$

$\Delta T_{\max}$ and $\Delta T_{\min}$, respectively, represent the maximum and minimum one between ΔT_1 and ΔT_2.

In the experiments, the tube-wall temperature was not measured directly. The heat transfer coefficient of the tube side (h_i) is determined from:

$$\frac{1}{U} = R_t = \frac{1}{h_i} + \frac{A_i \ln(r_o/r_i)}{2\pi kL} + \frac{A_i}{A_o h_o} \tag{7}$$

where r_i and r_o are the inner radius and outer radius of the test tube, respectively. A_{i} and A_{o} are the inner and outer surface area of the tube, respectively. k is the thermal conductivity of tube material, L is the length of the heat exchange tube, and h_i and h_{o} are the heat transfer coefficients for inside and outside flows, respectively.

The Nusselt number can be calculated as

$$Nu = \frac{h_i \cdot D_i}{k} \tag{8}$$

where D is the characteristic diameter; the thermal conductivity k is calculated from the fluid properties at the local mean bulk fluid temperature.

The Reynolds number is based on the average flow rate of the test section.

$$Re = \frac{D \cdot u \cdot \rho}{\mu} \tag{9}$$

where μ is the dynamic viscosity of the working fluid, and u is the mean velocity.

The friction factor (f) can be written as

$$f = \frac{\Delta p}{\frac{L}{D} \cdot \frac{\rho u^2}{2}} \tag{10}$$

where Δp is the pressure drop in the test section.

The performance evaluation criterion (PEC) is a dimensionless ratio, which is used for the evaluation of the overall performance of the enhanced tube and defined as follows:

$$PEC = (Nu_c/Nu_s)/(f_c/f_s)^{1/3} \tag{11}$$

When $PEC > 1$, it indicates that the enhanced tube has an advantage over the smooth tube; otherwise, the corrugated heat transfer component compares unfavorably with the smooth tube.

2.4 Heat and mass transfer performance

For the engineering applications and to design exchangers, the prediction of heat and mass transfer performance is important. We presented experimental data on the Nusselt numbers for turbulent regimes. In our experimental study, the hot fluid is at the tube side, and the cold fluid is at the shell side.

The heat transfer and resistance performance of corrugated tube are compared to smooth tube, aiming to reflect the superior of the corrugated tube. Ratio of *Nu* in the corrugated tube to that in the smooth tube (Nu_c/Nu_s) and ratio of f in the corrugated tube to that in the smooth tube (f_c/f_s) are adopted to indicate the enhancement degree of heat transfer and flow resistance performance.

Figure 6 shows the effect of Re_c (*Re* of the cold fluid) on Nu_c/Nu_s, f_c/f_s, and *PEC*, along with the changing Re_h (*Re* of the hot fluid). The figure exhibits that with the increase of *Re*, Nu_c/Nu_s, f_c/f_s, and *PEC* decline. The decreasing rate of Nu_c/Nu_s and *PEC* is almost linear, but f_c/f_s is decelerated.

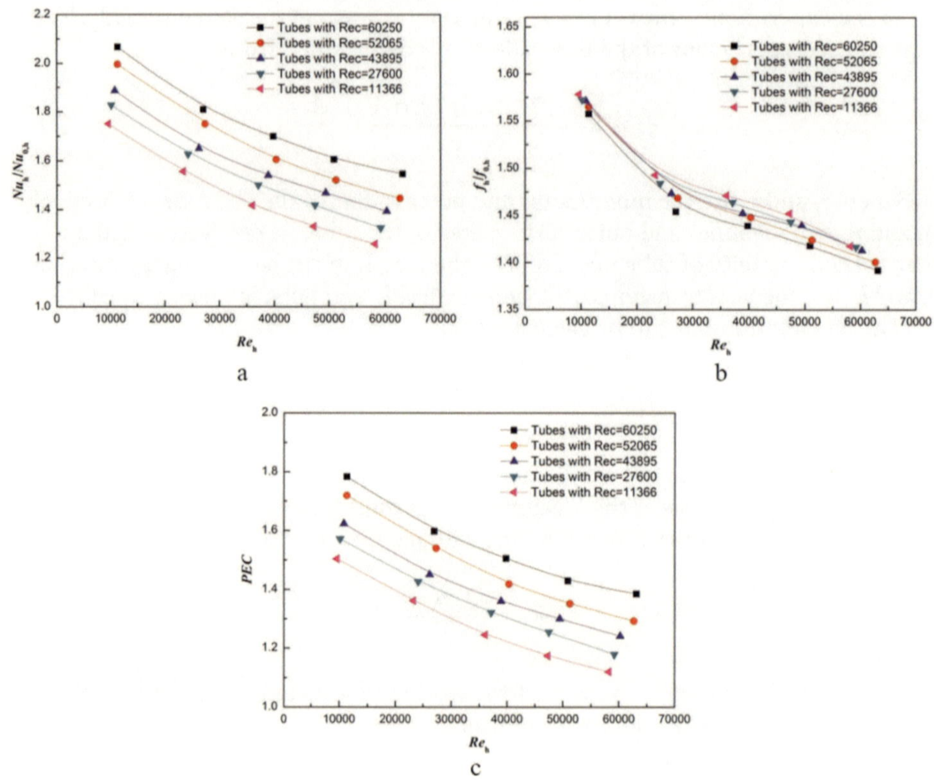

Figure 6.
Flow and mass transfer performance. (a) Nuc/Nus; (b) f/fs; (c) PEC.

3. Heat and mass transfer at outward convex corrugated tube heat exchangers

3.1 Numerical methods

3.1.1 Physical model and meshing system

The first task to accomplish in a numerical simulation is the definition of the geometry followed by the mesh generation. The geometry of the design needs to be created from the initial design. Any modeling software can be used for modeling and shifted to other simulation software for analysis purpose.

Figure 7 shows a schematic view of the structural parameters for corrugated tube investigated in this chapter, which include inner diameter (D), tube length (L), corrugation height (H), corrugation pitch (P), corrugation crest radius (R), and corrugation trough radius (r). Since the investigated corrugated tubes are used in tube-shell type heat exchanger, the flow region inside of tube is named "tube side" and out of tube is named "shell side."

Mesh generation is the process of subdividing a region to be modeled into a set of small control volumes. In general, a control volume model is defined by a mesh network, which is made up of the geometric arrangement of control volumes and nodes. Nodes represent points at which features such as displacements are calculated. Control volumes are bounded by set of nodes and also defined by the number of mesh. One or more values of dependent flow variable (e.g. velocity, pressure, temperature, etc.) will be contained in each control volume. Usually, these represent some type of locally averaged values. Numerical algorithms representing approximation to the conservation law of mass, momentum, and energy are then used to compute these variables in each control volume.

Mesh generation is often considered as the most important and most time consuming part of CFD simulation [13]. The quality of the mesh plays a direct role on the quality of the analysis, regardless of the flow solver used. In this work, a 3D non-uniform mesh system of hexahedral elements was established via the professional mesh generation software ICEM to accurately control the size and number of cells in the domain, as illustrated in **Figure 8**. The near-wall vicinity should be present

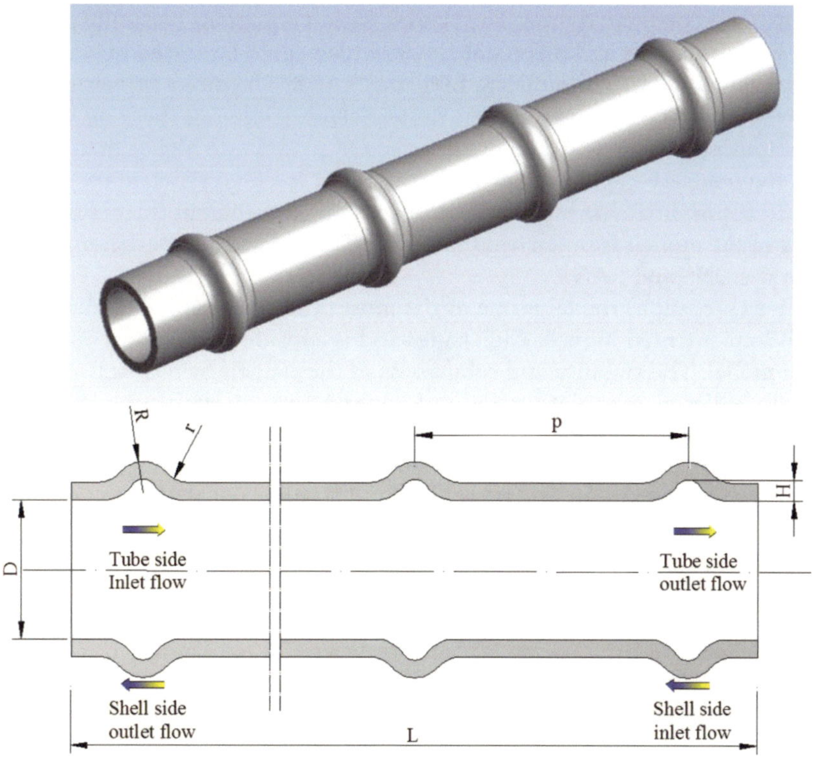

Figure 7.
Structure parameters of outward convex corrugated tube.

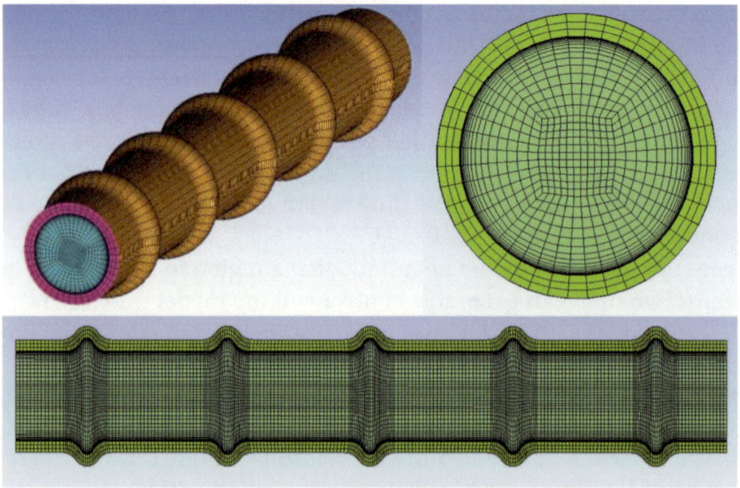

Figure 8.
Schematic diagram of meshing system for the simulated corrugated tube.

drastic velocity and temperature gradients, so a high density of gradient elements was applied in this region. Nevertheless, the remaining domain was modeled with relatively sparse elements. The first layer of thickness should satisfy $y^+ \approx 1$.

3.1.2 Mathematical model

Mathematical model should be constructed to numerically describe flow and heat transfer of corrugated tube. The Navier-Stokes equations generally are adopted to describe the laminar and turbulent flows, which could be solved by various kinds of simulation model including DNS, LES, and RANS. The direct numerical simulation (DNS) can solve accurately the turbulent fluctuation, but these models require huge computing power, which is many orders of magnitude higher than other models. Reynolds-averaged Navier-Stokes (RANS) is a high efficient model that can be used to approximate turbulence by time-averaged turbulent fluctuation, but the accuracy of the models is much less than DNS. The accuracy and efficient of LES are between the DNS and RANS.

The k-ε (k-epsilon) model is one of the most prominent RANS models, which has been implemented in most CFD codes and is considered the most common industry model. The stability and robustness of the models have a well-established regime of predictive capability, satisfying general purpose simulation by offering a comparative good accuracy. In our research work for outward convex corrugated tube, we use standard k-ε model for numerical simulation research.

The governing equations in a RANS (Reynolds Averaged Navier-Stokes) manner are given below.

Continuity equation:

$$\frac{\partial(\rho u_i)}{\partial x_i} = 0 \tag{12}$$

Momentum equation:

$$\frac{\partial}{\partial x_j}\left(\rho u_i u_j\right) = -\frac{\partial P}{\partial x_i} + \frac{\partial}{\partial x_j}\left[\mu\left(\frac{\partial u_i}{\partial x_j} + \frac{\partial u_j}{\partial x_i} - \frac{2}{3}\delta_{ij}\frac{\partial u_k}{\partial x_k}\right)\right] + \frac{\partial}{\partial x_j}\left(-\rho\overline{u_i' u_j'}\right) \tag{13}$$

Energy equation:

$$\frac{\partial}{\partial x_i}[u_i(\rho E + P)] = \frac{\partial}{\partial x_j}\left[\left(\lambda + \frac{c_p \mu_t}{Pr_t}\right)\frac{\partial T}{\partial x_j} + \mu_{eff} u_i\left(\frac{\partial u_i}{\partial x_j} + \frac{\partial u_j}{\partial x_i} - \frac{2}{3}\delta_{ij}\frac{\partial u_k}{\partial x_k}\right)\right] \quad (14)$$

The standard k-ε model is adopted here to close governing equations:

$$\frac{\partial}{\partial x_i}(\rho k u_i) = \frac{\partial}{\partial x_j}\left[\left(\mu + \frac{\mu_t}{\sigma_k}\right)\frac{\partial k}{\partial x_j}\right] + G_k - \rho\varepsilon \quad (15)$$

$$\frac{\partial}{\partial x_i}(\rho \varepsilon u_i) = \frac{\partial}{\partial x_j}\left[\left(\mu + \frac{\mu_t}{\sigma_\varepsilon}\right)\frac{\partial \varepsilon}{\partial x_j}\right] + C_{1\varepsilon}\frac{\varepsilon}{k}(G_k) - C_{2\varepsilon}\rho\frac{\varepsilon^2}{k} \quad (16)$$

where μ_t is the turbulent or eddy viscosity, and G_k represents the generation of turbulent kinetic energy due to the mean velocity gradients. The model constants $C_{1\varepsilon}$, $C_{2\varepsilon}$, C_μ, σ_k, and σ_ε are chosen for the default values 1.44, 1.92, 0.09, 1.0, and 1.30, respectively.

3.1.3 Boundary and initial condition

The next step in preprocessing is setting up the boundary conditions. Boundary conditions refer to the conditions that the solution of the equations should satisfy at the boundary of the moving fluid. Boundary condition will be different for each type of problem. In our research work, the initial and boundary conditions of the outward convex corrugated tube heat exchangers are shown as follows:

1. The inlet conditions at the shell side are as follows: velocity inlet $U = U_{in}$, $T_{in} = 563.15$ K, and the inlet turbulence specifications are a turbulence intensity of $I = 5\%$ and hydraulic diameter $D = 20$ mm.

2. The outlet conditions at the shell side are as follows: pressure outlet, $P_o = 7$ MPa, and the outlet turbulence specifications are a turbulence intensity of $I = 5\%$ and a viscosity ratio $\mu_t/\mu_{lam} = 5\%$.

3. The wall conditions are as follows: the outer wall temperature boundary condition is constant, $T_w = 700$ K, and the inner wall-coupled boundary condition was set as a no-slip boundary, $u = v = w = 0, T = T_w$, and $q = q_w$.

3.1.4 Numerical procedure

The final step in preprocessing is setting up the numerical procedure, which includes solver, discretization, and convergence criterion. In our work, the governing equations are discretized by the finite volume method and solved by the steady-state implicit format. The SIMPLE algorithm is used to couple the velocity and pressure fields. The second-order upwind scheme is applied herein. The convergence criterion for energy is set to be 10^{-7} relative error and 10^{-4} relative error for other variables.

3.2 Heat and mass transfer in tube and shell side of corrugated tube

The variable distribution exhibits the opposite similar tendency at the shell side compared with that at the tube side. In this chapter, we mainly analyze the distribution of velocity, temperature, and turbulence kinetic energy.

3.2.1 Flow structure

Figure 9(a) shows the velocity vector distribution in the tube side of outward convex corrugation tube. As shown from this figure, when fluid flow starts to cross the corrugation section bended from the straight segment, the flow boundary layer separates into two parts: one is the wall boundary layer developed at the near wall region; the other is shear layer associated with an inflection point of large velocity gradient developed away from the wall, which moves away from the surface at the separation point and forms a free shear layer. When the fluid flows through the upstream of the corrugation, the flow velocity decreases and the pressure increases due to the narrowing of the flow cross section. The fluid layer near the wall is gradually difficult to overcome the rising pressure due to the small amount of momentum, resulting in a reflow of the original flow direction. The recirculating zone between the separating streamline and the free boundary streamline is generated at the upstream of the corrugation.

Figure 9(b) indicates the velocity vector distribution in the shell side of outward convex corrugation tube. As shown in this figure, the upstream side boundary of the corrugation is influenced by the accelerating outer-flow, that is, a favorable gradient. As the boundary layer thickens, instabilities occur when the near-wall fluid begins to decelerate as shown in **Figure** 7. The flow separates at the downgrade of the corrugation crest, which is associated with an inflection point of the large velocity gradient developed away the wall.

3.2.2 Temperature distribution

Figure 10(a) shows the temperature distribution in the tube side of outward convex corrugation tube. As shown in **Figure 8**, the wall velocity boundary layer becomes thicker at the upstream side of the corrugation accompany gradually, while the temperature boundary layer gets thicker along the flow direction, due to the eddy generating. Then it goes into thinner at the downstream side of the corrugation with the velocity boundary layer getting thinner, due to the scouring action of the fluid.

Figure 10(b) shows the temperature distribution in the tube side of outward convex corrugation tube. As shown in **Figure 9**, the wall velocity boundary layer becomes thicker at the downstream side of the corrugation accompany gradually, while the temperature boundary layer gets thicker along the flow direction, due to the eddy generating. Then, it goes into thinner at the upstream side of the corrugation with the velocity boundary layer getting thinner, due to the scouring action of the fluid. The thinnest temperature boundary layer occurs at the corrugation crest.

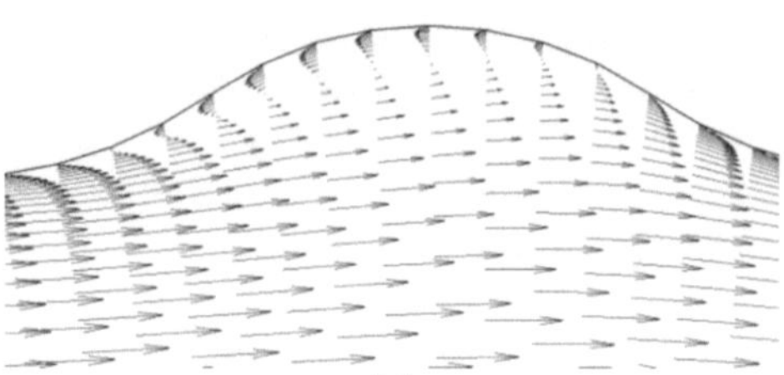

(a)

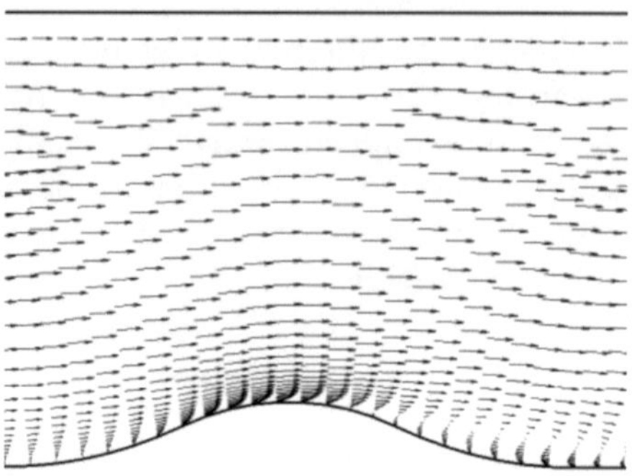

(b)

Figure 9.
Velocity vector distribution at tube side and at shell side. (a) tube side; (b) shell side.

Figure 10.
Temperature distribution at tube side and at shell side. (a) tube side; (b) shell side.

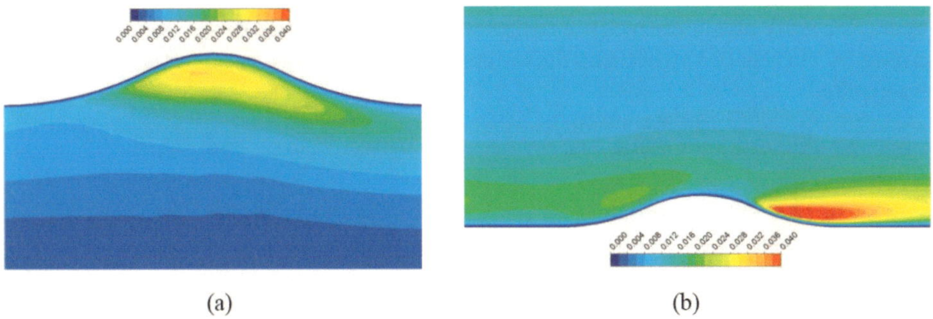

Figure 11.
TKE distribution at tube side and at shell side. (a) tube side; (b) shell side.

3.2.3 Turbulence kinetic energy

Turbulence kinetic energy (TKE) is one of the most important variables in boundary layer since it is a measure of the turbulence intensity, which is tightly related to the velocity profile. **Figure 11(a)** shows the turbulence kinetic energy distribution in the tube side of outward convex corrugation tube. As shown in this figure, the magnitude of the TKE gradient increases past upstream side section of corrugation with a noticed reduction after the flow reattaches as it enters downstream side section of corrugation. The location of the maximum turbulence kinetic energy extends over most of the corrugation, before descending when passing the downstream section of the wave trough.

Figure 11(b) shows the turbulence kinetic energy distribution in the tube side of outward convex corrugation tube. As shown in this figure, the magnitude of the high TKE extends fairly constant past most of the corrugation with a noticed reduction after the flow reattaches. The location of the high TKE extends over most of corrugation at a height, which roughly equals to the maximum corrugation height, before subsiding toward the corrugation trough.

4. Heat and mass transfer enhancement of outward convex corrugated tube heat exchangers

4.1 Enhancement methods

Heat transfer enhancement methods are classified into three classifications: active, passive, and compound. The active methods include electrostatic and magnetic fields, induced pulsation, mechanical aid, vibration, and jet impingement. These methods require external activating power to enhance the heat transfer [3–6]. Passive methods modify the geometrical structure to expand the effective surface area to disturb the actual boundary layer. Compound methods combine the two heat transfer augmentation methods to increase heat transfer performance. In the above-mentioned methods, passive methods have attracted significant attention from researchers and engineers since they are user-friendly and affordable. Extensive research has been devoted to develop highly efficient heat transfer components to better understand the physical mechanisms and optimal parameters of passive heat transfer augmentation methods.

The heat transfer enhancement mechanism in the corrugated tube is described as follows. The periodically corrugated structure on the tube wall arouses periodic alteration of velocity gradient, leading to adverse and favorable pressure gradient locally. The recurrent alternation of axial pressure gradient induces the secondary disturbance, and then the produced intensive eddy destroys the flow boundary layer. The eddy also increases the turbulence intensity of the flow. The disturbance caused by corrugated structures thus increases the heat transfer coefficient drastically.

4.2 Tube side enhancement

Figure 12 shows the effect of *Re* on Nu_c with various *p/D* and *H/D*. The Nu_c tends to increase linearly with the increasing *Re* with a fixed structure of the corrugated tube. This behavior occurs because the increases of flow velocity break wall thermal boundary layer and could obtain higher convective heat transfer coefficient. Moreover, with the decreasing *p/D* and increasing *H/D*, the values of the Nu_c increase.

In order to compare the performance between corrugated tube and smooth tube, the ratio of *Nu* in the corrugated tube to that in the smooth tube (Nu_c/Nu_s) is adopted to indicate the relative grow rate of heat transfer performance. **Figure 13** shows the effect of *Re* on Nu_c/Nu_s with various *p/D* and *H/D*, and the figure exhibits that with the increase of *Re*, Nu_c/Nu_s declines deceleratedly. Moreover, the Nu_c/Nu_s increases with the decreasing *p/D* and increasing *H/D*.

4.3 Shell side enhancement

Figure 14 shows the effect of *Re* on Nu_c with various *p/D* and *H/D* in the shell side. Compared with **Figure 12**, the changing tendency of Nu_c along with *Re*, *p/D*,

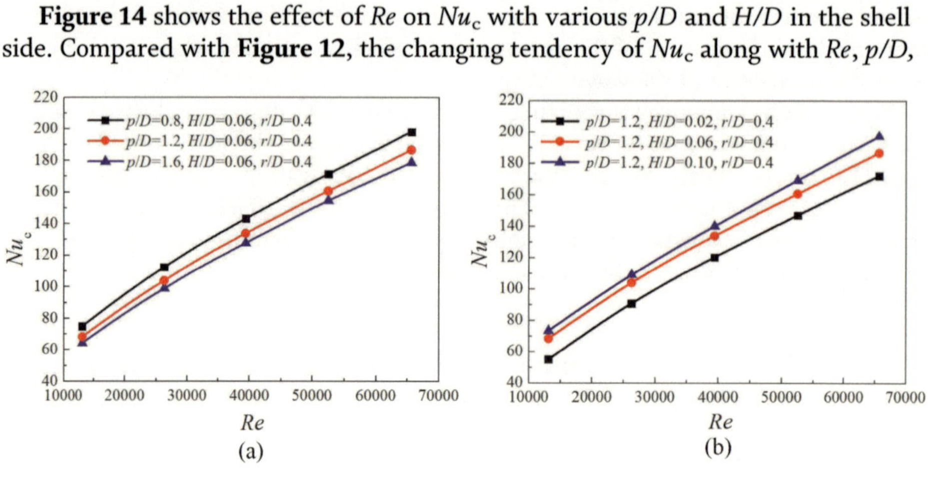

Figure 12.
Effect of Re on Nu with various p/D and H/D in the tube side. (a) various p/D (b) various H/D.

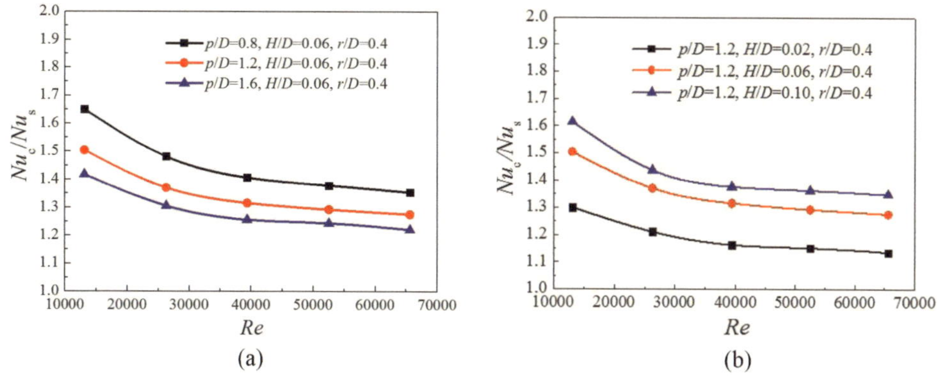

Figure 13.
Effect of Re on Nu_c/Nu_s with various p/D and H/D in the tube side. (a) various p/D (b) various H/D.

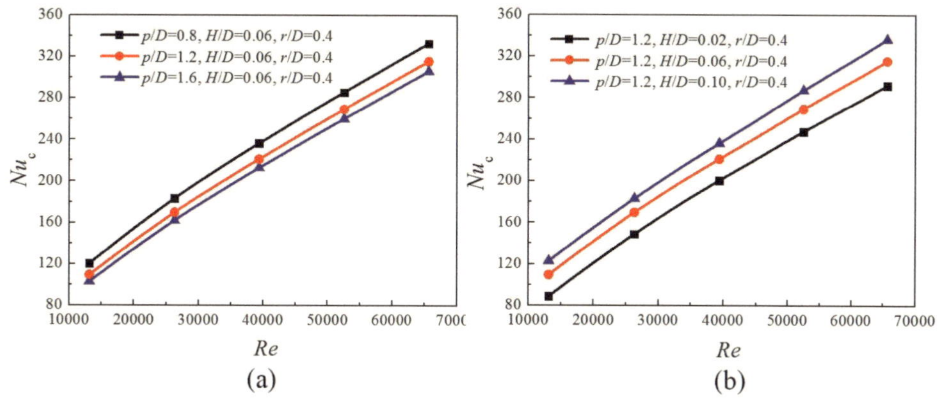

Figure 14.
Effect of Re on Nu with various p/D and H/D in the shell side. (a) various p/D (b) various H/D.

and *H/D* is consistent, but the Nu_c in the shell side is obviously higher than in the tube side.

Figure 15 shows the effect of *Re* on Nu_c/Nu_s with various *p/D* and *H/D* in the shell side. It can be found when compared with **Figure 13**, the changing tendency of Nu_c/Nu_s along with *Re*, *p/D,* and *H/D* is also consistent, but the Nu_c/Nu_s in the shell side is obviously higher than in the tube side.

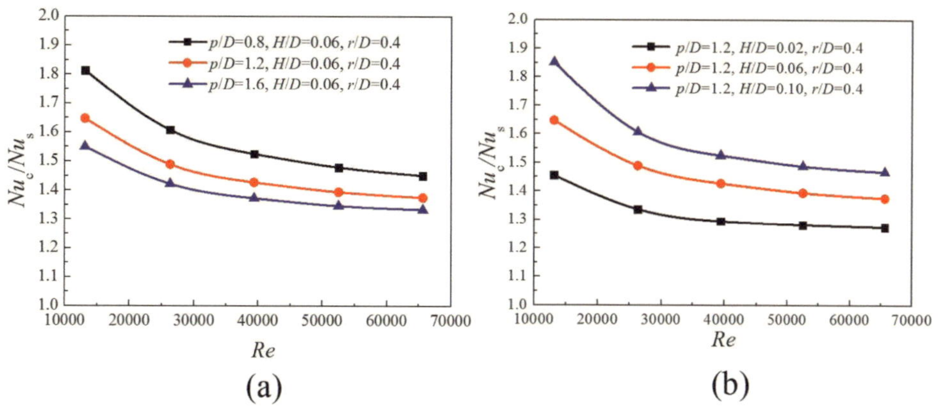

Figure 15.
Effect of Re on Nu with various p/D and H/D in the shell side. (a) various p/D (b) various H/D.

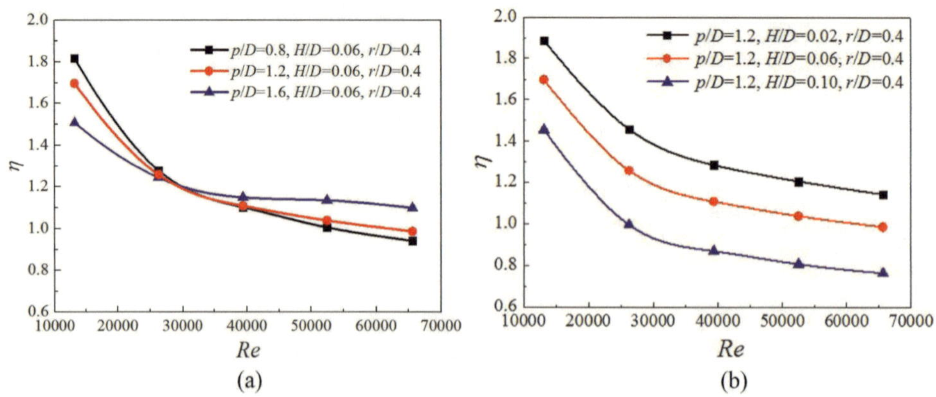

Figure 16.
Effect of Re on η with various p/D and H/D in the tube side. (a) various p/D (b) various H/D.

4.4 Overall enhancement

Generally, heat transfer enhancement accompanies with a penalty of flow resistance when a heat transfer enhancement component (corrugated tube in this paper) is utilized in a heat exchanger compared to the smooth tube. Therefore, an assessment criterion needs to be constructed to evaluate the overall heat transfer performance for the investigated corrugated tube. The function of overall heat transfer performance is adopted as follows:

$$\eta = (Nu_c/Nu_s)/\left(f_c/f_s\right)^{1/3} \quad (17)$$

Figure 16 indicates the effect of *Re* on overall heat transfer performance (η) with various *p/D* and *H*/D in the tube side of outward convex corrugated tube. The figure displays that with the increase of *Re*, η declines deceleratedly. This is because the Nu_c/Nu_s gradually decreases along with increasing *Re*. In addition, with the increase of *p/D*, η decreases when $Re < 30,000$, but increases when $Re > 30,000$. This can be explained from the fact that decreasing extent of Nu_c/Nu_s is larger than that of f_c/f_s with increase in *p/D* when $Re < 30,000$, but lower when $Re > 30,000$. Moreover, the η decreases obviously with the increasing

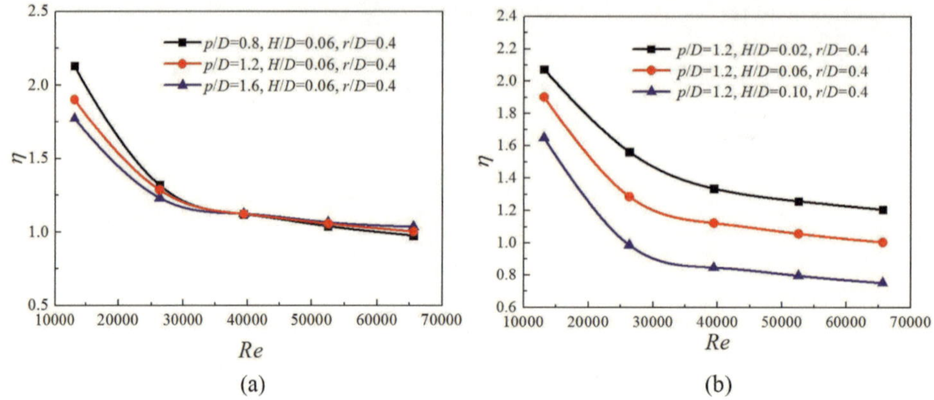

Figure 17.
Effect of Re on η with various p/D and H/D in the shell side. (a) various p/D (b) various H/D.

H/D, and the decreasing extent from H/D = 0.02 to H/D = 0.06 is more obvious than that from H/D = 0.06 to H/D = 0.10. This variation is quite intuitive because of the fact that increasing extent of Nu_c/Nu_s is larger than that of f_c/f_s along with increasing H/D.

It can be observed from **Figure 17** that the changing trend of η with various p/D and H/D in the shell side is almost the same from the tube side. However, the values of η in the shell side are larger than in the tube side. Therefore, the overall heat transfer enhancement in the shell side is superior to the tube side.

5. Optimization of outward convex corrugated tube heat exchangers

5.1 Optimization methods

Response surface methodology (RSM) is composed of a series of statistical and mathematical method for analyzing empirical results, which can construct connection between effect factors and objective functions. The sensitivity of each effect factor and the interactions between two factors can also be analyzed to the objective functions. Recently, RSM has been extensively used to study on the optimal design of heat exchangers, which is able to efficiently and accurately provide the design consideration for heat exchangers [7–9].

RSM constructs the relationship between objective functions and design variables using a series of statistical and mathematical methodology. The function expression of the relationship could be written as follows:

$$G = f(X_1, X_2, \ldots, X_k) + \varepsilon \tag{18}$$

where G represents the objective functions and X_1, X_2, …, X_k stand for design variables, f represents an approximate function, and ε is the residual error between the real value and the approximate value. The approximate functions are described as a quadratic polynomial, aiming to reflect the nonlinear characteristic between objective functions and design variables. In this study, the quadratic polynomial function, including the linear, square, and interaction terms, can be expressed as follows:

$$G = b_0 + \sum_{I=1}^{N}(b_I X_I) + \sum_{I=1}^{N-1}\sum_{J=I+1}^{N}(b_{I,J} X_I X_J) + \sum_{I=1}^{N}(b_{I,I} X_I^2) + \varepsilon \tag{19}$$

where b_I represents the linear effect of design variable X_I, $b_{I,I}$ represents the quadratic effect of X_I, and $b_{I,J}$ represents the linear-linear interactions between X_I and X_J.

5.1.1 Optimization procedure

In our present work, we adopted the flow chart of optimization procedure as shown in **Figure 18**. Three objective functions including heat transfer, pressure drop, and overall heat transfer performance in a heat exchanger tube are selected for optimization. In this simulation plan, a most popular design method called the design of experiment (DOE) and central composite design (CCD) is applied. As shown in **Figure 19**, points including factorial points and center points augmented

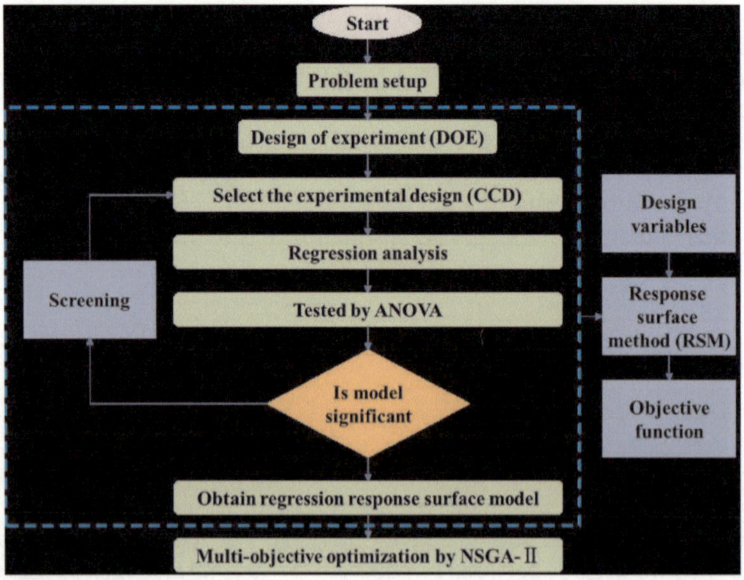

Figure 18.
Flow chart of optimization procedure.

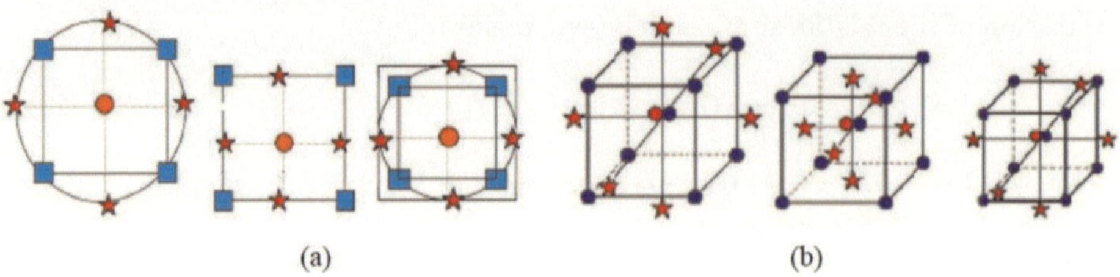

Figure 19.
CCD model. (a) Two factors. (b) Three factors.

by axial points are set in CCD. The numerical results for DOE runs are utilized in reflecting the behavior of responses with geometrical and flow parameters.

5.1.2 NSGA-II algorithm

Nondominated sorting genetic algorithm II (NSGA-II) combined with the multi-objective optimization is adopted in this study. The advantages of NSGA-II are a uniformly distributed Pareto-optimal front, which can suitably detect Pareto-optimal front for multi-objective problems, decrease time consuming, and present solutions with a single run.

Figure 20 shows the NSGA-II flowchart. As specified in **Figure 18**, the RSM has been employed to determine the fitness functions in the optimization algorithm. As well, cross over and mutation contained in genetic operators are used in order to generate a new population. Finally, the optimization process is wrapped up with condition of repetitions number.

This algorithm uses two functions including nondominated sorting function and crowding distance function, respectively. This subprogram takes population members as input, ranks them, and puts them into fronts in proportion to their ranks. Crowding distance function has been designed to avoid the accumulation of

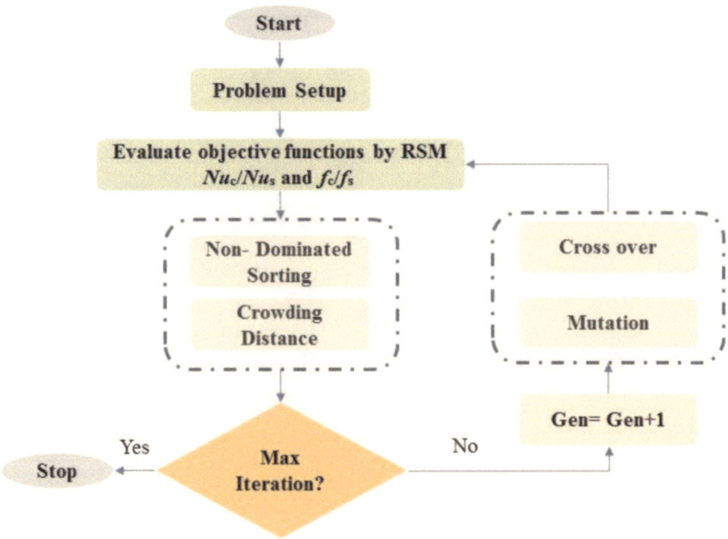

Figure 20.
NSGA-II flowchart.

population members in a limited distance. On the other hand, there are no blank intervals in the domain by using crowding distance function. The function is applied for comparison between members of a front that has equal ranks. Compared to the previous and the next member and also the first and the last member of the population, the normalization Euclidean distance of each solution of the front is for each reference point. Normalization is applied to avoid the problem that the objectives are in the different scale.

5.2 Analysis of variance

ANOVA is one statistical analysis method used to evaluate the fitness of regression models, perform significance testing, and construct simplified regression models between design factors and objective functions. **Tables 1** and **2** are ANOVA for Nu_c and f_c. According to the values of R^2, the fitting degree of the RSM is estimated. The F value and P value indicate the influencing significances of the model terms, judging the significant degree of each model term for the global sensitivity analysis. If the model term is the most significant, the corresponding P value is minimum, and F value is maximum. Generally, the terms having a P value >0.05 are considered insignificant and are removed from the models.

5.3 Regression model of responses

The regression response surface models are described in quadratic polynomial form. Coefficients in the models are determined based on a series of statistical and mathematical methods. The models evaluate the objective functions G including Nu_c/Nu_s, f_c/f_s, and η, which are expressed as:

$$G = b_0 + b_1 \cdot Re + b_2 \cdot Y + b_3 \cdot CR + b_{1,2} \cdot Re \cdot Y + b_{1,3} \cdot Re \cdot CR + b_{2,3} \cdot Y \cdot CR + b_{1,1} \cdot Re^2 + b_{2,2} \cdot Y^2 + b_{3,3} \cdot CR^2 \quad (20)$$

Source	DF	Seq SS	Adj SS	Adj MS	F	P
Regression	14	112756	112756	8054	1893.32	0.000
Linear	4	111873	111873	27968	6574.69	0.000
p/D	1	2398	2398	2398	563.65	0.000
H/D	1	8140	8140	8140	1913.54	0.000
r/D	1	186	186	186	43.80	0.000
Re	1	101149	101149	101149	23777.79	0.000
Square	4	63	63	16	3.73	0.042
$(p/D)\bullet(p/D)$	1	1	28	28	6.65	0.028
$(H/D)\bullet(H/D)$	1	37	17	17	4.03	0.073
$(r/D)\bullet(r/D)$	1	0	3	3	0.61	0.452
$Re\bullet Re$	1	25	25	25	5.94	0.035
Interaction	6	820	820	137	32.13	0.000
$(p/D)\bullet(H/D)$	1	585	585	585	137.46	0.000
$(p/D)\bullet(r/D)$	1	9	9	9	2.00	0.187
$(p/D)\bullet Re$	1	33	33	33	7.76	0.019
$(H/D)\bullet(r/D)$	1	11	11	11	2.64	0.135
$(H/D)\bullet Re$	1	178	178	178	41.90	0.000
$(r/D)\bullet Re$	1	4	4	4	1.01	0.339
Residual Error	10	43	43	4		
Total	24	112799				

Standard deviation = 2.06
Predicted residual error of sum of squares (PRESS) = 268.69
R^2 (Adequate) = 99.96% R^2 (Predicted) = 99.76% R^2 (Adjusted) = 99.91%

Table 1.
Analysis of variable (ANOVA) for Nu_c.

In our optimum work, the regression response surface models for evaluating Nu_c and f_c are expressed as:

$$Nu_c = 64.89 - 51.44p/D + 1026.68H/D - 34.95r/D + 0.0046Re - 377.83p/D \cdot H/D - 1.82(\text{E} - 04)p/D \cdot Re + 0.0042H/D \cdot Re + 20.83(p/D)^2 - 8.12(\text{E} - 09)Re^2 \quad (21)$$

$$f_c = 0.077 - 0.05p/D + 1.21H/D - 0.037r/D - 7.76(\text{E} - 07)Re - 0.57p/D \cdot H/D + 0.012p/D \cdot r/D - 0.38H/D \cdot r/D - 4.41\,(\text{E} - 06)H/D \cdot Re \quad (22)$$

5.4 Response surface analysis

We applied 2D response surface contour plots to describe the regression response surface model, in order to display the interaction influence of each pair of design variables on the required responses. From the 2D response surface contour plots, the regulation of objective functions with changing design variables can be clearly observed, distinguished by contour plot color. **Figures 21** and **22** show the 2D surface plots of the combined effects for the standard deviation of Nu_c and f_c. It

Source	DF	Seq SS	Adj SS	Adj MS	F	P
Regression	14	0.019961	0.019961	0.001426	157.04	0.000
Linear	4	0.017620	0.017620	0.004405	485.16	0.000
p/D	1	0.002109	0.002109	0.002109	232.28	0.000
H/D	1	0.014332	0.014332	0.014332	1578.59	0.000
r/D	1	0.000275	0.000275	0.000275	30.27	0.000
Re	1	0.000904	0.000904	0.000904	99.51	0.000
Square	4	0.000599	0.000599	0.000150	16.51	0.000
(*p/D*)•(*p/D*)	1	0.000380	0.000022	0.000022	2.40	0.152
(*H/D*)•(*H/D*)	1	0.000209	0.000131	0.000131	14.47	0.003
(*r/D*)•(*r/D*)	1	0.000004	0.000002	0.000002	0.18	0.681
Re•*Re*	1	0.000006	0.000006	0.000006	0.69	0.426
Interaction	6	0.001742	0.001742	0.000290	31.97	0.000
(*p/D*)•(*H/D*)	1	0.001342	0.001342	0.001342	147.76	0.000
(*p/D*)•(*r/D*)	1	0.000014	0.000014	0.000014	1.57	0.239
(*p/D*)•*Re*	1	0.000034	0.000034	0.000034	3.80	0.080
(*H/D*)•(*r/D*)	1	0.000147	0.000147	0.000147	16.18	0.002
(*H/D*)•*Re*	1	0.000193	0.000193	0.000193	21.28	0.001
(*r/D*)•*Re*	1	0.000011	0.000011	0.000011	1.25	0.290
Residual Error	10	0.000091	0.000091	0.000009		
Total	24	0.020052				

Standard deviation = 0.003
Predicted residual error of sum of squares (PRESS) = 0.00057
R^2 (Adequate) = 99.55% R^2 (Predicted) = 97.16% R^2 (Adjusted) = 98.91%

Table 2.
Analysis of variable (ANOVA) for f_c.

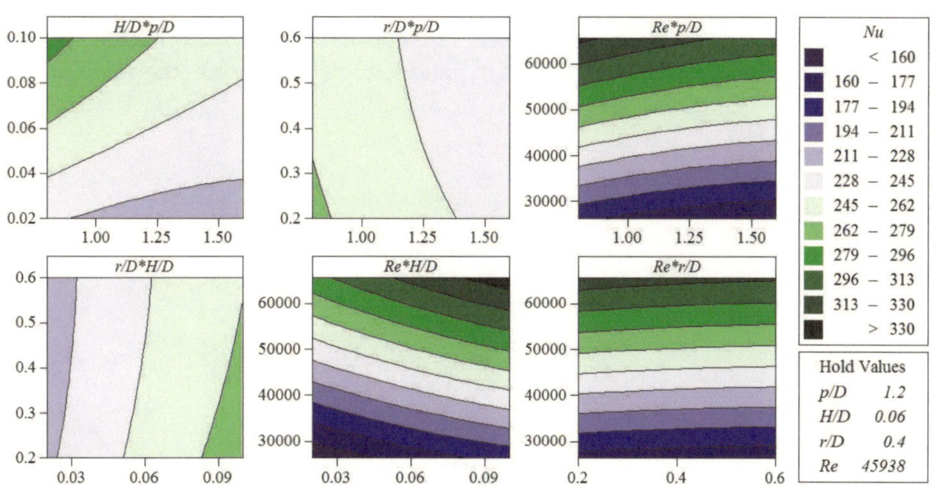

Figure 21.
Response surfaces contour plots of combined effects for Nu_c.

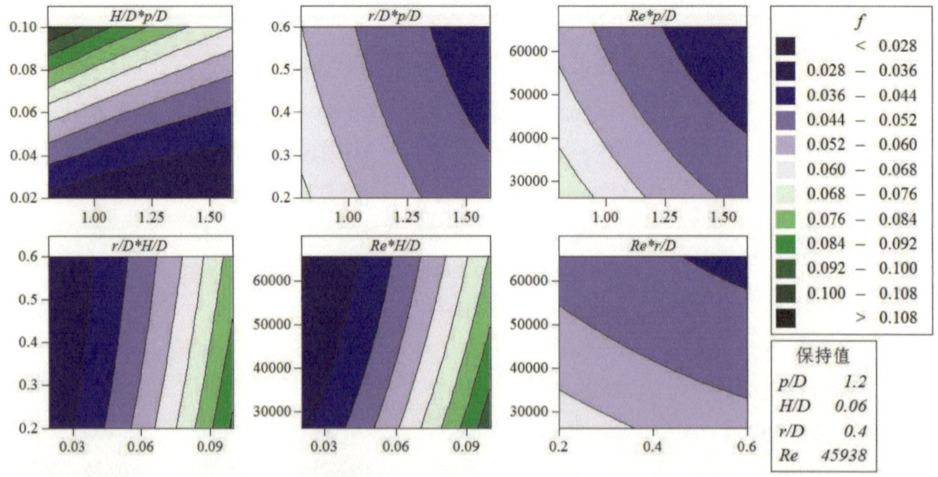

Figure 22.
Response surfaces contour plots of combined effects for f_c.

can be observed that the decrease of p/D, the increase of H/D, the decrease of r/D, and the increase of Re result in the augment of Nu. Moreover, it can be also seen that the decrease of p/D, the decrease of H/D, the increase of r/D, and the decrease of Re result in the weak of f_c.

5.5 Pareto front

By inspecting the numerical results of Nu_c/Nu_s and f_c/f_s, it is found that these two responses are varied with the changes of the design parameters. There must exist design parameters corresponding to the optimal objective functions. The goal of optimization for a corrugated tube subjected the design constrains of structural limitation in this study is to find the optimal values of designing parameters to maximize Nu_c/Nu_s and minimize f_c/f_s. In this study, the multi-objective optimization is executed by NSGA-II. The results for Pareto-optimal curve are shown in **Figure 23**, which clearly reveal the conflict between the two responses, Nu_c/Nu_s and f_c/f_s. Any changed design parameter that increases Nu_c/Nu_s leads to an increase of f_c/f_s. It is worth noting that the minimum values of f_c/f_s with Nu_c/Nu_s for various points on Pareto optimal front. Therefore, the reported results are applicable for a problem with one objective function (f_c/f_s) and specific constraint (the value of

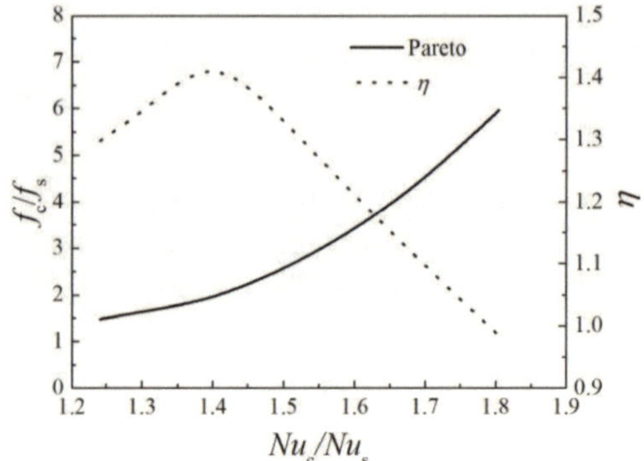

Figure 23.
Pareto-optimal curve.

selected or input Nu_c/Nu_s). This means that the presented multi-objective optimization method provides a general optimal solution in simplified form, and one may obtain an optimum design (minimum of f_c/f_s and maximum of η) with a specified Nu_c/Nu_s.

6. Conclusions

A deep investigation of the heat and mass transfer was given in outward convex corrugated tube heat exchangers in this chapter. The detailed structure of the novel tube has been introduced, in which the heat and mass transfer mechanism is different with the traditional tube type. A specific manufacturing procedure by hydraulic bulking system has been presented for the novel tube type. The experimental setup and measuring system for the novel tube type have been depicted. From obtained experimental data, we found that with the increase of *Re*, Nu_c/Nu_s, f_c/f_s, and *PEC* decline. The decreasing rate of Nu_c/Nu_s and *PEC* is almost linear, but f_c/f_s is decelerated. The numerical study on the heat and mass transfer at outward convex corrugated tube heat exchangers has been displayed. The distribution of velocity, temperature, and turbulence kinetic energy has been analyzed. The recirculating zone between the separating streamline and the free boundary streamline is generated, which breaks the thermal boundary layer to enhance the heat transfer performance. Turbulence kinetic energy is improved at the recirculating zone. Heat and mass transfer enhancement of outward convex corrugated tube heat exchangers has been revealed. Both on the tube side and shell side, with the decreasing p/D and increasing H/D, the values of Nu_c and Nu_c/Nu_s increase. Moreover, with the increase of p/D, η decreases when $Re < 30{,}000$, but increases when $Re > 30{,}000$; the η decreases obviously with the increasing H/D. The multi-objective optimization is executed by RSM combined with NSGA-II. ANOVA is used to evaluate the fitness of regression models, perform significance testing, and construct simplified regression models between design factors and objective functions. 2D response surface contour plots are applied to describe the regression response surface model. Multi-objective optimization method provides a general optimal solution in simplified form, and one may obtain an optimum design (minimum of f_c/f_s and maximum of η) with a specified Nu_c/Nu_s.

Acknowledgements

The authors gratefully acknowledge the support by the National Natural Science Fund (Grant No. 51506034).

Conflict of interest

The article has not been previously published, is not currently submitted for review to any other journal, and will not be submitted elsewhere before one decision is made.

Author details

Huaizhi Han[1,2]*, Bingxi Li[3], Yaning Zhang[3], Quan Zhu[1] and Ruitian Yu[2]

1 School of Chemical Engineering, Sichuan University, Chengdu, China

2 College of Power and Energy Engineering, Harbin Engineering University, Harbin, China

3 School of Energy Science and Engineering, Harbin Institute of Technology, Harbin, China

*Address all correspondence to: hanhz_hrbeu@163.com

References

[1] Han HZ, Li BX*, Yu BY, He YR, Li FC. Mint: Numerical study of flow and heat transfer characteristics in outward convex corrugated tubes. International Journal of Heat and Mass Transfer. 2012;**55**:7782-7802. DOI: 10.1016/j.ijheatmasstransfer.2012.08.007

[2] Yang LB, Han HZ*, Li YJ, Li XM. Mint: A numerical study of the flow and heat transfer characteristics of outward convex corrugated tubes with twisted-tape insert. ASME Journal of Heat Transfer. 2016;**138**:1-8. DOI: 10.1115/1.4031171

[3] Wang FQ, Tang ZX, Gong XT, Tan JY*, Han HZ, Li BX. Mint: Heat transfer performance enhancement and thermal strain restrain of tube receiver for parabolic trough solar collector by using asymmetric outward convex corrugated tube. Energy. 2016;**114**:275-295. DOI: 10.1016/j.energy.2016.08.013

[4] Wang FQ, Lai QZ, Han HZ, JY T. Mint: Parabolic trough receiver with corrugated tube for improving heat transfer and thermal deformation characteristics. Applied Energy. 2016; **164**:411-424. DOI: 10.1016/j.apenergy.2015.11.084

[5] Gong XT, Wang FQ, Wang HY, Tan JY, Lai QZ, Han HZ. Mint: Heat transfer enhancement analysis of tube receiver for parabolic trough solar collector with pin fin arrays inserting. Solar Energy. 2017;**144**:185-202. DOI: 10.1016/j.solener.2017.01.020

[6] Han HZ, Li BX*, Shao W. Mint: Effec t of flow direction for flow and heat tra nsfer characteristics inoutward convex a symmetrical corrugated tubes. International Journal of Heat and Mass Transfer. 2016;**92**:1236-1251. DOI: 10.1016/j.ijheatmasstransfer.2014.11.076

[7] Han HZ, Yu RT, Li BX, Zhang YN, Wang W. Mint: Multi-objective optimization of corrugated tube with loose fit twisted tape using RSM and NSGA-II. International Journal of Heat and Mass Transfer. 2019;**131**:781-794. DOI: 10.1016/j.ijheatmasstransfer.2018.10.128

[8] Han HZ, Li BX*, Wu H, Shao W. Mint: Multi-objective shape optimiza tion of double pipe heat exchanger with inner corrugated tube using RSM method. International Journal of Thermal Sciences. 2015;**90**:173-186. DOI: 10.1016/j.ijthermalsci.2014.12.010

[9] Han HZ, Li BX*, Shao W. Mint: Multi-objective optimization of outward convex corrugated tubes using response surface methodology. Applied Thermal Engineering. 2014;**70**:250-262. DOI: 10.1016/j.applthermaleng.2014.05.016.

[10] Shah RK, Sekulić Dušan P. Surface basic heat transfer and flow friction characteristics. In: R K S, Sekulić Dušan P, editors. Fundamentals of Heat Exchanger Design. Hoboken: John Wiley & Sons; 2017. pp. 425-562. DOI: 10.1002/9780470172605.ch7

[11] Karthikeyan D. Thermal analysis of Shell and tube heat exchanger using CFD. In: Parmar N, editor. Design and Analysis Shell and Tube Type Heat Exchanger: Taguchi Method and CFD Analysis. Balti: LAG LAMBERT Academic Publishing; 2017. pp. 26-43. DOI: 110.5281/zenodo.56040

[12] Russell TF, Robinson AS, Wagner NJ, editors. Mass and Heat Transfer: Analysis of Mass Contactors and Heat Exchangers. New York: Cambridge University Press; 2008

[13] ANSYS FLUENT User's Guide, Release 17.0. Ansys Inc.; 2016

Chapter 5

A Numerical Solution Model for the Heat Transfer in Octagonal Billets

Panagiotis Sismanis

Abstract

In the quest for high-quality steel products, the need of cast billets with minimum surface and internal defects is of paramount importance. On the other hand, productivity is required to be as high as possible in order to reduce production cost. Different billet shapes have been applied with emphasis upon square, rectangular, and circular cross-sections. It is obvious that the best billet shape that minimizes surface and subsurface defects is the circular one. Nevertheless, this shape creates some problems with respect to handling and safety reasons. One recent attempt is to produce normal octagonal-shaped billets that appear to approach the circular shape albeit easier to handle. In this study, a numerical solution for the heat transfer during solidification in the continuous casting of octagonal billets has been carried out. The developed model deploys an implicit scheme in order to solve the differential equations of heat transfer under the appropriate boundary conditions in a section of an octagonal billet, assuming fully axisymmetric cooling of the bloom. The geometry of the octagonal billet plays an interesting role in the development of the heat transfer analysis. Based upon fundamental principles, a computer program has been developed for this purpose. Consequently, results from the numerical solution are presented and discussed.

Keywords: octagonal billet, solidification, continuous casting, implicit, micro-segregation, prediction model, shell thickness, temperature distribution

1. Introduction

The mathematical analysis of heat transfer in a billet during the continuous casting process has been investigated from the early stages of the development of continuous casting machines. In this study, some fundamental works related to the numerical solution of heat transfer in solidifying billets are presented. A pioneer work is mentioned in [1] in which a mathematical model of heat transfer in continuously cast steel slabs is described. A computer program was developed to model the problem using a 1D transient conduction equation with the appropriate boundary conditions. Some years later, the design of the mold and spray sections of a continuous casting machine were examined in detail with the aid of a 2D heat-flow mathematical model [2]. Experimental data obtained from commercial casters were used for the validation of the deduced results. In that monumental work, the various aspects of heat transfer in the various sections of a casting machine were analyzed in detail. Later on, surface and subsurface defects in the cast billets became apparent, and research work included mechanical phenomena (like stress and strain) together with the heat transfer analysis; an early typical work on the subject was presented in [3]. It is worth mentioning that the caster automation includes most of the heat transfer formulation and analysis, albeit not published. Consequently, the present literature survey mostly presents indicative published works upon heat transfer analysis. In the 1990s, the continuing casting of steel started maturing, and this was reflected by two studies with practical implementations [4, 5]. In a review paper [6], the evolution of heat transfer and mechanical studies for the continuous casting were presented. Furthermore, an analysis on the ideal taper prediction for billet casting and a thorough analysis of thermomechanical behavior in billet casting were also presented [7, 8]. The salient features of heat transfer in the secondary cooling zone during the continuous casting of steel were examined in [9]. Primary cooling is considered the start-up of steel solidification process in a water-cooled copper mold, and secondary cooling zone is considered the region just after the mold until the end of the caster in which the billet is cooled by spray nozzles or air-mist systems. Radiation heat exchange between the billet surface and the environment plays also an important role in the secondary cooling zone. Fluid-flow phenomena in the mold captured the interest, and as of that, heat transfer studies were also reported on the subject later on; a typical work is given in [10]. In most similar works, general-purpose heat-transfer software has been deployed either as a direct prediction tool or as a verification one. A typical work is given in [11] in which a 1D heat-transfer simulation program (CON1D) was successfully verified against a 3D finite element model developed in Abaqus. Near-net-shape cast sections appeared in industrial practice in the early 2000, and therefore, their thermal-mechanical modeling during casting was developed as well [12]; in this type of analysis, the use of a package like Abaqus has been indispensable. In a relatively recent work [13], the 1D differential equation for heat transfer was solved in order to study the evolved microstructures during the solidification of round billets (so, the radial direction for heat transfer prevails). In the cases under study, the billet diameters were in the range from 210 up to 410 mm; the validation of their model was performed by deploying a semi-analytical model [14] for the prediction of the surface temperature of a cast round billet. The numerical model was used to calculate the local solidification time, the local gradient, and the local solidification time as a function of the distance from the round billet surface; furthermore, a simplified relation for the prediction of the columnar to equaxial transition was proposed. Deploying the finite volume method, a real-time 2D heat transfer and solidification model for continuous casting [15] were developed and tested online. The behavior was satisfactory with less than 10°C deviations on surface central temperatures. A dynamic heat transfer model was developed [16] in order to study the effect of the casting speed on the temperature field of continuously cast billets; various steel grades were taken under consideration. A real-time heat-transfer model and a heat-transfer coefficient identification method [17] were developed. The model validation was achieved by the measurement of surface temperatures by a CCD system that appeared to eliminate the impact of scales on the temperature measurement and keep the measured surface temperature fluctuation within the range of ±10°C. A finite volume method and the alternating direction implicit algorithm (ADI) were selected in order to develop a real-time heat transfer model [18] for the dynamic control of continuous cast billets. The system was applied online to control the electric current of the final electromagnetic stirring (FEMS) system. A simulation model of solidification and heat transfer of 150-mm-square billets was developed [19] for the continuous casting of grade 40 Cr. Nailing tests and temperature measurements were applied in order to fine-tune the model with maximum errors of less than 2%. In another study [20], different micro-segregation models coupled with fluid flow and heat transfer were run to study macrosegregation phenomena in a round billet with a 165-mm diameter. It was predicted that heavy centerline segregation occurs at the end of solidification when the central solid fraction reaches the value of 0.7. The need for high-speed casting under very controllable conditions has led the researchers to seek billet sections that can succeed in the demanding steelmaking environment. Recently, one of the leading manufacturers in the steelmaking sector, Danieli, has announced [21] the octagonal billet as a proven potential shape for successful high-speed casting. Consequently, the present author has taken the opportunity to study the heat transfer in a solidifying billet based on fundamental principles. The work has been carried out for a larger billet size than the one tested in [21] and for two special steel grades, the peritectic grade S355 J2 and the medium-high-grade 42CrMo4, that are both important for many end-user applications.

2. Formulation of the heat-transfer problem

The general three-dimensional (3D) heat-transfer equation that describes the temperature distribution inside a solidifying object is given by Eq. (1) according to [22]

$$\rho c \frac{\partial T}{\partial t} = \nabla \cdot k \nabla T + S \tag{1}$$

The source term S may be considered [23] to be of the form:

$$S = S_C + S_P\, T \tag{2}$$

Furthermore, T is the temperature, and ρ, c, and k are the density, heat capacity, and thermal conductivity, respectively. The 2D heat-transfer equation in Cartesian coordinates can be written as [22]

$$\rho c \frac{\partial T}{\partial t} = \frac{\partial}{\partial x}\left(k \frac{\partial T}{\partial x}\right) + \frac{\partial}{\partial y}\left(k \frac{\partial T}{\partial y}\right) + S \tag{3}$$

The boundary conditions that apply in the octagonal billet case are very similar to the ones for the rectangular billet that have been presented in detail in [24] and will not be repeated here. Nevertheless, two important boundary conditions apply in the case under study, which follow:

- Due to symmetry, the upper diagonal side (**Figure 1**, segment OB) of the domain of interest is supposed to be adiabatic, as well as the lower side (**Figure 1**, segment OK) is. In this way, the following formulation holds:

$$\frac{\partial T}{\partial s} = 0, \quad \text{where } s \text{ is the vertical axis on the line} \\ y = x\ \tan(\pi/8) \tag{4}$$

- The primary (mold) and secondary cooling systems are the ones applied in Stomana; this is presented in detail in a previous publication [24] and will not be repeated here. However, due to the potential of the octagonal mold, a surplus of water was used in favor of an enhanced cooling behavior for the solidifying billet; this was expressed as an extra percentage of water flow deployed for the octagonal billet.

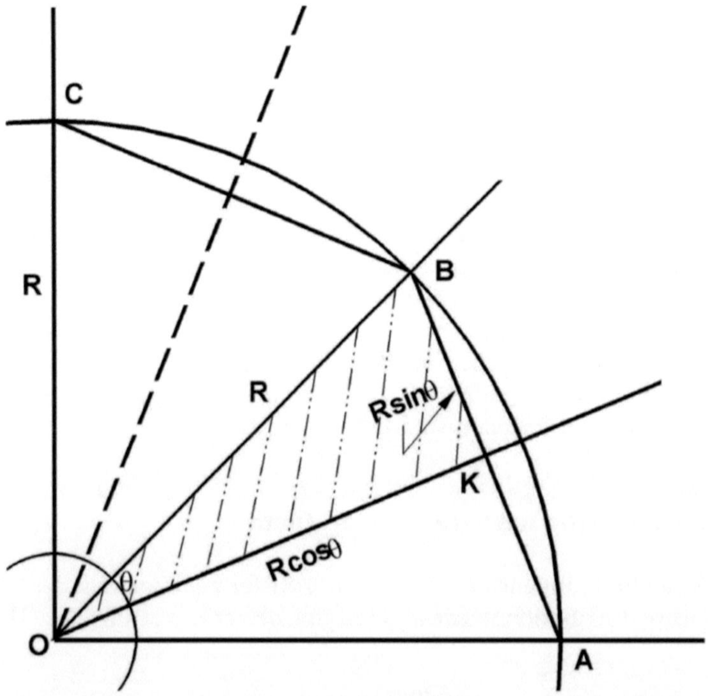

Figure 1.
The salient features of an octagon and its circumscribed circle.

3. Theoretical solution

There is no published work on the theoretical solution for the octagonal billet until now. However, a theoretical solution can be approximated by solving the DE of heat transfer on an equivalent circular geometry. According to **Figure 1**, the selected radius of that circle was R_e = 156.6 mm, as given by Eq. (5):

$$R_e = \frac{1 + \cos(\pi/8)}{2} R = 0.962\,162.84 = 156.6 \text{ mm} \tag{5}$$

This radius is actually the mean value between the circumradius and in-radius of the octagon [25]. The circumradius R being equal to 162.84 mm makes the currently studied octagonal billet equivalent in size to a 250 × 300 mm^2 billet produced in the Stomana plant, at Pernik, Bulgaria. Now the original problem can be converted into a heat-transfer problem of cylindrical geometry, and the DE together with the boundary conditions is formulated according to Eq. (6):

$$\begin{gathered} \frac{\partial T}{\partial t} = \alpha \left(\frac{\partial^2 T}{\partial r^2} + \frac{1}{r} \frac{\partial T}{\partial r} \right) \\ T = T_0,\ t = 0;\ \frac{\partial T}{\partial r} = 0,\ r = 0;\ -k \frac{\partial T}{\partial r} = h\left(T - T_f\right),\ r = R_e \end{gathered} \tag{6}$$

By variable transformation using Eq. (7)

$$\Theta = \frac{T - T_f}{T_0 - T_f} \tag{7}$$

the boundary condition at $r = R_e$ becomes:

$$\frac{\partial \Theta}{\partial r} + \frac{h}{k}\Theta = 0, \quad r = R_e \tag{8}$$

In order to fulfill the boundary condition at $r = R_e$ according to Eq. (8), Eq. (9) has to be solved for β:

$$\beta J_1(\beta) = A\, J_0(\beta), \quad A = \frac{R_e\, h}{k} \tag{9}$$

where J_0 and J_1 are the Bessel functions of zero and first order, respectively.

This theoretical solution is presented in the book of Carslaw and Jaeger [26] and is followed here. There are an infinite number of values for β, which solve Eq. (9); nevertheless, the first six roots [26] are enough for the computations. In this way, the derived solution arrives in the form of Eq. (10):

$$\begin{gathered}\Theta = \frac{T - T_f}{T_0 - T_f} = \sum_{n=1}^{\infty} \exp\left(-\beta_n{}^2 \xi\right) \frac{2A\, J_0(\beta_n\, r/R_e)}{\left(\beta_n{}^2 + A^2\right) J_0(\beta_n)} \\ \xi = \frac{\alpha\, t}{R_e{}^2}, \quad A = \frac{R_e\, h}{k}\end{gathered} \tag{10}$$

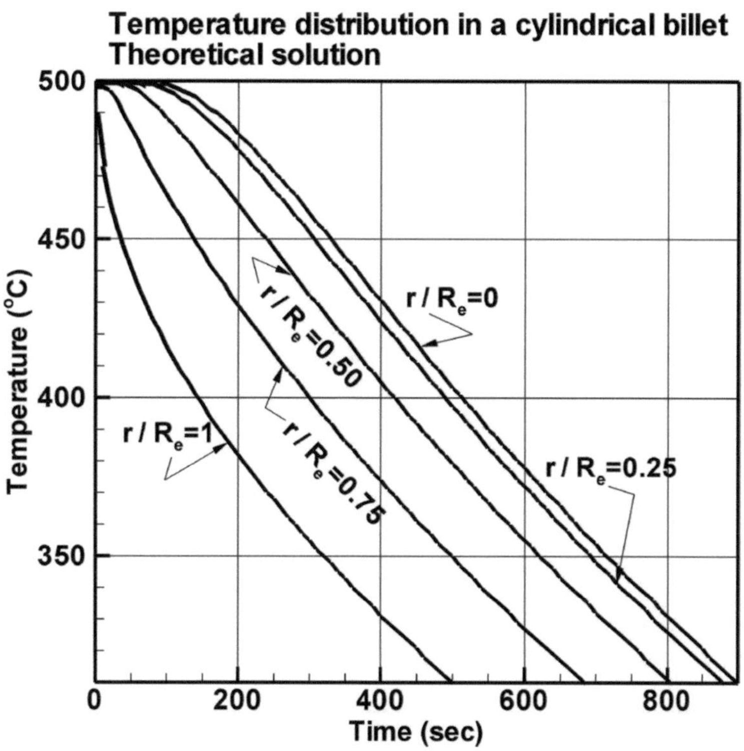

Figure 2.
Theoretical predictions of the temperature distribution in various positions in a cylindrical billet.

Eq. (10) was applied in order to compute the temperature distribution $T(r,t)$ inside the equivalent to the octagonal geometry cylindrical billet; some results are depicted in **Figure 2**. In fact, the following parameters were used in these computations: thermal conductivity, k = 50 W/m/K; density, ρ = 7600 kg/m^3; heat capacity, c = 400 J/kg/K; thermal diffusivity, $\alpha = k/(\rho c) = 1.645 \times 10^{-5}$ m^2/s; heat transfer coefficient, h = 200 W/m/K; cooling water temperature, T_f = 30°C, and initial billet temperature, T_0 = 500°C.

4. Numerical solution

Figure 1 illustrates the important geometrical features of the octagonal-shaped billet. Considering that the flow of heat takes place only in the radial direction toward the center of the octagon, and certainly toward the center of its circumscribed circle, it appears that due to symmetry, only a small trigonal part becomes the important region (the shaded area in **Figure 1**) for the solution of the DE of heat transfer. Actually, when uniform cooling takes place around the solidifying billet, then symmetrical conditions prevail in the relevant heat transfer. Consequently, heat flow takes place only in the radial direction. The strongly implicit scheme as presented by Patankar [23] was deployed for the solution of the DE of heat transfer. **Figure 3** illustrates the control-volume selection for a small number of nodal points.

The five-point formulation for the discretization equations was applied. Special attention was paid in the boundary condition formulation of the discretization equations, and a more detailed analysis on this is presented in Appendix I. One may realize that once the boundary conditions are properly written, then a finer grid will ultimately permit the convergence to the engineering solution. In **Figure 3**, only the lower trigonal part is necessary to be examined. The vertical side, MZ, is the boundary from which the main heat flow (cooling) takes place; the horizontal side OZ and the diagonal side OM, with an inclination of $\pi/8$, or 22.5°, are considered to be adiabatic to heat flow due to the aforementioned symmetry reasons. This study is part of a series of published works with respect to the numerical solution of the heat transfer equation in 2D (square and rectangular) and 3D domains [24, 27–29]. Although the core of the existing program remained intact, due to the nature of the present trigonal domain, the part of the program related to the

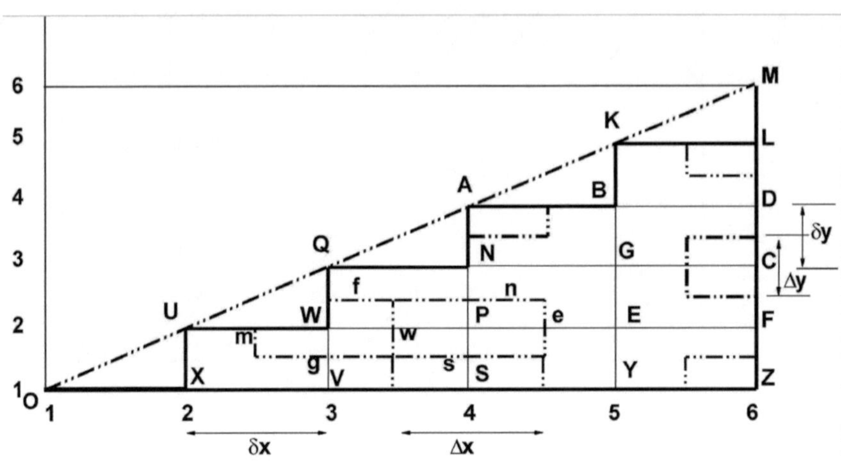

Figure 3.
Presentation of control volumes and nodal points for a coarse grid.

boundary conditions had to be developed from scratch. The octagonal billet with $R = 162.84$ mm can be discretized with a fine grid of 200×200 nodal points. In this way, the space intervals (δx, δy) were about $\delta x = R\cos(\pi/8)/199 = 0.756$ mm and $\delta y = R\sin(\pi/8)/199 = 0.313$ mm, respectively. Actually, keeping the ratio, $\delta y/\delta x = 0.313/0.756 = 0.414 = \tan(\pi/8)$, exhibited a relatively good convergence; the time interval was $\Delta t = 1$ sec.

5. Running the computer program

The Gauss-Seidel algorithm was applied for the iterative solution of the matrix equations in the case of the 42CrMo4 grade. Over-relaxation was used for the fastest possible convergence, and the over-relaxation parameter used was $\omega = 1.870$, which has exhibited good behavior for this type of studies [30]. The applied maximum error (tolerance) for the computed temperature at each nodal point was 5×10^{-3}. In addition to this, the Strongly Implicit Procedure (SIP) [31] was deployed for the iterative solution of the matrix equations in the case of the S355 J2 grade. In this case, however, the same tolerance (5×10^{-3}) was used for the maximum residual value and not the temperature difference at each nodal point. The C++ language was selected for the development of the program, and the deployed compiler was the Intel Parallel Studio XE 2015 Composer Edition for C++; the open source, cross-platform Code::Blocks, version 17.12 software was used as an integrated development environment for program edition and compiler invoking. The software was run in a DELL Alienware laptop with the Intel i7-6700HQ CPU (8 cores) at 2.6 GHz, 16GB RAM, running under a 64-bit Windows 10 Professional O/S.

6. Results and discussion

In order to validate the new developed program, a solution against the aforementioned theoretical one had to be sought. For this reason, an octagonal billet initially at 500°C, cooled afterwards by an hypothetical fluid at 30°C and with a heat transfer coefficient $h = 200$ W/m^2/K, was simulated. The values for the properties of density, thermal conductivity, and heat capacity were the same with the ones used in the theoretical solution. In total, 38 sets of data were randomly selected at various positions inside the cylindrical billet and were compared with similar results deduced from the numerical solution described here in order to verify its validity. **Figure 4** presents the compared results of these two sets of computed data. An analysis of variance (ANOVA) was performed for these two sets of results, and the following statistical results were derived: residual standard error, 3.133 on 36 degrees of freedom (DF); multiple-R squared, 0.9957; and F-statistic, 8333 on 1 and 35 DF, with a p-value $<2.2 \times 10^{-16}$. Consequently, a standard error of 3.1°C was found to represent the mean statistical error between the results obtained from the theoretical and current numerical solutions, respectively.

The advantage of studying the heat transfer in an octagonal billet marks the importance of such an experiment: the potential to test some results against theoretical ones gives the confidence about attaining the proper numerical solution; in fact, the octagonal geometry approximates the circular cross-section better than a square one for this comparison to be accomplished.

Table 1 presents the chemical analysis for the two selected grades under examination: 42CrMo4 and S355 J2. In the bloom caster of Stomana, in Pernik, the billet size of 250×300 mm^2 is normally used for the production of special steels. The equivalent billet to this size in octagonal shape has a circumradius of

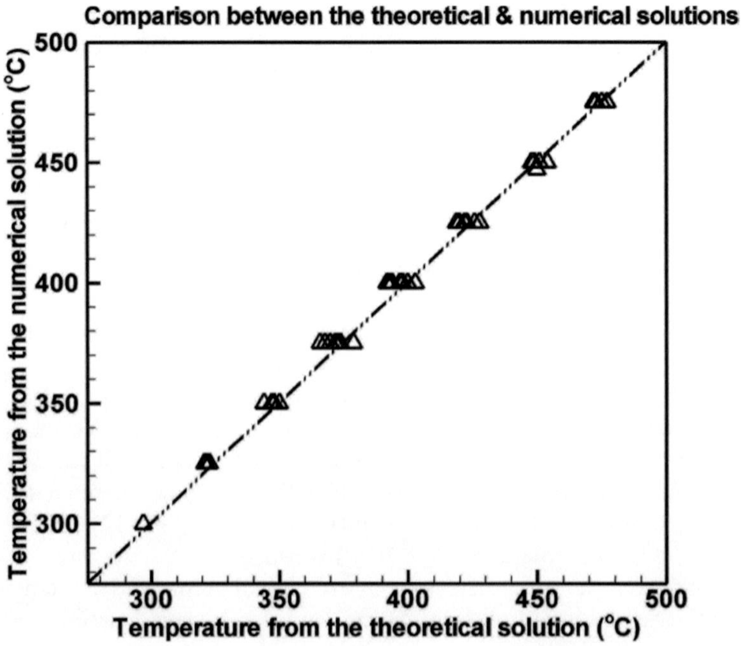

Figure 4.
Comparison of temperature values between the theoretical (for a cylindrical billet) and numerical (for an octagonal billet) solutions.

Grade	%C	%Si	%Mn	%P, S max	%Cr	%Mo	%Ni	%Cu	%V	%Al
1	0.42	0.22	0.78	0.015	0.98	0.20	0.07	0.22	—	0.03
2	0.17	0.21	1.06	0.015	0.10	0.02	0.09	0.22	0.03	0.03

Table 1.
Chemical analysis for grades 42CrMo4 (1) and S355 J2 (2).

R = 162.84 mm, as aforementioned. An interesting idea is that putting the octagonal billet in practice, there is a potential to increase productivity by using more cooling water, an advantage that has been tested in practice [21] for normal rebar grades. Consequently, in this study a surplus of mold water by 12% for both of the two grades 42CrMo4 and S355 J2 was used compared to the normal water-cooling process applied in Stomana for these grades, respectively. Furthermore, a surplus of water in the air-mist region by 3% for the high-carbon grade 42CrMo4 and 4% for the peritectic grade S355 J2 was applied, respectively.

Figure 5 illustrates the shell thickness (curve 3) increase as an octagonal billet of grade 42CrMo4 travels down Stomana's continuous casting machine. One may notice that the solidification completes at about 32.4 m from the liquid-steel meniscus in the mold.

At the same time, the centerline temperature (curve 2) drops appreciably at that point revealing the complete solidification at that point, as well; the surface temperature is also presented by curve 1. The casting speed was 0.70 m/min, and the superheat (SPH) was 30°C in the computation. The SPH is actually the excess temperature above liquidus temperature; the liquidus temperature is exclusively a function of the chemical composition of a steel grade.

Figure 6 depicts the temperature distribution in the domain of interest of a 42CrMo4-grade octagonal billet at a position of 19.2 m from the meniscus. Carbon

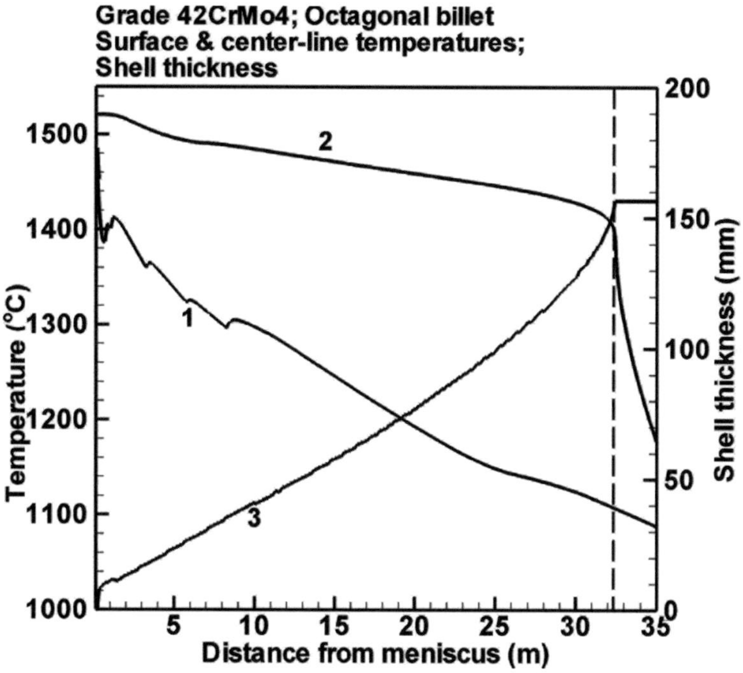

Figure 5.
Surface (1) and centerline (2) temperature values in an octagonal billet of the 42CrMo4 grade; thickness evolution (3) as the billet solidifies downwards the continuous caster.

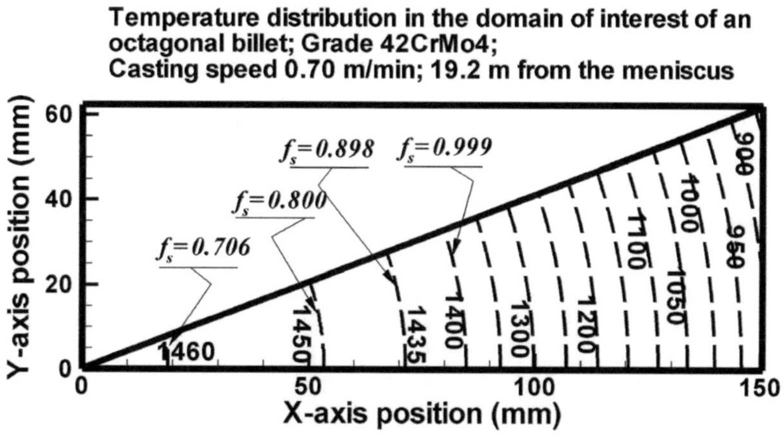

Figure 6.
Temperature distribution in the selected part of a 42CrMo4-grade octagonal billet. Solid fraction values (f_s) are presented in the mushy zone.

steels normally solidify gradually from liquid to solid phase. The degree of solidification is generally described by a parameter that is called solid fraction (f_s) and represents the percentage of the solid phase in the mixture of solid and liquid phases. When f_s = 0, then we have 100% liquid phase, and the steel temperature at this point is the liquidus temperature, T_l. When f_s = 1, then we have 100% solid phase, and the steel temperature at that point is the solidus temperature, T_s. Although the liquidus temperature is always a function of the chemical composition of a steel grade, this is not the case with the solidus temperature. The solidus temperature is also affected by the local cooling rate at solidification (C_R), which is expressed as

$$C_R = \frac{T_P - T_P{}^0}{\Delta t} \tag{11}$$

Eq. (11) shows that the temperature difference between a temperature T_P and the previous one $T_P{}^0$ at point P inside a solidifying billet within a time interval Δt, divided by this time interval, defines that local cooling rate. Consequently, the solidus temperature may be given by a formula of the type:

$$T_S = G_1(chemical\ analysis, C_R) \tag{12}$$

On the other hand, the solid fraction may be considered a function of the local cooling rate and temperature:

$$f_S = G_2(T, C_R) \tag{13}$$

Therefore, during solidification of carbon steels, there is always a range between solidus and liquidus temperatures in which the solid fraction is in the range from 0 to 1; this physical range inside a solidifying steel body is called mushy zone, and the whole phenomenon is related to micro-segregation. The simple micro-segregation model [32] has been adopted in this work, and the analysis has been described in similar previous studies [24, 29, 33]. It appears that the larger the carbon content in a steel grade, the larger the mushy zone gets, and central porosity becomes inevitable in the final stages of solidification.

The local cooling rate distribution for the case of a 42CrMo4-grade at the caster position presented in **Figure 6** is illustrated in **Figure** 7.

A short analysis showed that for the data presented in **Figure** 7, the C_R (cooling rate in °*C/s*) has the following statistics: average value μ = 0.106, standard deviation σ = 0.128, minimum = 0.023, and maximum = 0.494. More than 99% of the C_R population is within $\mu + 3^*\sigma = 0.490$.

Figure 8 depicts the surface (1) and centerline (2) temperatures in an octagonal billet as the billet solidifies downstream the Stomana billet caster; the shell thickness (3) is presented as a function of the distance from the meniscus of liquid steel in the mold. It is worth mentioning that a S355 J2-billet cast at a speed of 0.85 m/min solidifies even faster than a similar 42CrMo4-grade billet at a lower speed.

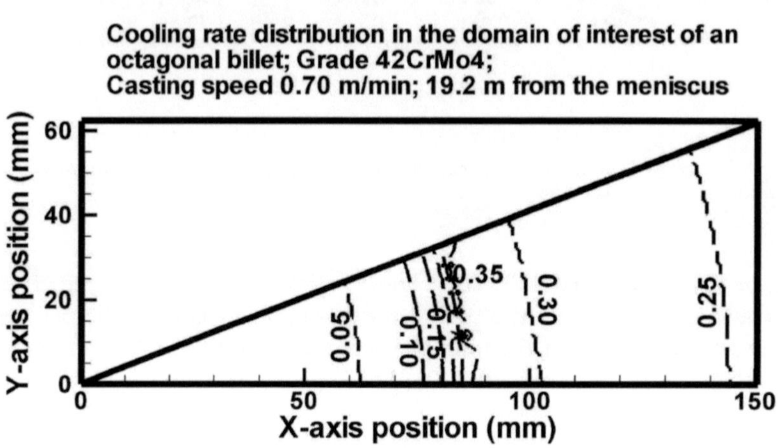

Figure 7.
Cooling rate distribution in the selected part of a 42CrMo4-grade octagonal billet.

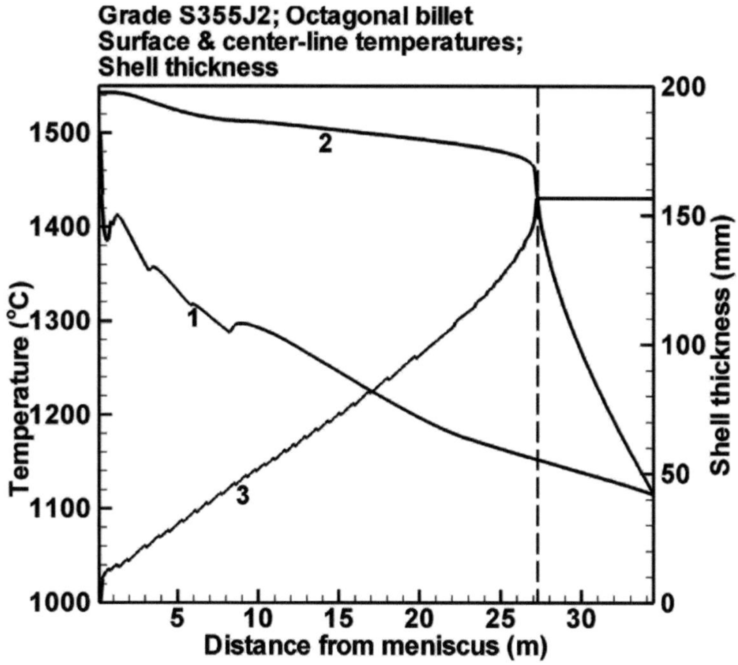

Figure 8.
Surface (1) and centerline (2) temperature values in an octagonal billet of the S355 J2 grade; thickness evolution (3) as the billet solidifies downwards the continuous caster.

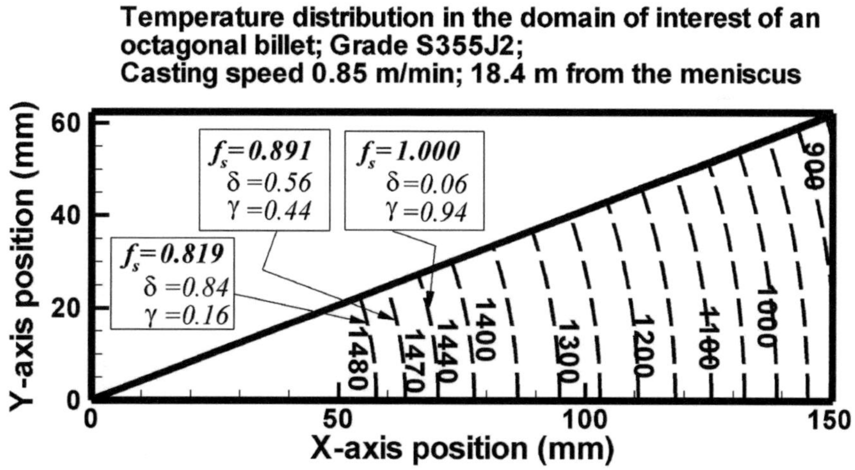

Figure 9.
Temperature distribution in the selected part of a S355 J2-grade octagonal billet. In addition to the solid fraction values (f_s) that are presented in the mushy zone, the percentages of δ- and γ- phases are shown, respectively.

Figure 9 illustrates the temperature distribution in the critical geometrical region of a S355 J2-grade octagonal billet at a distance of 18.4 m from the meniscus, cast at a speed of 0.85 m/min and a SPH = 30°C.

Peritectic grades have the tendency to crystallize in a mixture of delta (δ) and gamma (γ) phases of iron in an intermediate temperature range, before continuing to a 100% γ-phase solidification. For this reason, the percentages of δ- and γ-phases are also presented in **Figure 9** at the selected solid fractions.

All these calculations come also from the simple micro-segregation model [32] that is adopted in the developed program. **Figure 10** depicts the local cooling rates at the same position from the meniscus as for the aforementioned temperature distribution presented in **Figure 9**.

For the cooling rate distribution (C_R in °C/s) presented in **Figure 10**, a short statistical analysis derived the following results: average value μ = 0.138, standard deviation σ = 0.156, minimum = 0.027, and maximum = 0.490. Almost 99% of the population is within $\mu + 3^*\sigma$ = 0.606.

The minimum average number of iterations required for convergence by the Strongly Implicit Procedure (SIP) was attained at the value of 0.95 for the iteration

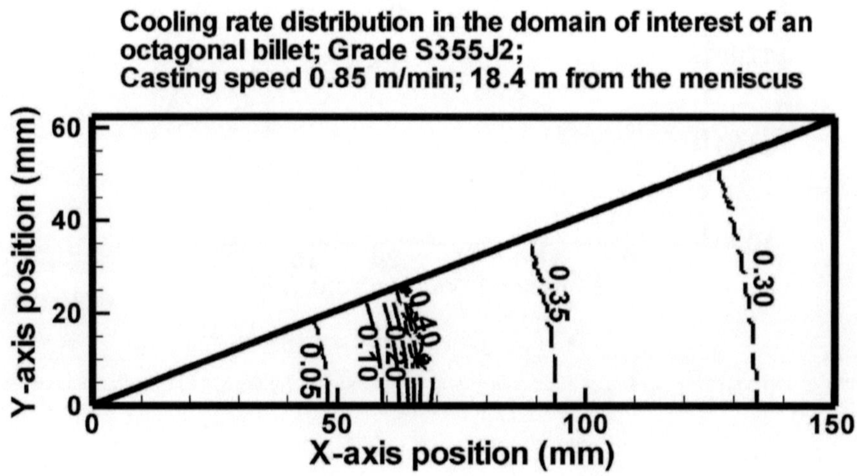

Figure 10.
Local cooling rate distribution in the selected part of a S355 J2-grade octagonal billet.

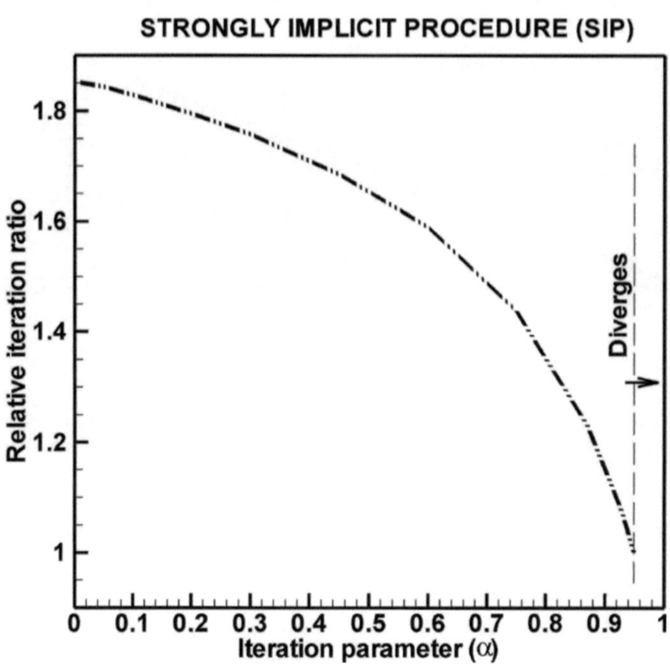

Figure 11.
Effect of the iteration parameter on the convergence behavior of the strongly implicit procedure.

parameter α as explained by Stone [31] in order to speed up the convergence process. Actually, this value (α = 0.95) was used in order to solve the derived system of matrix equations in the case of the S355 J2 grade. **Figure 11** illustrates the effect of this iteration parameter on the ratio of the required average number of iterations to the minimum number of iterations required at α equal to 0.95. In general, the convergence behavior improved as the selected iteration parameter increased. It is worth noting that convergence succeeded within an extensive range of the iteration parameter although above the limiting value of α = 0.95, the solution procedure started to diverge. For verification purposes, two sets of temperature results were generated by the Gauss-Seidel and Strongly Implicit Procedure, respectively, taken after a long simulation time (at the same time instant t = 2460 sec, equivalent to a distance of 34.85 m from meniscus). Using R [34], a Pearson's product–moment correlation test gave a coefficient of 0.9999344 for these two sets of results with a *t-Student* value equal to 17,455, 39,998 degrees of freedom and a *p-value* $< 2.2 \times 10^{-16}$.

7. Conclusions

The 2D differential equation of heat transfer was numerically solved for the derivation of the temperature distribution inside an octagonal billet. The deduced model was applied for the casting case of octagonal billets for an equivalent shape to the size of 250×300 mm^2 billets that are normally cast at the Stomana billet caster. Furthermore, the simulated octagonal billets were considered to be from the two special grades of 42CrMo4 and S355 J2. Two different iterative methods were applied for the solution of the resulting system of equations, the Gauss-Seidel method and the strongly implicit procedure, both exhibiting satisfactory behavior. A surplus of cooling water especially at the mold may result in a potential productivity increase. Computational results from the simple micro-segregation model were also included in the present study.

Acknowledgements

The author would like to express his gratitude to the top management for the continuous support on this kind of studies.

Nomenclature

c	heat capacity in J/kg/K
f_s	solid fraction
h	heat transfer coefficient in W/m^2/K
k	thermal conductivity, in W/m/K
S, S_c	source term, constant term of S, in W/m^3
S_p	source term compensating for the temperature dependence of S, in W/m^3/K
SPH	superheat (casting temperature—liquidus temperature) in °C
t, Δt	time, time interval in sec
T	temperature in a position inside a billet, or at a nodal point, in °C
T_l, T_s	liquidus, solidus temperature, in °C
x, y, z	spatial coordinates, in m
δx, δy	distance between adjacent nodal points along the x- and y-axis, in m
Δx, Δy	control-volume dimensions along the x- and y-axis, in m

ρ	density, in kg/m^3
ω	over-relaxation factor
Subscripts	
0	referring to an initial value
f	fluid (referring to a cooling fluid)
Superscripts	
0	referring to a value at a previous time step

A. Appendix I

The applied solution is based on the implicit scheme as described by Patankar in 1980 [23]. **Figure 3** shows the control volumes and the nodal points upon which the numerical solution was based; in the graph, only a very rough discretization was considered with 6 × 6 nodal points just for illustrative purposes. Normally, for nodal points in the interior of the geometry under investigation, the implicit scheme works on five adjacent points; in this way, for every internal point P, the four adjacent nodal points of interest are nominated as E (east), W (west), N (north), and S (south). The numerical discretization equation for the heat transfer equation at P is then given by Eq. (14)

$$a_P T_P = a_E T_E + a_W T_W + a_N T_N + a_S T_S + b_P \tag{14}$$

where the temperature coefficients and the constant term are given by Eq. (15):

$$\begin{aligned} a_E &= \frac{k_e\,\Delta y}{(\delta x)_e},\quad a_W = \frac{k_w\,\Delta y}{(\delta x)_w},\quad a_N = \frac{k_n\,\Delta x}{(\delta y)_n},\quad a_S = \frac{k_s\,\Delta x}{(\delta y)_s},\\ a_P{}^0 &= \frac{\rho c\,\Delta x\,\Delta y}{\Delta t},\quad b_P = S_C\,\Delta x\,\Delta y + a_P{}^0\,T_P{}^0,\\ a_P &= a_E + a_W + a_N + a_S + a_P{}^0 - S_P\,\Delta x\,\Delta y \end{aligned} \tag{15}$$

The described formulation is proven to satisfy the energy balance within the control volume that is included by the sides "n," "e," "w," and "s." Extra care is required in the derivation of the discretization equations at the borders of the studied geometry. For example, Eq. (16) describes the formulation for the temperature of a typical point on the outer area of the octagonal billet that is cooled by the water-cooled copper mold; the heat transfer is driven by a lower water temperature (T_f), by a rate given by a heat transfer coefficient h:

$$\begin{aligned} a_C T_C &= a_G T_G + a_D T_D + a_F T_F + b_C\\ a_G &= \frac{k_w\,\Delta y}{(\delta x)_w},\quad a_D = \frac{k_n\,(\Delta x/2)}{(\delta y)_n},\quad a_F = \frac{k_s(\Delta x/2)}{(\delta y)_s}\\ a_C{}^0 &= \frac{\rho c\,\Delta vol}{\Delta t},\quad b_C = S_C\,\Delta vol + a_C{}^0\,T_C{}^0 + h\,T_f\,\Delta y\\ a_C &= a_G + a_D + a_F + a_C{}^0 - S_P\,\Delta vol + h\,\Delta y\\ \Delta vol &= (\Delta x\,\Delta y/2) \end{aligned} \tag{16}$$

Similarly, another very interesting point of analysis on the derivation of the discretization equation is at a diagonal point A; the formulation is presented by Eq. (17); in this case, there is no heat transfer (due to symmetry) in the area above the diagonal (adiabatic formulation):

$$
\begin{aligned}
&a_A T_A = a_B T_B + a_N T_N + b_A \\
&a_B = \frac{k_e\,(\Delta y/2)}{(\delta x)_e}, \quad a_N = \frac{k_s\,(\Delta x/2)}{(\delta y)_s}, \\
&a_A{}^0 = \frac{\rho c\,\Delta vol}{\Delta t}, \quad b_A = S_C\,\Delta vol + a_A{}^0\,T_A{}^0 \\
&a_A = a_B + a_N + a_A{}^0 - S_P\,\Delta vol, \quad \Delta vol = (\Delta x \Delta y/4)
\end{aligned}
\tag{17}
$$

Author details

Panagiotis Sismanis
Sidenor Steel Industry SA, Athens, Greece

*Address all correspondence to: psismanis@sidenor.vionet.gr

References

[1] Mizikar EA. Mathematical heat transfer model for solidification of continuously cast steel slabs. Transactions of The Metallurgical Society of the American Institute of Mining, Metallurgical and Pet. 1967;**239**: 1747-1753

[2] Brimacombe JK. Design of continuous casting machines based on a heat-flow analysis: State-of-the-art review. Canadian Metallurgical Quarterly. 1976;**15**(2):163-175

[3] Samarasekera IV, Brimacombe JK. Thermal and mechanical behavior of continuous-casting billet molds. Ironmaking & Steelmaking. 1982;**9**(1): 1-15

[4] Brimacombe JK. Empowerment with knowledge – Toward the intelligent mold for the continuous casting of steel billets. Metallurgical and Materials Transactions. 1993;**24B**:917-935

[5] Dittenberger K, Morwald K, Hohenbichler G, Fleischl U. DYNACS cooling model – Features and operational results. In: Proceedings of the 7th International Conference on Casting; 1996; Linz, Austria. 1996. pp. 44.1-44.6

[6] Thomas BG. Modeling of the continuous casting of steel – Past, present and future. In: Proceedings of the Electric Furnace Conference; 2001; Phoenix, Arizona. Warrendale, PA: ISS; 2001. pp. 3-30

[7] Park JK, Li C, Thomas BG, Samarasekera IV. Analysis of thermo-mechanical behavior in billet casting. In: Proceedings of the Electric Furnace Conference; 2002; San Antonio, Texas. Warrendale, PA: ISS; 2002. pp. 669-685

[8] Li C, Thomas BG. Ideal taper prediction for billet casting. In: Proceedings of the ISSTech Conference; 2003; Indianapolis, Indiana. Warrendale, PA: ISSTech; 2003. pp. 685-700

[9] Schwerdtfeger KJ. Chapter 4: Heat withdrawal in continuous casting of steel. In: Cramb AW, editor. The Making, Shaping, and Treating of Steel. 11th ed. Pittsburgh, PA: The AISI Steel Foundation; 2003. Casting Volume

[10] Morales RD, Ramirez-Lopez P. Developments on fluid flow and heat flow theories in the continuous casting mold. In: Proceedings of the AISTech 2006 Conference; 2006; Cleveland, Ohio. Warrendale, PA: AIST; 2006. pp. 1005-1015

[11] Meng Y, Thomas BG. Heat Transfer and solidification model of continuous slab casting: CON1D. Metallurgical and Materials Transactions. 2003;**34B**(5): 685-705

[12] Hibbeler LC, Koric S, Xu K, Thomas BG, Spangler C. Thermo-mechanical modeling of beam blank casting. In: Proceedings of the AISTech 2008 Conference; 2008; Pittsburgh, PA. Warrendale, PA: AIST; 2008. Vol. II

[13] Straffelini G, Lutterotti L, Tonolli M, Lestani M. Modeling solidification microstructures of steel round billets obtained by continuous casting. ISIJ International. 2011;**51**(9):1448-1453

[14] Alizadeh M, Jahromi AJ, Abouali O. A new semi-analytical model for prediction of the strand surface temperature in the continuous casting of steel in the mold region. ISIJ International. 2008;**48**(2):161-169

[15] Ma J, Xie Z, Jia G. Applying of real-time heat transfer and solidification model on the dynamic control system of

billet continuous casting. ISIJ International. 2008;**48**(12):1722-1727

[16] Klimes L, Stetina J, Bucek P. Impact of casting speed on the temperature field of continuously cast steel billets. Materials Technology. 2013;**47**(4): 507-513

[17] Jiaocheng M, Jun L, Qiang Y, Liangyu C. The temperature field measurement of billet based on multi-information fusion. Materials Transactions – The Japan Institute of Metals & Materials. 2014;**55**(8): 1319-1323

[18] Yang J, Xie Z, Ji Z, Meng H. Real-time heat transfer model based on variable non-uniform grid for dynamic control of continuous casting billets. ISIJ International. 2014;**54**(2):328-335

[19] Chen Y, Peng Z, Wu L, Zhao L, Wang M, Bao Y. High-precision numerical simulation for effect of casting speed on solidification of 40Cr during continuous billet casting. La Metallurgia Italiana. 2015;**1**:47-51

[20] Dong Q, Zhang J, Qian L, Yin Y. Numerical modeling of macrosegregation in round billet with different microsegregation models. ISIJ International. 2017;**57**(5):814-823

[21] Entesano L, Isera M, De Luca A, Koblenzer H, Boldrin N. High-speed casting: A new concept in billet shape for rebar production. In: Proceedings of the AISTech 2018 Conference; 2018; Philadelphia, PA. Warrendale, PA: AIST; 2018. pp. 1707-1714

[22] Incropera FP, DeWitt DP. Fundamentals of Heat Transfer. New York: John Wiley & Sons; 1981. 49 p

[23] Patankar SV. Numerical Heat Transfer and Fluid Flow. New York: Hemisphere Publishing Corporation; 1980. 60 p

[24] Sismanis P. Evaluation of solidification times for medium and high carbon steels based upon heat transfer and solidification phenomena in the continuous casting of blooms. In: Kazi SN, editor. Heat Transfer. Rijeka, Croatia: IntechOpen; 2015. DOI: 10.5772/60706

[25] Wikipedia. Octagon [Internet]. 2018. Available from: https://en.wikipedia.org/wiki/Octagon [Accessed: November 15, 2018]

[26] Carslaw HS, Jaeger JC. Conduction of Heat in Solids. 2nd ed. New York: Oxford University Press; 1986. 202 p

[27] Sismanis P. Heat transfer analysis of special reinforced NSC-columns under severe fire conditions. International Journal of Materials Research. 2010;**101**: 417-430. DOI: 10.3139/146.110290

[28] Sismanis P. Modeling solidification phenomena in the continuous casting of carbon steels. In: Ahsan A, editor. Two Phase Flow, Phase Change and Numerical Modeling. Rijeka, Croatia: IntechOpen; 2011. DOI: 10.5772/19699

[29] Sismanis P, Bouzouni M, Papaefthymiou S. Modeling of crucial process parameters for the continuous improvement of special steels at the stomana plant. Journal of Materials Engineering and Performance. Novelty, Ohio: Springer, Published Online. ASM International; 2018;**27**(10):5130-5135. DOI: 10.1007/s11665-018-3616-z

[30] Anderson DA, Tannehill JC, Pletcher RH. Computational Fluid Mechanics and Heat Transfer. New York, NY: Hemisphere Publishing Corporation; 1984. 133 p

[31] Stone HL. Iterative solution of implicit approximations of multidimensional partial differential equations. SIAM Journal on Numerical Analysis. 1968;**5**(3):530-558

[32] Won YM, Thomas BG. Simple model of micro-segregation during solidification of steels. Metallurgical Transactions. 2001;**32A**:1755-1767

[33] Sismanis P. The effect of local cooling rates upon solidification phenomena in the continuous casting of carbon steels. In: Proceedings of the 8th European Continuous Casting Conference and Symposium in Numerical & Physical Modeling; 2014; Graz, Austria. ASMET; 2014. pp. 1462-1471

[34] R Core Team. R: A Language and Environment for Statistical Computing. Vienna, Austria: R Foundation for Statistical Computing; 2014. Available from: http://www.R-project.org/ [Accessed: December 19, 2018]

Chapter 6

CFD Simulation of Heat and Mass Transfer for Climate Control in Greenhouses

Cruz Ernesto Aguilar Rodriguez and Jorge Flores Velazquez

Abstract

Greenhouse plant production involves a number of processes such as transpiration, condensation, photosynthesis, and climate control. Such processes, in turn, set off mass and heat transfer phenomena that influence not only the quality and quantity of crop production but also its environmental cost. While these processes have considerably been analyzed in separate, they strongly interact with one another. For instance, increased radiation (mainly thermal infrared) increases temperature, reduces humidity, consequently increases transpiration, and affects CO_2 exchange as well as other reaction rates. Computational fluid dynamics (CFD) is a numerical tool with a solid physical basis which allows, through the construction of a computational model, to simulate the fluid flow environment. Heating, ventilation, and condensation have been analyzed in the greenhouse environment with CFD techniques. The current challenge is the interaction of these processes and their impact on the production system. The present work summarizes some CFD investigations carried out in this topic, in order to analyze the processes of heat and mass transfer in a greenhouse for agronomic purposes.

Keywords: ventilation, heating, crop production, numerical simulation, climate control

1. Introduction

The fast expansion of greenhouse technique around of world, as a means to supply food and produce, has posed emerging challenges in the operation and management of greenhouse climate. While such challenges have not changed in essence ever since the onset of agriculture, they have been considerably reshaped by the access to new technologies and information.

In semiarid regions, the main problems are the high temperatures that take place in daily and annual cycles. The same is true for cold temperatures. There are many options to auxiliary climate control systems, the implementation of which depends on many factors such as their cost, crop, location, and management, to name a few.

Greenhouse is an advantageous production system which realistically allows us to produce crops from all over the world during the whole year. Consequently, environment interior conditions such as temperature and humidity have to be controlled at a certain plant-specific level regardless of environmental conditions.

In greenhouse crop production, climate has peculiar considerations, because the most important data is the impact of the environmental factors on the crop cycle. The cultivation of plants requires a sufficient amount of light, a specific range of temperature, humidity, and CO_2, among other requirements. These requirements are primarily influenced by the greenhouse design and size and vary according to the local climate conditions. For instance, the radiation quantity inside the greenhouse depends on whether greenhouses are built with PVC or glass, because the surface material is the element to optimally use solar radiation for the required lighting [1].

Recently, heating, ventilation, and air conditioning (HVAC) systems have been extensively used in urban spaces, such as offices or stores, at agriculture area, and specifically in greenhouses. For instance, HVAC has been used in buildings in order to analyze the optimization and comfort inner Officine's and several uses. In greenhouses, the concept is incipient, even though the application of the ventilation and calefaction systems as a method to climate control in cold and warm regions is nothing new.

The climate produced in a greenhouse is the result of complex mechanisms involving the processes of heat and mass exchange. The internal climate is strongly dependent on the outside conditions, especially in unheated greenhouses. In greenhouse climate models, the parameters of the internal climate such as air, soil, and crop temperatures as well as air humidity are calculated using energy and water vapor balances for the various components of the system [2].

Climate in the greenhouse is a consequence of radiation crossing the cover material, usually plastic. After that, climate condition is a strong relationship between several factors, temperature, humidity, wind velocity, and the solar radiation. Environment conditions in the greenhouse depend on the energy management, by the auxiliary calefaction and ventilation systems. The equilibrium of these climate variables is a function of the efficiency of air exchange generating losses and gains of heat (temperature, radiation) and mass (humidity, gases) [3].

The automation in a greenhouse environment involves climate control, light-level control, and shade curtain management; gases inside the greenhouse are due to the plant reactions with the environment, requiring control of carbon dioxide (CO_2) concentration, irrigation and chemical treatment. Greenhouse automation is a modern, efficient, and accurate disruptive agriculture, which utilizes data collected within the system, to obtain better quality and higher yields, thereby increasing productivity [3].

Every process inside the greenhouse consumes energy and involves a change of mass between the sink or the source. The objective of this work was to show some results on greenhouse mass and energy transfer, using CFD.

2. Computational fluid dynamics (CFD) in a greenhouse simulation

Computational fluid dynamics (CFD) is an analysis tool based on numerical methods that show graphically the general and localized air movement inside the greenhouse owing to natural ventilation. Also, it is possible to determine spatial temperature distributions arising from such air movement, all this for any greenhouse type and open/closed configuration of the roof and side windows.

CFD modeling of different parameters in greenhouses has been used to examine various features such as vent configuration [4]; natural and mechanical ventilation [5, 6]; ventilation in screenhouses [7]; condensation, transpiration, and heat and mass transfer [8–10]; and, more recently, calefaction and HCVA [11] and their interactions [12, 13]. The analyses of these systems allow for climate control,

Figure 1.
Natural and mechanical climate control in greenhouses.

thereby offering the possibility to provide large numbers of high-quality crops with greater predictability.

CFD modeling has been used as a tool to get major details in facilities, for instance [14], uses CFD to analyze ventilation system in greenhouses. Based on CFD, simulation is possible to optimize some characteristics of ventilation systems, such as relationship between volume and vent area of greenhouses [15].

The performance of ventilation in enclosed spaces is affected by the flow of outside air [16], type of cover, height of the installation, and the ventilation opening [18]. Computational parametric studies on greenhouse structures can aid to identify design factors that affect greenhouse ventilation under specific climatic conditions [5, 19].

Modern auxiliary systems used for climate control demand new approaches of study, e.g., to quantify the effect of the back-wall vent dimension on solar greenhouse cooling. Traditional solar greenhouse (**Figure 1**) uses radiation to store energy and get advantages of its use naturally. Some studies [10] showed that it is possible to reduce averaged air temperature by approximately 1.7°C and the highest temperature drop by approximately 5.8°C, in comparison to a traditional solar greenhouse with brick back wall (TG). These authors also suggest that a back-wall vent of 1.4 m increased internal ventilation efficiency in a solar greenhouse by installing removable back walls [10].

On the other hand, modeling of climate systems is necessary for studying and regulating energy consumption and quality of indoor environment. In urban semi-closed spaces, modeling approaches are used in HVAC systems [20]. Physics-based models offer adequate capabilities for first-hand assessments but suffer from poor accuracy; data-driven models have very high accuracy on training data but suffer from lack of generalization beyond the training domain.

Numerical methods have also been implemented for analyses of crop production in semi-closed spaces. Santolini et al. [7] reviewed the effect of mass transfer in a screenhouse structure with CFD. Alternative computer-based simulation models have been used for examining typical greenhouses with alternative energies such as dynamic photovoltaic (PV) and plant cultivation [21, 22].

3. Heat and mass transfer equations and CFD simulations

Heat and mass transfer is investigated using CFD tools. A numerical model is built based on the solution of the governing equations for momentum, energy, and continuity within the greenhouse domain. General equations can be written as the convection-diffusion equation to simulate mass, velocity, temperature, or other variables inside the greenhouse (Eq. 1):

$$\frac{\partial \rho \varnothing}{\partial t} + \nabla(\rho \acute{v} \varnothing) = \nabla(\Gamma \nabla \varnothing) + s_{\phi} \quad (1)$$

where ∅ transport variable; u_j velocity vector (m s^{-1}); Γ_{ϕ} diffusion coefficient; s_{ϕ} source term to variable ∅ (temperature, CO_2, etc.).

Specific energy balance simulation is based on the solution of heat and mass balance equations applied to the whole greenhouse system [8].

For the heat balance of greenhouse air, the general equation is shown in Eq. (2):

$$\frac{\rho V C_p \Delta T}{\Delta T} = \sum_1 q_i A_i + \phi(C_p T_{out} - C_p T_{int}) \quad (2)$$

Mass balance of greenhouse air is described in Eq. (3):

$$\frac{M_w}{\partial T} = \Gamma_{crop} - \Omega_{cov} + \phi(W_{air} - W_{air_{ou}}) \quad (3)$$

where ρ is the density (kg cm^{-3}); t is the time (s); T is the temperature (°C); C_p is the heat capacity at constant pressure; Φ is the ventilation rate (kg s^{-1}); W (kg m^{-3}); $\sum_1 q_i A_i$ (W) is the sum of the convective contribution; Γ_{crop} (kg s^{-1}) is the transpiration rate; w_{air} (kg kg^{-1}) is the inside humidity ratio; W_{air_out} (kg kg^{-1}) is the outside humidity ratio; M_w is the water vapor mass.

The energy transfer process can occur basically in three physical phenomena: radiation, convection, and conduction. In greenhouse inner, convective heat transfer is the main source of temperature and energy. The conduction of energy occurs from the soil layers, and the flow is displaced depending on the quantity, always from higher to lower.

3.1 Conduction energy transfer process

Heat conduction is based on Fourier's law, in which one direction is a simple Eq. 4:

$$q_x = -kA\frac{\delta T}{\delta x} \quad (4)$$

where k material conductivity (W m^{-1}); A cross-sectional perpendicular area (m^2); $\frac{\delta T}{\delta x}$ thermal gradient (°C).

The convective effect is calculated using the cooling Newton's law (in Eq. 5):

$$Q = hA_s(T_s - T_{\infty}) \quad (5)$$

where h convective heat transfer coefficient (W m^{-2}); A_s area (m^2); T_s surface temperature (°C); T_{∞} fluid temperature (°C).

In a greenhouse, Eqs. (6) and (7) give the convective fluxes:

$$q_{conv,i} = \alpha_i(T_i + T_{air}) \quad [Wm^{-2}] \quad (6)$$

$$Q_{conv,i} = A_i \alpha_i(T_i + T_{air}) \quad [Wm^{-2}] \quad (7)$$

3.2 Radiation: energy transfer process

Outgoing radiation from a surface with nonzero transmissivity cover and sidewalls can be described in Eq. 8 [23]:

$$j_{i-in} = \varepsilon_i \sigma T_i^4 + \zeta_i g_{in} + \tau_{in} g_{ext} \quad [\mathrm{Wm}^{-2}] \tag{8}$$

where ε_i is the emissivity, σ is the Boltzmann constant, ζ_i is the reflectivity, and τ is the transmissivity. The outgoing radiation from opaque surface, soil, external soil, and sky is calculated with Eq. 9:

$$j_i = \varepsilon_i \sigma T_i^4 + \zeta_i \quad [\mathrm{Wm}^{-2}] \tag{9}$$

Incident radiation on a surface is (Eq. (10)):

$$g_i = \sum\nolimits_{j-sup} F_{i \to j} J_j \quad [\mathrm{Wm}^{-2}] \tag{10}$$

where $F_{i \to j}$ is the view factor between surfaces j and i (Eq. (11)).

Several factors are involved in these calculations. For instance, the net radiation balance could be simulated with Eqs. (11) and (12):

$$q_{rad} = j_i - g_i \quad [\mathrm{Wm}^{-2}] \tag{11}$$

$$Q_{rad-i} = A_i (j_i - g_i) \quad [\mathrm{W}] \tag{12}$$

In addition, the ideal black-body radiation is shown in Eq. 13:

$$E_{b\lambda} = \frac{C_1}{\lambda^5 \left(e^{\frac{C_2}{\lambda T}} - 1 \right)} \tag{13}$$

where $E_{b\lambda}$ spectral emissivity power (W m^{-2}); λ wave longitude (m); T absolute surface temperature (K); C_1 3.7405 $\times 10^{-16}$ (W m^2); C_2 0.0143879 m K.

The power surface emissivity is

$$E_b = \sigma T_s^4 \tag{14}$$

where T_s absolute surface temperature (K) and σ Boltzmann constant (5.6697×10^{-8} W m^{-2} K^{-1}).

The simulation of radiative heat exchange between black surfaces is based on Eq. 15:

$$Q_{1-2} = \sigma A_1 F_{12} \left(T_1^4 - T_2^4 \right) = \sigma A_2 F_{21} \left(T_1^4 - T_2^4 \right) \tag{15}$$

where F is the fraction of radiant energy that leaves the area A (m^2).

3.3 Mass transfer process in a greenhouse

In a greenhouse, the mass balance between inflows and outflows must be preserved. In general, Eq. 16 represents the mass balance:

$$\frac{\partial \acute{\rho}}{\partial t} + \frac{\partial}{\partial x_j}(\acute{\rho}\acute{u}_j) = 0 \tag{16}$$

where ρ air density (kg m^{-3}) and u_j wind velocity in j direction (m s^{-1}).

3.3.1 Condensation

Crop transpiration increases the percentage of water vapor in the environment, generating the possibility of obtaining saturated air. Environment saturation is an undesirable effect over the crops. There are some approximations in order to know condensation rate [8], which can be estimated as a difference between former quantity and the latter. Eq. (17) is used to estimate it:

$$\Omega_{cov} = \max(0, M_{w,air} - M_{w,\mathrm{cov}})\ [\mathrm{kg}] \tag{17}$$

where $M_{w,air}$ the humidity content of the greenhouse air (%) and M_{w,cov_in} the saturated humidity content of air at the cover temperature (%).

3.3.2 Water vapor

Water vapor transport is simulated with Eq. 18:

$$\frac{\partial(\acute{c})}{\partial t} + \frac{\partial}{\partial x_j}\left(\acute{u}_i\acute{c}\right) = \frac{\partial}{\partial x_i}\left(D_w + \frac{\mu_t}{\acute{\rho} s_{ct}}\right)\frac{\partial(\acute{c})}{\partial x_i} + S_w \tag{18}$$

where C mass concentration of component in air (kg kg^{-1}); u_i wind velocity in j direction (m s^{-1}); D_w water vapor diffusivity (m^2 s^{-1}); μ_t turbulence air viscosity (kg m^{-1} s^{-1}); ρ density (kg m^{-3}); S the average velocity module in the deformation (m s^{-2}) which is calculated with Eq. 19:

$$S_w = \frac{ET}{Lv}LAD \tag{19}$$

where ET latent heat flux density (W m^{-2}); Lv evaporation latent heat (J kg^{-1}); and LAD leaf area density (m^{-1}).

4. CFD simulations in greenhouses

4.1 Natural and mechanical ventilation

In a greenhouse crop production, the ventilation system is the most important auxiliary equipment for climate control. Natural or mechanical ventilation design accounts for the size of greenhouse to determine the vent dimension and position (**Figure 1**). Furthermore, new complementary devices have been adapted to enhance the efficiency of air renewal rates. For instance, the use of the back-wall vent dimension on solar greenhouse cooling was investigated by He et al. [10] using CFD. In this study, the average air temperature in a solar greenhouse with removable back walls (RG) was reduced by approximately 1.7°C with a back-wall vent of 1.4 m, thereby increasing ventilation efficiency.

The presence of screens in the lateral and roof windows reduces the ventilation rate. However, according to [7], screens promote uniform velocity distributions inside the greenhouse compared to no-screened greenhouses, especially near the crops. **Figure 2** shows the results of a CFD simulation in a screenhouse, more specifically the exchange of air inside/outside near the screenhouse roof.

The advantages of numerical simulation are the possibilities to observe details in specific zones of the greenhouses (**Figure 2A**) and to convert a discrete phenomenon continuously. **Figure 2B** shows the mass air that enter and exit from a screenhouse under five exterior velocities simulated, when crop is simulated and empty. When the screenhouse is empty, mass balance is very similar; however, the crop reduces this flow until 200 kg s^{-1} when exterior wind velocity is 5 m s^{-1}.

In a greenhouse with combined mechanical and natural ventilation (**Figure 3**), the velocities' patron is marked different. For instance, when only mechanical ventilation (fist one) is simulated, temperatures' distribution is basically due to mechanical convection as a consequence of these air movements. In the second one, roof windows are 30% open and the wind patron changed. If just mechanical ventilation is working, under vents, the velocity is near to zero, but if the roof windows are open, the wind distribution is better than only mechanical ventilation.

CFD simulation of the ventilation systems, natural, mechanical, or combined, allows to observe the distribution of the air in a problematic zone and infer process of mass and energy transfer due to the interaction with the external climate conditions.

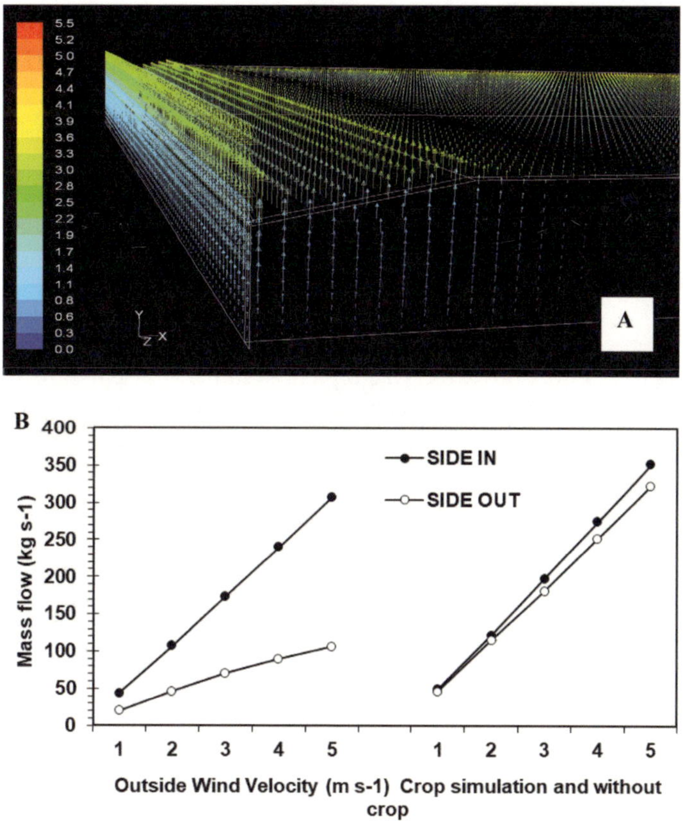

Figure 2.
(A) Top view of mass exchange in the roof of screenhouse and (B) scalar mass flow (kg s^{-1}) side exchange (inside/outside) by CFD simulation with and without crop.

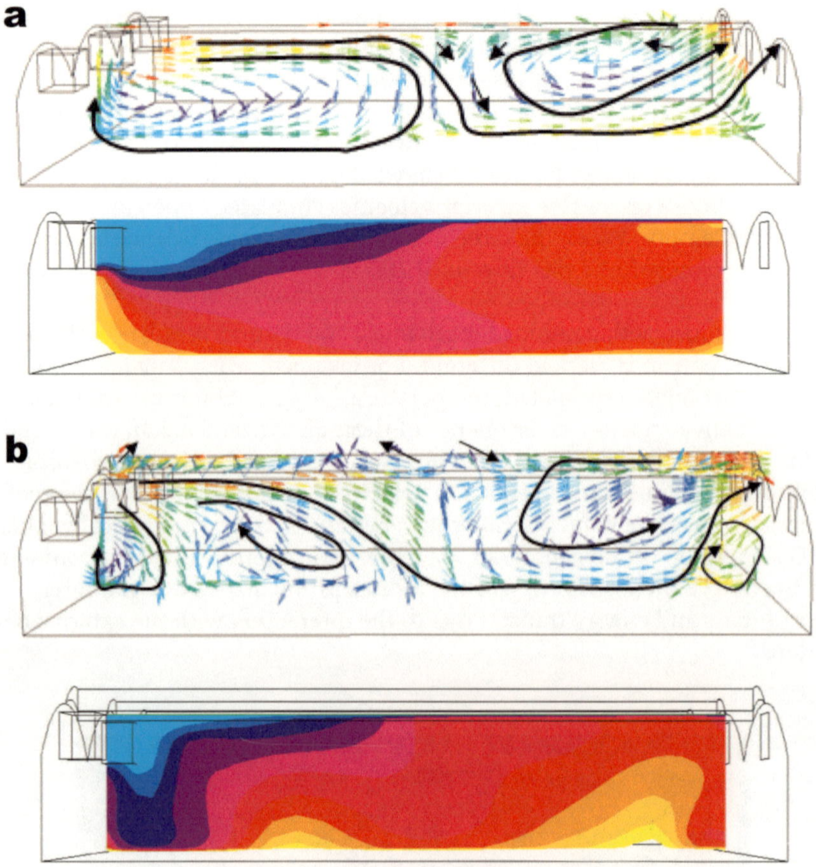

Figure 3.
Wind velocity vectors' distribution in the central spans of greenhouse closed (A) and all-open (B).

4.2 CFD heating pipe tube simulation

Heating in greenhouses strongly influences crop yields [17], energy consumption, and operation costs; however, this type of systems is essential to achieve sustainable production. A method to prevent low temperatures below a threshold makes use of the forces arising from a temperature or convection gradient.

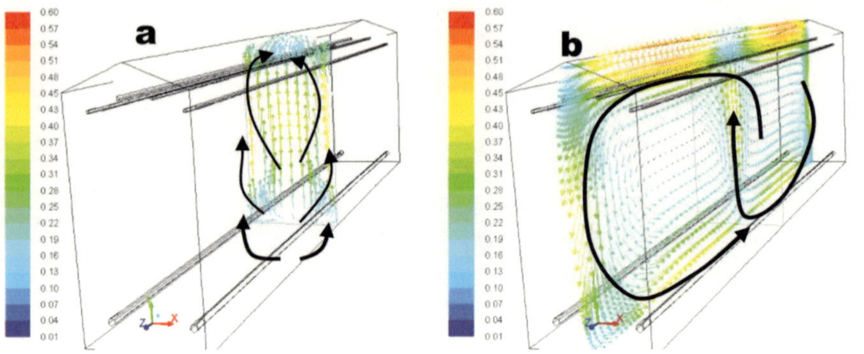

Figure 4.
Distribution of wind speed vectors (m s^{-1}) to the center of the module in a frontal (a) and longitudinal (b) plane.

The systems that most cold-climate greenhouses use are a collector wall and a heating system based on water or gas driven by a pipe. The heating pipes (pipe heating) is an effective means of keeping the greenhouse warm by promoting convection and radiation of heat. The layout of these tubes and the heating power determine the spatial distribution of temperature and the flow patterns induced by the movement of air due to the convective effect (**Figure 4**).

Teitel et al. [24, 25] mentioned that the best way to place the tubes is at medium height and under the crop, with the tubes as close as possible to the leaves. Other configurations have been analyzed by various researchers [24, 26, 27], which highlight the influence of the heating system with crops and radiative aspects. These investigations have unveiled the advantages of installing hot water pipes (pipe heating) in the lower part of the crop without promoting excessive evaporation [28]. Such pipe heating systems also favor the removal of humidity, which is known to negatively influence air quality. Moisture transport has been analyzed using computational fluid dynamics (CFD) to address various aspects such as condensation [8] and refrigeration [18], especially in closed greenhouses [5].

Numerical methods have been widely used to study climate variable inner greenhouses [29]. In 2007 and recently 2017 [30, 31] analyzed the heat distribution by three pipes and perforate polyethylene ducts to manage low temperature in tomato crop greenhouses. CFD gets observed as strong thermal gradients near to the ground and roof and well conditions in the crop zone. In this study, the effect of determining the flow and temperature patterns is the location and power of heating devices [31].

Figure 4 shows the air movement in a small greenhouse, with heating system based on five heat water tubes. The air movement and energy transference are due to the convection method, because temperature in the low tubes is higher than the upper pipe tubes. Normally wind velocities in greenhouse oscillation are between 0.1 and 0.5 m s^{-2}, due to pressure effect. In this system, wind velocity, just for convection effect, is 0.2–0.3 m s^{-1}.

A homogeneous temperature distribution is observed throughout most of the day (**Figure 5**). In a greenhouse almost all of the management processes need energy; in fact, in cool regions, the cost due to the climatic control is nearly 40% of the total production cost or more, depending on the automation grade of sensors and controls.

4.3 Transpiration

Transpiration is a special component with enormous importance in the balance of energy and transfer process in the greenhouse system. Crop transpiration is an important process useful not only in the production process but also in the climate

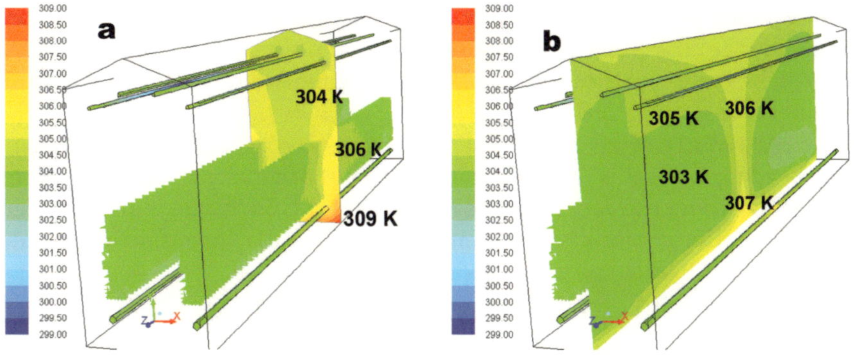

Figure 5.
Temperature gradient (K) to the center of module: frontal plane view (a) and longitudinal plane view (b).

control. Actually, transpiration is the first cooling natural system; when the high temperature is increasing, transpiration occurs very fast, and temperature is controlled. In CFD it is possible to simulate this phenomenon as a source term from the crop, as a flow of water. To speed up energy transport calculus use the model Penman-Monteith (Eq. 20) with some simplification.

Simulation in Fluent is based on Eq. 20, and for the simulation of transpiration, it is necessary to make a balance of energy between the plant and the environment, creating a system of equations implemented in the simulation as a "user-defined function" (UDF) so that terms such as transpiration, the consumption of CO_2, etc. can be calculated [4]. Nowadays most of the factors and estimated values of latent heat of vaporization in the energy balance equation can be measured using data of density, thermal conductivity and psychometric constant.

$$ET = \frac{\Delta(Rn - G) + \rho_a c_p \frac{(e_s - e_a)}{r_a}}{\Delta + \gamma\left[1 + \frac{r_c}{r_a}\right]} \tag{20}$$

where ET is the potential evapotranspiration (kg m^{-2} s^{-1} o mm s^{-1}); Rn is the net radiation (kW m^{-2}); G is the heat flux in soil (kW m^{-2}); $(e_s - e_a)$ is the vapor pressure deficit (kPa); r_c is the crop resistance (s m^{-1}); r_a is the aerodynamic resistance (s m^{-1}); Δ is the slope of the vapor pressure saturation (es/T) (Pa °C^{-1}); ρ_a is the air density (kg m^{-3}); c_p is the specific heat of the air (MJ kg^{-1}°C^{-1}); and γ is the apparent psychometric constant (kPa °C^{-1}).

In the case of stomatal resistance, it is possible to measure it directly and relate it to the environmental variables involved (solar radiation, VPD, temperature and CO_2 concentration). For each crop, the resistance will be different, but in general an average resistance in the canopy can be estimated according to the foliar area index [33]. To estimate external leaf resistance, it has been assumed that temperature of the leaf and air is the same, so it is possible to estimate a coefficient r_c with Eq. (21):

$$r_c = \frac{r_i}{L} \tag{21}$$

where R_c is the internal resistance of the leaf canopy to the transfer of water vapor (s m^{-1}), L is the leaf area index, and r_i internal resistance of the leaf (s m^{-1}). **Figure 6** shows the simulated results of the distribution of humidity and mass fraction along the greenhouse using the simplified model of [33]. Numerically it was demonstrated that the Penman-Monteith transpiration model is not particularly sensitive to the variables with the simplification of the model mentioned, which can be an indication of a good result.

Transpiration of the crop is directly affected by the foliar area (**Figure** 7), and consequently the strict relationship between this and the vapor pressure deficit (VPD) will be the variable to follow for an approximation to the transpiration of greenhouse crops.

The largest source of variation between the models compared is based on the leaf area of the crop; while it is true that transpiration is originally associated with the amount of radiation, the dependence of stomata in this exchange is also founded. **Figure** 7 shows the variation of the transpiration of the crop as a function of the leaf area index (LAI), in this case a tomato crop.

4.4 Gas simulation (ammonia)

Mass transfer in semi-closed spaces is an important process. Ventilation is the primary mechanism for gas removal. Air movement assumes a mixture of liquid,

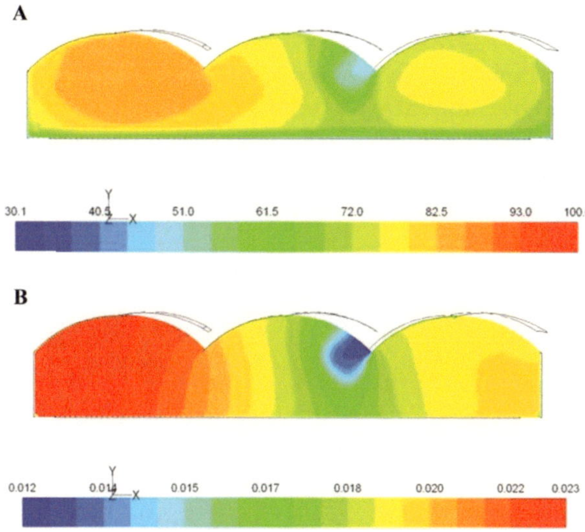

Figure 6.
Contour of relative humidity (%) (A) and transpiration as a mass fraction of H_2O (B) using CFD simulation [33].

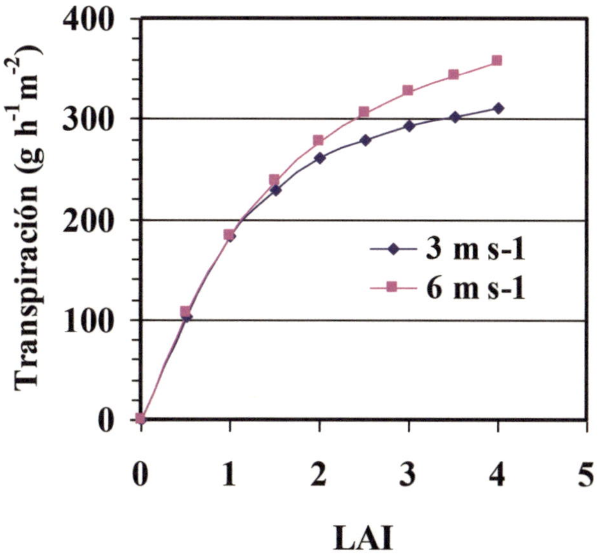

Figure 7.
Variation of transpiration (g m^{-2} h^{-1}) of a tomato crop as a function of the leaf area (IAF), when 3 and 6 m s^{-1} speed is simulated outside of the wind [33].

vapor, and nonconsumable gases. In this case, the species transport model available in ANSYS Fluent was used to simulate the mass transport, beginning from the diffusion flux of species i, which arises due to gradients of concentration and temperature. Such species model uses the dilute approximation (Flick's law) to model mass diffusion. For turbulent flows, mass diffusion can be written as in Eq. 22 [32]:

$$\bar{J}_i = \rho D_{i,m} \nabla Y_i - D_{T,i} \frac{\nabla T}{T} \tag{22}$$

In Eq. (19), J_i is the diffusion flux of species i (m^2 s^{-1}), ρ is the density of the mixture (kg m^{-3}), $D_{i,m}$ is the mass diffusion coefficient for species i in the mixture

m (m^2 s^{-1}), and $D_{T,\,i}$ is the turbulent diffusion coefficient (m^2 s^{-1}). Yi is the mass fraction of specie i, and T is the temperature of the flow (K). CFD can simulate this process and visualization of the movement as shown in **Figures 8–10**.

The discretization of components in semi-closed facilities can better depict fluxes under different scenarios. **Figure 8** shows the air movement along the barn and how the temperature is changing. In addition, air exchange promotes an efficient distribution of gas concentration by the effect of ventilation system.

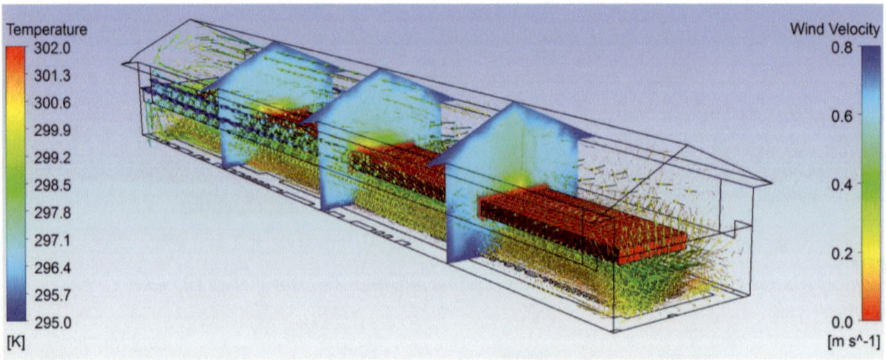

Figure 8.
Wind velocity vectors (m s^{-1}) under cages and surrounding temperature profiles (°C).

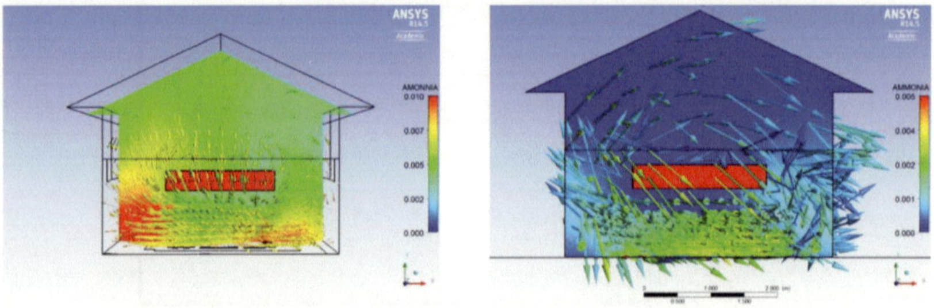

Figure 9.
Relative gas concentration by air exchange effect.

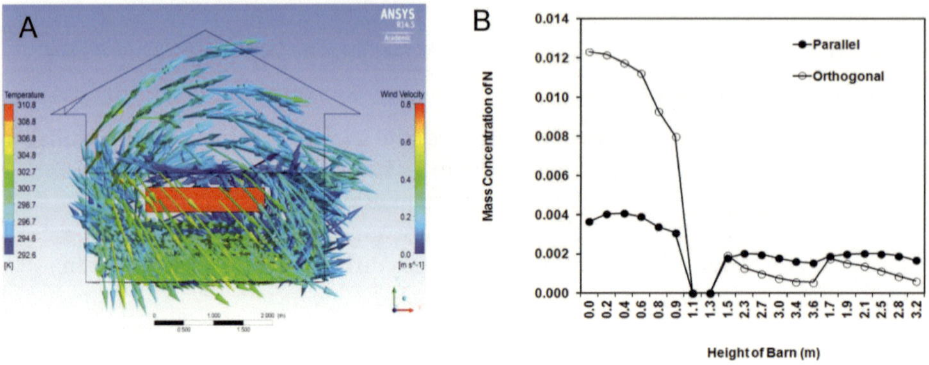

Figure 10.
CFD results of (A) profile of wind velocity and temperature and (B) relative gas concentration of nitrogen in a vertical profile, under two wind direction configurations (cages are in the 1.1 m height).

Performance of the vents is a function of their size, position, and proportion to the whole ground area.

In this study mass and energy transfer was revised to get reduced the negative effect of the ammonia gas in the rabbit barn development. Two climatic variables are responsible to the rabbit's health: temperature and humidity. Both climate variables were got better when the position of windows was changed. These results are consistent with CFD simulations, where the effective renovation rate depends on the position of the window. In some cases more than 50% of the air cannot get in through the inlet vent, producing a ventilation rate of 5.4 kg s^{-1}. As a consequence, a greater dispersion of toxic gases and lower temperature gradients (5 K) are produced.

The air exchange rate is an indicator of gas movement, because it is similar for both the air and the gas being simulated such as the ammonia (**Figure 9**) with a wind direction normal to the ridge. When the wind is parallel to the vents, the air that enters the vents by pressure difference produces a higher ventilation rate at the zone beneath the cages (**Figure 10**), even when the ventilation rate is close to zero. In contrast, when the wind flows normal to the ridge, ventilation rates increase.

Numerical models show a representative environmental dynamics, which can supply information for manage and control of several climate factors.

Continuity equation indicates that mass quantity entrance must be the mass in exit; however, with the change in the configuration of orientation of barn, the gas concentration can be better. Using CFD simulation, the concentration of gas under/over cages is calculated. **Figure 10** shows the mass transfer due to natural ventilation systems and the wind direction with respect to the size of the windows. In this case the position of the size of the windows was enough to reduce the mass transference under cages. Results indicated that the rate of mass change is the same, but distribution of gases (mass exchange) can be managed using different configuration of windows.

5. Conclusions

Numerical tools applied at predictive models of heat and mass transfer are helpful to better manage water-climate-soil inputs to plants in greenhouses. Computational fluid dynamics models are used to describe the greenhouse microclimate and the behavior of the plant-environment interaction in greenhouses. CFD is a powerful tool, to get the analysis of interactions between components of biosystem. Cover material, soil, and crop with other components must be included in the model. The crop can be considered as a porous medium and measured transpiration and sensible heat transfers. CFD models and auxiliary programming tools have been widely used to measure the interactions between the mass and energy transfer processes within the greenhouse and other biosystems, with excellent results.

Author details

Cruz Ernesto Aguilar Rodriguez and Jorge Flores Velazquez*
Mexican Institute of Water Technology, Mexico

*Address all correspondence to: jorge_flores@tlaloc.imta.mx

References

[1] Subin MC, Savio JL, Karthikeyan R, Periasamy C. Analysis of materials used for greenhouse roof covering-structure using CFD. Materials Science and Engineering. 2018;**346**: 012068. DOI: 10.1088/1757-899X/346/1/012068

[2] Baptista FJ, Bailey BJ, Meneses JF, Navas LM. Greenhouses climate modelling. Tests, adaptation and validation of a dynamic climate model. Spanish Journal of Agricultural Research. 2010;**8**(2):285-298

[3] Shamshiri R, Wan Ismail WI. A review of greenhouse climate control and automation systems in tropical regions. Journal of Agricultural Science and Applications. 2014;**2**(3):176-183. Available from: www.j-asa.org

[4] Baeza EJ, Perez-Parra JJ, Lopez JC, Montero JI. CFD simulation of natural ventilation of a parral greenhouse with a baffle device below the greenhouse vents. Acta Horticulturae. 2008;**801**: 885-892

[5] Flores-Velázquez J. Análisis del clima en los principals modelos de invernaderos en México mediante CFD [PhD Diss]. Almería, Spain: Universidad de Almería; 2010

[6] Fidaros D, Bexevanou C, Bartzanas T, Kittas C. Numerical study of mechanically ventilated broiler house equipped with evaporative pads. Computers and Electronics in Agriculture. 2018;**149**:101-109. DOI: 10.1016/j.compag.2017.10.016

[7] Santolini E, Pulvirenti B, Stefano B, Barbaresi L, Torreggiani D, Tasianari P. Numerical study of wind-driven natural ventilation in a greenhouse with screens. Computers and Electronics in Agriculture. 2018;**149**: 41-53. DOI: 10.1016/j.compag.2017.09.027

[8] Piscia D. Analysis of night-time climate in plastic-covered greenhouses [Tesi Doctoral]. Terrasa, Barcelona, Spain: Departament de Màquines i Motors Tèrmics E.T.S.E.I.A.T., Universitat Politecnica de Catalunya; 2012

[9] Bouhoun H, Bournet PE, Cannavo P, Chantoiseau E. Development of a CFD crop submodel for simulating microclimate and transpiration of ornamental plants grown in a greenhouse under water restriction. Computers and Electronics in Agriculture. 2018;**149**:26-40. DOI: 10.1016/j.compag.2017.06.021

[10] He X, Wang J, Guo S, Zhang J, Wei B, Sun J, et al. Ventilation optimization of solar greenhouse with removable back walls based on CFD. Computers and Electronics in Agriculture. 2018; **149**:16-25

[11] Amanowicz Ł, Wojtkowiak J. Validation of CFD model for simulation of multi-pipe earth-to-air heat exchangers (EAHEs) flow performance. Thermal Science and Engineering Progress. 2018;**5**:44-49. DOI: 10.1016/j.tsep.2017.10.018

[12] Ghani S, Bakochristou F, ElBialy E, Gamaledin S, Rashwan M. Design challenges of agricultural greenhouses in hot and arid environments—A review. Engineering in Agriculture, Environment and Food. 2018. In press

[13] Zhang G, Choi C, Bartzanas T, In Bok L, Kacira M. Computational fluid dynamics (CFD) research and application in agricultural and biological engineering. Computers and Electronics in Agriculture. 2018; **149**:1-2. DOI: 10.1016/j.compag.2018.04.007

[14] Sase S. Air movement and climate uniformity in ventilated greenhouses. Acta Horticulturae. 2006;**719**:313-324

[15] Impron I, Hemming S, Bot GPA. Simple greenhouse climate model as a design tool for greenhouses in tropical lowland. Biosystems Engineering. 2007; **98**:79-89

[16] Rico-García E, López-Cruz IL, Herrera-Ruiz G, Soto-Zarazua GM, Castaneda-Miranda R. Effect of temperature on greenhouse natural ventilation under hot conditions: Computational fluid dynamics simulations. Journal of Applied Sciences. 2008;**8**:4543-4551

[17] Bakker JC. Greenhouse climate control: Constraints and limitations. Acta Horticulturae. 1995;**399**:25-37

[18] Kim T, Kato S, Murakami S. Indoor cooling/heating load analysis based on coupled simulation of convection, radiation and HVAC control. Building and Environment. 2001;**36**:901-908. Available from: www.elsevier.com/locate/buildenv

[19] Romero-Gómez P, Choi CY, Lopez-Cruz IL. Enhancement of the greenhouse air ventilation rate under climate conditions of central Mexico. Agrociencia. 2010;**44**:1-15

[20] Afram A, Janabi-Sharifi F. Review of modeling methods for HVAC systems. Applied Thermal Engineering. 2014;**67**:507-519. DOI: 10.1016/j.applthermaleng.2014.03.055

[21] Taki M, Rohani A, Rahmati MA. Literature study for solar thermal simulation and applications in greenhouse. Engineering in Agriculture, Environment and Food. 2018. DOI: 10.1016/j.eaef.2018.12.007. Available online 24 December 2018. In Press, Accepted Manuscript

[22] Gao Y, Dong J, Isabella O, Santbergen R, Tan H. Modeling and analyses of energy performances of photovoltaic greenhouses with sun-tracking functionality. Applied Energy. 2019;**233-234**:424-442. DOI: 10.1016/j.apenergy.2018.10.019

[23] Teitel M, Segal I. Net thermal radiation under shading screens. Journal of Agricultural Engineering Research. 1995;**61**(1):19-25. DOI: 10.1006/jaer.1995.1026

[24] Teitel M, Tanny J. Radiative heat transfer from heating tubes in a greenhouse. Journal of Agricultural Engineering Research. 1998;**69**:185-188

[25] Teitel M, Segal I, Shklyar A, Barak M. A comparison of pipe and air heating methods for greenhouses. Journal of Agricultural Engineering Research. 1999;**72**:259-273

[26] Popovski K. Location of heating installations in greenhouses for low temperature heating fluids. In: Industrial Thermal Effluents for Greenhouse Heating. European Cooperative Networks on Rural Energy. Proceedings of CNRE Workshop, Dublin, Ireland. CNRE Bulletin No. 15. 1986. pp. 51-55

[27] Roy JC, Boulard T, Bailley Y. Characterisation of the heat transfer from heating tubes in a greenhouse. 2000. E-Proceedings, AGENG 2000; Warwick, UK

[28] Kempkes FLK, van de Braak NJ. Heating system position and vertical microclimate distribution in chrysanthemum greenhouse. Agriculture and Forest Meteorology. 2000;**104**:133-142

[29] Boulard T, Haxaire R, Lamrani MA, Roy JC, Jaffrin A. Characterisation and modelling of the air fluxes induced by natural ventilation in a greenhouse. Journal of Agricultural Engineering Research. 1999;**74**:135-144

[30] Tadj N, Draoui B, Theodoridis G, Bartzanas T, Kittas C. Convective heat

transfer in a heated in a greenhouse tunnel. Acta Horticulturae. 2007;**747**: 113-120

[31] Tadj N, Nahal MA, Draoui B, Constantinos K. CFD simulation of heating greenhouse using a perforated polyethylene ducts. International Journal of Engineering Systems Modelling and Simulation. 2017; **9**(1):3-11

[32] ANSYS Inc. CFX Manuals. ANSYS Incorporated; 2003

[33] Montero JI, Antón A, Muñoz P, Lorenzo P. Transpiration from geranium grow under high temperature and low humidities in greenhouses. Agricultural and Forest Meteorology. 2001;**107**:323-332

Section 3

Heat Transfer in Power Generation

Chapter 7

Advances in Concentrated Solar Power: A Perspective of Heat Transfer

Fadi Alnaimat and Yasir Rashid

Abstract

Solar energy has the potential to reduce the dependence on the dwindling supply of fossil fuels through concentrated solar power (CSP) technology. CSP plants utilize solar thermal energy to produce electrical energy based on different thermodynamic power cycles. Solar collectors, reflectors, receivers, thermal fluid, and turbines are the main components of each CSP plant and involve intensive heat transfer at all stages. This chapter illustrates the thermal characteristics of the main components used in CSP technology. In addition, the solar thermal fluid characteristics and its stable operational ranges are discussed in this chapter. Heat capacity, vapor pressure, volume expansion, density and viscosity of the thermal fluid should not differ significantly at different temperatures during various operation stages because these variations can cause failure in the system, which is designed at the fixed material properties. Currently, CSP technology is associated with a higher cost compared to the electricity generated through gas power plants. Many efforts are made to search for sustainable and inexpensive materials to minimize the cost of CSP. One critical issue faced by CSP technology is the intermittent nature of the sun. Modern CSP plants integrate thermal energy storage (TES) unit to smoothen the power production or to shift the production from peak sunshine hours to peak demand hours.

Keywords: concentrated solar power, thermal energy storage, sensible and latent heat storage materials, thermochemical materials

1. Introduction

Solar energy is the most abundantly available source of energy theoretically. It can be used to produce electricity by thermodynamic process and by photovoltaics conversion [1]. Initially, photovoltaic technology was incepted to convert solar energy into electricity. In this technology, solar cells produce electricity when exposed to irradiance. The amount of voltage produced per unit cell is very small; hence, hundreds of such cells are connected in series and parallel junctions to finally produce high voltage for daily use applications. Such systems are called PV arrays which are available from domestic scale to grid-connected commercial plants. In some cases, solar irradiance is concentrated at a small PV surface to generate more electricity from small PV surface; however, such systems have certain implications like accelerated thermal degradation [2]. Such systems are called concentrated

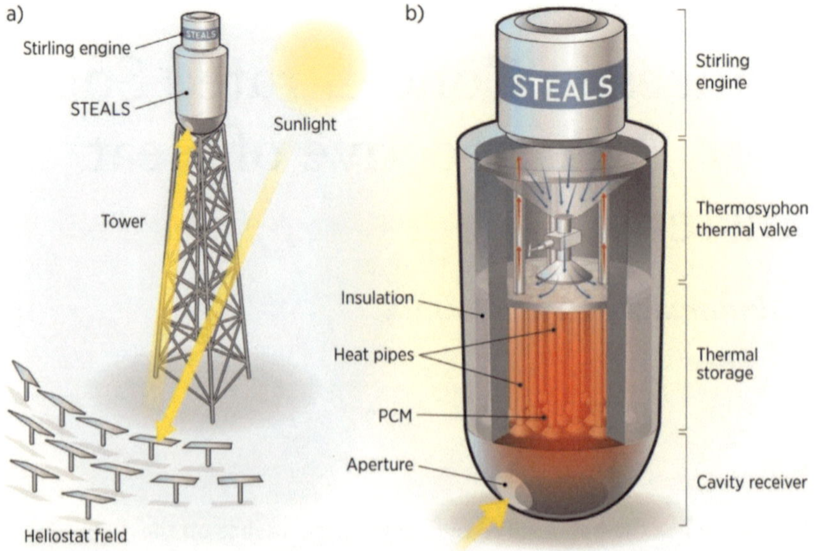

Figure 1.
A novel compact design of heliostat field type CSP with integrated latent heat storage; a) overall schematic and b) close-up view of the heat storage unit and Stirling engine [5].

photovoltaics (CPV). In further advancements, PV cells are integrated onto the buildings to produce electricity by decentralized means [3].

Concentrated solar power (CSP) is another technology to generate electricity from solar energy; however, it works on the principles of thermal energy. Solar radiations are concentrated at a point from where thermal fluid is passed. This fluid at high temperature and high pressure is passed through the power-generating unit (turbine or engine) to produce electricity [4]. The process of CSP plant is illustrated in **Figure 1** [5]. The components of a CSP plant are shown including reflectors, receiver, thermal fluid, and turbine/engines. Below is the comparison of photovoltaics with concentrated solar power. **Figure 1** shows a recent design of heliostat type CSP plant with a latent heat storage system coupled with the power generation unit. In this design, solar radiations are concentrated using heliostat mirrors on a point at the bottom of thermal energy storage tank containing phase change material (PCM), namely, aperture [5]. The tank is equipped with vertically arranged sodium pipes so that heat can be supplied passively to the entire part of the tank from receiving point. Above the tank is a thermal valve at the interface of sodium pipes. Sodium in vapor phase (at very high temperature) rises through the thermal valve (red arrows) and strikes at the head of Stirling engine where it passes its thermal energy to the energy, get condensed, and flow downward through the funnel (blue arrows). Opening and closing the thermal valve controls the flow of sodium vapors, and hence thermal energy flows to the power block and in a way regulates the power generation of the plant. The concept is proposed and tested for a plant in the capacity range of 0.1–1 MWelectric (MWe) [5]. In the case where multiple Stirling engines are needed, all will be integrated at the top of the tower in a parallel arrangement.

2. Comparison of photovoltaics with concentrated solar power

The subsequent section covers the comparison of photovoltaics technology with the concentrated solar power in terms of system energy efficiency and systems sustainability.

2.1 System efficiency

Annual electricity production in CSP plants is higher than the PV plants. The main reason behind this difference in electricity production is that PV plants work only during sunshine hours. In the case of cloudy days or during night time, outputs of such plants are zero. CSP plants can function for prolonged hours because such plants may have additional storage for thermal energy which is exploited when sun is not available like night time [6]. Other than this feature, efficiency of PV plants decreases with time due to thermal degradation and cracks in the cell. However, in terms of land use, PV plants are better in producing more electricity from the same area as compared to CSP. It is because PV modules occupy less space comparatively with reference to the solar collectors and reflectors in CSP. It is estimated that contemporary PV plants are functioning at the solar conversion efficiency of 14–22% [7].

2.2 System sustainability

Sustainability of a system is a vital measure to decide the future of any technology. Generally, cost of the systems, environmental impacts, and social acceptance define the sustainability of the system. Cost is the primary indicator to choose the most economical technology among all available options. It includes all types of costs including start-up, installation, operation, and maintenance cost. Environmental impact is another important parameter which considers the effects of any technology on the environment from its initiation to disposal. Severity of this parameter in decision-making is increasing continuously because of greenhouse gas emissions and consequent global warming [7].

2.2.1 Cost

As described earlier, the cost of any system is the most important factor for the sustainability of the system. It is quite logical that any renewable energy technology will be able to penetrate market if the cost associated with its unit production is less than the unit cost of electricity through grid station operated on fossil fuels. It is defined as grid parity, and it is based on the levelized electricity cost (LEC) which can be estimated using Eq. (1):

$$LEC = \frac{f_{cr} IC + C_{O\&M}}{E_{el}} \tag{1}$$

where f_{cr} is the annuity factor, IC is the investment cost, $C_{O\&M}$ is the annual operation and maintenance cost, and E_{el} is the annual net electricity output, and annuity factor can be calculated using Eq. (2):

$$f_{cr} = \frac{k_d(1+k_d)^n}{(1+k_d)^n - 1} + k_{ins} \tag{2}$$

where k_d is the real debt interest rate, k_{ins} is the annual insurance rate, and n is the depreciation period in years [7].

In both cases, the primary cost component is the initial cost of the system. According to the International Energy Agency (IEA), estimated initial cost for PV plant ranges from 2000 to 5200 US$/kW, while it lies in the range of 4200–8400 US $/kW for CSP plants. Further, the maintenance cost for PV is 1% of the initial cost. This value is approximately 2% of the initial cost in CSP. The difference in such costs is due to the complexity of the CSP system. Interestingly, even after the initial and maintenance cost difference, economic returns and incentives of CSP plant are higher as compared to PV plants. PV technology works only during daytime and totally unavailable during peak consumption hours of electricity. CSP technology has the capability to shift its production to peak consumption hours and also takes advantage of higher tariff rates during peak consumption hours [7].

2.2.2 Environmental impact

Impacts of any technology on the environment throughout the life cycle of the technology are an important measure to define sustainability of the system. The approach usually considered from cradle to grave is called life-cycle assessment (LCA). In terms of PV and CSP plants, most of the carbon footprints are during the manufacturing/installation and decomposing. During operational stage, there is almost no impact on the environment [7, 8].

In the case of PV plants, primary reason is the manufacturing of PV cells and modules, and the measure of greenhouse gas emissions is almost 50 g CO_2 eq/kWh for PV plants throughout the life of the plant according to NREL. It involves such ways and materials that pose severe threat to the environment [7, 8]. Particularly, the cell materials in second generation are more hazardous, and breathing in silicon dust is dangerous for workers according to NREL claims. In CPV, the issue is mainly due to mirrors and collector tube. The institute also reported that greenhouse gas emissions for CSP plants range from 22 to 23 g CO_2 eq/kWh during the life cycle of such plants. Overall, environmental impact of CPV is higher throughout life as compared to CSP [7, 8].

2.2.3 Social acceptance

Adoption of a new technology by the customers and end user is extremely important for the success of the technology. In the case of PV and CSP, the technology is warmly welcomed by all societies around the globe. Mainly, such plants are developed in Spain, the United States, and India. Solar energy is used in street lights, for pumping water and solar cooking, domestic level applications, and grid level mega power plants. The growth in both sectors is continuously increasing as the cost of the technology is getting lower than the fossil fuel-based electricity. Further, the public is aware of the clean and noise-free generation of electricity. Mainly, CSP is used for large-scale applications; however, the scale of PV is extremely large spanning from domestic use to the grid-integrated mega power plants [7].

Next sections deal with the concentrated solar power (CSP) technology from different design, thermal fluid and heat transfer characteristics.

3. Heat transfer in concentrated solar power plants

Basic principle of CSP lies on thermal energy capturing, transportation, and harnessing it. In all stages, heat transfer is involved. All modes of heat transfer in CSP are detailed in the subsequent section.

3.1 Heat transfer by radiation

Radiation is energy emitted by a solid object, liquid, or gas that is at a finite temperature. This matter can also dissipate radiant energy impinging on it by reflection and can be capable of absorption. It is classified into two, short wavelength of high energy originating from the sun and long wavelength of low energy originating from lower-energy sources [9]. The radiation energy can be balanced in CSP plants as [10]

$$(1 - r)S + L = L + H + \lambda E + G \tag{3}$$

where $(1 - r)S$ represents short radiations, L represents long radiations, H is sensible heat, λE is latent heat, and G is heat exchange by conduction.

Therefore, the net radiations are

$$R_n = (1 - r)S + L - L = H + \lambda E + G \tag{4}$$

The total energy absorbed into the system is converted into sensible heat, latent heat, and heat conductions [10].

Sunlight includes different spectra of wavelength starting from very short (extreme energy) to very long (low energy). The relation of wavelength and frequency is given as

$$\lambda = \frac{c}{\nu} \tag{5}$$

where λ is the wavelength, ν is the frequency, and c is the speed of light.

For a radiative-emitting surface of an arbitrary element of area dA_1, a differential solid angle $d\omega$ may subtend a point onto a differential area dA_2; the differential solid angle can be defined as

$$dw = \frac{dA_2}{r^2} \tag{6}$$

where dA_2 is the differential area.

Spectral radiation (G) can be calculated as

$$G_\lambda(\lambda) = \int_0^{2\pi}\int_0^{2\pi} I_\lambda(\lambda, \theta, \varnothing) cos\theta dw \tag{7}$$

where

$$dw = sin\theta d\theta d\varnothing \tag{8}$$

Considering total radiations as the rate of radiation from all sides per unit area from all directions and at all wavelengths

$$G = \int_0^{\propto} G_\lambda(\lambda) d\lambda \tag{9}$$

For the case of diffuse radiations, $I_\lambda(\lambda)$ is free of θ and φ as

$$G_\lambda(\lambda) = \pi I_\lambda(\lambda) \tag{10}$$

The amount of reflected radiation radiations is the rate at which radiation of wavelength λ leaves a unit area of the surface per unit wavelength interval $d\lambda$. It can be calculated as

$$J_\lambda(\lambda) = \int_0^{2\pi}\int_0^{2\pi} I_{\lambda, e+r}(\lambda, \theta, \varphi) cos\theta sin\theta d\theta d\varphi \tag{11}$$

Finally, in general, most engineering heat transfer applications are employed to characterize the radiative heat transfer interaction between an object and its environment with respect to wavelength:

$$\dot{Q}_{rad} = \sigma \varepsilon_{\lambda} A \left(T^4 - T_{\infty}^4\right) \tag{12}$$

where σ is the Stefan-Boltzmann constant, 5.67 x 10^{-8} W/m^2 K^4, A is the object surface area, and T_{∞} is the ambient temperature.

3.2 Heat transfer by conduction

Heat transfer by conduction is occurred due to lattice vibration in solids and particle collision in stationery fluids. In a solid with temperature gradient onto surfaces, the one with higher temperature experiences higher level of collision among particles, so it transfers its energy to the neighboring particles which transfer it further [11]. In the same way, heat is transferred from hot to cold surface until steady state is achieved. The rate of heat transfer by conduction is given by Fourier's law as

$$\dot{Q}_{cond} = kA\frac{dT}{dx} \tag{13}$$

where A is the surface area, k is the thermal conductivity, dT is the temperature gradient across surfaces, and dx is the length between the surfaces between heat transfers that are measured.

In CSP plants, heat transfer due to conduction is always changing with time; that's why it is necessary to understand the transient nature of conduction heat transfer. In certain scenarios, lumped systems are assumed while calculating heat transfer at the interface of solid and liquid. The reason is that heat conduction within the solid is comparatively negligible as compared to the heat transfer from solid to the fluid at its interface. It is further necessary to confirm that there are no spatial temperature changes within the solid to assure its lumped natured. After a specific time, the system will reach a steady state when high thermal energy of solid is transferred to the low-temperature fluid to eventually reaching the same level of temperatures [12]. The time can be calculated as

$$\frac{T - T_{\infty}}{T_i - T_{\infty}} = \exp\left(-\frac{hA_s}{\rho V c_p}\tau\right) \tag{14}$$

where T is the temperature of hot-surfaced solid, T_{∞} is the ambient temperature, T_i is the initial temperature, h is the convective heat transfer coefficient, A_s is the surface area at the solid-fluid interface, ρ is the density of the fluid, c_p is specific heat capacity, and τ is the thermal time constant. The constant can be calculated as

$$\tau = \frac{\rho V c_p}{hA_s} = RC \tag{15}$$

where R is the convective resistance and C is the lumped capacitance of the solid. To confirm the validity of lumped capacitance, Biot number (Bi) can be calculated as

$$Bi = \frac{hL_c}{k} = \frac{R_{cond}}{R_{conv}} = \frac{L_c/kA}{1/hA} = \frac{T_{s,1} - T_{s,2}}{T_{s,2} - T_{\infty}} \tag{16}$$

For the value of Biot number less than 0.1, it is assumed that temperature within the solid is nearly the same, and the maximum temperature difference is at the interface of solid-fluid. Rearrangement of equations provides us

$$\frac{T - T_\infty}{T_i - T_\infty} = \exp\left(-Bi.F_0\right) \tag{17}$$

where F_0 is a dimensionless number which characterizes transient problems and can be calculated as

$$F_0 = \frac{\left(\frac{k}{\rho c_p}\right)t}{L_c^2} \tag{18}$$

where L_c is the characteristic length which is different for rectangular, cylindrical, and spherical coordinate systems [12].

For rectangular coordinates, $L_c = L$

For cylindrical coordinates, $L_c = \frac{r}{2}$

For spherical coordinates, $L_c = \frac{r}{3}$

3.3 Heat transfer by convection

Heat transfer through convection occurs between fluid in motion and its bounding surface. The heat transfer can happen naturally due to buoyancy effect in which medium moves due to the density difference caused by the variation in temperature. It is called natural convection. On contrary, a fluid can be forcefully pushed or pumped through mechanical means like fans or pumps. The heat transfer through this way is called forced convention. Heat transfer through convection can be measured as

$$\dot{Q}_{conv} = hA_s(T_s - T_\infty) \tag{19}$$

where h is the convective heat transfer coefficient, A_s is the surface area, T_s is the temperature of hot surface, and T_∞ is the ambient temperature. The range of convective heat transfer coefficient for gases is 2–5 and 25–250 W/m^2 K for natural and forced convections, respectively [12]. The ranges 50–1000 and 100–20,000 W/m^2 K are for liquids in natural and forced convection cases, respectively [12]. For materials that involves phase change (boiling and condensation), the range is 2500–100,000 W/m^2 K [12]. This extraordinary difference in the coefficient value is exploited in CSP plants using heat pipes involving boiling and condensations. In CSP applications, Newton's law of cooling (dt = $T_s - T_\infty$) can overpredict the heat transfer rate. More accuracy to calculate heat transfer rate per unit length can be obtained using [12]

$$\Delta T_{lm} = \frac{(T_s - T_\infty) - (T_s - T_0)}{\ln\left(\frac{T_s - T_i}{T_s - T_\infty}\right)} \tag{20}$$

and the outlet temperature of the fluid passing through the pipe can be calculated using

$$\frac{T_s - T_i}{T_s - T_\infty} = \exp\left(-\frac{\pi D N \bar{h}}{\rho V N_T S_T c_p}\right) \tag{21}$$

where S_T is the transverse pitch, D is the tube diameter, N is the number of tubes, and $(\bar{h})$ is the average value of heat transfer coefficient. Finally, heat transfer can be obtained with the following

$$\dot{Q}' = N\left(\bar{h}\pi D\Delta T_{lm}\right) \tag{22}$$

4. Thermal energy storage in concentrated solar power

The prime difference between PV and CSP plant is an optional storage of solar energy in CSP and to harness it during the most demanding hours to produce electricity. This feature makes the CSP capable of electric power production during evening, the time after closing of offices, when the demand of electricity is at its peak [13]. At that time, output of CPV plant is zero, and it creates a big mismatch between supply and demand. Although a solution for such mismatch is to store electricity in large-scale batteries during sunshine hours so as to supply during evening and night, the option of batteries is very costly and environmentally unacceptable. Solar energy in the form of thermal energy storage (TES) is comparatively much better option as compared to the electrical energy storage in batteries because it is inexpensive and has minimal environmental impacts [14]. Few of the CSP plants are equipped with TES systems, while others do not have this feature.

Performance of a CSP plants with TES systems is dependent on the design of integration of TES into the power generation cycle (thermodynamic process) of the plants [15]. Generally, TES is classified into two main categories based on the motion state during charging and discharging, i.e., active systems and passive systems. Charging is the process in which heat is fed to the medium to raise its energy, while discharging is the process in which heat is extracted from the medium to bring it back to the original state. In active systems [16], thermal energy is stored into a medium which also acts as energy carrier. The medium with high thermal energy is pumped, and it transfers its energy to the thermodynamic cycle through forced convection. In passive system [17], thermal energy is stored into a stationery and motionless medium through which thermal fluid is circulated to extract heat, and the thermal fluid is circulated through the thermodynamic cycle to deliver thermal energy there. The following sections contain the further classifications of such systems with visual illustrations for clear understandings.

4.1 Active direct concept for TES

Active systems are further categorized into direct and indirect systems. In direct active systems, energy materials serve the dual functions of energy storage and transportation. In such systems, TES materials absorb heat from solar absorber during charging and stored in a hot tank as shown in **Figure 2** [13]. During discharging, the TES material is pumped from the hot tank to the thermodynamic cycle where it transfers its energy to the system and return back to cold storage. For the next charging cycle, the TES material is again pumped to solar absorber to gain energy for the next cycle. In this design, heat exchanger is not required; however, the TES material should have properties of heat storage capacity and good flowability [15].

4.2 Active indirect concept for TES with two tanks

Active indirect systems have further two designs for TES integration into CSP plants with the options of two tanks and single tank for storage. In such systems, materials to store thermal energy are different than the heat transfer fluid (HTF). An example of active indirect system with two-tank storage is shown in **Figure 3** [13]. At the start of the charging/discharging cycle, TES material is stored in a cold tank, while HTF is moving in the flow path of solar collector, thermodynamics cycle, and heat exchanger. During charging, TES material is pumped from the cold tank to a hot tank through the heat exchanger where it absorbs heat from HTF and stored into the hot tank. The TES is pumped back to the cold tank during discharging in which the materials return heat to the HTF through heat exchanger. Hence, TES material and HTF are totally different and are not mixed with each other during the process.

4.3 Active indirect concept for TES with single tank

Active indirect design with a single storage tank is also possible which is quite efficient than the two-tank storage system. In this design, TES material in cold and hot

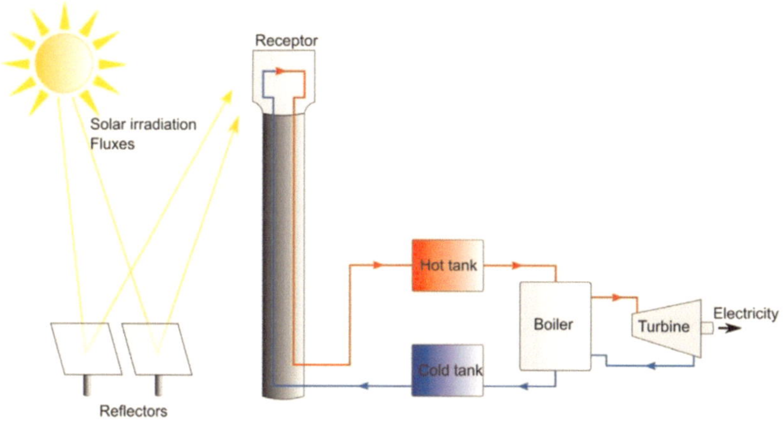

Figure 2.
Active direct concept for TES [13].

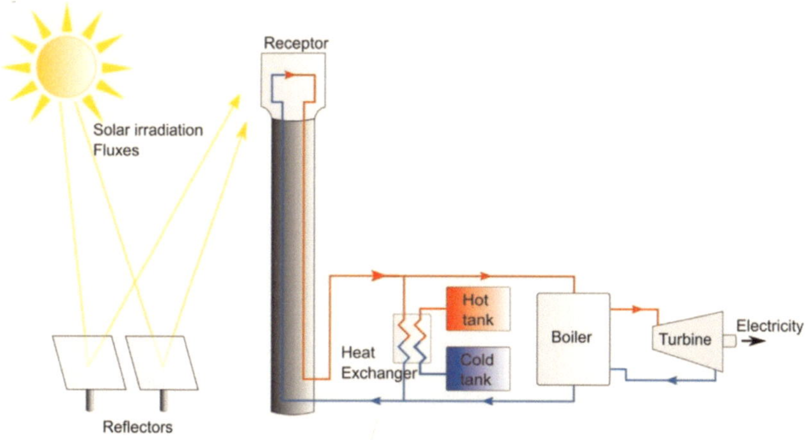

Figure 3.
Active indirect concept for TES using two tanks [13].

state is stored in the same tank as illustrated in **Figure 4**. The materials get separated by itself due to different material properties at different temperature. Thermal stratification causes the hot material to be stored in the upper part of the tank, while cold material is stored in the bottom. This system is also known as thermocline system, which is approximately 35% cheaper than the two-tank storage system.

In thermocline storage systems, a filler material like rocks, concrete, or sand is used so that less quantity of TES materials is required to enhance the thermocline effect. In this system, handling accuracy during charging and discharging of the TES is required using controlled methods and devices to avoid mixing of materials. Among differently investigated structures during discharging of thermocline, packed-bed configuration is reported to be the best performing [13].

4.4 Passive concept for TES

This is another concept of thermal energy storage in which TES materials are fixed and do not move during entire processes. The concept is demonstrated in **Figure 5**. The TES material is kept in a tank through which HTF is circulated. During charging phase, hot HTF is passed through TES and it transfers its heat to the TES. During discharging, cold HTF is passed from TES which is already at very

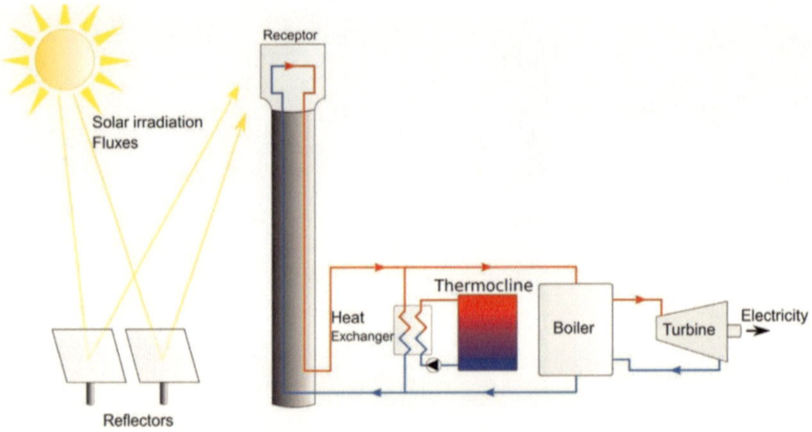

Figure 4.
Active indirect concept for TES integration, single tank [13].

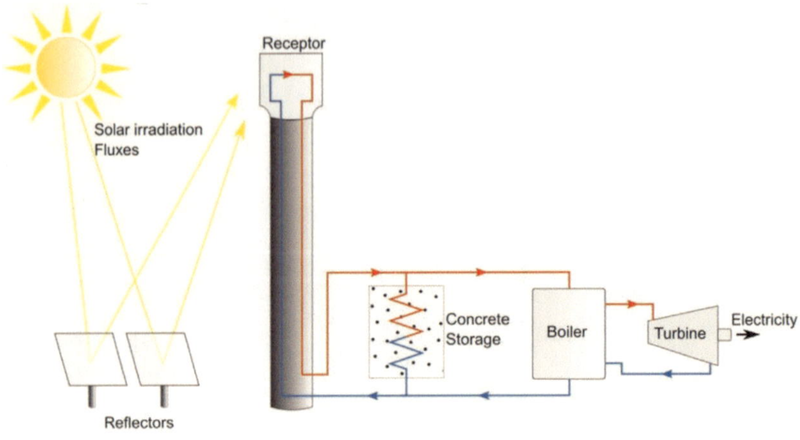

Figure 5.
Passive concept for TES [13].

high temperature. In such systems, thermal conductivity of the storage material and surface contact area defines the rate of heat transfer from and to the HTF during charging and discharging, respectively. In this type of design, rocks and concrete are used as TES materials.

5. Materials for thermal energy storage

Thermal energy storage materials are very specific in terms of physical and thermal properties for the best performance of the CSP plants. These materials are generally categorized into three, namely, sensible storage, latent storage, and thermochemical storage. Explanation of these categories is given in the subsequent sections. There are few properties of TES materials which are common for all materials. Energy storage density is very basic thing that defines the size of the TES tanks as well as associated cost with it. The higher is the energy storage density of a material, the less of its quantity is required to store a specific amount of thermal energy [14]. Similar is the case of mass density of the material. In the case of latent heat and thermochemical storage, equivalent terms are heat of fusion of the material and heat of reaction of the materials, respectively [14]. Thermal conductivity and operating temperatures are very important parameters in defining the overall efficiency and performance of the system. If a material is a good thermal conductor, it takes less time during charging and discharging. In case thermal conductivity is low, conductivity enhancers, nanofillers, and enhanced contact surface geometries are utilized for better results. In general, the materials should be inexpensive, readily and widely available, less corrosive, and less hazardous to the environment and to human health [14].

5.1 Sensible TES storage materials

These materials store heat in the form of rise in temperature. The property of the material that is associated with this phenomenon is the heat capacity. A material with higher heat capacity is good for energy storage. Examples of such materials are sand, rocks, concrete, etc. [18]. The amount of thermal energy stored in TES materials as sensible storage can be calculated using Eq. (23):

$$Q_T = V \times \rho \times C \times \Delta T \tag{23}$$

where V is the volume, ρ is the average density, C is the specific heat capacity of the sensible energy storage material, and ΔT is the difference of temperature from initial to final stage.

5.2 Latent heat storage

In these types of materials, energy is stored in the form of change of phase. This phase change may be in solid-liquid, liquid-gas, solid-gas, and solid-solid [19]. Commonly known materials lie in the categories of liquid-gas and solid-liquid phase change materials for CSP applications. Water-steam is an example of the former type, and binary salt is an example of the latter type. For latent heat storage materials, parameter of interest is the phase transition temperatures and latent heat of fusion during this phase transitions. It is highly desirable that materials are stable within a long temperature range, and its solidification temperature is as low as possible. The reason behind the low preferred solidification temperature is that it should not be deposited as a solid in the circulation pipes in active systems. The

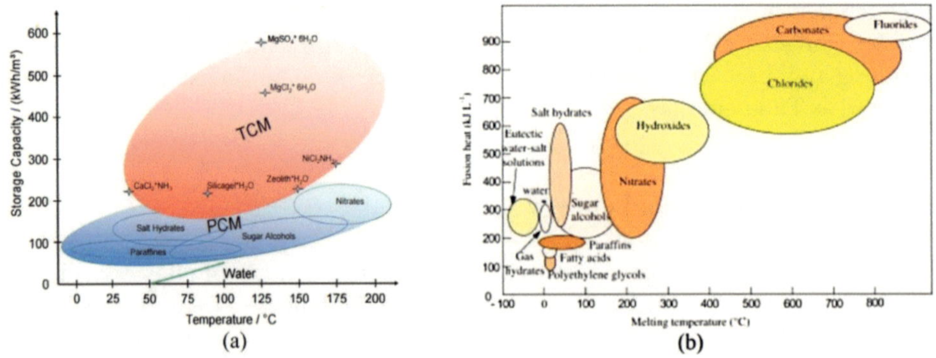

Figure 6.
(a) PCM with respect to storage capacity and (b) PCM with respect to heat of fusion [21].

amount of energy in this type of material is the sum of sensible energy storage from initial temperature to the final temperature and the energy storage during phase transition as a latent heat [18, 19]. Energy stored as a latent heat of fusion can be calculated using Eq. (24):

$$Q_{\text{latent}} = V \times \rho \times L \tag{24}$$

where L is the latent heat of fusion of the material.

5.3 Thermochemical energy storage

This type of energy storage is based on the chemistry of endothermic-exothermic reversible reactions. Surplus heat energy is used to initiate a reaction which is highly endothermic. During charging, the heat is taken by the reactants, and due to reactions occurrence, the reactants are converted into products. These products are stored for days, weeks, and seasons. Interestingly, the storage is at ambient conditions, and energy losses in this storage are minimal. During discharging, these products are converted back to the reactants with the release of huge amounts of heat. That heat is transported to the thermodynamic cycle [20]. The reversible $CaO/CaCO_3$ carbonation reaction (CaL) is one of the most promising since CaO natural precursors are affordable and earth-abundant. However, CaO particles progressively deactivate due to sintering-induced morphological changes during repeated carbonation and calcinations cycles.

Figure 6(a) is the representation of different types of materials based on the heat storage capacity [21]. As it is evident from the figure, thermochemical materials (TCM) possess the most storage density in the range of 170–600 kWh/m^2. Energy storage density of latent heat storage called PCM comes lower than TCM ranging from 70 to 250 kWh/m^2. Sensible storage materials are the lowest in terms of energy storage density. The only advantage in sensible storage is the absence of degradation and corrosion and very low cost. **Figure 6(b)** is the classification of TES materials based on the melting points [21]. This analysis gives an indication about the selection of materials for specific ranges of melting points.

6. Characteristics of thermal fluids

According to the 2050 vision of the International Energy Agency (IEA), energy production share by CSP is 630Gwe. Keeping in view the high future targets, scientists and researchers are working on different designs of CSP. Among all parameters, thermal fluid is a key component because overall performance of the CSP is dependent on the thermal energy. Thermal fluid is the transport material that carries thermal energy from solar receiver/hot storage and delivers it to the thermodynamic cycle [22]. In the context of thermal fluids, required characteristics are:

1. High heat capacity

2. Low viscosity

3. Uniform thermophysical properties in the operational temperature range

4. Enhanced heat transfer at heat exchanger

Thermal fluids are categorized into two classes based on the behavior of the materials. The classification of the thermal fluids is represented in **Figure 7**.

In certain cases, thermal property enhancer nanoparticles agglomerate and form clusters after a limited operational life. This agglomeration of nanoparticle declines the performance of thermal fluids. A quaternary salt is developed recently with low melting point (85.4°C), wide operating range (600°C), reduced risk of blockage, and less corrosive effect with the system [23]. Correlation of heat transfer with nanofluid is described in the subsequent section.

6.1 Heat transfer with nanofluids

Thermal characteristics of nanofluids are different than solid-liquid mixtures as these fluids contain suspended particles (metallic or nonmetallic) in the liquid base. Heat transport properties are altered because of the suspended ultra-fine particles [24]. Generally, the volume content of these particles is below 10% in the fluid. Addition of nanoparticles increases thermal conductivity and heat transport properties of the fluid as compare to the pure fluid. For instance, Xuan and Li reported an increase in the thermal conductivity ratio from 1.24 to 1.78 with the increase in particles from 2.5 to 7.5% [25]. The change in properties of the fluid is dependent on the particle shape, dimensions, quantity, and characteristics. However, micrometer- and millimeter-sized particles are reported to settle down quickly producing clogs in the channels, eroding pipelines, and causing huge pressure drop [26]. Heat

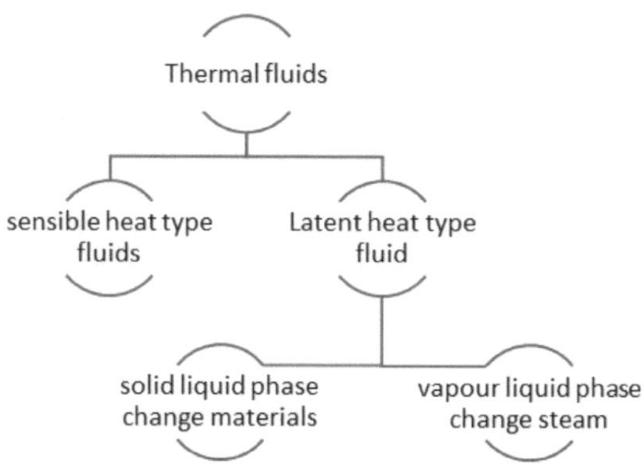

Figure 7.
Classification of thermal fluids.

transfer correlations, fundamentals, and theory can be read through the literature presented in [24, 27, 28].

7. Current issues in CSP

One critical issue in TES for high-temperature applications is corrosion of the TES materials with its containment. Most of the molten salts used for TES are highly corrosive. The reactions of salts and formation of corrosion is not completely understood, and the problem still needs attention especially in the context of stress corrosion cracking in molten salts [29]. The corrosion is either in the form of oxide layer formation on the container or degradation of the container material. Standard structural materials (stainless steel or carbon) degrade after coming into interaction with salt chlorides by the chloridation. Protective coatings are under investigations to mitigate the effects of corrosion [29].

Currently, energy production from CSP technology is costly [30]. To circumvent the issue of high cost, the US Department of Energy launched the SunShot Initiative in 2011 and has put forth an aggressive research and development (R&D) plan to make CSP technologies cost competitive with other energy generation sources on the grid by the end of the decade [31]. The goal of the SunShot Initiative includes the levelized electricity cost (LEC) from solar power plants less than 6 ¢/kWh without subsidies, a round-trip annualized exergetic efficiency greater than 95%, storage cost less than $15/kW ht, by 2020, which would pave way for rapid, and large-scale adoption of solar electricity [31].

8. Conclusion

Among different available options of solar energy, concentrated solar power is considered the most efficient technology available contemporarily and trending in future developments. Its conversion efficiency from solar thermal energy to electricity is very high because it can achieve as high temperature as 800°C. The higher-achieved temperature leads to higher efficiency because it enhances the exergy level of the system. For future developments, molten salts and liquid metals are considered better options for energy storage as well as for thermal fluids. Although many combinations of salts in different mixing ratio are available in literature and rapid research is undergoing on the developments of further mixtures, the problem of corrosion is still existing. It is anticipated that liquid metals can solve the problem of corrosion and their stability and energy storage density is even higher than salt hydrates. Research in the similar lines for HTF is also undergoing at a high pace. Most of the developed molten salt HTFs are based on nitrates/nitrites. However, the annual nitrate/nitrite salts production is limited due to their reserves. Therefore, carbonate or chloride-based salts are proposed and evaluated in the most recent studies. One of the major issues of the molten salts is their relatively high corrosive nature to metal alloys. The corrosion issues must be resolved completely before commercial application of the molten salts as the HTF in the CSP Technology.

Acknowledgements

The authors would like to express their appreciation to the United Arab Emirates University (UAEU) for funding the research through grants (31N265 and 31R153).

Conflict of interest

The authors declare no conflict of interest.

Nomenclature

$C_{O\&M}$	annual operational and maintenance cost
CPV	concentrated photovoltaics
CSP	concentrated solar power
E_{el}	annual electricity output
f_{cr}	annuity factor
HTF	heat transfer fluid
IC	investment cost
IEA	International Energy Agency
K_{id}	real debt interest rate
K_{ins}	annual insurance rate
LOC	levelized electricity cost
MWe	megawatt electricity
n	depreciation period
PV	photovoltaics
PCM	phase change material
TES	thermal energy storage

Author details

Fadi Alnaimat* and Yasir Rashid
Department of Mechanical Engineering, United Arab Emirates University, Al Ain, UAE

*Address all correspondence to: falnaimat@uaeu.ac.ae

References

[1] Settino J, Sant T, Micallef C, Farrugia M, Spiteri Staines C, Licari J, et al. Overview of solar technologies for electricity, heating and cooling production. Renewable and Sustainable Energy Reviews. 2018;**90**:892-909

[2] Radwan A, Emam M, Ahmed M. Comparative study of active and passive cooling techniques for concentrated photovoltaic systems. In: Exergetic, Energetic and Environmental Dimensions. Imprint: Academic Press, Elsevier; 2018. pp. 475-505. Available from: https://linkinghub.elsevier.com/retrieve/pii/B978012813734 5000275

[3] Hasan A, Alnoman H, Rashid Y. Impact of integrated photovoltaic-phase change material system on building energy efficiency in hot climate. Energy and Buildings. 2016;**130**:495-505

[4] Kabir E, Kumar P, Kumar S, Adelodun AA, Kim K-H. Solar energy: Potential and future prospects. Renewable and Sustainable Energy Reviews. 2018;**82**:894-900

[5] Rea JE, Oshman CJ, Olsen ML, Hardin CL, Glatzmaier GC, Siegel NP, et al. Performance modeling and techno-economic analysis of a modular concentrated solar power tower with latent heat storage. Applied Energy. 2018;**217**:143-152

[6] Cocco D, Migliari L, Petrollese M. A hybrid CSP-CPV system for improving the dispatchability of solar power plants. Energy Conversion and Management. 2016;**114**:312-323

[7] Khan J, Arsalan MH. Solar power technologies for sustainable electricity generation—A review. Renewable and Sustainable Energy Reviews. 2016;**55**: 414-425

[8] Desideri U, Zepparelli F, Morettini V, Garroni E. Comparative analysis of concentrating solar power and photovoltaic technologies: Technical and environmental evaluations. Applied Energy. 2013;**102**:765-784

[9] Duffie JA, Beckman WA. In: Duffie JA, Beckman WA, editors. Solar Engineering of Thermal Processes. 4th ed. Hoboken: John Wiley; 2013. 910 p

[10] Bonan G. Ecological Climatology: Concepts and Applications. 3rd ed. Cambridge: Cambridge University Press; 2016. Available from: http://ebooks.cambridge.org/ref/id/CBO9781107339200

[11] Tada H. Gourmet Engineering Lecture Notes [Internet]. 2002. Available from: http://emerald.tufts.edu/as/tampl/en43/lecture_notes/ch4.html

[12] Incropera FP, editor. Fundamentals of Heat and Mass Transfer. 6th ed. Hoboken, NJ: John Wiley; 2007. 997 p

[13] Pelay U, Luo L, Fan Y, Stitou D, Rood M. Thermal energy storage systems for concentrated solar power plants. Renewable and Sustainable Energy Reviews. 2017;**79**:82-100

[14] Hassan A, Shakeel Laghari M, Rashid Y. Micro-encapsulated phase change materials: A review of encapsulation, safety and thermal characteristics. Sustainability. 2016; **8**(10):1046

[15] Crespo A, Barreneche C, Ibarra M, Platzer W. Latent thermal energy storage for solar process heat applications at medium-high temperatures—A review. Solar Energy. 2018. Available from: https://linkinghub.elsevier.com/retrieve/pii/S0038092X18306534 (In press)

[16] Pointner H, Steinmann W-D. Experimental demonstration of an

active latent heat storage concept. Applied Energy. 2016;**168**:661-671

[17] Gil A, Medrano M, Martorell I, Lázaro A, Dolado P, Zalba B, et al. State of the art on high temperature thermal energy storage for power generation. Part 1—Concepts, materials and modellization. Renewable and Sustainable Energy Reviews. 2010; **14**(1):31-55

[18] Bhale PV, Rathod MK, Sahoo L. Thermal analysis of a solar concentrating system integrated with sensible and latent heat storage. Energy Procedia. 2015;**75**:2157-2162

[19] Joemann M, Oezcan T, Kauffeld M, Pollerberg C. Process steam and chilled water production with CPC-collectors, steam jet ejector Chiller and latent heat storages. Energy Procedia. 2016;**91**: 767-776

[20] Dizaji HB, Hosseini H. A review of material screening in pure and mixed-metal oxide thermochemical energy storage (TCES) systems for concentrated solar power (CSP) applications. Renewable and Sustainable Energy Reviews. 2018;**98**:9-26

[21] Khan MMA, Saidur R, Al-Sulaiman FA. A review for phase change materials (PCMs) in solar absorption refrigeration systems. Renewable and Sustainable Energy Reviews. 2017;**76**:105-137

[22] Singh T, Hussien MAA, Al-Ansari T, Saoud K, McKay G. Critical review of solar thermal resources in GCC and application of nanofluids for development of efficient and cost effective CSP technologies. Renewable and Sustainable Energy Reviews. 2018; **91**:708-719

[23] Chen X, Wu Y, Zhang L, Wang X, Ma C. Experimental study on the specific heat and stability of molten salt nanofluids prepared by high-temperature melting. Solar Energy Materials and Solar Cells. 2018;**176**: 42-48

[24] Xuan Y, Roetzel W. Conceptions for heat transfer correlation of nanofluids. International Journal of Heat and Mass Transfer. 2000;**43**(19):3701-3707

[25] Xuan Y, Li Q. Heat transfer enhancement of nanofluids. International Journal of Heat and Fluid Flow. 2000;**21**(1):58-64

[26] Choi SUS, Eastman JA. Enhancing Thermal Conductivity of Fluids with Nanoparticles. ASME-Publications-Fed 231; 1995. pp. 99-106

[27] Abelman S, Parsa AB, Sayehvand H-O. Nanofluid flow and heat transfer in a brinkman porous channel with variable porosity. Quaestiones Mathematicae. 2018;**41**(4):449-467

[28] Mahian O, Kolsi L, Amani M, Estellé P, Ahmadi G, Kleinstreuer C, et al. Recent advances in modeling and simulation of nanofluid flows-part I: Fundamentals and theory. Physics Reports. 2018;**790**:1-48. Available from: https://linkinghub.elsevier.com/retrieve/pii/S0370157318303302

[29] Fernández AG, Muñoz-Sánchez B, Nieto-Maestre J, García-Romero A. High temperature corrosion behavior on molten nitrate salt-based nanofluids for CSP plants. Renewable Energy. 2019; **130**:902-909

[30] Du E, Zhang N, Hodge B-M, Kang C, Kroposki B, Xia Q. Economic justification of concentrating solar power in high renewable energy penetrated power systems. Applied Energy. 2018;**222**:649-661

[31] The SunShot Initiative. Department of Energy [Internet]. 2019. Available from: https://www.energy.gov/eere/solar/sunshot-initiative

Chapter 8

Numerical Simulation of the Accidental Transient of an Industrial Steam Boiler

Amina Lyria Deghal Cheridi, Abla Chaker and Ahcene Loubar

Abstract

Numerical simulation allows a better understanding of thermal-hydraulic phenomena that can take place in thermal installations. It is a capital contribution, especially in accident situations. The code RELAP5/Mod3.2 makes it possible to predict the thermal-hydraulic behavior of these installations during the normal and accidental operations. The present chapter focuses on accidental transient modeling and simulation of an industrial steam boiler by the code RELAP5/Mod3.2. This steam boiler is radiant type, high power, natural circulation, and a single drum. The model of the boiler developed for the RELAP5/Mod3.2 code encompasses the entire installation. The control loop of the water level in the steam drum and the superheated steam temperature are also included in the model. The qualification process of the steam boiler model is based on the steam boiler operation data under steady-state operating conditions. The comparative study shows that the theoretical results of the code RELAP5 are in good agreement with the operating data of the installation. To evaluate the behavior and response of the boiler in accident situations, the loss of feedwater following pump power loss, with and without protective operations, was simulated.

Keywords: industrial steam boiler, natural circulation, safety, modeling and simulation, RELAP5/Mod3.2, accidental transient

1. Introduction

Electricity production largely depends on the production of steam using coal, gas, or nuclear fission of uranium as heat sources. To produce steam, it is necessary to heat the water to its boiling point and then to provide a sufficient amount of heat to change the boiling water into steam. Steam production and utilization techniques are therefore important aspects of engineering technology. The steam generator is one of the means used to produce steam. A steam boiler plays an important role in all types of industries; it is one of the key components of a thermal installation. The main function of the steam boiler is to produce steam for the purpose of using it for industrial reasons such as the production of electrical energy, petro-chemistry, district heating, and others [1]. In general, the steam boilers can be classified in two categories: water-tube and fire-tube steam boilers. The choice of the type of industrial steam boiler to be selected can be made according to several criteria, the main one being the thermal power to be supplied or its equivalent in production of steam.

In the steam boiler, several problems can occur during its service because it works in severe conditions (high pressure, high temperature, corrosive environment, and continuous operation). These problems have an influence on steam boiler operation and sometimes lead to serious consequences such as explosions. Indeed, accidental transient was already observed during normal operation of the installation [2]; the most important and the most frequent are loss of feedwater, loss of flow, pipeline ruptures, loss of electrical power, equipment failures, and others. Early detection of such faults under operation is of great importance. Therefore, it is very necessary to perform an accident analysis to evaluate causes and make an assessment of the accidents' consequences [3]. Finally, it is important to consider the safety aspects and analysis of the steam boiler to guarantee the reliability and stability.

Steam boiler is a complex equipment considering the nonlinear, phase change, and inverse response behavior (shrink and swell). However, the operating conditions of the steam boiler are very difficult to control because all the parameters are interrelated. In addition, the steam boiler has very high manufacturing, operating, and maintenance cost. Hence, it is very difficult to take measurement and carry out tests directly on steam boiler. However, modeling and simulation are also effective tools for safety assessment and prediction of installation behavior of real process under transient conditions. The usefulness of numerical simulation tools is mainly based on the development of numerical methods, the progress of programming, and the provision of powerful computing resources [4].

The power plants' safety is largely based on simulation [5]. Nowadays, the best-estimate nuclear system codes such as TRAC [6], RETRAN [7], RELAP5 [8, 12], ATHLET [9], CATHARE [10], and APROS [11] are widely used to investigate the thermal-hydraulic characteristics of nuclear power plants either during steady-state operation or accidental transients and simulate the overall behavior of the installation (pumps, piping, heat exchangers, tanks, valves, control loops, etc.).

They are mainly produced to simulate the behavior of nuclear installations, but they can also be used to study the normal and accidental operation of conventional thermal, industrial, and solar installations [13, 14, 1, 2].

RELAP5 code used to carry out the present study is a thermal-hydraulic analysis system code of a realistic estimation level (best estimate). It is used to simulate the thermal-hydraulic transient of light water systems during postulated accidents [15]. RELAP5 is widely used in nuclear safety studies; its scope extends to energy systems using water and its vapor. Research work in this direction is very limited to the nuclear field. Extrapolation of the code scope is possible for the thermal-hydraulic behavior study of an industrial boiler [1, 2, 16].

In this chapter, realistic simulation of the global behavior of an industrial natural circulation steam boiler during normal and accidental operation is performed using RELAP5/Mod3.2 system code with a thermal-hydraulic performance analysis of the main equipment of the installation. A better understanding of the physical phenomena occurring during all phases of a hypothetical accident is necessary for the safety of an installation. The accidental transient simulated in this chapter is the loss of feedwater (pump stop) with and without protective operations. The chapter is divided into the following sections:

- Presentation of the steam boiler
- The RELAP5/Mod3.2 computer code
- The simulation application by the RELAP5 code

- The transient simulation of the steam boiler

- Conclusion

2. Steam boiler

Steam is used in energy production and in many industrial processes. Its production and utilization techniques are therefore important aspects of engineering technology. Among the means of producing steam, there are steam generators, which are vital to power and processing plants.

2.1 Steam boiler history

The use of two-phase systems accompanied by a phase change to transform thermal energy into mechanical energy is old. It dates back to the first century with the invention of the aeolipile by the Greek mathematician Héron d'Alexandrie [17]. However, no practical system was built until the Italian architect and inventor Giovanni Branca designed a boiler. But, it is really only from the end of the seventeenth century that engineers developed modern steam machines. The first real steam machine was built by English engineer Thomas Savery in 1698; this machine was used for pumping water. The James Watt boiler, built in 1785, who was one of the first engineers to achieve the thermodynamic properties of steam, used the safety valve and valves to control the flow of water and steam in its boilers [18]. At the beginning of the nineteenth century, British engineer Richard Trevithick and American inventor Oliver Evans developed machines without condenser using high-pressure steam. Trevithick used this model steam engine to equip the first railway locomotive. Trevithick and Evans built road vehicles powered by steam [18]. The French engineer Marc Seguin (1781–1875) developed a fire-tube boiler, which in 1827 equipped George Stephenson's famous "Rocket" locomotive. The first improvement in Evans' boiler was the "Lancashire" fire-tube boiler patented in 1845 by British engineer William Fairbairn, in which the flue gases circulated through tubes inserted in the water tank, increasing the area through which heat could be transmitted. Fire-tube boilers, however, had limited capacity and pressure and sometimes presented a risk of explosion [17]. The first boiler with water tubes (**Figure 1**) patented in

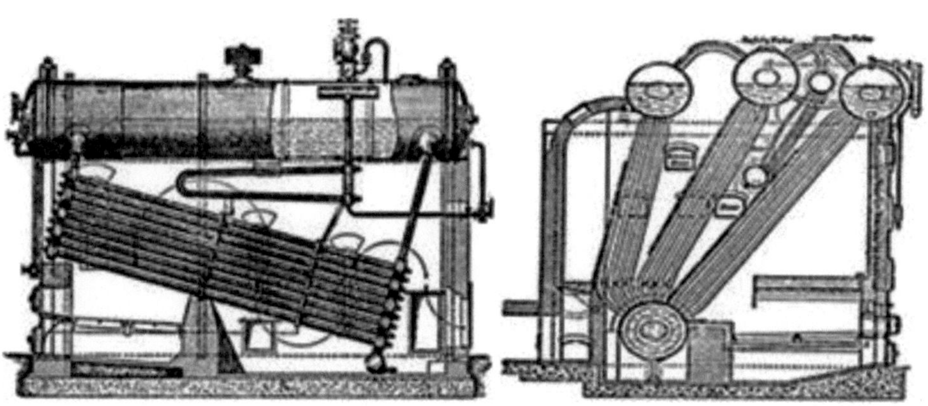

Figure 1.
First boilers with water tubes (Babcock and Wilcox).

1867 by American inventors George Herman Babcock and Stephen Wilcox allowed a higher pressure than that of the fire-tube boiler [19]. In this boiler, the water passed through tubes heated from the outside by the combustion gases, and the steam was collected in a top drum. In the twentieth century, the water-tube boiler found wide applications due to advances such as high-temperature steel alloys and modern welding techniques, which made the water-tube boiler the standard boiler type for all high-capacity boilers.

2.2 Steam boiler classification

Steam boilers can be classified according to various parameters such as design (fire-tube or water-tube), depending on the support, circulation method of water, steam and water/steam mixture (natural circulation, forced circulation), and thermal power. These are fuel-steam generators; they consist of two separate compartments, one in which the fuel burns and the other in which the water circulates. But generally, they are classified in two categories: water-tube and fire-tube steam boilers.

2.2.1 Fire-tube steam boiler

In this type of boiler, the flue gas passes inside submerged tubes in the water (**Figure 2**) [20]. These steam generators are widely used in industrial and commercial facilities, especially in the locomotives and marine applications. Modern fire-tube steam boiler can produce steam pressure up to 25 bars (low and medium pressure) and a flow rate of 1–25 t/h [19]. They can use natural gas, oil, or solid fuel.

The fire-tube steam boiler consists of a cylindrical tank, which contains tubes inside. These tubes collect the hot gases at the exit of the burner. Hot gases, accumulated in a first pass at the back of the steam boiler, are carried by a group of tubes submerged in water to a second pass at the front of the boiler. A second group of submerged tubes take the combustion gases to a third pass at the rear of the steam boiler; this third pass opens on the chimney for the evacuation of fumes to the outside. The heat transfer between the tubes and the combustion gases is mainly done by the convection mode. A typical example of this boiler is illustrated by **Figure 3**.

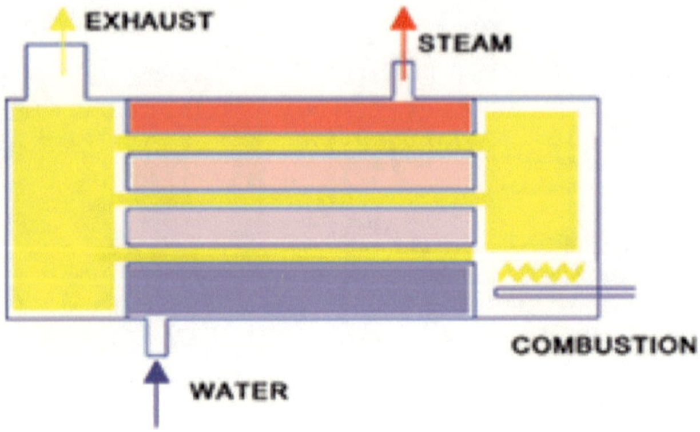

Figure 2.
Principle of fire-tube steam generator.

Figure 3.
Typical fire-tube steam boiler.

2.2.2 Water-tube steam boiler

It is a type of steam generator in which water circulates in tubes that are externally heated by flue gases (**Figure 4**) [19, 20]. They represent the vast majority of steam generators in service.

These steam boilers are used in industrial and power plants to produce high steam pressure. They use gas, oil, or solid combustible as fuel [19]. A typical water-tube steam boiler is illustrated by **Figure 5**. Generally, water-tube steam boilers have two or more tanks, the upper tank called collecting tank (drum) and lower tank called distributor tank. The hot gases produced by the burner are directly in contact with the evaporating tubes; inside of these, vaporization occurs. The steam thus generated is collected in the drum, and the excess water is returned to the bottom tank by non-heated pipes (downcomer). The heat transfer between the tubes and the combustion gases is mainly done by radiation. The flue gases can also be used in the preheating of combustion air and the feedwater.

The performance comparison of the two types of steam boiler is presented in **Table 1**.

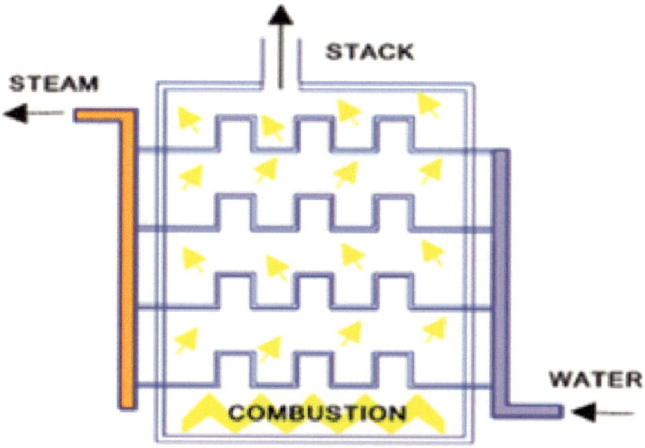

Figure 4.
Principle of water-tube steam boiler.

Figure 5.
Typical water-tube steam boiler.

Properties	Fire-tube steam boiler	Water-tube steam boiler
Start-up (equivalent power)	Slow (large volume of water to heat)	Quick
Adaptation to regime changes	Mediocre (significant inertia)	Good
Heating surface	Medium	High
Security	Mediocre	Good
Congestion	Low	Strong
Price	Limit	High
Usual applications		
• Power	• Moderately high	• Important
• Flow rate	• 1.5–25 t/h	• Higher
• Max working pressure	• 10–20 bars	• 70–225 bars

Table 1.
Comparison of the two types of steam boiler.

2.3 Water circulation mode

The role of the water circulation or the emulsion of water and steam in the steam boiler tubes is to ensure, on the one hand, the correct cooling of the tubes located in the hottest areas or exposed to radiation and that receives at this part the maximum heat flow and, on the other hand, to ensure the generation of saturated steam, that is to say, the passage of the heated fluid from the water state to the water and vapor emulsion state. There are two main types of circulation, natural circulation and forced circulation.

2.3.1 Natural circulation

It establishes itself in the circuits of the steam boiler. It is ensured by the difference of weight between two columns of fluids, one containing only water and the other a mixture of water and steam. The effect of natural circulation decreases when approaching the critical pressure of 221 bars; the practical limit of the use of natural circulation is 180 bars [21]. In steam boilers with natural circulation, the boiling is done in nucleated form, which guarantees a good cooling of the vaporizing tubes [20, 21]. When the power decreases, the void fraction varies relatively little, and consequently the flow in circulation decreases much less quickly than the steam flow of the steam boiler; this guarantees the cooling of the tubes. Natural circulation is therefore relatively more active at low power. It should be noted that the natural circulation is better when the pressure is low and the heating part is located at the bottom of the furnace.

2.3.2 Forced circulation

The circulation of water in such boiler is provided by the feed pumps [21]. This circulation allows great liberty in circuit design, since circulation is always assured. In addition, the diameter of the tubes may be smaller than in natural circulation boilers. In forced circulation boilers, the water vaporized in totality by the unstable film evaporation regime and then by stable film is established necessarily in the final parts of the circuit; it must then be controlled that the speed of the emulsion is sufficient to ensure the cooling of the tube. Forced circulation boiler can be used for all subcritical and supercritical pressures.

2.4 Different components of a steam boiler

We distinguish mainly:

- Drums
- Combustion chamber
- Heat exchangers (economizer, superheater, desuperheater)
- Integrated control
- Valves and flappers
- Feedwater and steam piping
- Pumps

3. Modeling and simulation using RELAP5/Mod3.2

This section describes the modeling and simulation of an industrial steam boiler using RELAP5/Mod3.2 in the steady-state and accidental transients.

3.1 Presentation of the steam boiler used for this study

The steam boiler used in this chapter is a water-tube, radiant type, and high power, with natural circulation from an ABB ALSTOM brand. It is installed in the natural gas liquefaction (NGL) complex which is operated by SONATRACH Company; is located at 5 km east side of Skikda, Algeria; and has been in production since 1970 [22]. The complex contains six units, each one equipped with a steam boiler, to provide superheated steam primarily for driving a turbine, therefore making energy available to the unit [22].

This boiler operates at high heat flux density to produce 374 t/h of superheated steam at 73 bars and 487°C with a design thermal efficiency around 92% [2, 22]. It is composed of three main parts: the steam generator, the superheated steam line, and the feedwater line. A schematic representation of the steam boiler installation is illustrated by **Figure 6** [1, 2].

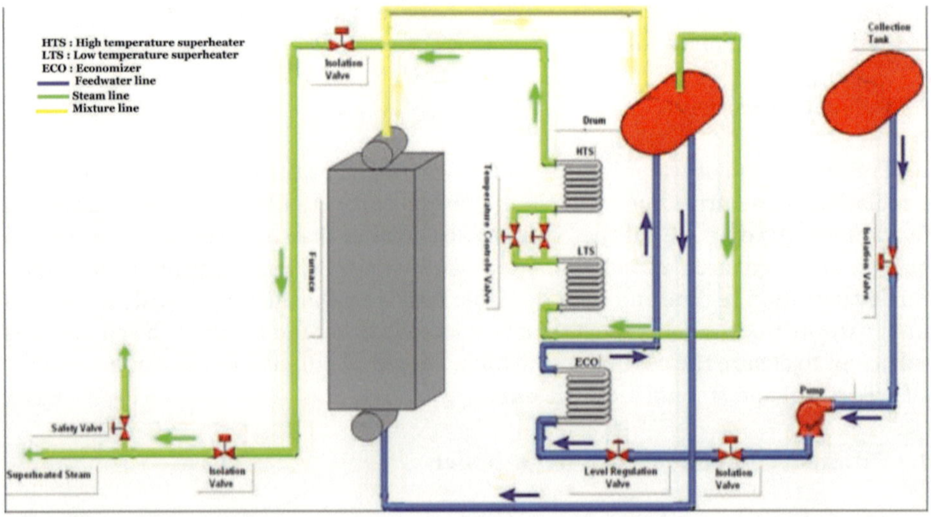

Figure 6.
Steam boiler installation [1].

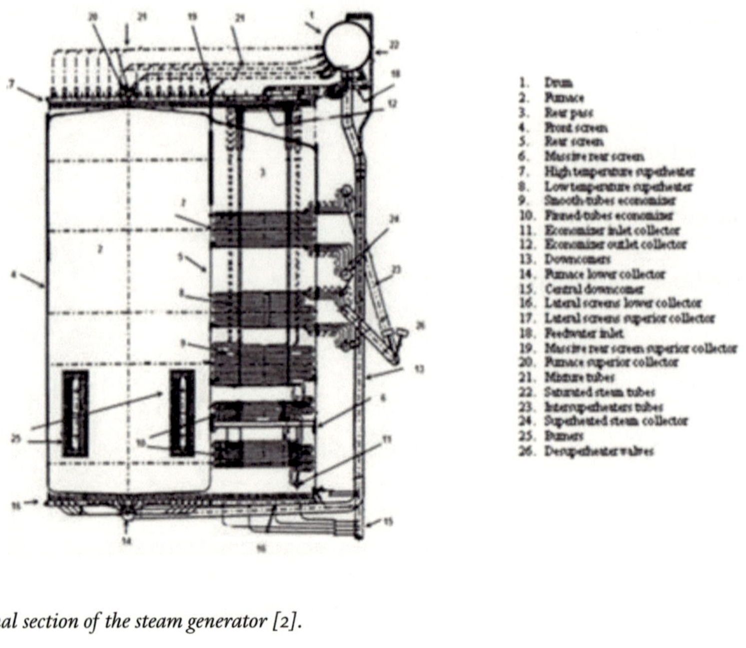

Figure 7.
Longitudinal section of the steam generator [2].

The entire plant can be subdivided into three main parts, the feedwater line, which refers to the saturated liquid phase, the steam generator, and, finally, the main steam line and its transformations. The steam generator consists of one drum and two main parts; the first concerns the combustion chamber, and the second

Technical parameters	Unit	Values
Steam flow rate	t/h	374
Drum pressure	Bar	76.9
Feedwater inlet temperature	°C	118
Outlet furnace gas temperature	°C	1147
Airflow rate	Nm^3/h	344,800
Natural gas flow rate	Nm^3/h	45,699
Air excess in the furnace	%	1.3
Efficiency	%	92
Estimated heat flux densities		
Furnace	kW/m^2	162
Economizers	kW/m^2	34.35
Primary superheater (HTS)	kW/m^2	60.54
Secondary superheater (LTS)	kW/m^2	34.97

Table 2.
Steam boiler operating parameters.

is the rear pass materialized by the water walls that form the evaporating tubes. The rear pass receives the superheaters at high and low temperatures at the top and the economizers below. The steam boiler is designed to operate by combination of automatic and manual operation. The main feedwater line includes the collection tank, two feed pumps, tree economizers, control and isolation valves, and feed piping. The main steam line is constituted by high and low superheaters, steam piping, pipeline of desuperheater, and control and safety valves. The steam generator shown in **Figure 7** [2] and the main operating characteristics of the steam boiler are given in **Table 2**.

The heat transfer between the wall of the tubes and the combustion gases is generally done by two modes, radiative and convective [23]; in radiant steam boilers, as the name suggests, it receives almost the heat by radiation: convection and conduction represent only 5% [19, 20]. The heat received by the water walls is conducted through the membranes and walls of the tubes and transferred by forced convection to nucleate boiling to the water/vapor mixture in the vaporizer tubes.

The installation contains two control loops: water level control in the drum and superheated steam temperature control, in order to maintain the stable operation of the steam boiler. A detailed description of the steam boiler plant can be found in Ref. [1, 2].

3.2 Adopted code and nodalization

3.2.1 RELAP5/Mod3.2 code presentation

The Reactor Excursion and Leak Analysis Program (RELAP5) is a best-estimate nuclear system code; it was developed at Idaho National Engineering Laboratory (INEL) at the request of the US Nuclear Regulatory Commission (NRC) [15]. It is mainly used for the transients' analysis of light-water reactor (LWR); however, the generalization of the RELAP5 code allowed its application to the nuclear and nonnuclear fields [1, 4, 11]. It has been designed to simulate the thermal-hydraulic behavior of installations during accidental or incidental transients. RELAP5 is based on a nonhomogeneous and nonequilibrium hydrodynamic model for the two-phase system. It solves the unstable and one-dimensional equation of mass, energy, and momentum for each phase using the semi-implicit finite difference numerical method [15, 24].

The series of RELAP codes are started by Reactor Leak and Power Safety Excursion (RELAPSE). Previous versions of the RELAP code are RELAP2 and RELAP3, where the name RELAPSE has been changed to RELAP. All these versions are based on equilibrium homogeneous model for two-phase flow [15]. The development of a model of nonhomogeneous nonequilibrium was undertaken for RELAP4. In 1976, the last version (RELAP4/MOD7) of this series of codes has been released. It is clear that a complete rewrite of the code was required to effectively accomplish this goal. The result of this effort was the beginning of the RELAP5 project [15]. RELAP5/MOD3 is the third major release of the RELAP5 thermal-hydraulic system code which was realized in 1985. It is written in FORTRAN 77 for a variety of 64-bit and 32-bit computers. The latest version of the RELAP5 (RELAP-3D) code simulates three-dimensional thermal-hydraulic and neutron phenomena.

RELAP5 is designed in a modular way, using an ordered structure. The procedures and the models are separated into sub-programs and constitute the basis of thermal, hydraulic, and neutronic treatment. An option introduced makes it possible to perform the various calculations related to the steady state, by using the following algorithms:

- Algorithm for kinetics
- Algorithm for the control system
- Algorithm for the hydrodynamic transient
- Algorithm for the thermal transient

Parameters such as pressure, flow rates, and densities would adjust quickly, but the thermal effects evolve more slowly. The accelerated transient technique is therefore used to reduce the transient computation time required to reach steady state. The transient calculation is characterized by the temporal variation of one or more variables related to the studied problem. Usually, the transient regime must be preceded by a well-established steady state in which the initial conditions of the simulated accident are completed. The introduction of the initial values is necessary for the execution of a problem either in the steady state or in the transient state. These values are provided by the user in the input for each component [15].

The RELAP5/MOD3.2 code includes many generic component models for the modeling of various systems and physical phenomena such as pipe, pump, turbines, separators, valves, accumulator, point kinetics of reactors, heat structure, control system component, etc. [25]. In addition, other special process models are introduced for the different form losses, flows in pipes with variable surfaces, branching and choked flow, and others. The programming of the various hydrodynamic calculations is based on a concept of volumes and junctions. System simulation consists to subdivide the plant into components connected by flow junctions. The main component models that are introduced in the RELAP5/MOD3.2 code are grouped in **Table 3**.

The code allows the calculation of the heat transfer through the solid walls, delimiting the hydrodynamic volume. Heat structures are solid elements that generate heat or not, put in contact with the fluid volume. Each heat structure is defined by the indices of the left and right control volumes, the solid volume, its thickness, and the type of the material. The heat transfer modeling of metal structures usually

Component	Label	Schematic	Definition and scope
Single volume	SNGLVOL	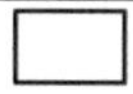	Represents a fluid volume in the system
Pipe or annulus	PIPE	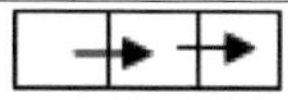	Represents a pipe in the system
	ANNULUS		Special pipe, used to simulate an annular flow
Branch	BRANCH	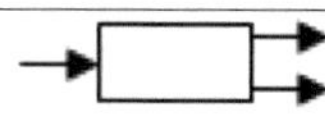	Represents a stream pipe flow juncture
	SEPARATR		Used to simulate the separator in a steam boiler
	TURBINE		Used to simulate a steam turbine
Single junction	SNGLJN	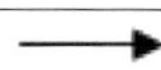	Designed to connect one component to another
Time-dependent volume	TMDPVOL	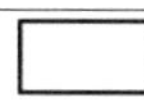	Imposes the thermodynamic conditions at the system boundary
Time-dependent junction	TMDPJUN		Connect some components to another and imposes the circulation flow
Valve	VALVE		A special junction used to simulate an action and the presence of different valves
Pump	PUMP		Simulates the centrifuge pump
Accumulator	ACCUM		Simulates a PWR accumulator

Table 3.
Main thermal-hydraulic components of the RELAP5/Mod3.2 code [25].

includes fuel rods and plates (source of electrical heat or nuclear), heat transfer through the tubes of the steam generator, and the heat transfer to the walls of pipes and tanks in the case of a reactor. The temperature distribution in the heat structures is represented by one-dimensional heat conduction in spherical, rectangular, or cylindrical coordinates. The thermal conductivity and the heat capacity can be simulated by a series of tabulated values according to the temperature or a given function. The integral form of the heat conduction equation is given by expression (1), and finite differences are used for solving this equation [26].

$$\iiint_v \rho C_p(T,\bar{x}) \frac{\partial T}{\partial t}(\bar{x},t) dV = \iint_s k(T,\bar{x}) \bar{\nabla} T(\bar{x},t) \cdot d\bar{s} + \iiint_v S(\bar{x},t) dV \qquad (1)$$

The heat transfer model of the RELAP5 code divides the thermal transfer between the two phases—liquid and vapor (**Figure 8**). The total heat flux Q takes the following expression [26]:

$$Q = h_g(T_w - T_{refg}) + h_f(T_w - T_{reff}) \qquad (2)$$

where h_g: coefficient of heat transfer to steam; h_f: coefficient of heat transfer to liquid; T_w: wall temperature; T_{refg}: vapor reference temperature; T_{reff}: liquid reference temperature.

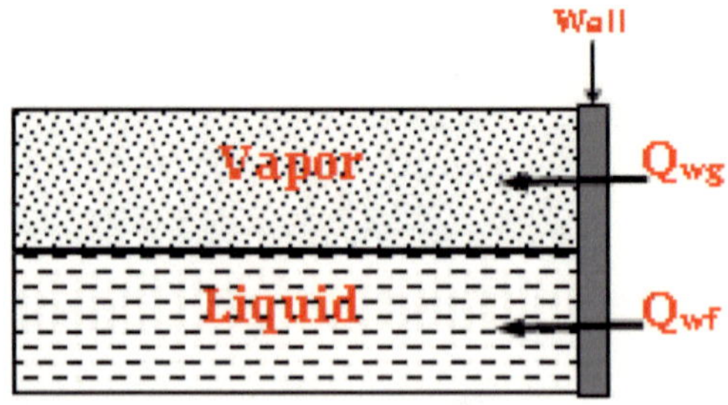

Figure 8.
Heat transfer process.

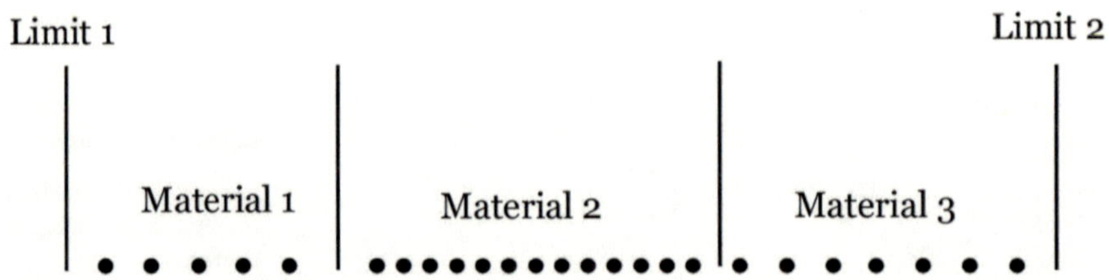

Figure 9.
Discretization scheme.

The reference temperature can be the local temperature of liquid and vapor or the saturation temperature, all depending on the heat transfer correlation used. The wall temperature is calculated implicitly, and the reference temperature can be variable during the calculation. Wall-fluid heat transfer is subdivided into three regimes: condensation, convection, and boiling [26].

Figure 9 illustrates the position of the different nodes (mesh points) for the temperatures' calculation. Each interval may contain different spacing between nodes, different materials, or both. The interval between nodes takes an axial direction for a rectangular structure and a radial direction for a cylindrical or spherical structure. Heat sources can be simulated by the kinetics of the reactor (nuclear source), a series of tabular values as a function of time, or by a control variable.

The code permits the introduction of different boundary conditions such as isolation conditions of tubes, surface temperature tables as a function of time, and atmospheric losses. These boundary conditions can be simulated in different ways: imposed heat flow, imposed temperature, and convection coefficient. A heat transfer correlation series is used to calculate the heat transfer between the circulating fluid and the metal structures connected to the hydrodynamic volumes. This series covers the different modes of heat transfer, convection, radiation, nucleate boiling, transient boiling, and boiling by film.

Boiling curves are used to select correlations of heat transfer. Modeled heat transfer regimes are classified as nucleate boiling, critical heat flow point (CHF), and dispersed flow regime. The heat transfer of condensation is also modeled. The pre-boiling regimes concern the liquid monophasic convection, subcooled nucleate boiling, and nucleate boiling at saturation [15].

3.2.2 Steam boiler nodalization

Knowledge of all the components and parts of the installation as well as all the physical phenomena that may occur in the system is essential for the modeling of any thermal installation. Preparing data to access this type of work using the RELAP5 code requires considerable effort because of the large amount of

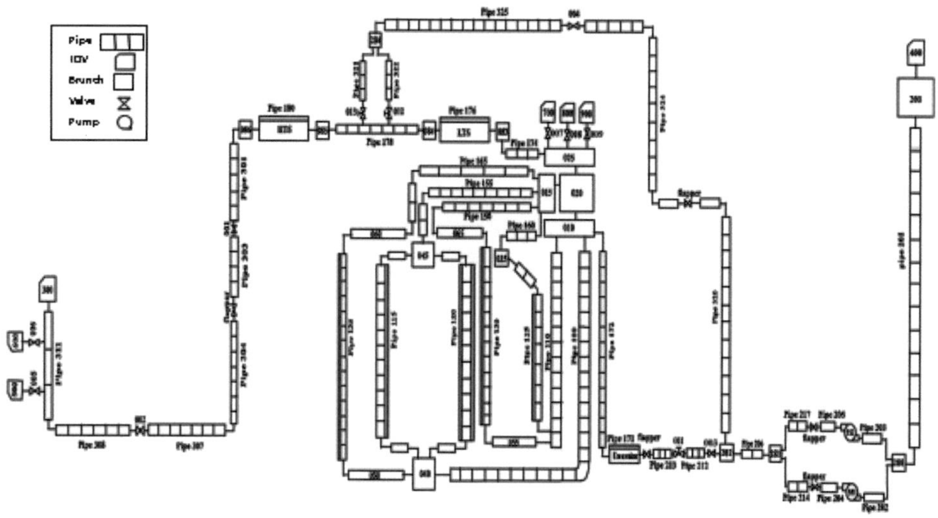

Figure 10.
Nodalization of the steam boiler installation.

information required for the entire installation and its associated components. The information and data of the modeling of the steam boiler plant were obtained from the installation documentation and staff [22], that is, the RELAP5 steam boiler model is based on geometrical and technical data.

The philosophy of using the RELAP5 code is to subdivide the hydrodynamic system into control volumes connected by flow junctions. The thermal behavior of the metal wall of the boiler tubes, such as heat transfer with the fluid, is modeled by heat structures that are connected to vaporizer tubes and heat exchangers. The heat densities between the combustion gases and the external surfaces of the vaporizer tubes are calculated from the energy balance performed on the fumes at each exchanger.

The thermal-hydraulic conditions at the inlet and outlet of the installation represent the condensed feedwater that enters the collection tank and the superheated steam flowing to the turbine. Regulation plays a very important role in the operation of the steam boiler; the RELAP5 code includes the possibility to model the regulation system by components specific to the code. The installation of the entire steam boiler is modeled in 582 control volumes, 589 junctions, and 142 heat structures. The thermodynamic conditions at the system boundaries are imposed by "time-dependent volumes" component **Figure 10** shows the nodalization diagram of the entire installation. More details on the steam boiler nodalization are given in Ref. [1, 2]. The modeling of the steam boiler using the RELAP5 code will be followed by a qualification at the steady-state level.

3.3 Validation at steady-state level

Numerical simulation allows a better understanding of thermal-hydraulic phenomena that could take place in industrial installations; they are of a capital contribution especially in accident situations. The RELAP5 code allows the prediction of the thermal-hydraulic behavior and response of the steam boiler during normal and accidental operations. Prior to the transient accident analysis, it is essential to check the establishment of the steady state in different points of the steam boiler installation. The steady state is reached after running the RELAP5/Mod3.2 code for 5000 seconds in our case study. To demonstrate the establishment of the steady

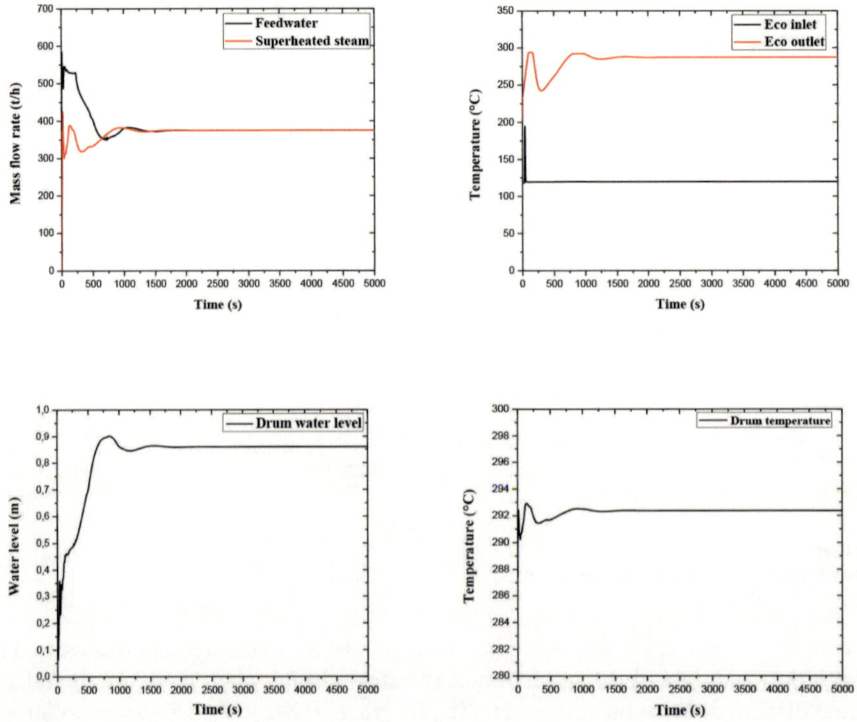

Figure 11.
Main steam boiler parameters during steady state.

Boiler parameters	Units	Experimental data	Simulation data
Feedwater flow rate	t/h	374	374.121
Steam flow rate	t/h	374	374.357
Desuperheater flow rate	t/h	25	26.043
Inlet economizer temperature	°C	118	119.0
Outlet economizer temperature	°C	287	287.030
Outlet drum steam temperature	°C	292	292.368
Inlet LTS temperature	°C	292	292.316
Outlet LTS temperature	°C	370	370.848
Inlet HTS temperature	°C	322	320.141
Outlet HTS temperature	°C	487	487.338
Drum water level	mm	860	860.003
Pressure at collection tank	Bar	1.89	1.89
Drum pressure	Bar	76.9	77.2
Inlet steam generator pressure	Bar	82	78.2
Outlet steam generator pressure	Bar	73	73.199
Outlet pump pressure	Bar	91.93	94.150

Table 4.
Comparison between operating and calculated data at steady state.

state, we selected some steam boiler operating parameters (**Figure 11**). The analysis of the curves representing the evolution of these parameters showed that the regime is stationary and well established, and the set-point values of the regulation system were reached.

The qualification of the steam boiler RELAP5 model is based on available operating data, and it aims to verify that the steady state is well reproduced. In order to validate the plant nodalization under steady-state condition, the simulation results are compared with the experimental data; it provides precious information on the quality of the nodalization, the selection of the appropriate code options, and the appropriate choice of the boundary and initial conditions (1). The comparison between the RELAP5/Mod3.2 results and operating data at steady state is summarized in **Table 4**. As it could be seen, the simulation results are in good agreement with the operating data of the steam boiler, proving the adequacy of the model and expressing the capacity and reliability of the RELAP5/Mod3.2 code in simulating thermal-hydraulic behavior of industrial installations. At this level, it should be noted that the present model could potentially be used for further transient analysis.

3.4 Transient calculation

For a steam boiler, loss of feedwater is the most severe incident that can occur and that may potentially end with serious consequences because water flow rate decreases suddenly leading to a decrease in drum water level and the walls of the tubes are overheated. Various factors can produce this accident; it can be caused by pump power loss, failure of the feedwater pump, ruptures and leakages from pipes located in the main feedwater line, feedwater control valve closing, or failure of the water level regulation [27].

In this chapter, the numerical simulation of the steam boiler thermal-hydraulic behavior and response during loss of feedwater accident caused by the pump power loss is discussed. The transient was performed including protected and unprotected scenarios. In the first one (protected scenario), it is assumed that all control systems are functioning properly to mitigate the sequences of the accident; in the second one (unprotected scenario), it is assumed that there is a failure in the security and control system. Prior to the accident, the steam boiler was operating under steady-state condition. The accidental transient is initiated when the feedwater pump costs down accidentally leading to a sudden decrease in feedwater flow rate. The burners' shutdown is actuated immediately following the triggering of the pump stopping alarm signal. **Table 5** groups the main events describing the accidental scenario as a function of time.

Occurrence instants	Sequence	
	Protected scenario	Unprotected scenario
−500 to 0 seconds	Steady-state regime	X
At 0.0 seconds	Feedwater pump costs down	X
After 0.25 seconds	Alarm signal generation	
At 5 seconds	Burners' shutdown	
At 100 seconds	Closing the steam isolation valve	
At 1000 seconds	End of transient	

Table 5.
Main accidental sequences of the transient.

3.5 Transient analysis

To illustrate the behavior and response of the steam boiler during this transient, the main parameters are presented in curves showing their evolution with time. The curves from **Figures 10–15** show the behavior of each selected parameter even during protected or unprotected scenarios. The transient simulation is preceded by a steady-state period equal to 3000 seconds (this time corresponding to the stability of the entire installation at nominal steam boiler load) with a time step size equal to 10^{-3} seconds.

The temporal variation of feedwater and superheated steam mass flow rate for both scenarios are illustrated by **Figure 12**. Before the accident, the two flow rates are at the same initial value of 374 t/h. After the accident occurrence, the feedwater flow rate decreases instantly, and the superheated steam flow rate vanishes gradually for 90 seconds after the burner shuts down, due to the stopping of steam generation inside the vaporizer tubes. In the second scenario (unprotected), an increase in superheated steam flow rate is observed up to 513.78 t/h which is due to the continuous water vaporization in the vaporizing tubes. At the instant t equals 318 seconds, the flow rate begins to decrease until there is more water in the tubes to vaporize.

Figure 13 shows the time variation of the pressure in the drum. In the first scenario, following the accident and burners' shutdown, the pressure drops rapidly until 72.44 bars at 58 seconds. From this instant, it continues to decrease but more slowly until the end of the transient. This is due to the cooling of the boiler by the ventilation air. In the second case (unprotected), after the accident, the pressure increases to a value of 81.58 bars resulting from the vaporization. Then at time 313 seconds, the pressure starts to drop, and when it reaches 72.69 bars, it stabilizes at that value until the end of transient.

The water level in the steam boiler is a key parameter since it indicates the mass of water in the boiler. So, for safety reasons it must be kept in a limited range [28]. The behavior and response of the water level in the drum are shown in **Figure 14**. It is maintained before the accident at the value of 860 mm (set-point). For the case of the protected scenario, and after stopping the feedwater pump, the level drops suddenly to the value of 359 mm due to the decrease in pressure (**Figure 13**) which generates an intense vaporization of water in the drum. Then it continues to decrease but more slowly until reaching the value of 206 mm at the end of the transient. In the unprotected case, the level decreases slowly and almost linearly contrary to the first case, due to the presence of vapor bubbles in the drum.

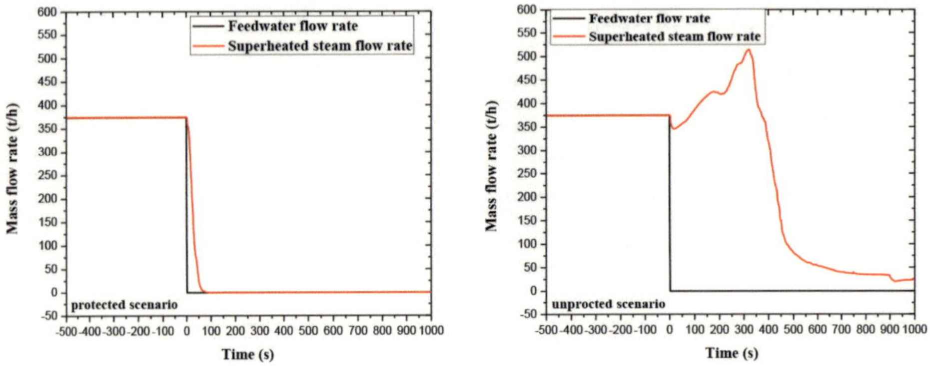

Figure 12.
Time variation of feedwater and steam flow rates.

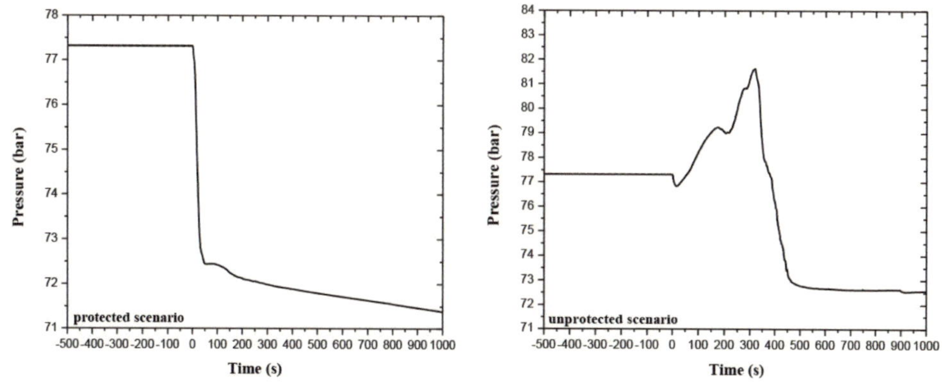

Figure 13.
Time variation of drum pressure.

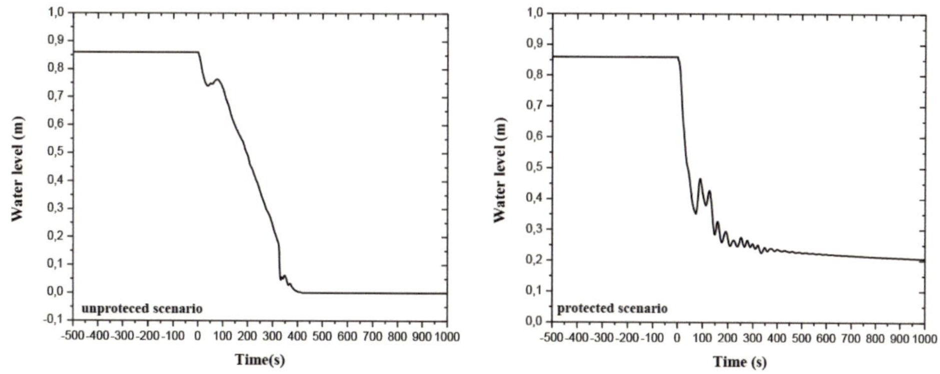

Figure 14.
Time variation of the drum water level.

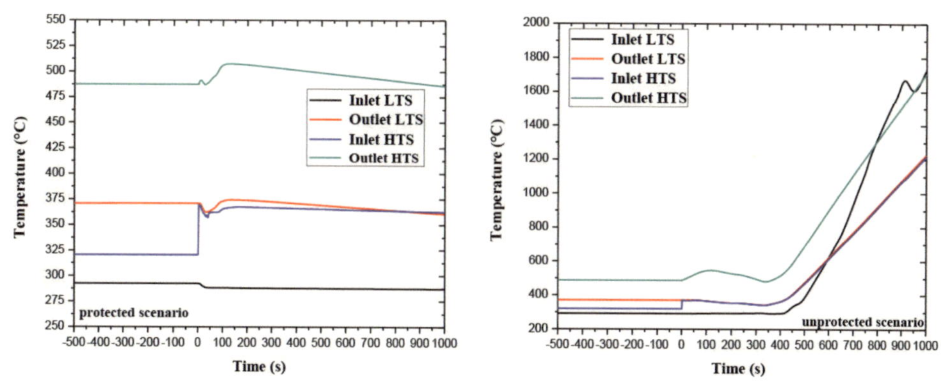

Figure 15.
Temporal variation of the superheated steam temperature inlet/outlet of the superheaters.

Figure 15 shows the time variation of the inlet and outlet temperatures of the superheated steam in low-temperature superheater (LTS) and high-temperature superheater (HTS). At steady state, the temperature values at the inlet and outlet of the two superheaters are, respectively, 292.6 and 370.8°C (LTS) and 321.2 and 487.3°C (HTS). After stopping the pump, temperatures decrease after stopping the burners. The protected scenario shows that the inlet temperature of SHT increases to the value of 370.4°C due to the lack of the desuperheating flow following the feed pump stop. Then it decreases to the value of 359°C. The temperatures increase again due to the heat inertia of the flue gases and then begin to decrease linearly. This decrease is caused by the cooling of the superheater by the ventilation airflow.

For the unprotected case, after stopping the pump, temperatures remain stable and then increase rapidly, reaching very high values of the order of 1720°C. This rise is due to the nonstop of the burners and the decrease of the steam flow rate.

In the steam boiler, wall tubes are designed to operate under highest heat transfer condition (1), where heat is supplied to the outer tubes' surfaces by the fumes. Therefore, and from the safety point of view, it is very important to know the evolution of the wall temperature of the vaporizer tube of the combustion chamber under accidental conditions. It is a key parameter in the safety analysis of the thermal installation. In natural circulation steam boilers, the vaporization regime is in every way in the form of nucleate boiling in order to ensure the continuous cooling of the wall heated by water [29]. As long as this vaporization regime is maintained, the inner wall temperature remains higher than that of the saturation.

The temporal evolution of the evaporator tube inner wall temperature and the heat transfer coefficient during transient for protected and unprotected scenarios is shown in **Figures 16** and **17**, respectively. Before the accident occurrence, the heat transfer inside the tubes is ensured by the nucleate boiling regime, which is characterized by a moderate internal wall temperature, of the order of 303°C, and a good

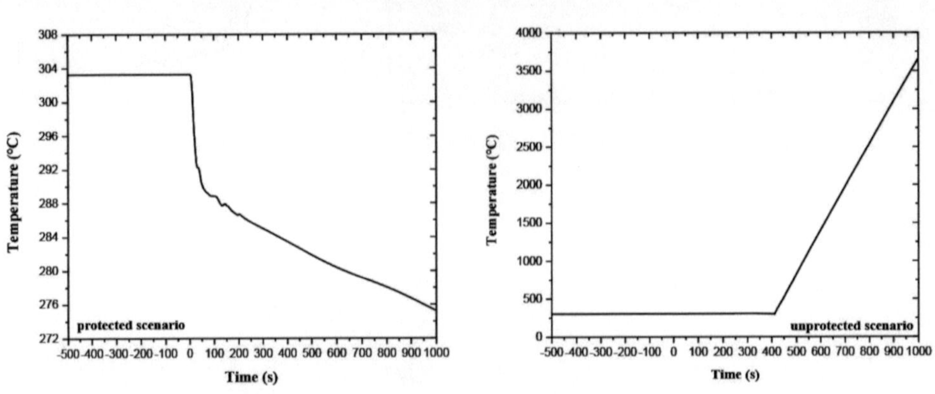

Figure 16.
Temporal variation of the inner wall temperature of the vaporizing tubes.

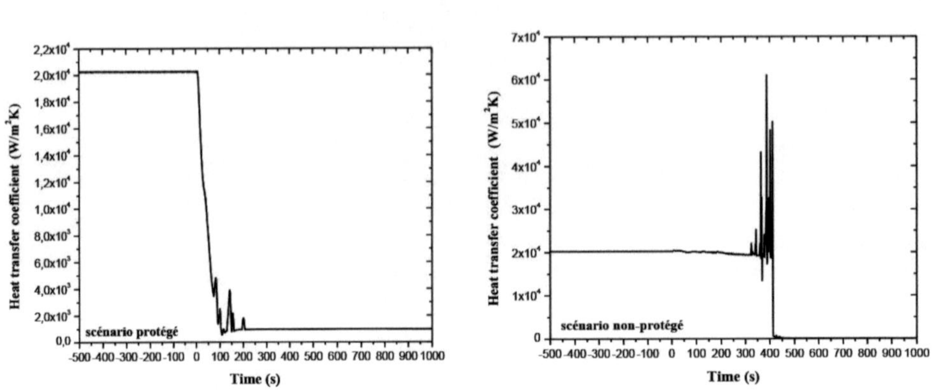

Figure 17.
Temporal variation of the heat transfer coefficient in vaporizing tubes.

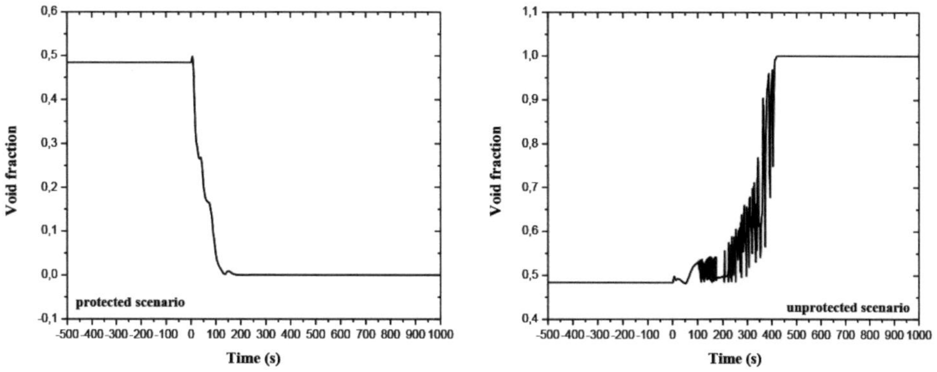

Figure 18.
Temporal variation of the void fraction in the vaporizer tubes.

heat transfer coefficient equal to 20.25 kW/m^2 K. Just after stopping the pump, the wall temperature drops from its initial value to 289.38°C, and the heat transfer coefficient drops from 20 to 5 kW/m^2 K in a time interval equal to 66 seconds. This drop is caused by stopping the burners; nucleate boiling is therefore stopped. Thereafter, the wall temperature decreases linearly until the end of the transient, and heat transfer is achieved by simple convection.

In the second scenario, instabilities in the heat transfer coefficient are observed, which implies that there is a poor heat transfer inside the vaporizing tubes, and the inner wall temperature is quasi-constant. From 410 seconds, the boiling crisis appears, leading to the dryout phenomenon. In fact, the liquid film becomes unstable and is depleted under the effect of intense vaporization. Hence, the wall surface dries out, the heat transfer coefficient drops sharply to 45.5 W/m^2 K, and the temperature of the inner wall increases rapidly to very high values (3900°C) due to the appearance of the boiling crisis. This temperature is higher than the allowed maximum operating value of the plant (500°C) [2, 22], which leads the melting of the vaporizer tube in the combustion chamber.

It is very important to study the void fraction variation during the transient to understand the flow behavior in both phases. **Figure 18** shows the temporal variation of the void fraction in the vaporizer tubes. For the protected scenario, we can see that before the accident (at steady state), the void fraction is maintained at the value 0.4837. After the accident, an instantaneous increase in the void fraction up to 0.4988 resulting from loss of feedwater is observed. After burner's shutdown, the void fraction becomes almost null, and the flow regime is characterized by liquid-phase convection.

During the unprotected case, the void fraction increases to reach unit, between the moment of the accident and the moment of the boiling crisis appearance; as it is shown, there are instabilities in the void fraction during its increase. These are probably caused by poor circulation inside the vaporizer tubes.

4. Conclusions

Modeling and thermal-hydraulic behavior simulation of an industrial water-tube steam boiler during the accidental transient using the RELAP5/Mod3.2 code are presented in this chapter. The transient investigated in this study is the loss of feedwater following the cost down of the feedwater pump. The transient was performed in two steps: the first one concerns the simulation of the protected scenario where the protection systems are operational, and the second one is the simulation of the unprotected scenario.

The results obtained make it possible to analyze and better understand the behavior and response of the installation to the accidental transients by the evolution of the steam boiler thermal-hydraulic parameters. Furthermore, the study clearly demonstrates the protective systems' role in preserving the structural integrity of the steam boiler.

This study has shown that the basic models of RELAP5 code give the possibility of reproducing the main thermal-hydraulic phenomena that may occur in the installation. Thus, it was possible to develop a basic model that can simulate steam boiler operation during normal and accidental transients. In addition, the capacity and reliability of the RELAP5/Mod3.2 code for thermal-hydraulic analysis of conventional thermal installations such as industrial steam boilers have been demonstrated.

Finally it was possible to demonstrate, using RELAP5 modeling capabilities, that in the case of safety and protection system failure, the critical phenomenon of the boiling crisis is established in the combustion chamber which is undoubtedly the cause of the frequent explosion of the steam boilers.

Author details

Amina Lyria Deghal Cheridi[1*], Abla Chaker[2] and Ahcene Loubar[1]

1 Nuclear Safety Department, Nuclear Research Centre of Birine, Algeria

2 Energetical Physics Department, Constantine 1 University, Algeria

*Address all correspondence to: al_cheridi@yahoo.fr

References

[1] Deghal Cheridi AL, Chaker A, Loubar A. Numerical simulation of a 374 tons/h water-tube steam boiler following a feedwater line break. Annals of Nuclear Energy. 2016;**97**:27-35

[2] Deghal Cheridi AL, Chaker A, Loubar A. Thermal-hydraulic simulation of a radiant steam boiler using RELAP5/Mod3.2 computer code. Computers and Chemical Engineering. 2013;**52**:168-176

[3] Dias SP. Modeling and simulation of an industrial steam boiler with ECOSIMPRO. In: Proceedings of the 1st Meeting of Ecosimpro Users; Madrid: UNED. 2001

[4] Serraro-Aguilera J, Valenzuela L, Parras L. Thermal-hydraulic RELAP5 model for a solar direct steam generation system based on parabolic trough collectors operating in once-through mode. Energy. 2017;**133**:796-807

[5] Smrekar J, Assadi M, Fast M, Kustrin I. Development and artificial neural network model for a coal-fired boiler using real plant data. Energy. 2009;**34**:144-152

[6] Watanabe T, Ishigaki M, Hirano M. Analysis of BWR long-term station blackout accident using TRAC-BF1. Annals of Nuclear Energy. 2012;**49**:223-226

[7] Tsai C et al. The pressurization transient analysis for Lungmen advanced boiling water reactor using RETRAN-02. Nuclear Engineering and Design. 2010;**240**:3037-3045

[8] Groudev P, Stefanava A. Validation of RELAP5/Mod3.2 model of VVER 440 on reactor scram transient. Progress in Nuclear Energy. 2006;**48**:146-154

[9] Hainoun A, Schaffrath A. Simulation of subcooled instability of high flux research reactor using the extended code ATHLET. Nuclear Engineering and Design. 2001;**207**:163-180

[10] Micaelli JC, Barre F, Bestion D. CATHARE code development and assessment methodologies. In: Trans. of the ANS, Winter Meeting; San Francisco. Vol. 73. 1995. pp. 509-510

[11] Tuunanen J et al. Analysis of PACTEL passive safety injection experiments with APROS, CATHARE and RELAP5 codes. Nuclear Engineering and Design. 2000;**198**:261-286

[12] Bokhari IH. Steady-state thermal hydraulic and safety analysis of a proposed mixed fuel (HEU& LEU) core for Pakistan research reactor. Annals of Nuclear Energy. 2004;**31**:1265-1273

[13] Rahmani A et al. D-type steam boiler simulation using RELAP5/Mod3.2 computer code. International Journal of Heat and Technology. 2006;**25**(2):119-126

[14] Ferri R, Cammi A, Mazzei D. Molten salt mixture properties in RELAP5 code for thermodynamic solar applications. International Journal of Thermal Sciences. 2008;**47**:1676-1687

[15] RELAP5/Mod3.2 code Manual Volume: Code Structure, System Models, and Solution Methods; 1998

[16] Rahmani A et al. Assessment of boiler tubes overheating mechanisms during a postulated loss of feedwater accident. Applied Thermal Engineering. 2009;**29**:501-508

[17] Berles AE et al. Two Phase and Heat Transfer in the Power and Process Industries. New York: McGraw-Hill; 1981

[18] Burstall AF. A History of Mechanical Engineering. Cambridge, UK: The MIT Press; 1965

[19] Truelove JS. Furnaces and Combustion Chamber. Philadelphia, USA: Hemisphere Publishing Corporation; 1983

[20] El-Wakil MM. Power Plant Technology. New York: McGraw-Hill; 1985

[21] Engineering Technique B 1461, B 1462, B124, B1480; 1993

[22] Technical Operations Manual of the Industrial Steam Boiler ABB ALSTOM; 2000

[23] Tong LS, Weisman J. Thermal Analysis of Pressurized Water Reactors. Heat Transfer in Steam Generator. 3rd Edition, USA: American Nuclear Society; 1996

[24] IAEA-TECDOC-1395. Inter Comparison and Validation of Computer Codes For Thermal-Hydraulic Safety Analysis of Heavy Water Reactors; 2004

[25] RELAP5/MOD3.2 Code Manual Volume V: User's Guidelines. The Thermal-Hydraulic Group; 1998

[26] RELAP5/Mod3.2 Code Manuals, NUREG/CR-5535 Report. Vol. IV; 1998

[27] Juslin K, Companion Model A. Approach to modeling and simulation of industrial processes [Ph. D thesis]. Finland: Helsinki University of Technology; 2005

[28] Kim H, Choi S. A model on water level dynamics in natural circulation drum type boiler. International Communications in Heat and Mass Transfer. 2005;**32**:786-796

[29] Payan LA et al. Critical heat flux prediction for water boiling in vertical tubes of a steam generator. International Journal of Thermal Sciences. 2005;**44**:179-188

Chapter 9

Exhaust Gas Heat Recovery for an ORC: A Case Study

Armando Gallegos-Muñoz, Fabián Luna-Cabrera, Martín Picón-Núñez, Francisco Elizalde-Blancas and Juan Manuel Belman-Flores

Abstract

This work aims at developing a heat exchanger (HEX) sizing approach considering the need to maximize the heat recovery within the limitations of pressure drop and space. The application consists in the recovery of the energy contained in exhaust gases coming from an internal combustion engine (ICE). Two heat exchanger geometries are selected as case studies. The design approach involves the application of design of experiments (DOE) techniques and computational fluid dynamics (CFD) simulations. DOE techniques are used to observe the influence of some selected parameters (factors) in the design of the heat exchangers, and CFD simulations are carried out to determine the performance of the heat exchanger. The information obtained is used to determine local Nusselt number correlations that are used for the design of the heat exchangers.

Keywords: heat exchanger, heat waste, CFD, neural network, optimization

1. Introduction

Industrial applications of waste heat recovery require several types of heat exchangers. The correct selection and optimization of the heat exchangers are critical for heat transfer. Several papers have been published that deal with the selection of the most suitable heat exchanger technology for a specific application. Hatami et al. [1] developed a numerical study to model two types of heat exchangers (HEXs) used to recover the exhaust waste heat from internal combustion engines (ICEs). In the work, authors aimed at finding the best viscous model to fit experimental data. One of the exchangers belongs to a compression ignition (CI) engine with water as cold fluid, while the second exchanger belonged to a spark ignition (SI) engine with a mixture of 50% water and 50% ethylene glycol (EG) as cold fluid. From the study, authors concluded that the heat recovery can be improved by increasing the number of fins and length, where maximum heat recovery occurs with high engine load and speeds. On a different work, Hatami et al. [2] applied a response surface methodology (RSM) based on central composite design (CCD) to derive an optimization approach of finned-type heat exchangers to recover waste heat from the exhaust of a diesel engine. The design is performed for a single-point operation (1600 rpm and 60 N m) of an OM314 diesel engine. Based on the CCD principle, 15 exchangers with different fin heights (FH), fin numbers,

IntechOpen

and fin thicknesses (FT) were numerically modeled, and optimization was carried out to maximize heat recovery and minimize pressure drop along the heat exchanger. The results showed that the height of the fins has a higher impact on pressure drop than fin number and thicknesses. On the other hand, fin number enhances heat recovery.

Bari et al. [3] performed a study on pancake-shaped heat exchangers to be fitted in a vehicle. The heat exchangers studied were of the shell-tube and U-tube type. CFD simulations were carried out to optimize the design and calculate the additional power that could be achieved by using these optimized heat exchangers. The effectiveness of pancake-shaped heat exchanger is on average 3% higher than that of the optimized round-shaped heat exchanger. Bari et al. [4] conducted experiments using water as the working fluid to estimate the exhaust waste heat recoverable from a diesel engine using two available heat exchangers. Two identical shell and tube heat exchangers were fitted into the exhaust of the engine, and experiments were conducted to estimate the additional energy that could be gained with this setup. Simulation tools were used to compare the performance of the heat exchangers with experimental data. Then the effects of changing important parameters such as length, diameter of shell, and number and diameter of tubes on the heat recovery were investigated. It was found that the effectiveness was higher for smaller shell diameters. After optimization, the additional power increased from 16 to 23.7%.

Tan et al. [5] reported the use of artificial neural network (ANN) models to simulate the thermal performance of a compact fin-tube heat exchanger with air and water/ethylene glycol antifreeze mixtures as the working fluids. They demonstrated that, once trained, an artificial neural network could predict the overall heat transfer rate between the liquid and air steams with a high degree of accuracy. The neural network predictions were in much closer agreement to the experimental data than corresponding predictions derived using a conventional nonlinear regression model.

Shivakumar et al. [6] tested the applicability of neural networks in order to correlate the experimentally determined heat transfer parameters of a multi-pass cross-flow heat exchanger. The waste heat from an internal combustion (IC) engine was used to heat the water in a cross-flow heat exchanger. The experimental results were used to train the ANN model. A multilayer perceptron (MLP) with back-propagation algorithm was used for training the network. The predicted results by the ANN model were compared with experimental data. They concluded that an MLP network can be used to predict the thermal performance characteristics of multi-pass cross-flow heat exchanger using a limited number of experimental data.

Hatami et al. [7] used a multi-objective optimization approach based on ANN and genetic algorithm (GA) to the numerical outcomes of a finned-tube heat exchanger in a diesel exhaust heat recovery application. The results confirm that the optimized case widely increased the recovered heat and exergy while keeping the pressure drop at low levels. Although the optimized case exhibited higher irreversibility, its second law efficiency is significantly greater than the non-optimized case, especially at high engine loads. The average efficiency of the proposed HEX is about 8% for the exergy recovery from the exhaust of a light diesel engine.

Aly et al. [8] investigated the 3D turbulent flow and heat transfer of coiled tube-in-tube heat exchangers. Heat exchangers are analyzed considering conjugate heat transfer from the hot fluid in the inner-coiled tube to the cold fluid in the annulus region. After simulations, the Taguchi method was used to find the optimum condition for some design parameters in the range of coil diameter from 0.18 to 0.3 m and tube and annulus flow rates from 2 to 4 and 10 to 20 l/min, respectively. Results showed that the Gnielinski correlation (used extensively for predicting Nusselt number for turbulent flow in ducts) can be used to predict Nusselt number for both the inner-coiled tube and the annular coiled tube using the friction factor correlation for helical tubes. The application of the Taguchi method showed that the annulus side flow rate, the tube side flow rate, the coil diameter, and the flow configuration are the most important design parameters in coiled tube-in-tube heat exchangers.

Hossain et al. [9] optimized heat exchangers used in the recovery of exhaust heat from a 40-kW diesel generator. With the available experimental data, computer simulations were carried out to optimize the design of the heat exchangers. The optimized heat exchangers were then used to estimate additional power gained considering the turbine isentropic efficiency. The proposed heat exchangers could produce 11% additional power using water as the working fluid at a pressure of 15 bar. The effects of the working fluid pressure were also investigated to maximize the additional power production. The pressure was limited to 15 bar which was constrained by the exhaust gas temperature. However, higher pressure is possible for higher exhaust gas temperatures from higher capacity engines.

This work aims at showing a stepwise approach for the sizing of a heat exchanger for waste heat recovery and subsequent use in an Organic Rankine Cycle (ORC). For maximum power production and minimum pressure drop, the exchanger must be optimized. Besides, space limitation poses an additional constraint to the design. The approach introduced in this work allows the designer to simultaneously achieve all these design objectives.

2. Method description

The proposed method seeks to maximize heat transfer and minimize pressure drop. Besides, within the exchanger, overheated areas (to avoid evaporation of the cold fluid) and overcooled areas (to avoid condensation and corrosion on the hot side) must also be minimized [10]. A combination of different tools is used to solve the complex problem. To avoid overdesign, accurate Nusselt correlation must be developed. Given the space limitations, two exchanger geometries, namely, a finned heat exchanger and a helical heat exchanger, are analyzed to have an additional degree of freedom between heat recovery and pressure drop. For the same heat load, the finned tube will exhibit lower heat transfer area but higher pressure drop, while the helical tube will have larger surface area but lower pressure drop. Given the constraints in terms of mass flow rate, pressure drop, heat transfer, and space, maximum and minimum values for these parameters must be fixed. Since a very large possible combination of operating conditions can result, it is important to discriminate between them. One way of doing this is by designing the experiments or identifying the most representative set of design variables that allow to reduce the search space. Once this is done, in principle the geometry should be constructed and tested to see which of the designs exhibit overheating and overcooling areas. Computational fluid dynamics techniques can be used to this end. Besides, CFD can also provide local heat transfer coefficients which can be correlated for design purposes. As mentioned earlier, the approach used in this work is by means of artificial neural networks.

Since maximum power production is the final desired outcome, the design with the maximum exergy recovery will lead to maximum power production. Thus, exergy analysis is included, and the Organic Rankine Cycle is modeled using the HYSYS simulator [11].

2.1 Process description

The steps followed in the analysis introduced in this work are detailed described below and graphically shown in **Figure 1**.

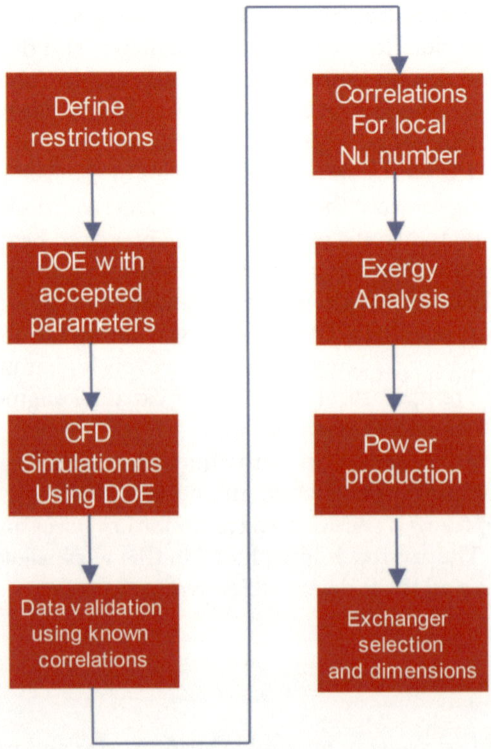

Figure 1.
Sequence of the proposed analysis and design method.

1. The restriction parameters (mass flow rate, pressure drop, heat transfer, power required, and geometrical constrains) are defined.

2. A design of experiments, [12] is established using the parameters (factors: heat flux, pressure drop, overheated and overcooled areas of both types of heat exchangers) accepted to select the conditions for the CFD simulation.

3. A CFD simulation for each of the resulting experiments above is carried out.

4. Local Nusselt numbers obtained from the CFD simulations are compared with published correlations for validation.

5. Once the results of the simulation are validated, these can be used to produce regression equations to predict the best combination of parameters to maximize heat transfer and minimize overheated and overcooled areas and pressure drop.

6. The neural network developed in [10] is used to fit complex relation emerging from CFD results.

7. Exergy analysis is applied to determine the exergy gain and exergy efficiency of the heat exchangers.

8. The heat recovered from the diesel engine is used in an Organic Rankine Cycle for power production.

9. Selection and design of final heat exchanger.

The diagram of the process is presented in **Figure 1**.

3. Case study

The proposed method is applied on two types of heat exchangers, namely, a finned-tube heat exchanger and a helical-tube heat exchanger. These configurations were selected because they are compact, of relatively simple geometry and of easy modification of parameters. The open literature indicates that these types of heat exchangers are typically used in heat recovery from exhaust gases. The geometry of both exchangers consists of two concentric tubes: the hot gas flows at the internal tube and ethylene glycol flows in the annular space between tubes. The outer surface of the exchangers is isolated. **Figures 2** and **3** show the helical-tube heat exchanger and the finned-tube heat exchanger, respectively. The only parameter that remains the same, as reference, for both types of heat exchangers is the linear length from inlet to outlet which is set to 1 m.

CFD techniques are used to analyze the performance of the units. The response variables of the CFD simulation are gas outlet pressure (PG), ethylene glycol outlet pressure (PEG), heat exchanger's surface area referred to the hot side (AG), heat exchanger's surface area referred to the cold side (AEG), heat flux from the hot side (QG"), and heat flux to the cold side (QEG"). For the design of experiments, the factors considered for the case of the helical-tube heat exchanger are internal

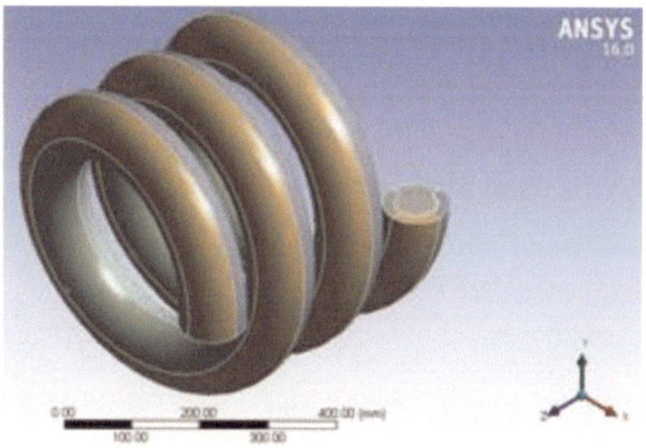

Figure 2.
Helical-tube heat exchanger.

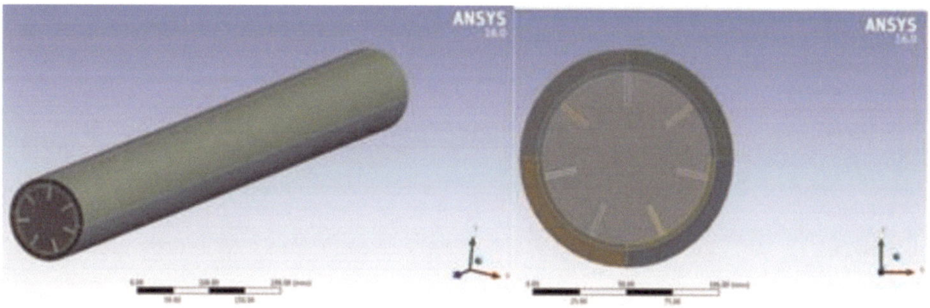

Figure 3.
Finned-tube heat exchanger.

diameter (ID), external diameter (ED), mass flow rate of the gas (mass), and inlet gas temperature (temp). The range of parameters are shown in **Table 1**.

In the case of the finned-tube heat exchanger, where the fins are assumed to be straight, the parameters considered are fin height, fin thickness, fin density (FD), mass ratio between gas and ethylene glycol (mass ratio), and inlet gas temperature (temp). The range of the parameters are shown in **Table 2**.

The variables were selected based on the impact they have on the dimensions of the heat exchanger, as well as on the operating conditions over the range where maximum heat transfer will be achieved. These variables will allow to find the optimal conditions of the heat exchanger when determining new correlations for the Nusselt number to eliminate the risk of oversizing.

The factors chosen for the finned heat exchanger were taken from the work of Hatami et al. [7]. The mass ratio factor is the ratio between the two fluids, namely, gas and ethylene glycol. After trying different mass ratios, the best fit between CFD results and Minitab [13] regressions was obtained. To select the factors for the case of helical heat exchanger, six parameters were initially considered: internal diameter, external diameter, helix diameter, helix pitch, mass of gas, and temperature of gas. Some parameters were eliminated to get the minimum number of factors that could exhibit a good fitting to the simulated results. The parameters eliminated were helix diameter and helix pitch. The minimum number of factors required to get a good fit were internal diameter, external diameter, and temperature and mass of gas.

The experiment design indicates that 27 configurations to simulate the helical-tube heat exchanger are needed, while 46 are required for the finned-tube unit. Each one of these configurations was simulated using Ansys Fluent 2016 [14]. The simulations were made under the following considerations:

a. The standard k-ε model with standard wall function turbulent model was used for the gas side.

b. A laminar model was used at the ethylene glycol side ($50 \leq Re \leq 250$).

Parameter	Low	High
ID [mm]	60	80
ED [mm]	100	110
Mass [kg/s]	0.07	0.135
Temp [K]	550	700

Table 1.
Design parameters for the helical-tube heat exchanger.

Parameters	Low	High
FH [mm]	17	32
FT [mm]	2	6
FD [mm]	4	10
Mass ratio	0.5	0.71
Temp [K]	550	700

Table 2.
Design parameters for the finned-tube heat exchanger.

c. There is no phase change on either side.

d. The gas pressure drop must be lower than 10 kPa.

e. The y+ value must be around y+ > 30 and y+ < 300 (wall treatment) [15].

The density of the gas was determined using the incompressible ideal gas model because the Mach number in all cases was lower than 0.3. The density and viscosity of ethylene glycol were calculated using a user-defined function (UDF). In this way, the variation of density and viscosity with temperature was considered applying the equations

$$\rho_{EG} = -0.9904 * temp + 1417 \quad (1)$$

$$\eta_{EG} = 3.724E4\ \exp\ (-0.05021 * temp) + 0.2811 \exp\ (-0.01356 * temp) \quad (2)$$

For the thermal conductivity and viscosity of the hot gas, the equations used are

$$\kappa_{GA} = 6.22 * 10 - 5 * temp + 0.008116 \quad (3)$$

$$\eta_{GA} = 3.755 * 10 - 5 * \exp\ (0.0002586 * temp) \\ -3.561 * 10 - 5 * \exp\ (-0.001614 * temp) \quad (4)$$

To validate the results of the CFD simulations, the following published correlations are used [10]:

Helical-tube heat exchanger

Turbulent regime

$$\mathrm{Nu_s} = \frac{\mathrm{PrRe}(\mathrm{f_s}/8)}{1 + 12.7\sqrt{\frac{\mathrm{f_s}}{8}}\left(\mathrm{Pr}^{\frac{2}{3}} - 1\right)}\left[1 + \left(\frac{\mathrm{d_h}}{\mathrm{L}}\right)^{2/3}\right] \quad (5)$$

Laminar regime

$$\mathrm{Nu} = \left[\left(4.364 + \frac{4.636}{\mathrm{x_3}}\right) + 1.816\left(\frac{\mathrm{De}}{\mathrm{x_4}}\right)^{3/2}\right]^{1/3} \quad (6)$$

Finned-tube heat exchanger

Turbulent regime

$$\frac{\mathrm{Nu_n}}{\mathrm{Nu_{D-n}}} = \left[\frac{\mathrm{d_i}}{\mathrm{d_{im}}}\left(1 - \frac{2\mathrm{e}}{\mathrm{d_i}}\right)\right]^{-0.2}\left(\frac{\mathrm{d_i d_h}}{\mathrm{d_{im}^2}}\right)^{0.5} \sec^3\beta \quad (7)$$

Laminar regime

$$\mathrm{Nu} = 1.86\left(\frac{\mathrm{Pe_b d_i}}{\mathrm{L}}\right)^{1/3}\left(\frac{\mu_\mathrm{b}}{\mu_\mathrm{w}}\right)^{0.14} \quad (8)$$

To determine the maximum exergy recovery, the expression used is

$$\frac{\mathrm{d}\phi_{\mathrm{CV}}}{\mathrm{dt}} = \sum\dot{\phi}_\mathrm{Q} + \sum\dot{\mathrm{m}}_\mathrm{i}\psi_i - \sum\dot{\mathrm{m}}_\mathrm{e}\psi_e + \dot{\mathrm{W}}_{\mathrm{act}} - \dot{\mathrm{I}}_{\mathrm{total}} \quad (9)$$

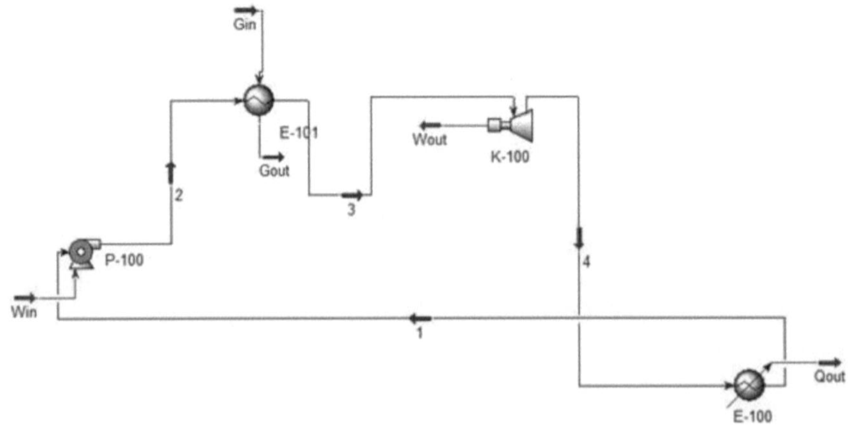

Figure 4.
Flow diagram for the simulation of the organic Rankine cycle.

The generation of power from a low temperature heat source can be achieved by means of an Organic Rankine Cycle. **Figure 4** shows the diagram of the ORC cycle used for the simulation using HYSYS [11]. The working fluid is butane and the main components of the cycle are:

Pump (P-100): adiabatic efficiency 75%.

Heat exchanger (heater) (E-101): tube passes 2, shell passes 1, $\Delta P = 0$.

Turbine (K-100): adiabatic efficiency 75%.

Heat exchanger (cooler) (E-100): $\Delta P = 0$.

4. Results

As described in the methodology applied (**Figure 1**), the first step consists in the validation of the CFD simulations according to the DOE results from where the input data are chosen. Next, from the CFD local Nusselt numbers obtained, correlations are obtained. Then an exergy analysis is applied to the heat exchangers, and finally the simulation in Aspen HYSYS [11] is carried out.

4.1 Validation of CFD simulation

For the CFD simulation, the sweep hex elements were used; 473,600 elements were applied to the helical-tube heat exchanger and 949,500 to the finned-tube heat exchanger. The boundary conditions used in the CFD simulations for both types of heat exchangers are inlet mass flow, outlet pressure and insulated external surface. The regime of flow is subsonic. The gas side exhibits turbulent flow and the ethylene glycol a laminar regime.

For the gas side, the conditions at the inlet are mass, temperature, turbulent intensity, and turbulent length scale. The conditions at the outlet are backflow temperature, turbulent intensity, and turbulent length scale. For the ethylene glycol side, the conditions at the inlet are mass and temperature. The CFD solution provides the following results for both exchangers: mean temperature, wall temperature, pressure drop, and heat flux.

For the grid independence, several meshes were tested for each of the 73 configurations; a total of 198 simulations were carried out with the aim of finding the meshes that exhibit less variation in the prediction of results. A finer mesh was used at the inner face of the gas cavity, around the fins and at the inner and outer face of ethylene glycol cavity to fulfill the required y + value. **Figures 5** and **6** show the refinement for the two types of heat exchangers. The parameters used in the solver of the CFD simulations are shown in **Table 3**.

The parameters in **Table 3** gave the best results regarding mass and energy balance. For turbulent flow, the physical model SIMPLEC is recommended [16]. For gradient, the least squares cell-based method was selected. This method is less expensive in terms of simulation time [17]. For pressure interpolation, the second-order scheme is recommended. Second-order upwind was used to get more accuracy in the solution of the momentum equations [18]. First-order upwind was used to calculate turbulent kinetic energy because it is less time-consuming [17]. First order upwind was used to calculate turbulent dissipation rate because is less time consuming [17]. Second-order upwind was used to get more accuracy in the solution of the energy equations [18].

The results obtained from the CFD simulations were validated using Eqs. (5)–(8). **Figures 7–10** show the comparison of Nusselt number. It can be observed that, for both heat exchanger geometries, the numerical results and the ones obtained from the correlation show similar tendency with a good approximation between them.

In the same way, local Nusselt number for both types of heat exchangers is presented in **Figures 11–14**. The most relevant configurations (experiments) were considered for each type of heat exchanger.

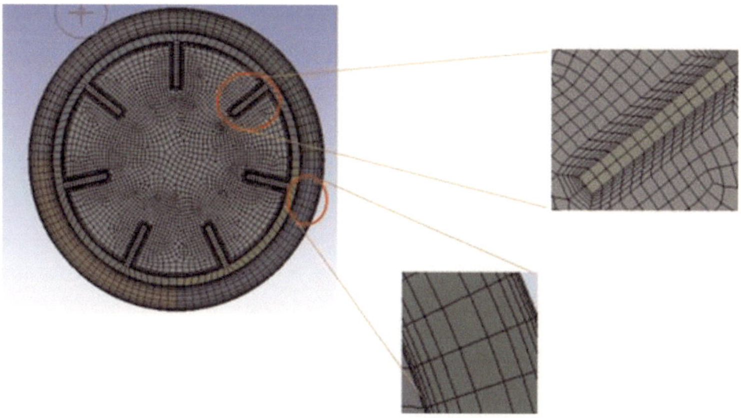

Figure 5.
Mesh refinement in the finned-tube geometry.

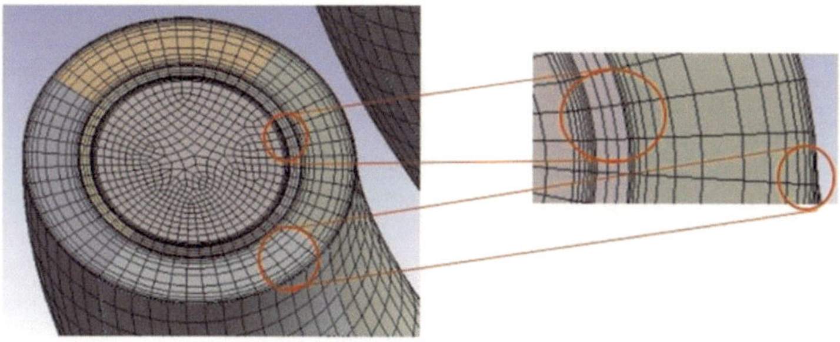

Figure 6.
Mesh refinement in the helical-tube geometry.

4.2 Design of experiments (DOE)

With the aim of obtaining regression equations for Nu number, the design of experiments is applied. A second-order regression model for each of the variables is used. The DOE used was the Box-Behnken response surface design. The advantages of this method are as follows: it is a second-order model, and the experimental

Time	Steady
Scheme	SIMPLEC
Gradient	Least squares cell-based method
Pressure	Second order
Momentum	Second-order upwind
Turbulent kinetic energy	First-order upwind
Turbulent dissipation rate	First-order upwind
Energy	Second-order upwind

Table 3.
Parameters used in the solver of the CFD simulations.

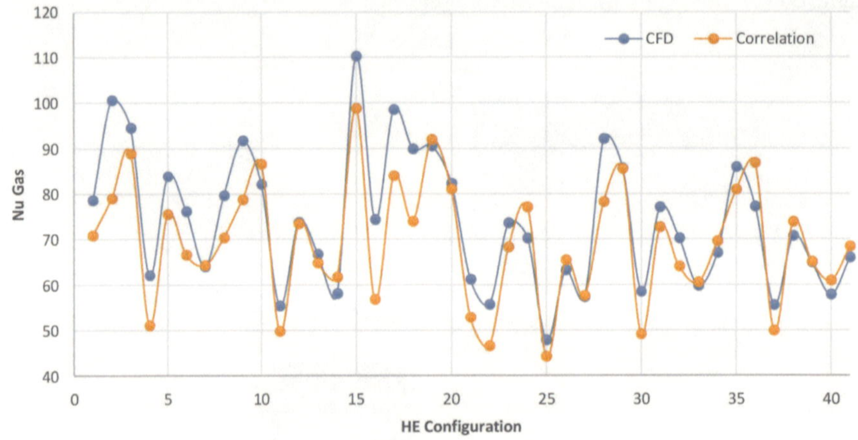

Figure 7.
Validation of nu number on the hot side (G) for the finned-tube geometry.

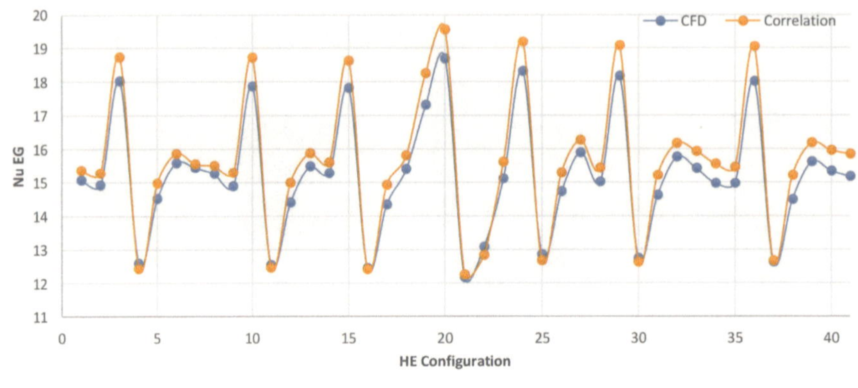

Figure 8.
Validation of nu number on the cold side (EG) for the finned-tube geometry.

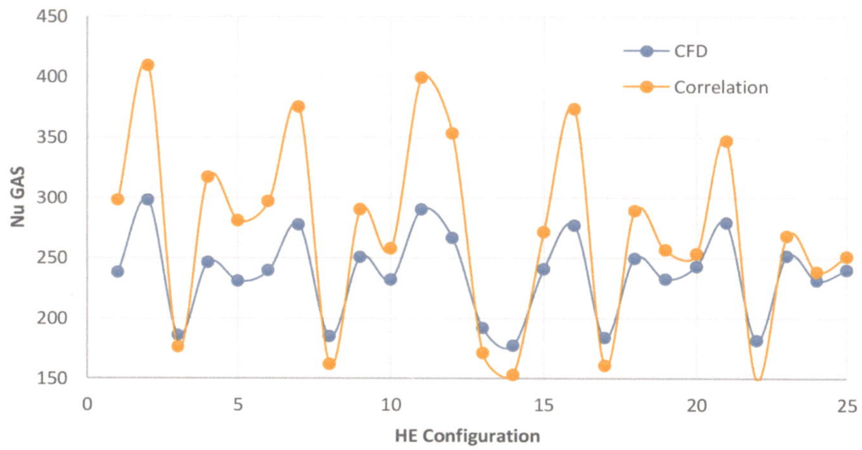

Figure 9.
Validation of nu number on the hot side (G) for the helical-tube geometry (G).

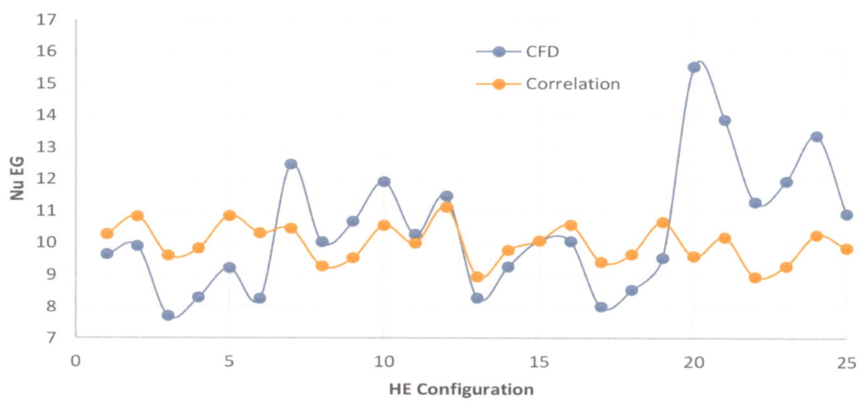

Figure 10.
Validation of nu number on the cold side (EG) for the helical-tube geometry.

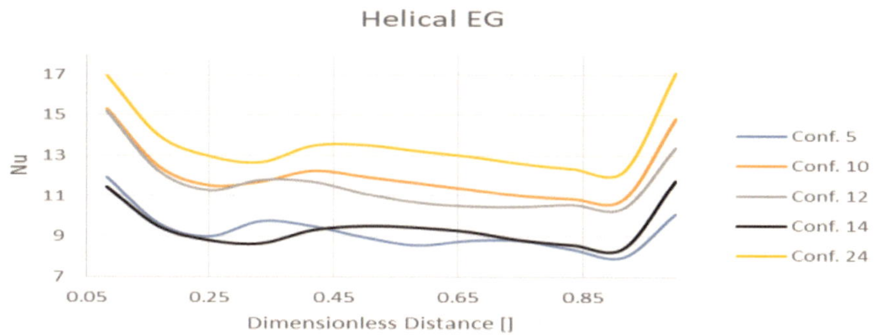

Figure 11.
Local Nusselt number for ethylene glycol side. Helical heat exchanger.

points are within the experimental space. Other methods like CCD have experimental points outside the experimental space which present divergence in the CFD simulations. Therefore, the Box-Behnken method was used in the computer simulations. The results are:

a. Regression equations that relate inputs of each heat exchanger with the listed response variables

b. Optimization plots which give the best parameter combination that maximizes some selected parameters and minimizes the rest of the parameters

The equations obtained are

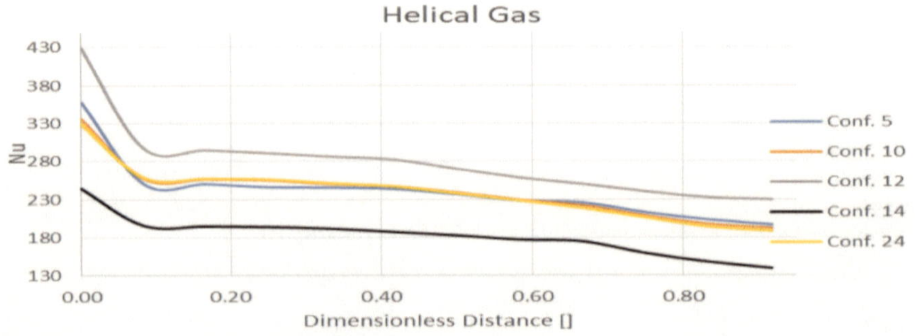

Figure 12.
Local Nusselt number for gas side. Helical heat exchanger.

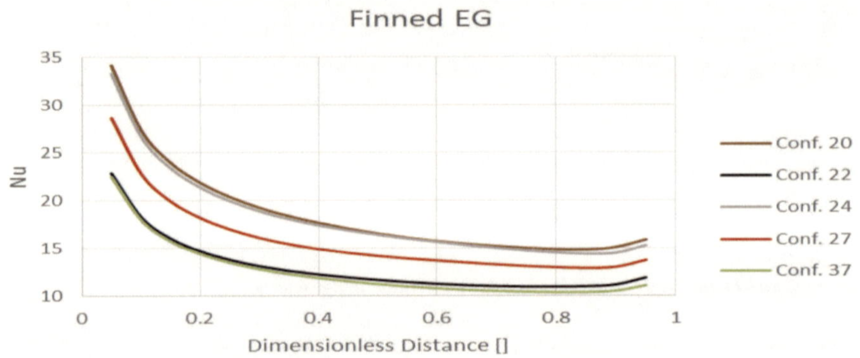

Figure 13.
Local Nusselt number for ethylene glycol side. Finned heat exchanger.

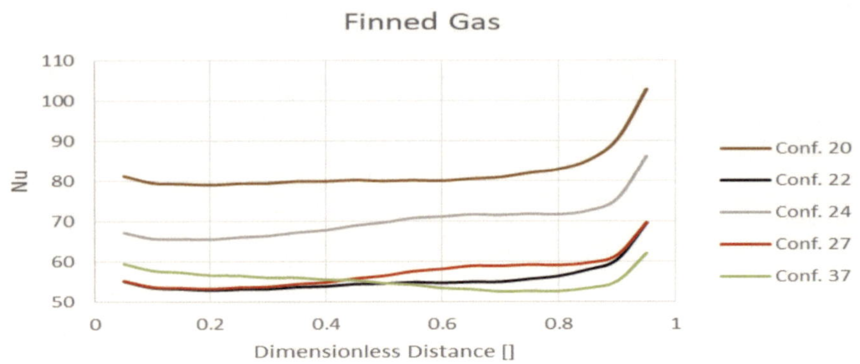

Figure 14.
Local Nusselt number for gas side. Finned heat exchanger.

$$QG'' = -19483 - 63ID - 92ED + 38412MASS + 86.9TEMP + 3.354ID^2 + 2.66ED^2 - 3.20ID * ED - 0.164ID * TEMP - 658ED * MASS - 0.471ED * TEMP + 225MASS * TEMP \quad (10)$$

$$QEG'' = -16212 - 111ED + 29547MASS + 73TEMP + 2.9ID^2 + 2.59ED^2 - 3.22ID * ED - 0.131ID * TEMP - 585ED * MASS - 0.403ED * TEMP + 212.9MASS * TEMP \quad (11)$$

$$PEG = 84 + 12.31ID - 9.4ED + 186MASS + 0.0657ID^2 + 0.0955ED^2 - 0.000053TEMP^2 - 0.1834ID * ED + 4.39ID * MASS - 0.00292ID * TEMP - 3.49ED * MASS + 0.00238ED * TEMP - 0.093MASS * TEMP \quad (12)$$

$$PG = 5.52 - 0.1187ID + 0.058ED + 26.86MASS + 0.00319TEMP + 0.000386ID^2 - 0.000328ED^2 + 0.000157ID * ED - 0.000014ID * TEMP + 0.00058MASS * TEMP \quad (13)$$

$$AG = 227 + 4.84ID - 2.05ED - 647MASS - 0.688TEMP + 0.02127ID^2 + 0.0092ED^2 + 588.4MASS^2 + 0.000430TEMP^2 - 0.03658ID * ED - 4.608ID * MASS - 0.004547ID * TEMP + 3.26ED * MASS + 0.00300ED * TEMP + 0.6390MASS * TEMP \quad (14)$$

Variable	S	R-sq	R-sq (adjusted)
QG	189.525	99.91	99.8
QEG	178.049	99.91	99.8
PG	0.0264185	99.93	99.86
PEG	4.01934	92.72	84.22
AG	0.924692	99.09	98.03
AEG	3.31958	98.23	95.3

Table 4.
Standard deviation and mean square error for the case of the helical-tube geometry.

Variable	S	R-sq	R-sq (adjusted)
QG	101.885	99.89	99.81
QEG	180.39	99.92	99.85
PG	3.23035	99.43	98.74
PEG	0.110461	99.98	99.96
AG	0.242708	95.76	92.37
AEG	1.40149	99.12	98.42

Table 5.
Standard deviation and mean square error for the case of the finned-tube geometry.

$$AEG = -2184 - 5.78ID + 15.3ED + 259MASS + 4.686TEMP - 0.0198ID^2 - 0.0837ED^2 - 1329MASS^2 - 0.003373TEMP^2 + 0.0611ID*ED + 0.00095ID*TEMP - 1.37ED*MASS - 0.00121ED*TEMP + 0.573MASS*TEMP \tag{15}$$

Tables 4 and 5 show the parameters that have a significant effect as well as the standard deviation(s) and mean square error (R) for the helical and finned geometries.

From the regression parameters, it is evident that the heat flux on the gas side (QG") and the heat flux on the ethylene glycol side (QEG") for both types of heat exchangers have a high standard deviation. However, the regressed expressions for these parameters seem to adjust very well with the results of CFD simulations as shown in **Figures 15–18**. The legend RS fitting stands for response surface fitting.

The curves for the gas side and ethylene glycol side for the case of helical heat exchanger show a similar behavior since the distance separating both gas side and

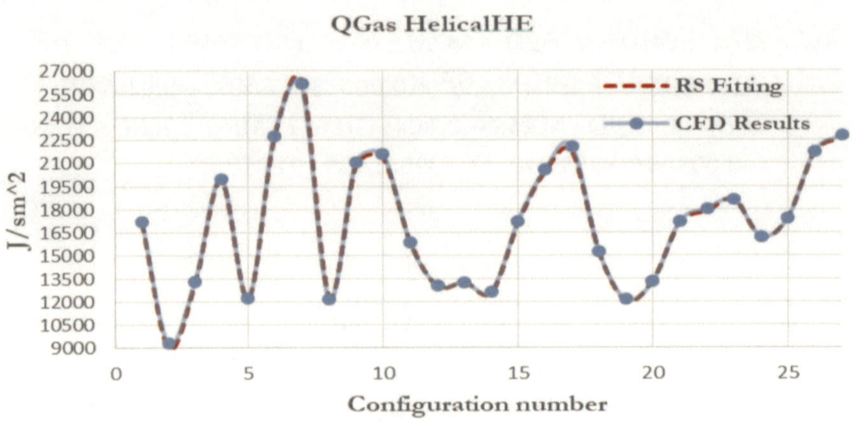

Figure 15.
Plot of QG" vs. exchanger configuration for the helical-tube geometry.

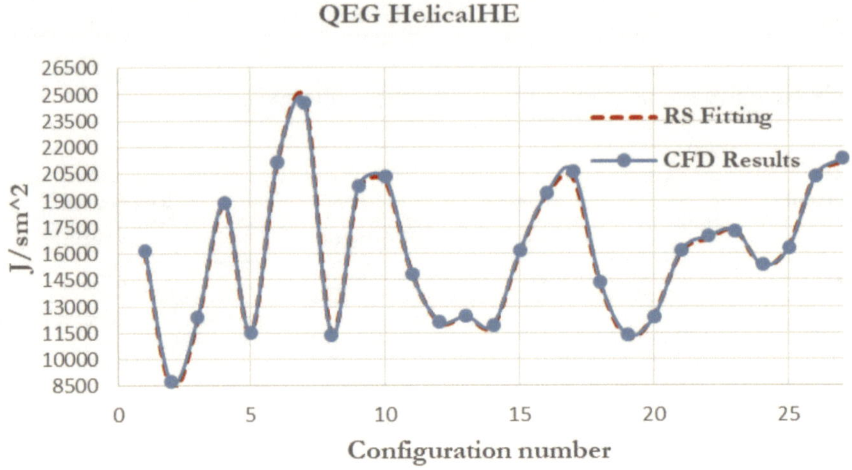

Figure 16.
Plot of QEG" vs. exchanger configuration for the helical-tube geometry.

ethylene glycol side is small. This length corresponds to the inner tube thickness. So, the heat transfer area for the gas side and the ethylene glycol is similar. However, a difference in value exists between QG" and QEG", and that difference can be observed in **Figures 15** and **16**.

In the case of the finned heat exchanger, the surface area of the gas side differs from that of ethylene glycol side. In this case the QG" and QEG" plots show a different behavior. This is shown in **Figures 17** and **18**.

In order to generate correlations for local Nusselt numbers exclusively for the bank of heat exchangers simulated, each one of the heat exchangers was divided in sections using the software Fluent [14]. These sections represent dimensionless distance from 0.05 to 0.95. In this way local Nusselt number can be obtained. The helical type of each exchanger was divided in 13 sections, and the finned type of each exchanger was divided in 19 sections. These divisions were done on each of the 41 configurations of the helical heat exchangers and 25 configurations of the finned heat exchangers. **Figure 19** shows the section on each heat exchanger.

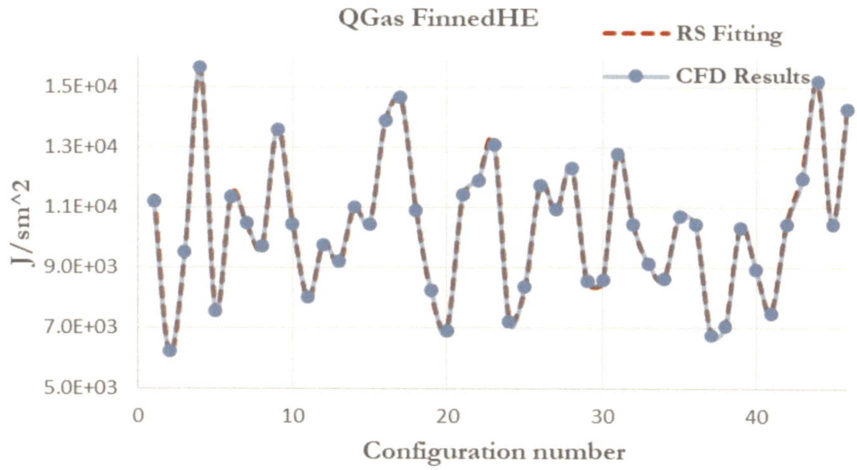

Figure 17.
Plot of QG" vs. exchanger configuration for the finned-tube geometry.

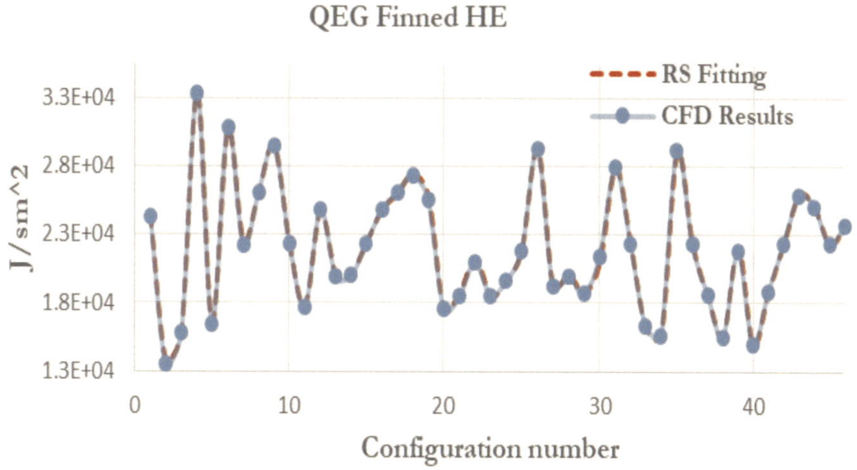

Figure 18.
Plot of QEG" vs. exchanger configuration for the finned-tube geometry.

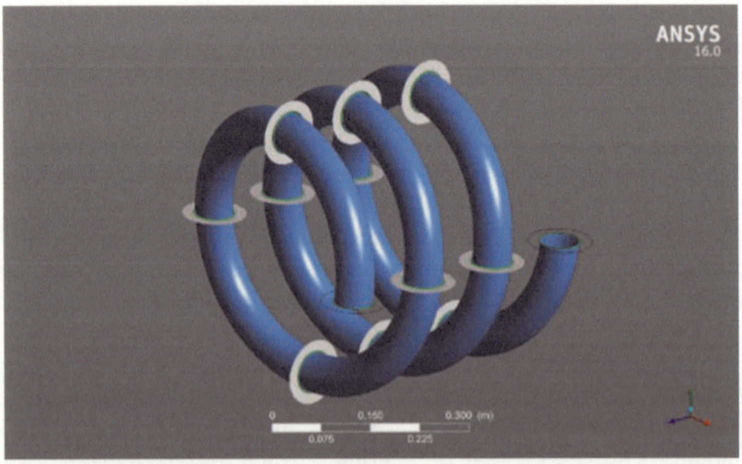

(a)

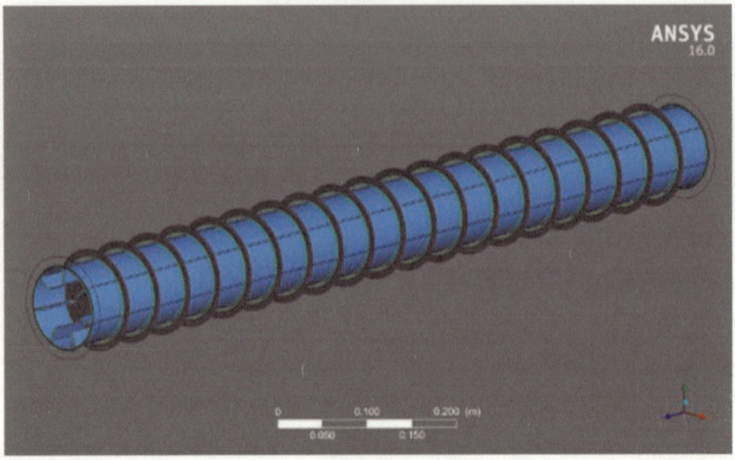

(b)

Figure 19.
Section for the determination of the local nu number. (a) Helical heat exchanger and (b) finned heat exchanger.

Next, correlations for each exchanger geometry to fit the CFD results are proposed. For the finned-tube heat exchanger, the ethylene glycol side exhibits a laminar flow regime, while the gas side exhibits a turbulent regime. The correlations for the hot side local Nu number at a dimensionless distance of 0.5 for all configurations are presented in **Figure 20**. **Figure 21** presents the local cold side Nusselt number for a dimensionless distance of 0.45 for all configurations. The correlations have the form.

$$\text{For ethylene glycol (EG): } Nu_x = mRe^n Pr^L \tag{16}$$

$$\text{For gas (G): } Nu_x = mRe^n [(A_f/A_t)Pr]^L \tag{17}$$

In the case of the helical-tube heat exchanger, a similar correlation for the ethylene glycol was proposed. On the gas side, a factor was proposed by dividing internal diameter of the gas side over the difference of the outside diameter of the annular side minus the internal diameter of the gas side. In the same way, a parameter (δ) was used. This parameter is the ratio between the helix diameter and the internal diameter of the gas side. The correlations for Nu number are presented in **Figure 22** for the hot side and **Figure 23** for the cold side for dimensionless distance of 0.7 and 0.1, respectively. The correlations have the form.

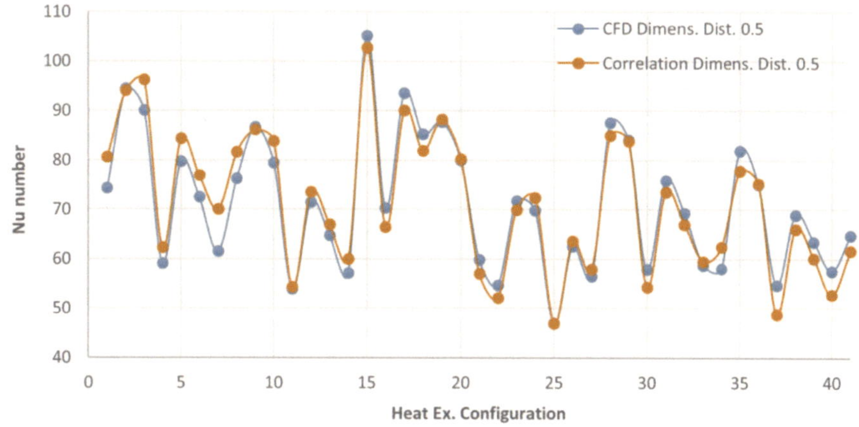

Figure 20.
Correlation for gas side (G) in the finned-tube geometry.

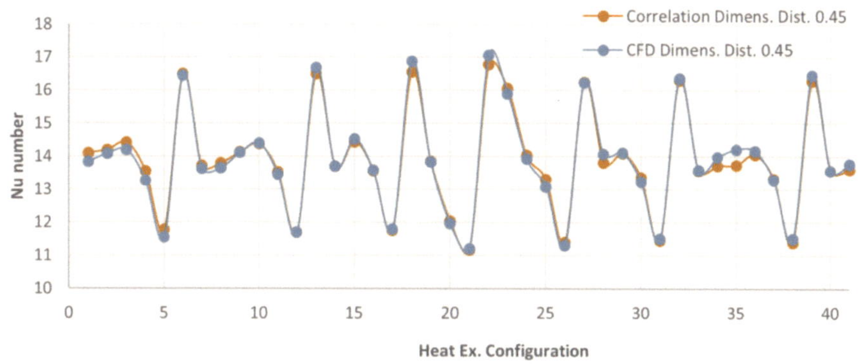

Figure 21.
Correlation for cold side (EG) in the finned-tube geometry.

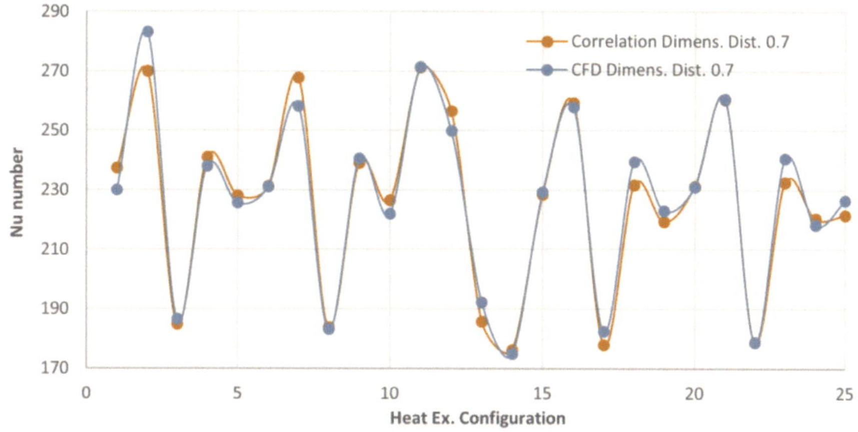

Figure 22.
Correlation for the hot side (G) of the helical-tube geometry.

$$Nu_x = mRe^n Pr^L \tag{18}$$

$$Nu_x = mRe^n[(D_i/(D_o - D_i))(1/(\delta)]^L \tag{19}$$

4.3 Neural network

An artificial neural network approach is proposed to fit the response variables of the DOE; these are the inputs given by the DOE (experiments) and CFD simulations. The object is to train the neural network using the input and the corresponding output data derived from the experimental measurements. This process is known as single training cycle or iteration. The cycle is repeated sequentially using a back-propagation algorithm so that training proceeds iteratively until the mean square error between the predicted outputs and corresponding measured values is reduced to an acceptable level. So, the results were introduced in the neural network, and the outputs of the network match very well with some results obtained from CFD. It is observed that the neural network can do a good fitting for the Nu number and heat flux for both types of heat exchangers. **Figures 24–27**

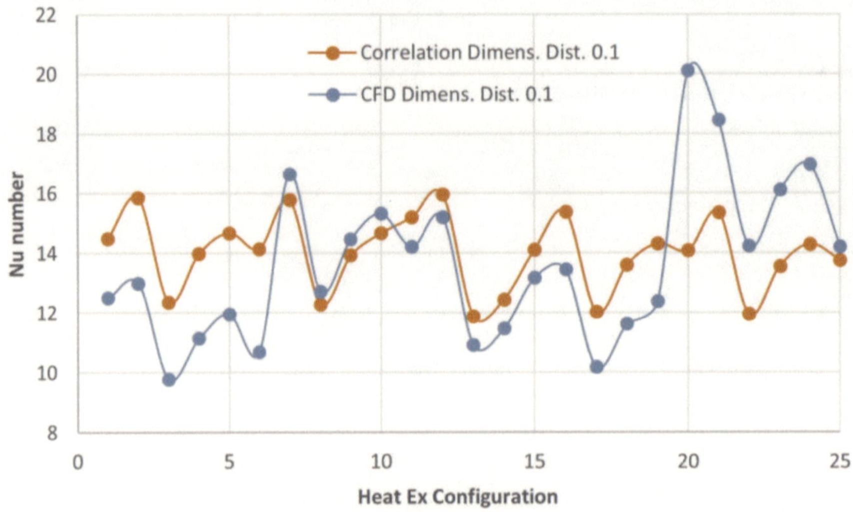

Figure 23.
Correlation for the cold side (EG) of the helical-tube geometry.

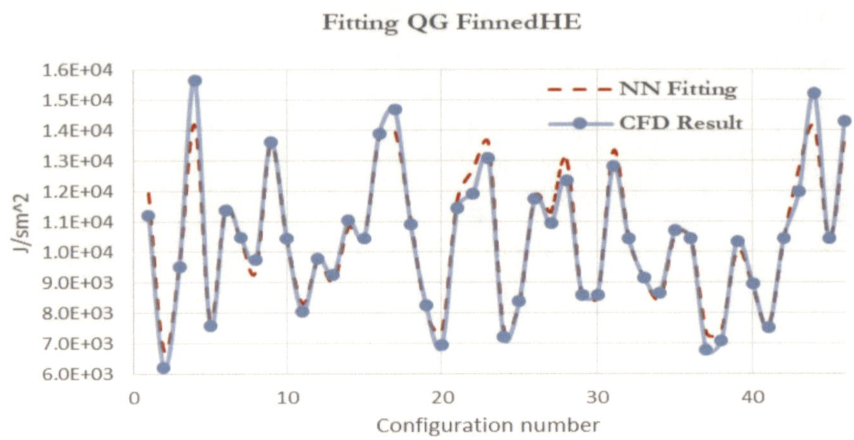

Figure 24.
Neural network fitting for heat flux on the hot side (gas).

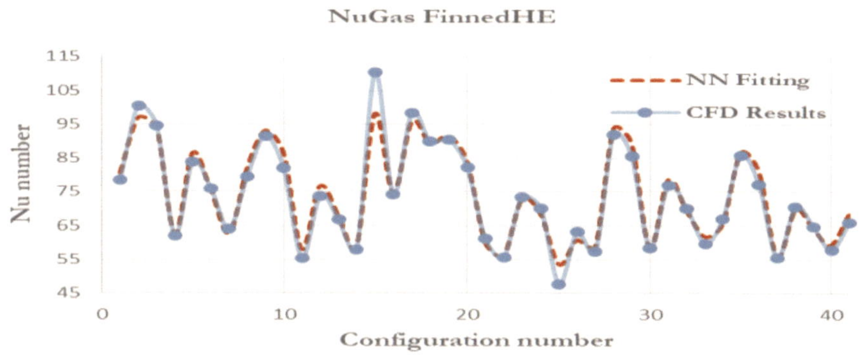

Figure 25.
Neural network fitting for the Nu number on the hot side of the finned tube exchanger.

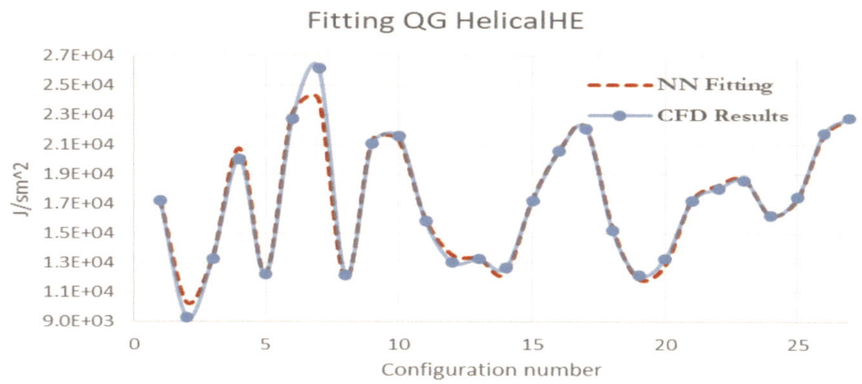

Figure 26.
Neural network fitting for the heat flux on the hot side (gas).

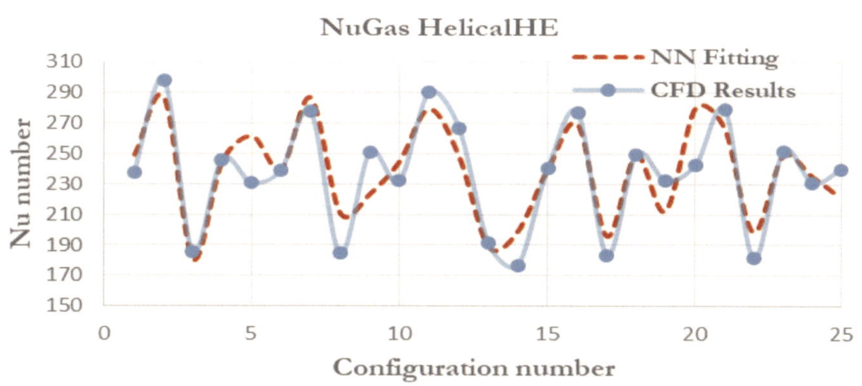

Figure 27.
Neural network fitting for the Nu number on the hot side of the helical exchanger.

show the results of the fitting for heat flux and Nu number. NN fitting stands for neural network fitting.

4.4 Exergy analysis

Exergy analysis allows to identify the components of the power cycle, whose parameters have greater influence on the maximum power generation of the

Organic Rankine Cycle. An optimization is done maximizing heat transfer and minimizing pressure drop and overheated and overcooled areas for each of the configurations simulated in CFD. **Tables 6** and **7** show the exergy analysis for each heat exchanger, where the last column shows the optimization results. From the results, it is seen that the helical heat exchanger is more efficient than the finned heat exchanger. The overheated area of the helical heat exchanger has similar values than the finned heat exchanger. The overcooled area of the helical heat exchanger is higher than the finned heat exchanger. Pressure drop for the ethylene glycol side for both types of heat exchanger has similar values. The same variable for the gas side is higher in the helical exchanger than in the finned exchanger. Finally, the heat transfer is higher in the helical heat exchanger than in the finned heat exchanger. The heat exchanger has similar dimensions, so it could be a good idea to use a helical heat exchanger to extract as much heat as possible. The pressure drop could be an important factor to consider as well. In that case, the finned heat exchanger could be considered.

	Conf 5	Conf 10	Conf 12	Conf 14	Conf 24	Helical O
Temp in gas [K]	700.0	700.0	700.0	700.0	700.0	700.0
Temp out gas [K]	582.4	559.2	591.0	514.3	545.8	560.6
Temp in EG [K]	300.0	300.0	300.0	300.0	300.0	300.0
Temp out EG [K]	389.9	406.2	391.7	409.9	415.6	404.0
ΔP gas [Pa]	3406.0	1806.4	3308.7	664.0	1062.7	882.9
ΔP EG [Pa]	5.7	24.0	17.5	9.3	45.2	25.2
Q gas [J/s]	−17011.3	−20296.6	−23368.1	−13803.7	−22188.2	−17717.6
Q EG [J/s]	16870.0	20086.6	22958.2	13880.6	21954.8	18076.9
Overheated area [%]	0.629	2.27	0.892	4.59	4.659	11.725
Overcooled area[%]	92.407	80.023	89.209	69.046	70.194	4.164
Second law effect [%]	22.7	27.2	22.7	29.9	29.7	9.9

Table 6.
Exergy balance for the helical-tube geometry.

	Conf. 20	Conf. 22	Conf. 24	Conf. 27	Conf. 37	Conf. 39	Finned O
Temp in gas [K]	700.0	700.0	625.0	700.0	625.0	700.0	697.0
Temp out gas [K]	605.9	575.8	542.7	584.2	520.5	590.4	617.8
Temp in EG [K]	300.0	300.0	300.0	300.0	300.0	300.0	300.0
Temp out EG [K]	319.6	337.0	317.1	330.2	331.1	328.5	320.5
ΔP gas [Pa]	318.6	107.8	443.7	324.1	131.0	271.7	109.1
ΔP EG [Pa]	56.7	18.4	59.3	31.6	19.3	31.9	42.1
Q gas [J/s]	−10111.7	−6647.4	−8680.5	−9303.4	−5498.7	−8815.0	−7662.0
Q EG [J/s]	9506.9	6321.3	8306.3	8811.6	5306.4	8336.1	7199.5
Overheated area [%]	0.52	0.764	1.277	0.322	0.961	0.481	1.066
Overcooled area [%]	26.4	27.601	9.623	44.097	6.839	37.083	9.53
Second law effect [%]	5.4	10.3	5.4	8.4	9.9	7.9	2.1

Table 7.
Exergy balance for the finned-tube geometry.

	Optimized	24	12
Mass [kg]	0.069	0.075	0.1
Temp [K]	404	415.6	391.7
Temp[°C]	130.85	142.45	118.55
W_{in} [kJ/s]	1.20E-02	1.44E-02	1.53E-02
W_{out} [kJ/s]	1.137	1.364	1.447
Eff %	6.28	6.21	6.3

Table 8.
Power output using the helical-tube heat exchanger.

	Optimized	22	20
Mass [kg]	0.145	0.07	0.2
Temp [K]	321	337	319
Temp[°C]	47	59	45.85
W_{in} [kJ/s	6.34E-03	4.15E-03	8.08E-03
W_{out} [kJ/s]	0.5994	0.3927	0.7647
Eff %	8.33	6.21	8.04

Table 9.
Power output using the finned-tube heat exchanger.

4.5 Power production

The commercial software Aspen HYSYS [11] is used to simulate an ORC thermodynamic cycle to determine the power obtained considering the operating conditions of the cycle. To determine the convenience of recovering heat from the combustion gases, it is essential to determine how much heat can be recovered. The output of the simulation model provides the maximum power obtained from the ORC. Butane is used as the working fluid and ethylene glycol as the heating fluid. Two of the best configurations and the optimized case were taken from each of the heat exchanger geometries. Configurations 12 and 24 were used for the case of the helical-tube heat exchanger and configurations 20 and 22 for the case of the finned-tube heat exchanger.

The results indicate that more power can be produced if the helical heat exchanger is used for the exhaust gas heat recovery. **Tables 8** and **9** show the results of the simulation of the Organic Rankine Cycle.

5. Conclusions

This work has introduced a methodology compounded of various techniques of analysis to solve a complex problem: to maximize the power production obtained through the operation of an Organic Rankine Cycle using the heat recovered from the exhaust gases of a diesel engine. Further complexity was imposed since the heat exchangers were required to fit in a fixed length dimension. Operating variables that need to be carefully maintained are heat exchanger pressure drop on the hot side to avoid operating problems in the engine; overheated and overcooled areas in the heat exchanger to avoid either evaporation of the cold fluid or condensation on the gas side. The approach followed to achieve the objective was composed of a set of tools such as design of experiments, computational fluid dynamics, artificial neural networks, exergy analysis, and process simulation. All these tools were required at some point of the design methodology. Although the overall approach seems to be rather complex and elaborated, it guided the results to the established objective.Two different types of heat exchanger technology were analyzed, resulting that for the objective of the design, the helical-tube heat exchanger, apart from fulfilling all the restrictions cited above, it also provides the larger power generation.

In terms of the results, additional conclusions can be drawn:

1. The optimized configuration for both types of heat exchangers does not exhibit the highest second law efficiency. This is so since several variables were considered in the optimization process, not only the heat transfer. The heat flux is maximized, but at the same time, the pressure drop and the overheated and overcooled areas are minimized.

2. A good prediction was obtained in the case of QG" and QEG" for the finned-tubed heat exchanger. This result is very important because the neural network could predict the complex behavior of the DOE and CFD results. This is so for three variables: Nu Gas, QG", and QEG" for the finned-tube geometry.

3. The power output obtained is in the order of 0.39 to 1.446 kW. So, this energy could be used to run several devices. An economic study is needed to determine the heat recovery rate at which the operation of a power generating engine becomes affordable.

Author details

Armando Gallegos-Muñoz[1*], Fabián Luna-Cabrera[1], Martín Picón-Núñez[2], Francisco Elizalde-Blancas[1] and Juan Manuel Belman-Flores[1]

1 Department of Mechanical Engineering, University of Guanajuato, Salamanca, Gto, México

2 Department of Chemical Engineering, University of Guanajuato, Salamanca, Gto, México

*Address all correspondence to: gallegos@ugto.mx

References

[1] Hatami M, Ganji D, Gorji-Bandpy M. A review of different heat exchangers designs for increasing the diesel exhaust waste heat recovery. Renewable and Sustainable Energy Reviews. 2014;**37**: 168-181. DOI: 10.1016/j.rser.2014.05.00450

[2] Hatami M, Jafaryar M, Ganji D, Gorji-Bandpy M. Optimization of finned-tube heat exchangers for diesel exhaust waste heat recovery using CFD and CCD techniques. International Communications in Heat and Mass Transfer. 2014;**57**:254-263. DOI: 10.1016/j.icheatmasstransfer.2014.08.015

[3] Bari S, Hossain S. Design and optimization of compact heat exchangers to be retrofitted into a vehicle for heat recovery from a diesel engine. Procedia Engineering. 2015;**105**: 472-479. DOI: 10.1016/j.proeng.2015.05.077

[4] Bari S, Hossain SN. Waste heat recovery from a diesel engine using shell and tube heat exchanger. Applied Thermal Engineering. 2013;**61**:355-363. DOI: 10.1016/j.applthermaleng.2013.08.02

[5] Tan C, Ward J, Wilcox S, Payne R. Artificial neural network modelling of the thermal performance of a compact heat exchanger. Applied Thermal Engineering. 2009;**29**:3609-3617. DOI: 10.1016/j.applthermaleng.2009.06.017

[6] Shivakumar KM, Srinivasa Pai P, Shrinivasa Rao BR. Application of neural networks for the prediction of heat transfer parameters in a multi pass cross flow heat exchanger. In: Proceedings of the 3rd World Conference on Applied Sciences, Engineering & Technology. Kathmandu, Nepal; 27–29 September 2014

[7] Hatami M, Ganji D, Gorji-Bandpy M. Experimental and numerical analysis of the optimized finned-tube heat exchanger for OM314 diesel exhaust exergy recovery. Energy Conversion and Management. 2015;**97**:26-41. DOI: 10.1016/j.enconman.2015.03.032

[8] Aly WI. Computational fluid dynamics and optimization of flow and heat transfer in coiled tube-in-tube heat exchangers under turbulent flow conditions. Journal of Thermal Science and Engineering Applications. 2014; **6**(3):031001. DOI: 10.1115/1.4026120

[9] Hossain SN, Bari S. Waste heat recovery from the exhaust of a diesel generator using Rankine cycle. Energy Conversion and Management. 2013;**75**: 141-151. DOI: 10.1016/j.enconman.2013.06.009

[10] Fabián LC. Methodology to design heat exchangers with limited space using engine exhaust gases to generate power [Thesis]. Salamanca, Gto: Guanajuato University; 2017

[11] Aspen Hysys. V10. USA: Aspen Technology, Inc. Available from: https://www.aspentech.com/en/whats-new-in-v10

[12] Minitab 17 Statistical Software [Computer Software]. State College, PA: Minitab, Inc; 2010. Available from: www.minitab.com

[13] Mathews PG. Design of Experiments with MINITAB. New Dehli: New Age; 2010

[14] Ansys-Fluent, Release 16.0. Available from: www.ansys.com/.../ansys-fluent-benchmarks-release-16

[15] Salim MS, Cheah SC. Wall y+ strategy for dealing with wall-bounded turbulent flows. In: Proceedings of the

International Multi Conference of Engineers and Computer Scientists (IMECS 2009). Hong Kong; 18-20 March 2009

[16] Ansys Fluent. Choosing the Pressure-Velocity Coupling Method. Retrieve from: https://www.sharcnet.ca/Software/Fluent6/html/ug/node1021.htm

[17] Cyklis P, Młynarczyk P. The influence of the spatial discretization methods on the nozzle impulse flow simulation results. Procedia Engineering. 2016;**157**:396-403. DOI: 10.1016/j.proeng.2016.08.382

[18] Ansys Fluent. Spatial Discretization. Retrieved from: http://www.afs.enea.it/project/neptunius/docs/fluent/html/th/node366.htm

Section 4

Heat and Mass Transfer in New and Emerging Technologies

Chapter 10

Heat and Mass Transfer of Additive Manufacturing Processes for Metals

Zhengying Wei and Jun Du

Abstract

Additive manufacturing (AM), a method in which a part is fabricated layer by layer from a digital design package, provides the potential to produce complex components at reduced cost and time. Many techniques (using many different names) have been developed to accomplish this via melting or solid-state joining. However, to date, only a handful can be used to produce metallic parts that fulfill the requirements of industrial applications. The thermal physics and weld pool behaviors in metal AM process have decisive influence on the deposition quality, the microstructure and service performance of the depositions. Accurate analysis and calculation of thermal processes and weld pool behaviors are of great significance to the metallurgy analysis, stress and deformation analysis, process control and process optimization etc. Numerical modeling is also a necessary way to turn welding from qualitative description and experience-based art into quantitative analysis- and science-based engineering branch. In this chapter, two techniques for producing metal parts are explored, with a focus on the thermal science of metal AM: fluid flow and heat transfer. Selective laser melting (SLM) is the one that is most widely used because it typically has the best resolution. Another is named metal fused-coated additive manufacturing (MFCAM) that is cost competitive and efficient in producing large and middle-complex components in aerospace applications.

Keywords: additive manufacturing, selective laser melting (SLM), metal fused-coated additive manufacturing (MFCAM), fluid flow, heat transfer

1. Introduction

As a kind of advanced manufacturing technology, additive manufacturing (AM) provides an effective and 'bottom up' manufacturing where a complex structure can be built into its designed shape by a 'layer-by-layer' approach, which can directly create geometric metal parts. AM is versatile, flexible, highly customizable and, as such, can suite most sectors of industrial production [1]. Even though metal additive manufacturing involves creating parts layer-by-layer, there are many different types, including material extrusion, material jetting, material droplet printing, binder jetting, sheet lamination, powder bed fusion, and directed energy deposition [2]. Most current metal AM systems are of the powder bed fusion type [3]. Due to the complexity of the physical process in the process of metal AM, it is very difficult to ascertain metal parts with high dimensional accuracy, no defects, small residual stress/deformation, compact microstructure and high mechanical properties [4–8]. At present, defect control and microstructure/composition control are key bottleneck problems that restrict the further development of metal AM technology. Both of these problems are closely related to the energy and mass transport process in deposition process, especially at the solid-liquid interface of molten pool [9–12]. How to understand and control the complex heat and mass transport in the molten pool is the key trend of the current research, and it is also the basis and prerequisite to break through the current technical bottlenecks and further improve the mechanical properties of the parts, such as strength, stiffness and fatigue. Therefore, this paper focuses on the scientific issues of transport phenomena and solidification behavior of molten pool during metal additive manufacturing.

2. Effect of processing parameters on forming defects during selective laser melting of AlSi10Mg powder

Selective laser melting is one of the most promising additive manufacturing processes. The randomly distributed packed powder particle is obtained using discrete element method (DEM) in PFC software. The correlation between the processing parameters (i.e., laser power, scanning speed, hatch spacing, and layer thickness) and the pores formation for the SLM-processed AlSi10Mg components was disclosed by simulations.

2.1 Numerical model

SLM process is complicated, involving heat transfer, evaporation, melting and solidification, re-melting and re-solidification, shrinkage and other thermophysical behaviors. In the numerical calculation model, in order to simplify the complicated physical process, the following assumptions need to be made: (1) The molten pool liquid is assumed as laminar and incompressible Newtonian fluid. (2) Mushy zone is treated as an isotropic permeability of porous medium in solid-liquid phase change. (3) Powder size is Gaussian distribution with sphere shape.

2.1.1 Establishment of randomly packed powder bed

A DEM-based randomly packed powder model was established by commercial platform PFC [13, 14]. In DEM, the contact between powder particles is regarded as a linear model while the mini deformations are allowable between the particles [15]. The linear force induced by the mini deformation could be regarded as the force exerted by a linear spring. The spring has a constant normal and shear stiffness, K_n and K_s [16]. In **Figure 1**, F_i, the contact force vector, is decomposed into two subvectors as F_n and F_s in the normal and shear direction, respectively. The stacking of the powder particles is accomplished by setting gravitational force until the powder particles reaches equilibrium. The powder bed porosity is set as 0.45, which is close to 0.5 which is the theoretical packing density. More details about the establishment of powder bed can be found in Refs. [13, 14].

As shown in **Figure 2**, two dense packing of Gaussian-sized spherical particles with dimensions of 500 × 200 × 35 and 600 × 300 × 35 μm were obtained by PFC with the parameters: the layer thickness of particles of 35 μm, the height of substrate of 30 μm. Then the powder bed model was converted to STL format for subsequent numerical simulation.

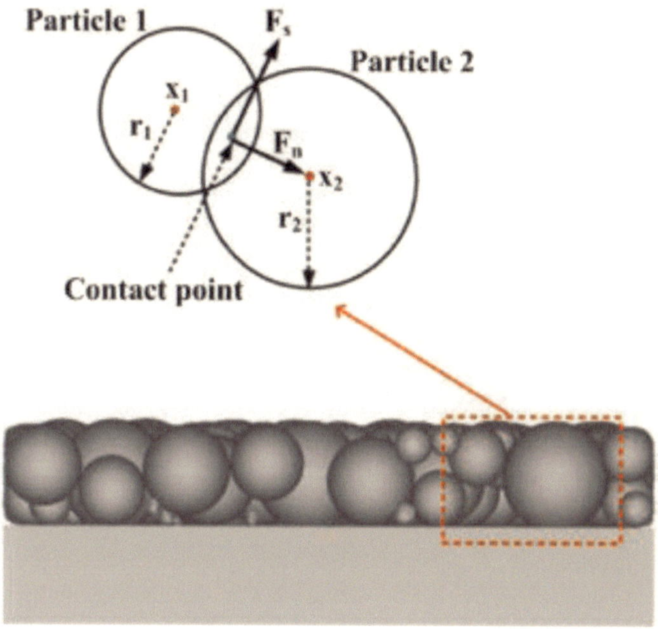

Figure 1.
Contact forces generated by interaction between two spherical particles x_1/x_2 and r_1/r_2 are the center and radius of particle 1 and particle 2, respectively.

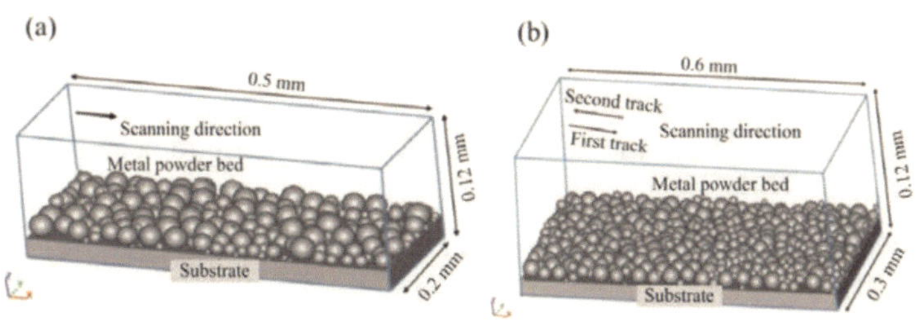

Figure 2.
Model of randomly packed AlSi10Mg powder bed. (a) Single track simulation, (b) multi-tracks simulation.

2.1.2 Conservation equations

The melt fluid flow is governed by N-S equations. The fluid free surface is captured employing the volume of fluid (VOF) method. The melt flow could be solved by the conservations of mass, momentum and energy, given by Eqs. (1)–(3), respectively.

Mass

$$\nabla \cdot \vec{V} = 0 \tag{1}$$

Momentum

$$\frac{\partial \vec{V}}{\partial t} + \left(\vec{V} \cdot \nabla\right) \vec{V} = -\frac{1}{\rho} \nabla P + \mu \nabla^2 \vec{V} + \vec{g}\, [1 - \beta(T - T_m)] \tag{2}$$

Energy

$$\frac{\partial H}{\partial t} + \left(\vec{V} \cdot \nabla\right) H = \frac{1}{\rho} (\nabla \cdot k \nabla T) + S_U \tag{3}$$

where $\vec{V}$ is velocity of the melt, ρ is the liquid metal density, P represents hydrodynamic pressure, μ is liquid viscosity, $\vec{g}$ is the gravitational acceleration, β is volumetric thermal expansion coefficient of the material, T represents the fluid temperature, T_m represents the melting temperature of AlSi10Mg.

The VOF method as shown in **Figure 3** was employed to track the free surface of the particles model obtained by DEM as they are melted, and it defined a function of the fraction of fluid by the following equation [17].

$$\frac{\partial F}{\partial t} + \nabla \cdot \left(\vec{V}F\right) = 0 \tag{4}$$

where F is the volume fraction of the liquid in a cell. When the cell is filled with liquid, $F = 1$; when the cell is void, $F = 0$. The value of F is between 0 and 1 when both the void and liquid are in the cell.

In this work, the laser energy of the AlSi10Mg powder bed was defined as 0.18 [18].

2.1.3 Boundary conditions

The heat-flux boundary condition at fluid free surface was given by [19].

$$\kappa \frac{\partial T}{\partial z} = q(r) - h_c(T - T_0) - \varepsilon_r \sigma_s \left(T^4 - T_0^4\right) - q_{ev} \tag{5}$$

where h_c is the heat transfer coefficient, ε_r is the emissivity, σ_s is Stefan-Boltzmann constant, T_0 is ambient temperature, and q_{ev} is the heat loss by melt evaporation.

In SLM process, the liquid metal evaporation is given by the equation [20]

$$q_{ev} = 0.82 \frac{\Delta H^*}{\sqrt{2\pi MRT}} P_0 \exp\left(\Delta H^* \cdot \frac{T - T_{lv}}{RTT_{lv}}\right) \tag{6}$$

where M represents the molar mass, R is the ideal gas constant, P_0 represents the ambient pressure, T_{lv} is the boiling point of the metal melt, and ΔH^* is the effective enthalpy of loss metal vapor.

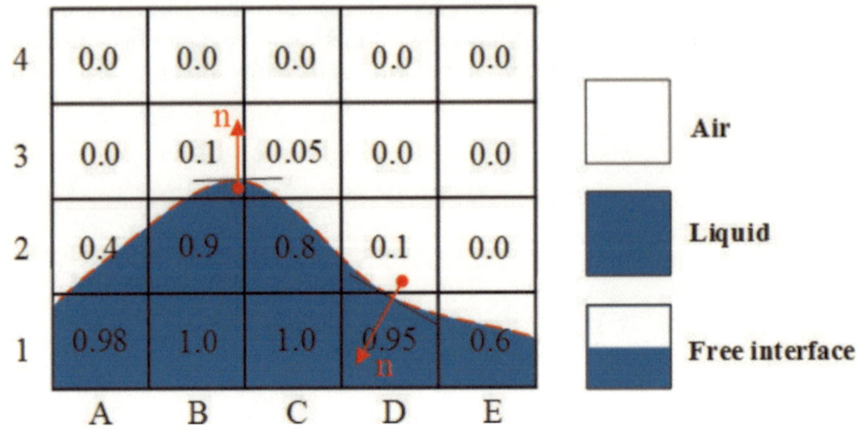

Figure 3.
Schematic diagram of VOF.

In order to simulate the Marangoni effect induced by the temperature gradient of the molten pool fluid, the shear stress should be balanced with boundary condition at fluid free surface, as given by [21]:

$$\begin{aligned} -\mu\frac{\partial u}{\partial z} &= \frac{\partial \gamma}{\partial T}\frac{\partial T}{\partial x} \\ -\mu\frac{\partial v}{\partial z} &= \frac{\partial \gamma}{\partial T}\frac{\partial T}{\partial y} \end{aligned} \tag{7}$$

where $\partial\gamma/\partial T$ is the surface tension gradient. The surface pressure boundary condition including the surface normal force was given by [22]:

$$-P + 2\mu\frac{\partial \vec{v}_n}{\partial n} = -P_r + \sigma\left(\frac{1}{R_x} + \frac{1}{R_y}\right) \tag{8}$$

where $\vec{v}_n$ represents the normal velocity vector, P_r is the recoil pressure, σ means the surface tension, R_x and R_y represent the principal radius of surface curvature.

$$P_r = 0.54P_0 \exp\left(L_{lv} \cdot \frac{T - T_{lv}}{RTT_{lv}}\right) \tag{9}$$

2.2 Material physical properties and numerical simulation

The AlSi10Mg powder (Felcon, China) used in SLM was produced by gas atomization. The chemical composition of the AlSi10Mg alloy is shown in **Table 1**. Drying the powder before laser melting by the drying ovens at temperature of 373 K can help reduce the humidity and the oxygen content within the powder. The scanning electron microscope (SEM) morphology of the AlSi10Mg powder is shown in **Figure 4a**, showing the morphology of powder particles is almost spherical. The powder particle size distribution was obtained by laser particle size analyzer (Sympatec, HELOS, Germany). In **Figure 4b**, the powder particle size is from 0 to 45 μm with the average size of 26.53 μm.

The thermophysical properties of AlSi10Mg and laser processing parameters are shown in **Figure 5** and **Table 2**. The temperature-dependent surface tension can be expressed as [23].

$$\sigma = 1000.726 - 0.152T \quad \text{When } T > T_l. \tag{10}$$

In this work, the final meshing of the model ensured the convergence of the simulation with the cell size of 2 μm. The minimum time step was defined as e^{-15} second while the maximum time step was defined as e^{-8} s. Implicit method was selected for the solvers of heat transfer, viscosity and surface tension. Explicit method was selected for free surface pressure solver. Numerical simulations were carried out on the commercial CFD platform Flow3D [13, 14].

Elements	Si	Mg	Fe	Cu	Ni	Zn	Pb	Sn	Ti	Al
wt.%	10.1	0.4	0.3	<0.05	0.03	0.05	0.03	0.03	0.01	Balance

Table 1.
Composition of AlSi10Mg (wt.%).

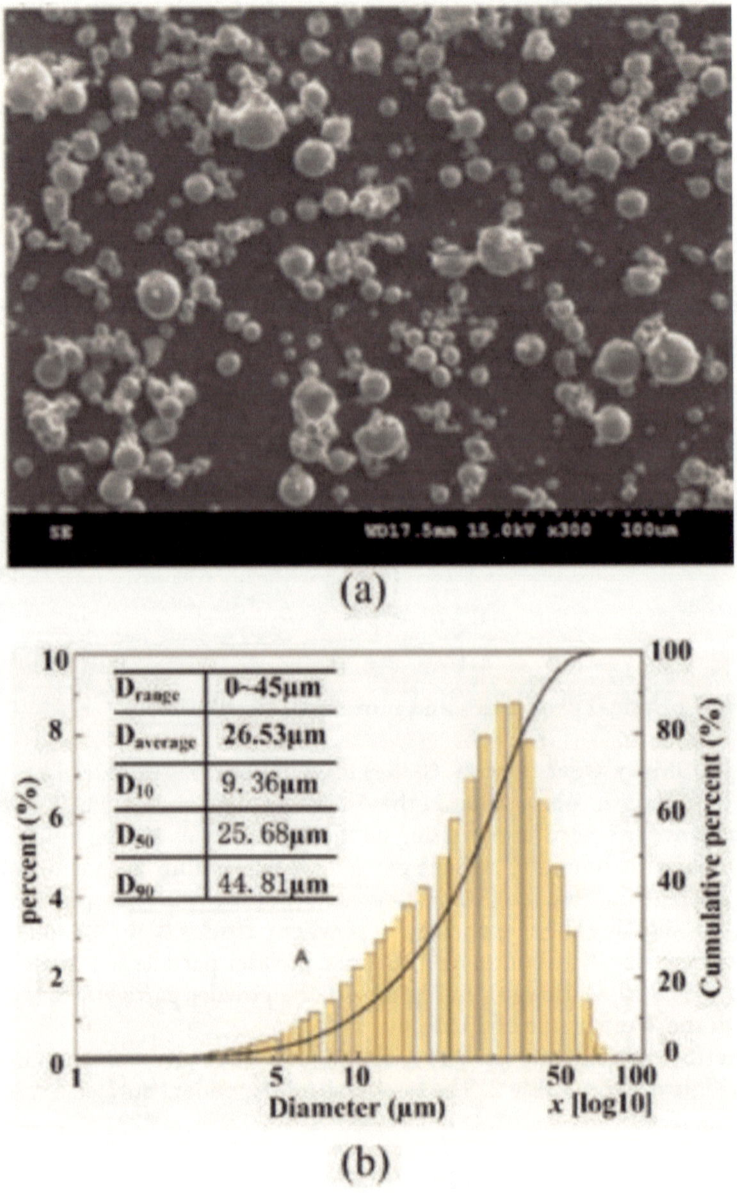

Figure 4.
(a) Microstructure of the AlSi10Mg powder, (b) the particles size distribution.

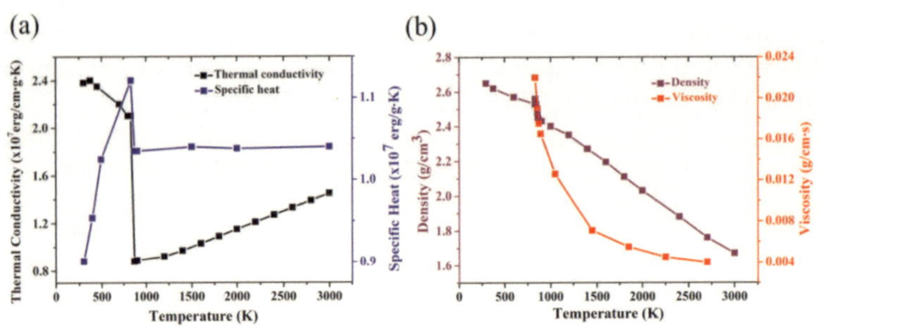

Figure 5.
Thermal material properties of AlSi10Mg: (a) thermal conductivity and specific heat, (b) density and viscosity.

Parameters	Value
Ambient temperature, T_0	293.15 K
Laser power, P	180 W
Scanning speed, v	100 cm/s
Laser beam spot size, D	70 μm
Stefan-Boltzmann constant, k_B	5.67×10^{-8} W/(m^{2}K^4)
Heat transfer coefficient, h_c	82 W/(m^{2}K)
Radiation emissivity, ε	0.4
Powder layer thickness, d	25–50 μm
Hatch spacing, H	50–70 μm
Powder size distribution	0–45 μm, Gaussian
AlSi10Mg solidus point, T_s	830.15 K
AlSi10Mg liquids point, T_l	870.15 K

Table 2.
SLM-processing conditions and material parameters used in this work.

2.3 Results and discussion

2.3.1 Model verification

Single melting track can used to validate the correction of the model and available of software. At experiment, the laser power and powder layer thickness were fixed at 180 W and 35 μm, and the laser scanning speed changed from 600 to 1600 mm. The experimental measurement criteria of the melt depth and melt width are as shown in **Figure 6a**. The melt depth is taken vertically from the free surface of the molten pool to the maximum depth of the melt boundary while the melt width is taken horizontally between the edges of the melt boundary. **Figure 6a** shows the micrograph of pool on cross section caused by Gaussian laser irradiation in the simulation and experiment for P = 180 W and v = 1000 mm/s. It is obviously that the calculated morphology of pool agree well with the experiment one. Meanwhile, **Figure 6b** shows the range and averaged experimental melt depth and width results at different laser scanning speed v with a fixed laser power P (P = 180 W,

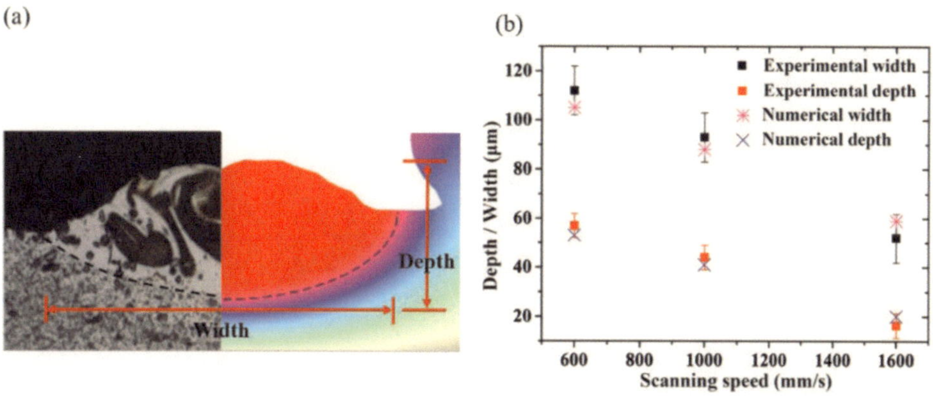

Figure 6.
(a) Morphology of molten pool at P = 180 W and v = 1000 mm/s, (b) depth and width of molten pool at different scanning speed with a fixed P = 180 W.

v = 600, 1000, 1600 mm/s) with their corresponding simulated values. Compared the numerical depths and widths of pool with experimental results, the simulated melt depths and widths all fall within their corresponding experimental range. Due to the existence of the inherent experimental error and numerical error, the errors between experiment and simulation are inevitable. So experimental results and numerical results demonstrate the proposed numerical model can provide good predictions on shape of molten pool and, can be provided to predict the effect of layer thickness and hatching spacing on the morphology of scan track.

2.3.2 Effect of laser power and scanning speed

Figure 7 shows the powder melting and solidification evolution process in SLM. First, laser beam is turned off, as shown in **Figure 7a**. When the laser is turned on, the beam energy is absorbed by the powder layer. With the heat accumulation of powder, powder layer melts. In **Figure 7b**, a depression area is made by the work of recoil force. Because the apply of laser beam in Gaussian distribution, the temperature gradient is created between the molten pool center and the molten pool edge, with the induced surface tension gradient and the Marangoni effect. Lastly, the depression area was filled with fluid by the Marangoni effect, as shown in **Figure 7c**.

Figure 8 shows the characteristic microstructure on the polished cross section of the SLM-processed AlSi10Mg samples. The cross-section of the samples produced

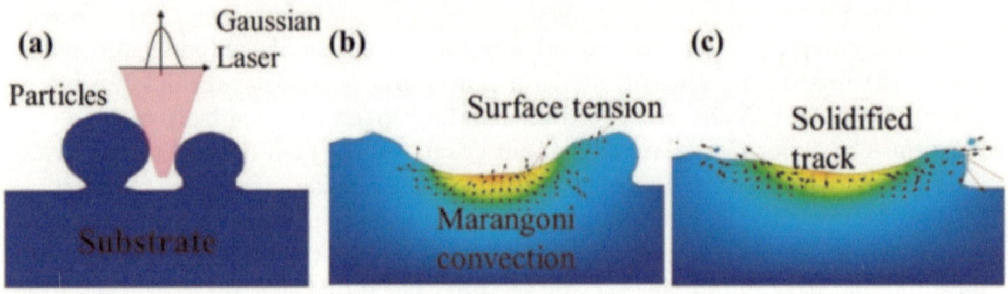

Figure 7.
Powder melting and solidification process evolution at the laser power of 180 W and scan speed of 1000 mm/s: (a) initial state (b) heating process (c) solidification.

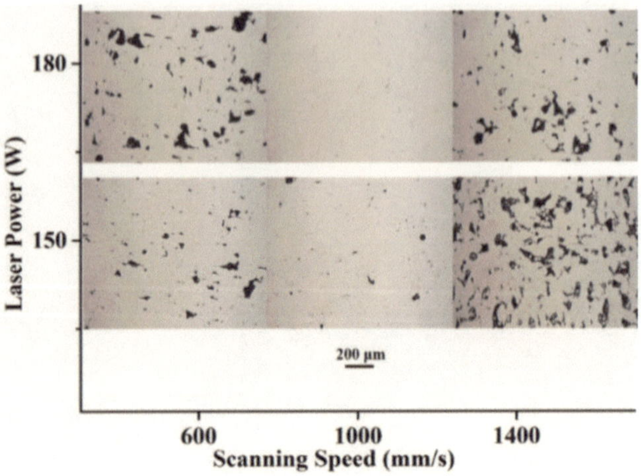

Figure 8.
Influence of laser power and scanning speed on the microstructure of SLM-processed samples.

by using the laser power of 180 W and scan speed of 1000 mm/s showed a fine microstructure without any apparent pores. At the low laser power of 150 W or scan speed of 600 mm/s, the cross-section of the as-built samples consisted of irregular-shaped pores were visible. However, at relatively high laser scan speed of 1400 mm/s, a large amount of the balling formation was present on the cross-section. It can be concluded that the balling phenomena was one of the typical metallurgical defects at high laser scan speed. In order to understand how the laser power P combined with the laser scan speed v affects the pores and balling defects, the line energy density (LED) is defined as [24]:

$$LED = \frac{P}{v} \tag{11}$$

Furthermore, the numerical studies were used to describe the forming mechanism of the metallurgical defects during laser melting and to provide a basis for the process optimization. The laser melting process of the AlSi10Mg powder at different scan times was shown in **Figure 9a**. In order to assess the effect of the LED on heating of the powder, the peak temperature and the interfacial velocity at the edge of the molten pool was rescored as the center of the laser beam moved to the point of X = 0.36 mm and Y = 0.1 mm. As shown in **Figure 9b**, the peak temperature and the interfacial velocity increased from 1810 K and 1.2 m/s to 2831 K and 5.1 m/s as the applied LED increased from 1.071 to 3 J/cm, respectively. At a relatively low LED, the insufficient laser energy result in insufficient melting of the powder. As

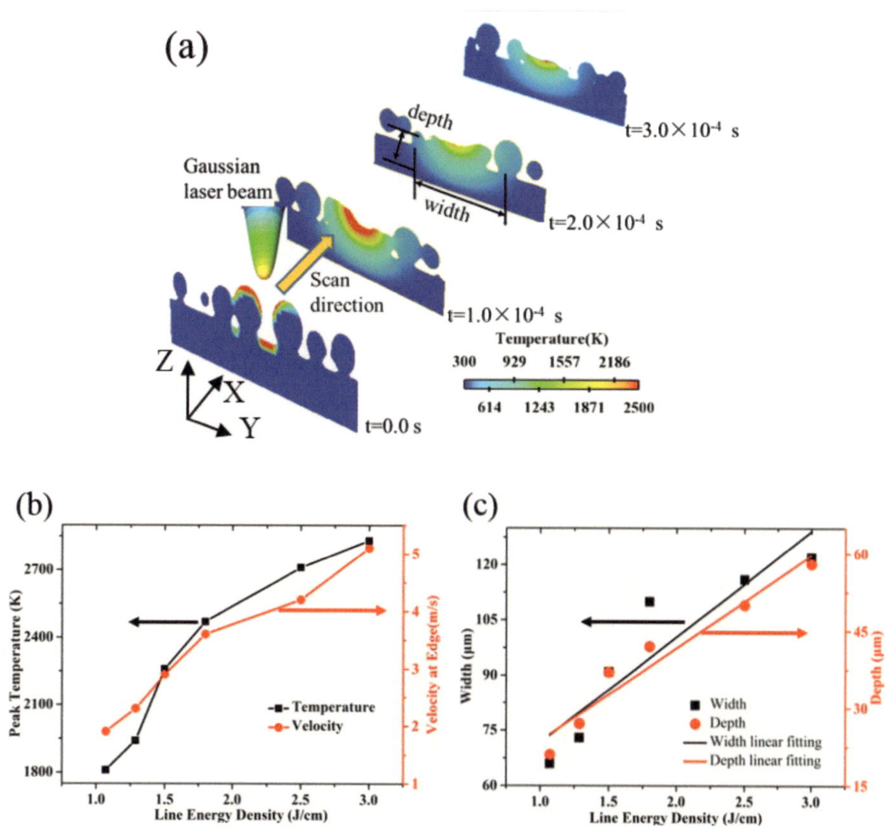

Figure 9.
(a) Temperature field and surface melt velocity obtained by simulation at condition of P = 180 W and v = 1000 mm/s. (b) Influence of line energy density on the peak temperature and surface melt velocity of molten pool. (c) Influence of line energy density on the width and depth of molten pool.

shown in **Figure 9c**, the depth of the molten pool is about 20 μm at LED of 1.071 J/cm. This implies that the low applied LED can lead to the insufficient melting of the powder and the resultant interlayer pores. In addition, the balling phenomena was severely occurred in this condition. On the contrary, at relatively high LED, the high temperature of the molten pool leads to a strong perturbation within the molten pool, thereby resulting in the instability of the free surface of the molten pool. In this case, a large amount of the pores was generated by the instability of the scan track [24].

When the laser moves to the point of X = 0.36 mm along scanning direction, the molten pool was chosen in this case.

2.3.3 Effect of hatching spacing

Figure 10 shows that at an interconnect pores will be formed in the overlap if the large interconnected gap has not been completely filled by re-melting liquid because of the high hatching spacing. An overlap in the multi-tracks is necessary to have continuity between two adjacent scan tracks leading to a dense solidification. Due to the overlap between the successive adjacent scans, the hatching spacing is always less than the laser beam radius. Hatch spacing is another parameter, which is highly affects the pore formation.

As revealed in **Figure 11**, at laser power of 180 W and scan speed of 1000 mm/s, the number of the inter-track pores was apparently reduced as the hatch spacing

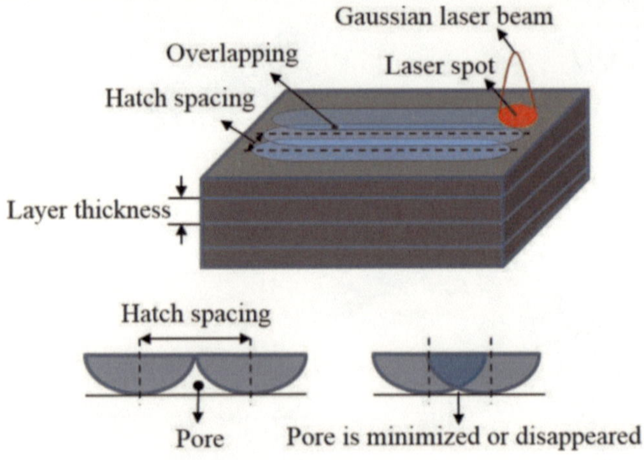

Figure 10.
The effect of hatching spacing on the formation of porosity in overlap between adjacent scan tracks.

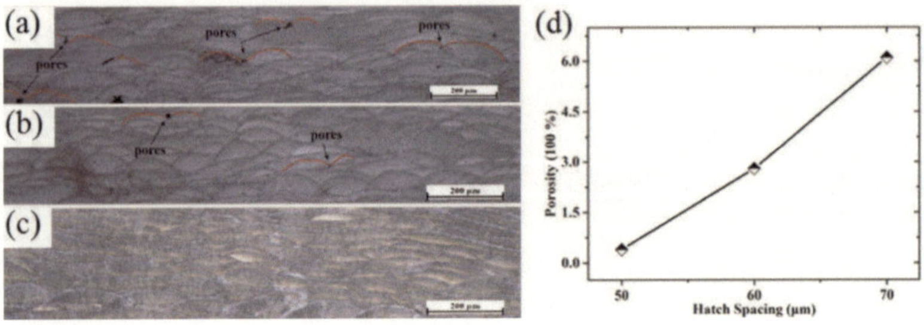

Figure 11.
Typical microstructure of the SLM-processed AlSi10Mg samples at hatch spacing of (a) 70 μm, (b) 60 μm, and (c) 50 μm. (d) Effect of the hatch spacing on the porosity of the as-built sample. The laser power of P = 180 W, scan speed of v = 1000 mm/s and layer thickness of d = 35 μm were fixed in these experiments.

decreased from 70 to 50 μm. As a result, the porosity of the SLM-processed sample was accordingly reduced from 6.1 to 0.3% (**Figure 11d**). Since at a relatively larger hatch spacing, the portion of the re-melted material was reduced, leading to insufficient overlap between two adjacent scan tracks.

As the case in **Figure 12**, the insufficient overlap will increase the possibility of the inter-track pores formation. In general, reducing the hatch spacing is beneficial for the reduction of the inter-track pores [3]. However, as the relatively low hatch spacing applied during SLM, although a fine bonding between two adjacent scan tracks can be obtained, the build efficiency was decreased under this condition. So it is important to choose a reasonable hatch spacing during SLM.

2.3.4 Effect of layer thickness

The powder layer thickness is one of the important factors in in producing the pores formation. **Figure 13** shows the topography of the second scan track formed by SLM using different powder layer thickness. It is clearly that a relatively continuous and stable scan track was generated at powder layer thickness of 35 μm. At low layer thickness of 25 μm or larger layer thickness of 50 μm, the scan track showed a severe irregular profile which will increase the possibility of the inter-layer pore

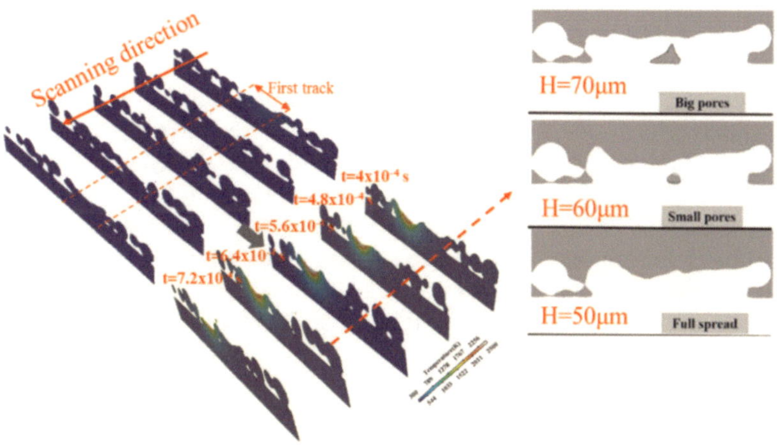

Figure 12.
Bonding between two adjacent tracks at laser power P = 180 W, scan speed v = 1000 mm/s, and hatch spacing of 50, 60, 70 μm.

Figure 13.
Influence of powder layer thickness on morphology of scan track in longitudinal view (left) and cross-sectional view (right). Laser power P = 180 W and scanning speed v = 1000 mm/s were fixed in simulations.

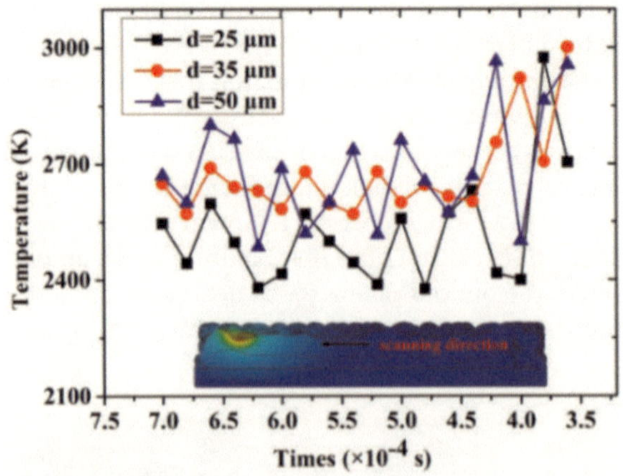

Figure 14.
Influence of powder layer thickness on the temperature of molten pool. Laser power P = 180 W and scanning speed v = 1000 mm/s were fixed in simulations.

formation. This is due to as the relatively larger powder layer thickness applied during SLM, the laser energy cannot melt previous layer, inducing inter-layer pores in the interface between two adjacent layers. However, decreasing the powder layer thickness increase the reflected radiation from the surface of the previous track, hence decreasing the peak temperature of molten pool, as shown in **Figure 14**. At powder layer thickness of 25 μm, the peak temperature of molten pool exhibits apparent fluctuation, and the following will produce the scan track with irregular surface. According to this numerical result, the thin powder layer thickness of 35 μm is recommended for AlSi10Mg in SLM process.

The pore and the balling defects lead to a decrease in the densification level of the SLM-processed samples. To evaluate the combined effect of the laser power P, scan speed v, hatch spacing H, and powder layer thickness d on the densification level of the as-built samples, a volumetric energy density (VED) is defined as:

$$VED = \frac{P}{vHd} \tag{12}$$

The relative density of the SLM-processed AlSi10Mg sample are shown in **Figure 15** as a function of the VED.

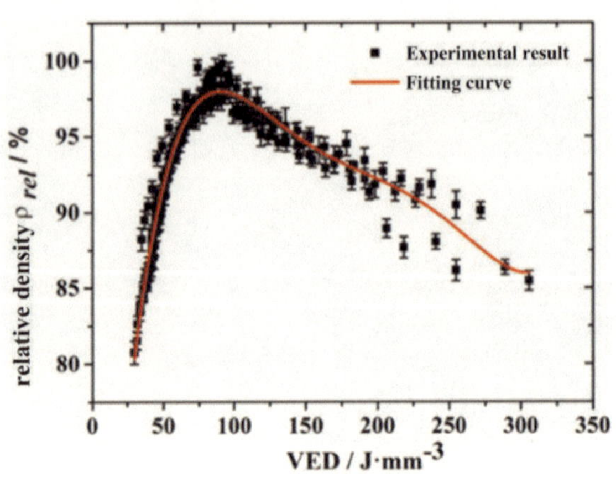

Figure 15.
Effect of the VED on the relative density of the SLM-processed samples.

As the applied VED within the range of 75–105 J/mm^3 during laser melting, the relative density of the as-built samples is larger than 97.5%. According to all the above research results, the VED of 102.86 J/mm^3 (P = 180 W, v = 1000 mm/s, H = 50 μm, d = 35 μm) was proposed during AlSi10Mg SLM process.

3. Numerical investigation of thermocapillary-induced deposited shape in MFCAM of aluminum alloy

A three-dimensional (3D) numerical model with volume of fluid method is developed for metal fused-coating additive manufacturing (MFCAM) process of aluminum (Al) alloys. It predicts the thermal flow field in the thermocapillary-induced melt, the surface deformation and solidified deposition geometry during MFCAM in successive depositing passes. Verification of the numerical model was performed by comparing the calculated results with metallography of deposited cross-sections, showing that there is a good qualitative agreement between the two, which indicates that the established numerical model is capable of simulating the complex heat and mass transfer phenomena in the varying polarity gas tungsten arc welding (VP-GTAW) based additive manufacturing. The effects of melt flow rate and the gap height between the substrate and fused-coating head on deposition geometry were studied. The results show that the deposition geometry is closely correlated to the melt flow rate. Increase in melt flow rate will lead to the obvious increase of deposition height, but the reverse is true in the gap height. These detailed physical insights facilitate the prediction of deposition defects in MFCAM of aluminum alloy.

3.1 Principle of MFCAM process

The schematic illustration of MFCAM process is shown in **Figure 16**. The experimental system is mainly composed of an induction heating, gas protection device, a pressure controller and a movable platform. The deposits of aluminum alloy can be created following the layer-by-layer approach by controlling the synchronization of the movable platform and the extrusion of liquid metal. A programmable multi-axis controller (PMAC) can be used to control the motion of

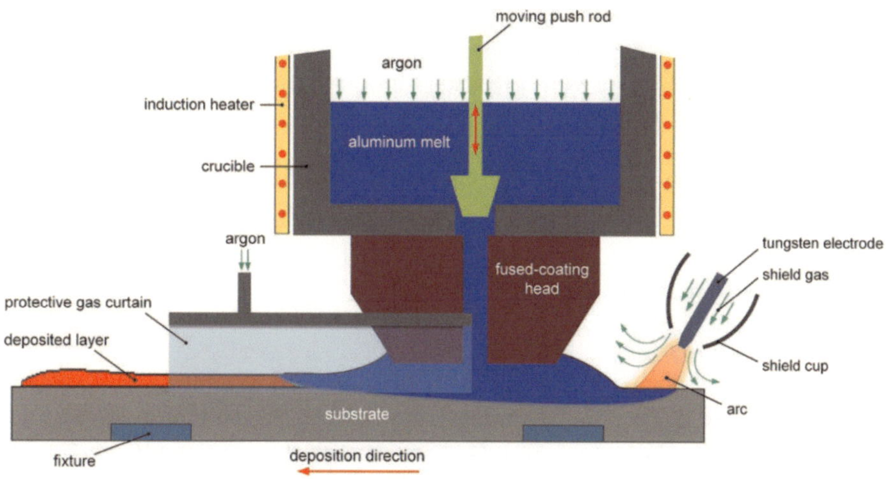

Figure 16.
Schematic diagram of MFCAM process.

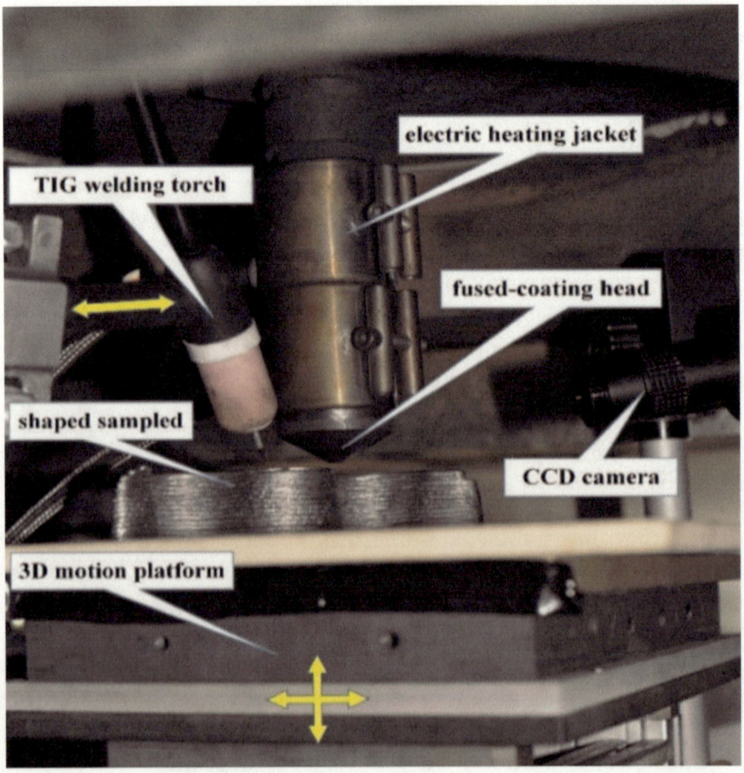

Figure 17.
Experimental setup for the MFCAM of Al alloys.

movable platform. The high-temperature liquid metal would flow through the inner flow channel of fused-coating head under the combined-action of moving push rod, surface tension and hydrostatic pressure. As the liquid metal contacts with the substrate surface or the previously deposited layers, a local thermocapillary-induced flow region will be created rapidly. To achieve a metallurgical bonding between the deposited layers, a pulsed variable polarity GTA welding arc was adopted to create a shallow molten pool in front of the thermocapillary-induced flow region. On the other hand, variable polarity GTAW arc can timely remove the aluminum oxide on the deposits. Last but not least, the local solidification conditions and thermocapillary-driven spreading motion of the melt can be tunable by arc heat input and the relative distance between welding arc and fused-coating head.

The corresponding experimental setup for the MFCAM of Al alloys is shown in **Figure 17**.

3.2 Materials and methods

3.2.1 Experimental procedure

The base material was 2024 aluminum alloy in initial T6 condition with a thickness of 6 mm. The material was cut into several pieces with $300 \times 60 \times 6$ mm^3 dimensions. The nominal chemical composition of the base metal is shown in **Table 3**. Pure argon (99%) was employed as shielding gas with flow rates of 15–18 L/min for VP-GTAW.

Cu	Si	Mn	Mg	Zn	Ti	Al
4.52	0.16	0.61	1.4	0.08	0.05	Remain

Table 3.
Chemical composition of the deposited 2024 aluminum alloy/wt%.

In VP-GTAW welding system, Fronius MagicWave 3000 welding power source was adopted. The VP-GTA welding process was used square wave AC mode. Before deposition, oil and other impurities were removed using acetone. The arc welding parameters applied in the experiments are presented in **Table 4**. Among them, the magnitude of DCEN welding current is I_{en}, the magnitude of DCEP welding current is I_{ep}, the welding speed is u, the pulse frequency is f. The DCEP duty ratio of VP arc was fixed at 50%.

After the experiments, the samples for metallographic observations were prepared by sectioning the deposits along the vertical direction using an electrical discharge wire cutting machine (Suzhou Simos CNC Technology Co., Ltd., Suzhou, China). Then, the samples were etched with modified Keller solution (50 ml H_2O, 1 ml HCL, 1.5 ml HF, and 2.5 ml HNO_3) after the processes of rough grinding, fine grinding, and polishing. The microstructure of the treated specimen was observed by Eclipse MA200 light microscope (OM) (Nikon Instruments (Shanghai) Co., Ltd., Shanghai, China).

3.2.2 Numerical modeling of MFCAM process

3.2.2.1 Physical model assumptions

A numerical model coupling electromagnetism force, heat transfer, and fluid flow in melt is derived in this section. The electric arc is modeled by an equivalent heat source applied to the upper surface of a workpiece. Gaussian distribution function can be used to describe the modeled electric arc quantities. It was assumed that the welding torch moves at a constant welding speed. The electromagnetic, continuity, momentum, and energy equations can be solved in the weld pool. The considered problem possesses symmetry with respect to the longitudinal vertical median plane, can therefore be calculated as one-half plate.

Up to now, there is no direct literature addressing the quantitative relation of shielding gas flow rate and weld bead dimensions during the VP-GTAW process. There is an optimum flow rate for weld shielding gases, but this is often decided by preference or experience. In this study, the thermocapillary-driven flow region is protected from atmosphere by pure argon (Ar) gas, and the flowing rate of the shielding gas was 15–18 L/min.

Test case	Adjustable parameters		Other fixed parameters
	f (Hz)	u (mm/s)	
T1	5	4	I_{en}/I_{ep} = 240/120 A U_{en}/U_{ep} = 12.0/11.2 V $\sigma_{j,\,EN}/\sigma_{j,\,EP}$ = 2.70/2.62 mm $\sigma_{q,\,EN}/\sigma_{q,\,EP}$ = 2.70/2.62 mm $\sigma_{p,\,EN}/\sigma_{p,\,EP}$ = 3.00/2.92 mm η_{en}/η_{ep} = 0.8/0.5
T2	5	5	
T3	5	6	

Table 4.
Parameters for VP-GTA welding of aluminum.

3.2.2.2 Governing equations

The numerical simulation of heat and mass transfer processes are governed by a set of equations in Flow3D model. In the present study, the liquid metal is assumed to be incompressible Newtonian fluid, and the flow should be laminar. The electric arc was assumed as an internal boundary condition can be described as Eq. (13).

$$k\frac{\partial T}{\partial \vec{n}} = \dot{q}_{arc} - \dot{q}_{conv} - \dot{q}_{rad} \tag{13}$$

where q_{arc}, q_{conv}, q_{rad} are the arc heat input, convective and radiative heat loss, respectively, $\vec{n}$ is the surface normal.

In this study, according to the actual processing conditions that the welding current in case of DCEP and DCEN (I_{en}/I_{ep} = 240/120 A) is high enough, arc stiffness and impact force exerted onto the weld pool surface is larger, and the arc column is perpendicular to the surface of the weldment. So, the Goldak's double-ellipsoidal heat source model was adopted [25], which can provide relatively accurate results, especially for the low penetration surface melting process. In the moving volumetric heat source model, the power density distributions of the front and rear quadrants can be described by Eqs. (14) and (15), respectively,

$$q_r = \frac{6\sqrt{3}q_{arc}f_r}{\pi a_r bc\sqrt{\pi}}\exp\left(-3[\frac{x^2}{a_r^2}+\frac{y^2}{b^2}+\frac{z^2}{c^2}]\right) \tag{14}$$

$$q_f = \frac{6\sqrt{3}q_{arc}f_f}{\pi a_f bc\sqrt{\pi}}\exp\left(-3[\frac{x^2}{a_f^2}+\frac{y^2}{b^2}+\frac{z^2}{c^2}]\right) \tag{15}$$

where f_f and f_r are the front and rear fraction of the heat flux; a_f, a_r, b and c are the parametric values obtained from the metallographic data and the weld bead profile; q_{arc} is the welding arc heat input.

The heat loss $\dot{q}_{conv}$ and $\dot{q}_{rad}$ can be calculated as follow:

$$\dot{q}_{conv} = h_{conv}(T - T_0) \tag{16}$$

$$\dot{q}_{rad} = \varepsilon\sigma_{sb}(T^4 - T_0^4) \tag{17}$$

According to the Vinokurov's empirical model [26], combined convection-radiation heat transfer coefficient was utilized as:

$$h_{vino} = 2.41 \times 10^{-3}\varepsilon T^{1.61} \tag{18}$$

The pressure boundary conditions on the weld pool surface can be described as Eq. (19).

$$P = P_{arc} + \frac{\gamma}{R_c} \tag{19}$$

where P_{arc} is the arc pressure, R_c is the curvature radius of the weld pool surface. The surface tension γ can be calculated as follow:

$$\gamma = \gamma_0 - \gamma_T (T - T_l) \tag{20}$$

The arc pressure distribution was assumed to follow the distribution of current density. It can be modeled by a Gaussian model with the same radius of arc drag force, as blow [27].

$$P_{arc}(x, y) = \frac{\mu_0 I^2}{4\pi^2 \sigma_r^2} \exp\left(-(\frac{r^2}{2\sigma_r^2}) \right) \tag{21}$$

where μ_0 is the magnetic permeability of free space, σ_r is the arc pressure parameter (DCEN phase: $\sigma_r = \sigma_{p,\ \mathrm{EN}}$; DCEP phase: $\sigma_r = \sigma_{p,\ \mathrm{EP}}$).

The arc drag force on the weld pool is greatly dependent on the current, the composition of shielding gas, and the tip angle of electrode. Here, the effect of arc drag force is considered as a spatial boundary distribution, which can be represented as follows [28].

$$P_{Drag}(r) = P_{Max} \sqrt{\frac{r}{r_{Shear}}} \exp - (\frac{r}{r_{Shear}})^2 \tag{22}$$

where r_{Shear} is the distribution parameter of arc drag force.

The body force mainly includes electromagnetic force (EMF), gravity, and buoyancy. The gravity acceleration is 9.81 m/s^2. The temperature-dependent properties were used for the density. The electromagnetic force, as an important body force, was considered by adopting the elliptically symmetric welding current density [29, 30]. The equations relating EMF are listed as below.

$$F_x = -J_z \times B_\theta \frac{x}{r_a} \tag{23}$$

$$F_z = -J_r \times B_\theta \tag{24}$$

$$J_z = \frac{I}{2\pi} \int_0^\infty \lambda J_0(\lambda r_a) \exp(-\frac{\lambda^2 \sigma_x}{2}) \frac{\sinh[\lambda(c-z)]}{\sinh(\lambda c)} d\lambda \tag{25}$$

$$J_r = \frac{I}{2\pi} \int_0^\infty \lambda J_1(\lambda r_a) \exp(-\frac{\lambda^2 \sigma_x}{2}) \frac{\cosh[\lambda(c-z)]}{\sinh(\lambda c)} d\lambda \tag{26}$$

$$B_\theta = \frac{\mu_m I}{2\pi} \int_0^\infty J_1(\lambda r_a) \exp(-\frac{\lambda^2 \sigma_x}{2}) \frac{\sinh[\lambda(c-z)]}{\sinh(\lambda c)} d\lambda \tag{27}$$

$$r_a = \sqrt{(x - x_0)^2 + [\frac{\sigma_x}{\sigma_y}(y - y_0)]^2} \tag{28}$$

where I is the arc current (DCEN phase: $I = I_{en}$, $\sigma_j = \sigma_{j,\ \mathrm{EN}}$; DCEP phase: $I = I_{ep}$, $\sigma_j = \sigma_{j,\ \mathrm{EP}}$), F_i are the components of the EMF force in the i-direction ($i = x, y, z$), B_θ is the angular component of the magnetic field, J_z and J_r are the axial and radial current density in the cylindrical coordinate system, J_0 and J_1 are the zero order and one order Bessel function, respectively, z indicates the vertical depth from top surface of workpiece, and c is the workpiece thickness.

3.2.2.3 Numerical model

A 2D longitudinal section of the 3D calculation model for MFCAM process is shown in **Figure 18**. The structural parameters of the model are: d is 1.2 mm, D is 4 mm; h is 1.2–2.4 mm; the substrate thickness is 6 mm, and the distance from arc to the rim of coating head is ~5.5 mm.

In the present study, peak current I_p (voltage) and base current I_b (voltage) are 220 A (18.2 V) and 183 A (16.7 V), respectively. The proportion of peak current time in one cycle is 0.5. Thermo-physical properties of material used in this study are listed in **Table 5**. The contact angle, defined as the angle at which the liquid metal meets the workpiece surface, is assumed to be 90°.

A grid size (200 μm) and computational time-step (2.5×10^{-6} s) for simulations are used. It took ~72 h to run 6 s of real-time simulation of a single-track deposition using high-performance computer of 48 × 1.6 GHz CPU.

3.3 Results and discussion

3.3.1 Experimental verification of the numerical model

To confirm the predictive accuracy of the established model, the corresponding deposition experiments were carried out. A cathode with 60° cone angle and a 3.2 mm radius float tip provides a good combination. The distance between tungsten electrode tip and workpiece surface was fixed at 4 mm. To avoid heat sinks, the workpieces need to be thermally insulated from the fixtures in VP-GTA welding process. Verifications of the numerical model were carried out by comparing the calculated results with the metallographic macro-sections. **Figure 19** presents the simulated and measured geometry of a single-track deposition by MFACM.

Calculated solidus isotherm 811 K—corresponds to the fusion line obtained in the experiments. It is found that the calculated deposition geometries and dimensions agree well with the experimental data. The average values of deposition width and penetration depth were measured to be 8.65 and 1.43 mm, respectively, which is in consistent with the calculated results. The average relative errors of the deposition width and height are never exceeded 5.7 and 11.5%. The fidelity of the models will be verified to better predict the surface evolution in MFCAM process and eliminate the deposition defects based on the understanding of its mechanism.

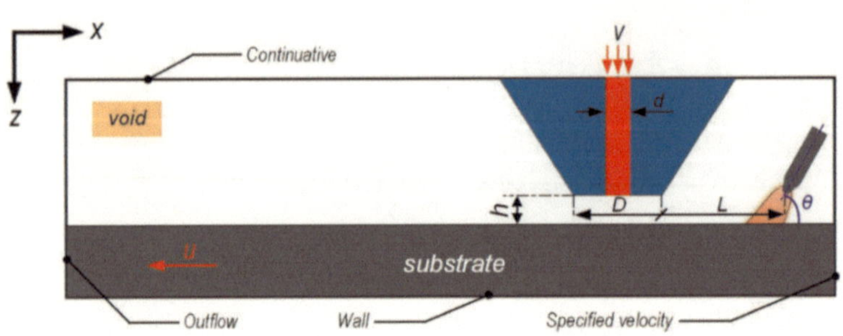

Figure 18.
Calculation model for FCAM process.

Nomenclature	Value
Density (kg m^{-3})	$f(T)$ [21]
Viscosity (N s/m^2)	$f(T)$ [21]
Surface tension (N m^{-1})	$f(T)$ [21]
Thermal conductivity (W m^{-1} K^{-1})	$f(T)$ [21]
Specific heat (J kg^{-1} K^{-1})	$f(T)$ [21]
Solidus temperature (K)	$T_s = 811$
Liquidus temperature (K)	$T_l = 905$
Latent heat for melting (J kg^{-1})	$h_{sl} = 3.97 \times 10^5$
Coefficient of thermal expansion (K^{-1})	$\beta = 1.5 \times 10^{-4}$
Radiation emissivity	$\varepsilon = 0.4$
Initial temperature of melt (K)	960
Substrate temperature (K)	298
Volume flow rate of the melt (mm^3 s^{-1})	$V = 30–70$
Substrate moving speed (mm s^{-1})	$U = 3–9$
Ambient temperature (K)	$T_\infty = 300$
Vaporization temperature (K)	$T_{ev} = 1163$
Arc thermal efficiency	$\eta = 0.7$ [22]
Distribution parameter of arc drag force (mm)	3.2

Table 5.
Thermo-physical material properties of 2024 aluminum alloy and calculation data used in the simulation.

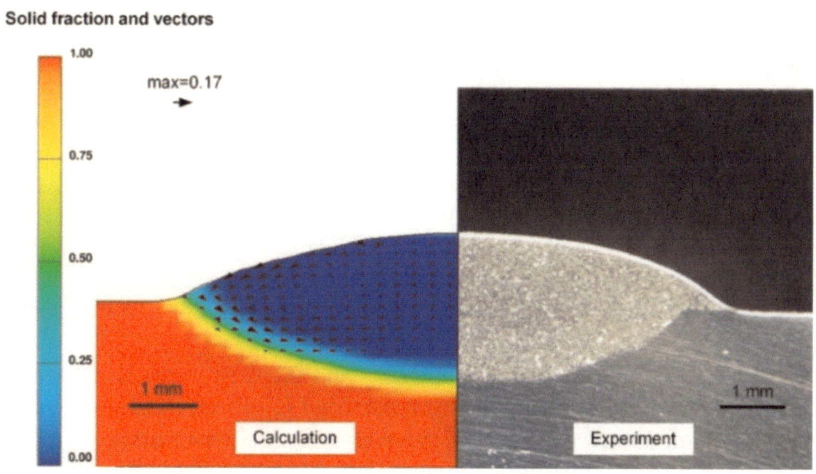

Figure 19.
Comparison between computed thermal profile (left) and experimentally determined (right) cross-section of a single-track deposition, all dimensions are in mm.

3.3.2 Thermophysical phenomena in MFCAM

During the fused-coating deposition process, the zone where the heat source is characterized by the temperature local sharp increase and being heated up to the state of melting, as shown in **Figure 20**. We need to how this evolutionary sequence of deposited layers. The development of deposition shape and the heat transfer and

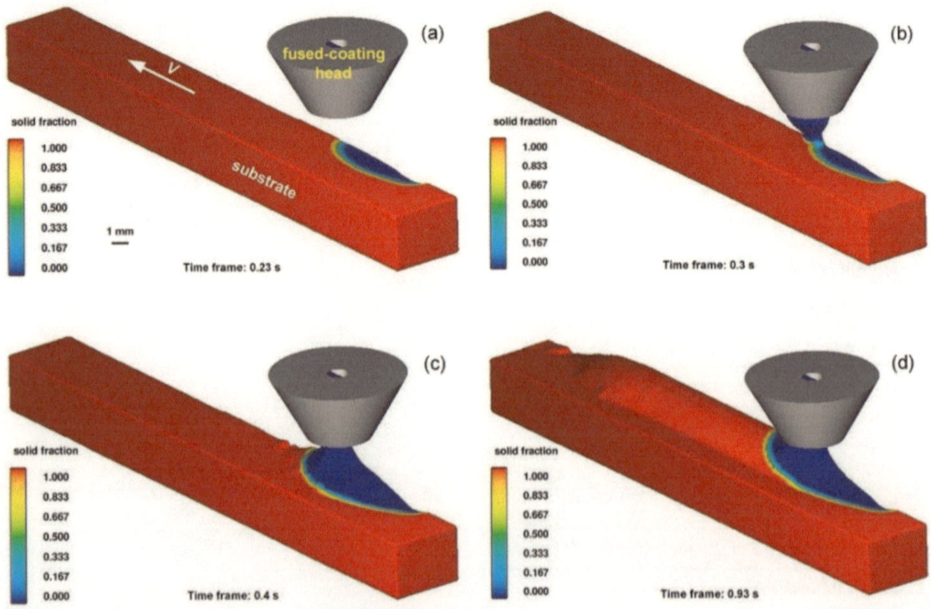

Figure 20.
Solid fraction distribution and the changes of surface morphology of the melt in four stages during single-track deposition. (a) formation of a shallow molten pool, (b) extrusion process of the melt, (c) extremely changes of surface morphology of the melt, and (d) relatively steady state.

flow characteristics of liquid metal were illustrated. The shape evolution of each single layer can be divided into four stages: (i) formation and growth of a shallow molten pool, (ii) extrusion process of the melt, (iii) extremely changes of surface morphology of the melt, and (iv) relatively steady state.

The shallow molten pool on a moving metal substrate is firstly created in the initial stage. The source of heat in welding is enough to melt the metal. According to the practical conditions, the welding torch needs to be tilted from the nominal center of the workpiece surface, which will slightly change the thermal flow field characteristics of the molten pool.

During the second stage of MFCAM process the hot liquid Al alloy will be extruded continuously toward the shallow molten pool. Meanwhile, the heat flux of the tilted welding arc still melts the base metal. Thus, the local thermal contact resistance of deposition regions can be reduced greatly, which is beneficial to the spreading of the melt. The corresponding thermal and flow fields in this stage is shown in **Figure 21**.

In the third stage, the extruded liquid alloy and the melt in the shallow molten pool trend to spontaneous fusion once the extruded liquid metal meets the surface of the molten pool. The free surface of the melt will experience extremely complex deformations. **Figure 22** illustrates the topology changes in the free surface, surface depression in the molten pool and the temperature distributions in the melt.

It can be seen from **Figure 22a** that there is a higher temperature region between the welding arc and fused-coating head. The teardrop-shaped gouging region is also found in the front of the molten pool. The causes of these phenomena can be attributed to the fact that it is relative difficult for the liquid metal located in the higher temperature region to dissipate heat because of the heat input by welding arc, at the same time, the melt continuously obtains latent heat from the being extruded liquid metal. The end-face of the coating head and the shallow molten pool will produce an obviously restriction effect to the spreading melt. As shown in **Figure 22b**, the local fluctuations of the melt's surface and solidification front are

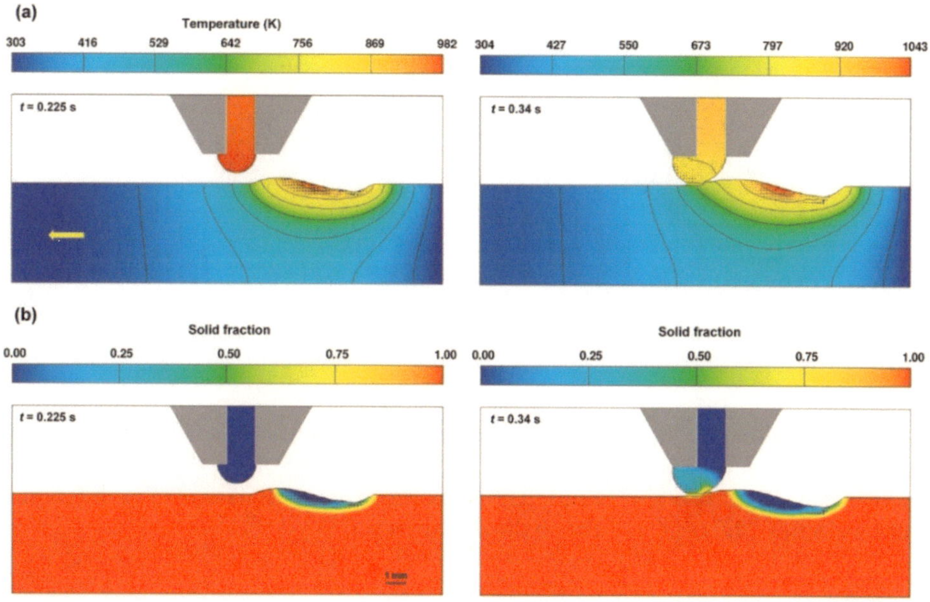

Figure 21.
Side views of (a) temperature distributions, and (b) solid fraction distributions in the second stage.

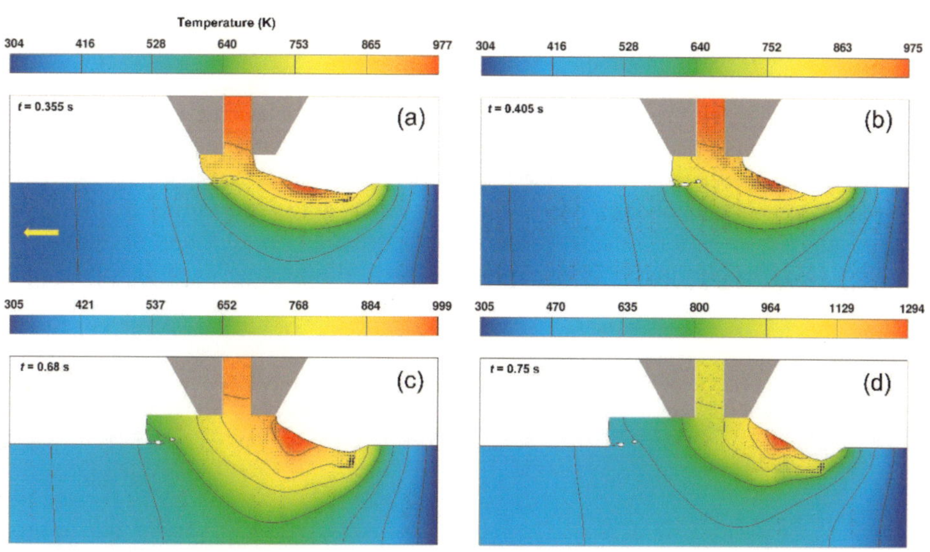

Figure 22.
Topology changes in the free surface, surface depression in the molten pool and the temperature distributions in the melt. (a) t = 0.355 s, (b) 0.405 s, (c) 0.68 s, and (d) 0.75 s.

always found, this could be attributed to the way in which a pulsed-current mode was adopted.

As time goes on, the various forces governing the fluctuation of molten pool surface and a macroscopic heat balance tend to equilibrium, therefore, the thermocapillary-driven deposited shape becomes to be continuous and stable, the 3D shape of the stable deposited layer can be observed in **Figure 23**. The calculated temperature field and velocity magnitude could be illustrated by **Figure 23a**. The dimensions associated to the solidified deposited layers will be characterized by *H* and *W*.

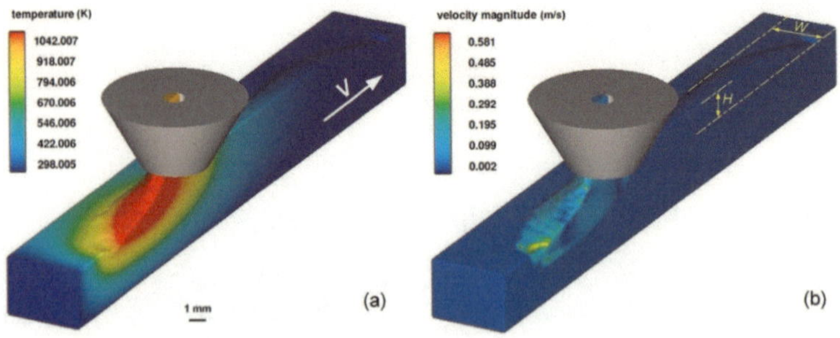

Figure 23.
(a) 3D shape of a single-track deposited layer and (b) surface velocity contour.

The calculated results show that the shape of deposited layers is flat and regular from the global aspect although there is a local fluctuation of melt's surface and solidification front. It is also found that a maximum liquid metal speed with 0.6 m/s is obtained near the center of the shallow molten pool.

3.3.3 Influence of melt flow rate

Figure 24 shows the influence of melt flow rate on the morphology characteristics of single-track deposits. The gap between substrate and fused-coating head is fixed at 1.8 mm, and the substrate moving speed is 6 mm/s.

As illustrated from **Figure 24**, an approximate linear increase in the deposition height was observed as the melt flow rate increase from 20 to 120 ml/min. However, in the respects of the deposition width, there obviously exist many highly nonlinear relationships. Therefore, the formation mechanism based on the thermal flow dynamics needs to be recognized. The principle of minimum enthalpy may be utilized to explain this phenomenon. The principle is described as follows: the thermocapillary-driven flow always has a characteristic to keep the minimum enthalpy value by automatically changing its size and adjusting its enthalpy. In

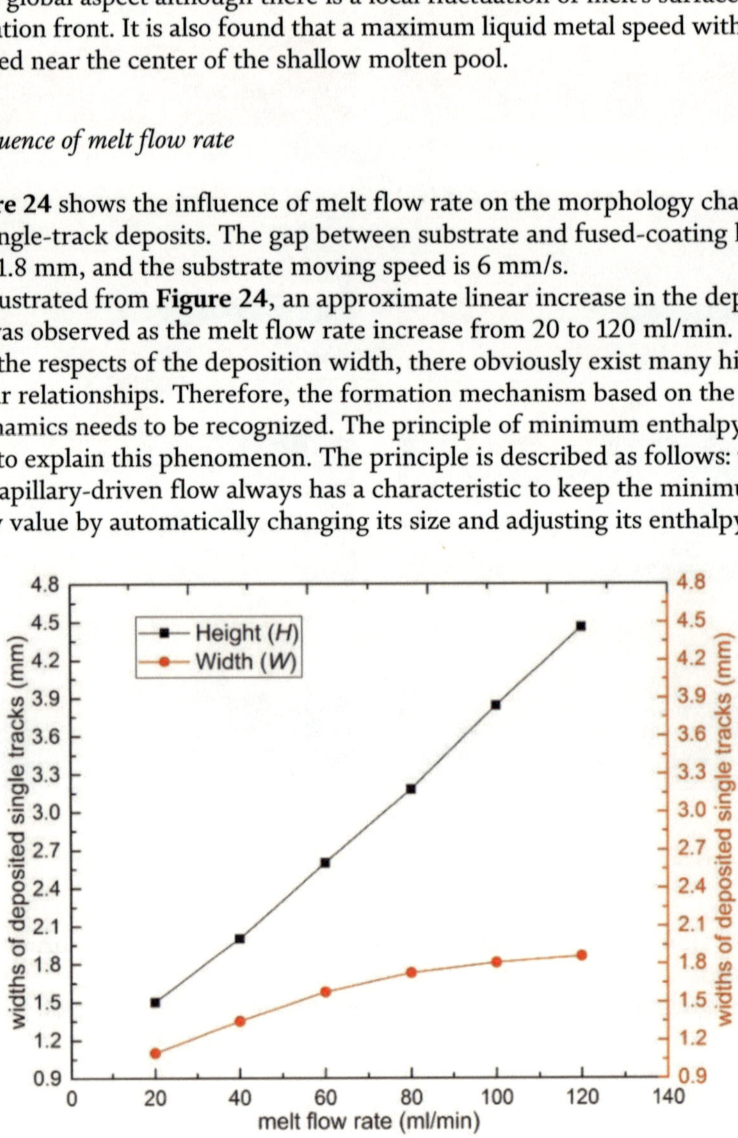

Figure 24.
Variation of the morphology characteristics of deposited single tracks with melt flow rate.

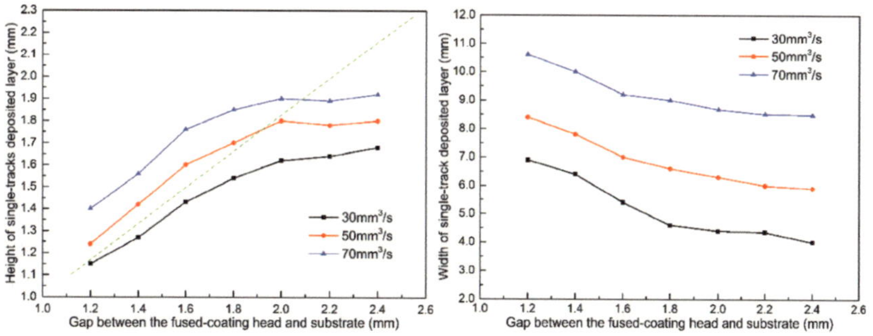

Figure 25.
Influence of the gap height on the morphology characteristics of single-track deposits.

MFCAM process, the area of the absorbed arc energy is heated alternatively. The residual heat in the material will be continuously accumulated from pulse to pulse. The heat accumulation, as a function of processing parameters, is proportional to the duration of arc heat input.

When the substrate moving speed and arc heat input keep invariant, the local high-temperature range and the molten pool size should remain unchanged. However, the local melting/solidification behaviors will be partially changed with the spontaneous fusion of extruded liquid metal and the melt in the molten pool. As the melt flow rate increases, heat sink effect on the penetration depth becomes more significantly since the additional heat energy from the being extruded melt is partially transferred to the molten pool, thus the additional thermal energy actually prolongs solidification time and makes the penetration depth increase.

3.3.4 Influence of the gap height

On the other hand, the influence of the gap height between substrate and fused-coating head on the morphology characteristics of single-track deposits is also discussed, as shown in **Figure 25**. The range of melt volume flow rate is 30–70 mm^3/s. The substrate moves at a constant speed of 5 mm/s.

As illustrated in **Figure 25**, it was proved that the deposition height increases with the increase in the gap height, but the reverse is true in the deposition width. The reason for this phenomenon can be explained by researching the morphology of free surface and the thermal-flow characteristics around the fused-coating head, two main causes are pointed out: (i) liquid metal state under the condition of being squeezed, (ii) the adhesion properties of the melt around the fused coating head.

Squeezed flow behaviors of the liquid metal within a narrower gap plays a dominant role in MFCAM process, the penetration may be deeper than that in a simple thermocapillary shear flow because of the presence of pressure gradients. In this situation, the flow direction of the melt transforms from the rear toward the lateral, thus the deposition widths achieve actually increase. Meantime, it is also found that a larger gap easily leads to a higher and narrower deposition layer. This is attributed to the fact that as the gap is sufficiently large, the thermocapillary force will become not evidence, while the gravity effect and the adhesion between liquid metal and fused-coating head become remarkable.

4. Conclusions

Two multi-physics numerical models were developed to investigate the three-dimensional transient dynamics of the melt in deposition processes associated with

SLM and MFCAM. The geometries of single-track depositions under both laser- and arc-heated modes were obtained by simulations and compared with experimental observations. The molten pool motion/geometry and the computed temperature distribution from simulated results are used to analyze the formation mechanisms of deposited layers. The major findings of the current work are summarized briefly summarized as follows:

1. A randomly distributed powder bed model was established to study the effect of layer thickness and hatching spacing on the thermal behavior of AlSi10Mg molten pool. The numerical calculation results were verified by experiments.

2. The pore and balling formation mechanisms during AlSi10Mg SLM were revealed by using numerical and experimental method. The cross-section of the part produced by SLM using the laser power of 180 W and scan speed of 1000 mm/s showed a fine microstructure without any apparent pores. The larger hatch spacing may result in poor inter-track bonding inducing the inter-track pores. Also, the larger powder layer thickness is one of the key factors in inducing the inter-layer pores. In order to produce a fully dense AlSi10Mg part without any apparent pore and balling defects, the laser power of 180 W, the scanning speed of 1000 mm/s, the powder layer thickness of 35 μm, and the hatch spacing of 50 μm are proposed during SLM.

3. The whole evolution processes of the deposited layer morphology in MFCAM process could be generally divided into four stages. There are local fluctuations of the profile of liquid-solid interface, which can be attributed to the spatial-temporal fluctuations of fluid momentum in pulsed current mode. The spatial variability of the liquid-solid interface results in the variability in the solidification parameters, which impact the deposited layer morphology.

4. The substrate moving speed can significantly affect the morphology of deposited layers. Width and height of deposited layers usually decrease with the increase of substrate moving speed.

Acknowledgements

The authors are grateful for the financial support for this research from National Natural Science Foundation of China (Grant No. 51775420), the State Key Development Program Research of China (Grant No. 2016YFB1100400) and the Fundamental Research Funds for the Central Universities (Grant No. xjj2016124).

Author details

Zhengying Wei* and Jun Du
The State Key Laboratory for Manufacturing Systems Engineering, Xi'an Jiaotong University, Xi'an, China

*Address all correspondence to: zywei@mail.xjtu.edu.cn

References

[1] Tofail SAM, Koumoulos EP, Bandyopadhyay A, Bose S, O'Donoghue L, Charitidis C. Additive manufacturing: Scientific and technological challenges, market uptake and opportunities. Materials Today. 2018;**21**:22-37

[2] Wohlers TT, Caffrey T. Wohlers Report 2014: Additive Manufacturing and 3D Printing State of the Industry Annual Worldwide Progress Report. Fort Collins, Colorado: Wohlers Associates, Inc.; 2014

[3] King WE, Anderson AT, Ferencz RM, Hodge NE, Kamath C, Khairallah SA, et al. Laser powder bed fusion additive manufacturing of metals; physics, computational, and materials challenges. Applied Physics Reviews. 2015;**2**:041304

[4] DebRoy T, Wei HL, Zuback JS, Mukherjee T, Elmer JW, Milewski JO, et al. Additive manufacturing of metallic components—Process, structure and properties. Progress in Materials Science. 2018;**92**:112-224

[5] Babu SS. Scripta Viewpoint Set: Materials science aspects related to additive manufacturing. Scripta Materialia. 2017;**135**:97-99

[6] O'Regan P, Prickett P, Setchi R, Hankins G, Jones N. Metal based additive layer manufacturing: Variations, correlations and process control. Procedia Computer Science. 2016;**96**:216-224

[7] Herzog D, Seyda V, Wycisk E, Emmelmann C. Additive manufacturing of metals. Acta Materialia. 2016;**117**: 371-392

[8] Hebert RJ. Viewpoint: Metallurgical aspects of powder bed metal additive manufacturing. Journal of Materials Science. 2016;**51**:1165-1175

[9] Yusuf SM, Gao N. Influence of energy density on metallurgy and properties in metal additive manufacturing. Materials Science and Technology. 2017;**33**:1269-1289

[10] Francois MM, Sun A, King WE, Henson NJ, Tourret D, Bronkhorst CA, et al. Modeling of additive manufacturing processes for metals: Challenges and opportunities. Current Opinion in Solid State and Materials Science;**21**:198-206

[11] Bikas H, Stavropoulos P, Chryssolouris G. Additive manufacturing methods and modelling approaches: A critical review. The International Journal of Advanced Manufacturing Technology. 2015;**83**:1-17

[12] Mukherjee T, Manvatkar V, De A, DebRoy T. Dimensionless numbers in additive manufacturing. Journal of Applied Physics. 2017;**121**:064904

[13] Wei P, Wei Z, Chen Z. Thermal behavior in single track during selective laser melting of AlSi10Mg powder. Applied Physics A. 2017;**123**:604

[14] Wei P, Wei Z, Chen Z. Numerical simulation and parametric analysis of selective laser melting process of AlSi10Mg powder. Applied Physics A. 2017;**123**:540

[15] Roux JN. Geometric origin of mechanical properties of granular materials. Physical Review E. 2000; **61**(6):6802

[16] Rong G, Liu G, Hou D, Zhou C. Effect of particle shape on mechanical behaviors of rocks: A numerical study using clumped particle model. Scientific World Journal. 2013;**2013**: 289215-589215

[17] Hirt CW, Nichols BD. Volume of fluid (VOF) method for the dynamics of free boundaries. Journal of Computational Physics. 1981;**39**:201-225

[18] Stacy SC, Zhang X, Pantoya M, Weeks B. The effects of density on thermal conductivity and absorption coefficient for consolidated aluminum nanoparticles. International Journal of Heat and Mass Transfer. 2014;**54**: 595-599

[19] Dai K, Shaw L. Finite element analysis of the effect of volume shrinkage during laser densification. Acta Materialia. 2005;**53**:4743-4754

[20] Semak V, Matsunawa A. The role of recoil pressure in energy balance during laser materials processing. Journal of Physics D: Applied Physics. 1999;**30**: 2541

[21] Min HC, Yong CL, Farson D. Simulation of weld pool dynamics in the stationary pulsed gas metal arc welding process and final weld shape. Welding Journal. 2006;**85**:271-283

[22] Masmoudi A, Bolot R, Coddet C. Investigation of the laser–powder–atmosphere interaction zone during the selective laser melting process. Journal of Materials Processing Technology. 2015;**225**:122-132

[23] Dou L, Yuan ZF, Li JQ, Wang X. Surface tension of molten Al-Si alloy at temperatures ranging from 923 to 1123 K. Chinese Science Bulletin. 2008;**53**: 2593-2598

[24] Gu DD, Wang Z, Shen Y, Li Q, Li Y. In-situ TiC particle reinforced Ti–Al matrix composites: Powder preparation by mechanical alloying and selective laser melting behavior. Applied Surface Science. 2009;**255**:9230-9240

[25] Goldak J, Chakravarti A, Bibby M. A new finite element model for welding heat sources. Metallurgical and Materials Transactions B. 1984;**15**: 299-305

[26] Desai RS, Bag S. Influence of displacement constraints in thermomechanical analysis of laser micro-spot welding process. Journal of Manufacturing Processes. 2014;**16**: 264-275

[27] Cho DW, Lee SH, Na SJ. Characterization of welding arc and weld pool formation in vacuum gas hollow tungsten arc welding. Journal of Materials Processing Technology. 2013; **213**:143-152

[28] Sreedhar U. Static GTAW: Experimental and numerical investigations and heat flux parameter estimation [PhD thesis]. Montpellier, France: Universit'e Montpellier-II; 2014

[29] Cho DW, Na SJ, Cho MH, Lee JS. Simulations of weld pool dynamics in V-groove GTA and GMA welding. Weld World. 2013;**57**:223-233

[30] Kou S, Sun DK. Fluid flow and weld penetration in stationary arc welds. Metallurgical and Materials Transactions A: Physical Metallurgy and Materials Science. 1985;**16**:203-213

Section 5

Multi-phase Flows and Phase Change

Chapter 11

Flow and Heat Transfer in Jet Cooling Rolling Bearing

Wei Wu, Jibin Hu and Shihua Yuan

Abstract

The real flow field inside a jet cooling ball bearing is an air-oil two-phase flow since the air entrainment in the lubricant is ineluctable. The flow heat transfer characteristics are thought to have significant impacts in cooling effect forecast and temperature control. To this end, the fluid flow and heat transfer characteristics of the air-oil two-phase flow inside a jet cooling rolling bearing have been investigated by simulation and experiment herein. A multiphase VOF numerical method for the flow field calculation has been proposed. The temperature distributions of the bearing system have been measured, and the flow pattern under different operation speeds has also been captured. The parameter effects on the jet cooling rolling bearing flow field and temperature distribution have been revealed accordingly. The research findings can be used for the structural optimization and the precise lubrication design of the rolling bearing.

Keywords: ball bearings, jet cooling, two-phase flow, VOF, flow pattern, temperature distribution

1. Introduction

Rolling bearings are the most widely used bearings in the vehicle transmission industry due to their advantages of less friction coefficient for rotatory movements [1]. With the future trends of vehicle transmission development shifting toward lighter and more compact designs with greater power density, the main shaft bearings are subjected to more severe demands of simultaneous shrinking size and excessive loading [2, 3]. These requirements enforce higher thermal load on the bearings. The rolling bearings under stress overload and poor heat dissipation conditions are prone to wear out easily due to metal-to-metal contact, which corrode surface of bearings.

Hence, lubrication of bearings plays a significant role in fatigue life assurance by reducing friction and wear, which ensure the safe and reliable operation of bearings. The choice of lubrication mode of rolling bearing mainly depends on load, working speed, and temperature. Grease and splash lubrication are capable enough for low-speed rolling bearings [4], while oil-jet lubrication could better meet the requirements under heavy load, high-speed, and high-temperature conditions [5].

Fluid jet heat transfer enhances the heat transfer rates in comparison with conventional methods. In an oil-jet cooling rolling bearing, a portion of the oil is utilized to lubricate the ball bearing and form lubricant film in ball raceway contacts. A larger film thickness is helpful to reduce the friction and the heat generation. However, for the high-speed operation, the total energy dissipated by friction is significant, leading to an excessive heat generation. To compensate for this problem, larger amounts of lubricant oil flow are required to provide sufficient cooling capability, while only a relatively small part of the oil is used to form a lubricant film [6, 7]. Besides, while a larger amount of oil mass flow may lead to a higher power loss due to the increased drag against the rotation of the rolling elements and, an insufficient oil mass flow may cause a lubrication failure.

Thus, both the power performance and the lubrication performance of a rolling bearing are considered to be significantly impacted by the flow around it, which is in typical forms of enclosed space flow or open space flow. The enclosed space flow relates to the flow inside a bearing chamber, which is a common flow state in the real operating conditions such as aero engines, etc. However, there is usually no bearing chamber design for the drivetrain applications with lightweight requirements. The flow state of these rolling bearings is an open space flow, in which the heat that determines the temperature of bearing and oil-out is generated by the drag among rolling elements.

Experimental investigations with regard to high-speed ball bearings with oil-jet lubrication have been reported by Zaretsky et al. [8], indicating a lower bearing temperature with a double-nozzle jet construction compared with a single-nozzle jet counterpart under the same given oil flow rate. Flow was simulated using the computational fluid dynamic method for acquiring working performance of the oil-mist lubrication, without and with the flow inside the rolling bearing [9, 10]. The results indicate that the opening near the inner raceway is expected to be the excellent location for supplying oil-mist or oil-jet for accomplishing better lubrication and cooling of ball bearing components. Both the flow pattern around one sphere and its drag coefficient are modified when placing another sphere in its vicinity.

The lubrication performance of the rolling bearing has been investigated a lot by single-phase method [11, 12]. The bearing temperature forecast model, considering the heat transfer of the oil flow and the air flow, respectively, is mainly presented by Harris [13, 14]. The convective heat transfer coefficients are given considering the lubricant oil flow and the air flow, respectively [15]. A computer program Shaberth has been used to calculate the thermal performance of ball bearings [16]. The oil volume fraction is required for input in Shaberth, and an empirical equation was derived for the oil volume fraction value inside the ball bearing as a function of oil flow rate, shaft speed, and bearing pitch diameter. However, the flow field inside a rolling bearing with oil-jet lubrication, in reality, is far more complicated since the air entrainment in the lubricant is ineluctable while the bearing is rotating. In addition, the air flow phase and the lubricant flow phase constantly interact with each other, and their interaction is strongly affected by the rotary speed of the bearing.

As stated above, while an accurate assessment of the flow around the ball bearing is of particular importance, the experiment measurements are not enough, and the numerical studies are mainly based upon single phase and empirical equation which are incapable in reflecting the real physics. This work herein investigates the air-oil two-phase transient flow inside the ball bearing. A CFD modeling method is presented for the air-oil two-phase flow inside the rotary ball bearing. It aims to increase the understanding of the temperature effects and flow pattern of the flow field inside the oil-jet lubricated ball bearing. The results can be used to optimize the precise oil-jet lubrication design to control the amount of lubricant oil inside the ball bearing at a later stage.

2. Oil-jet cooling mechanism and experimental setup

Lubrication and cooling are important for the rolling bearing fatigue life extension. It is good for the bearing temperature control. The fluid jet is an effective cooling method which has been used widely in heat dissipation. In the rolling bearing operation, the oil-jet method is used for very high-speed conditions. The number of nozzles, the jet velocity, and the oil flow rate are significant to satisfy the bearing lubrication requirement. In order to overcome the detrimental centrifugal effects in high-speed applications, the lubricant oil under pressure is directed at the side of the bearing. The jet velocity should be sufficiently high since the air surrounding the bearing causes the oil to be deflected from the inner ring. Further, a sufficient amount of lubricant oil should be supplied to the bearing. However, an excessive amount of lubricant oil will increase the bearing temperature unnecessarily.

Figure 1(a) shows the schematic configuration of the oil-jet lubrication ball bearing under investigation. The cooling oil is injected into the bearing raceway through a small-diameter nozzle to lubricate and cool the bearing and ensure the safe and reliable operation of the rolling bearing. The high-speed oil-jet atomizes into droplets which then mix with the surrounding air to from a two-phase mixture. Besides, the two-phase mixture is working in an open space environment that has a direct contact with air of the outlets. In this way, coexistence of oil and air two-phase flow is formed inside the ball bearing.

In the high-speed operation of bearings, a large amount of heat generated by friction between moving pairs will lead to bearings and their adjacent parts. A sharp increase of friction heat will lead to a significant increase in the working temperature. If the heat not effective discharged, it will inevitably lead to the failure of bearings, which will damage the service life of the bearings. The three major heat dissipation methods inside the ball bearing are shown in **Figure 1(b)**, including the conduction through solid structures, the convection from solid structures to fluids, and the radiation to surrounding media. The latter one is not taken into consideration since it dissipates a negligible part of heat in the bearing case. When the bearing is in a state of thermal equilibrium, the main factors that impact the bearing temperature are bearing load, speed, oil-in temperature, and oil volume fraction inside the bearing, which indicates that the flow field and heat transfer characteristics between the two-phase flow and solid components have a considerable influence in the bearing temperature. Thus, it is essential to clarify the correlations between the fluid flow and thermal behaviors in oil-jet cooling bearing, in order to

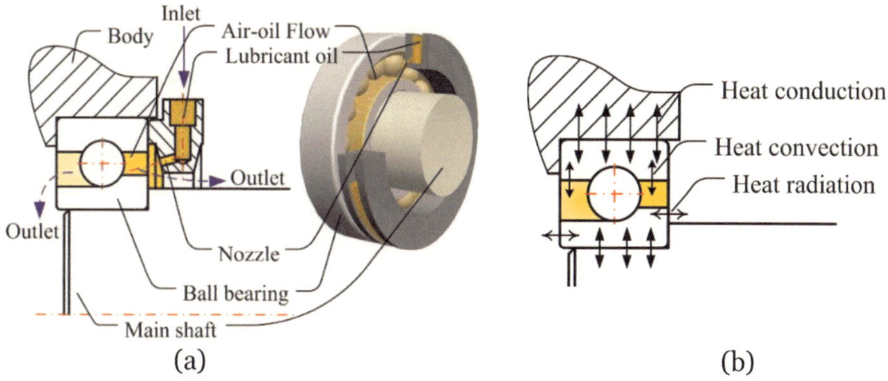

Figure 1.
Configuration of the oil-jet cooling ball bearing and heat transfer method: (a) configuration of jet cooling mechanism and (b) heat dissipation methods.

make precise assessment of cooling method and optimize the design of cooling devices.

Experiment apparatuses for testing the flow pattern and temperature distribution of the ball bearing have been built up, as shown in **Figure 2**. No load is applied in the flow pattern test, as shown in **Figure 2(a)**. The shaft of the tested ball bearing is horizontal. The maximum tested speed is 4500 r/min which is the rated speed of the driving motor. The parameters of the tested ball bearing are given in **Table 1**. The exact value of the oil volume fraction distribution is difficult to process, since the flow field inside the bearing is compact. Thus, the photograph of the flow pattern is presented to validate the simulated results.

Further, a tested apparatus for temperature test of ball bearing has been built up, as shown in **Figure 2(b)**. The outer ring temperatures of different positions are measured by three temperature sensors. r is the radial coordinates and ω is the circumferential azimuth angle. The technical data of the experimental apparatus is presented in **Table 2**. The measured bearing temperature distribution can be used for computational model validation.

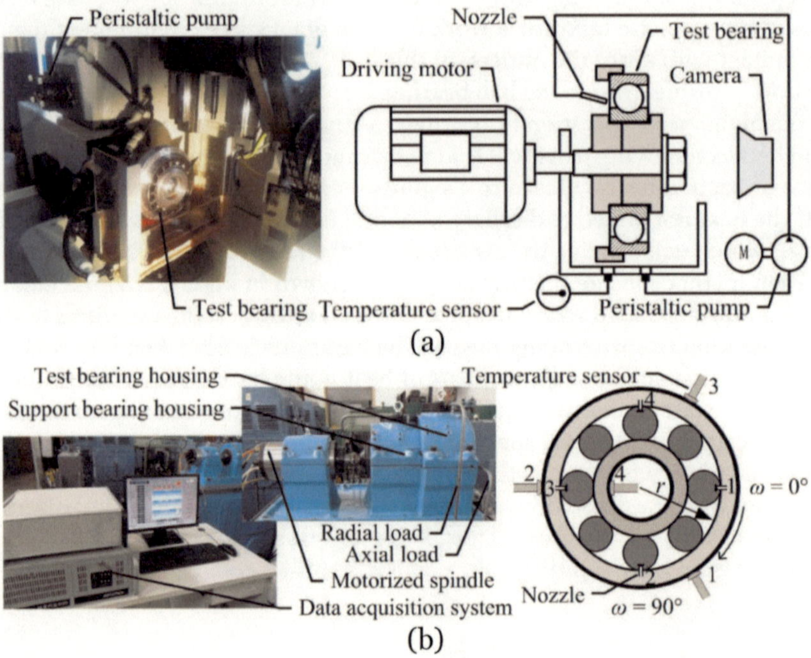

Figure 2.
Test apparatuses: (a) rolling bearing flow pattern test apparatus and (b) rolling bearing temperature distribution test apparatus.

Quantity	Conversion from Gaussian and CGS EMU to SI
Inner diameter (mm)	50
Outer diameter (mm)	90
Width (mm)	20
Ball diameter (mm)	12.186
Contact angle (deg.)	40

Table 1.
Specifications of the test ball bearing.

Apparatus and sensor	Technical data
Motorized spindle	0–15,000 r/min
Temperature sensor	Pt1000 (−70 to 500°C)
Oil flow transducer	FT-110 (1.0–10 L/min)
Vibration transducer	JHT-II-B (±15 g)
External radial force	Hydraulic loading (0–30 kN)
External axial force	Hydraulic loading (0–30 kN)

Table 2.
Technical data of the test apparatus.

3. Mathematical modeling

Since the two-phase flow changes constantly in high-speed rotating rolling bearings, the development of a computational fluid dynamics (CFD) model with a two-phase flow method to determine the amount of lubricant oil is of practical value as a substitution of empirical correlations. The volume of fluid (VOF) model is selected to model the two-phase interactions inside the bearing for its recognized ability in tracking the air-oil interface. The two-phase coexistence of the bearing is in the turbulent state. The RNG k-ε model is an economical turbulent model for the rotating or swirling flows, so it is chosen as the turbulent governing equations.

3.1 Governing equations

The oil is defined as the primary phase in the two-phase calculation, and its volume fraction in each cell is denoted as φ_{oil}, with $\varphi_{oil} = 1$ representing a pure oil phase and $\varphi_{oil} = 0$ representing a pure air phase. If $0 < \varphi_{oil} < 1$, the cell of interest is in the air-oil two-phase state. If the air volume fraction is similarly denoted as φ_{air}, then the sum of the volume fractions of the two phases in the flow domain should meet the follow the constraint:

$$\varphi_{\text{oil}} + \varphi_{\text{air}} = 1 \tag{1}$$

The properties of an air-oil two-phase flow in the VOF method are treated as the volumetric average of that of the individual phase. Thus, the density, dynamic viscosity, and thermal conductivity are expressed as

$$\rho = \varphi_{\text{oil}}\rho_{\text{oil}} + \varphi_{\text{air}}\rho_{\text{air}} \tag{2}$$

$$\mu = \varphi_{\text{oil}}\mu_{\text{oil}} + \varphi_{\text{air}}\mu_{\text{air}} \tag{3}$$

$$k = k_{\text{oil}}\alpha_{\text{oil}} + k_{\text{air}}\mu_{\text{air}} \tag{4}$$

where ρ_{oil} is the oil density, ρ_{air} is the air density, μ_{oil} is the oil dynamic viscosity, μ_{air} is the air dynamic viscosity, k_{oil} is the oil thermal conductivity, k_{air} is the air thermal conductivity, and k_f represents the effective thermal conductivity of the two-phase flow.

The governing equations for the entire computational domain are as follows:

Continuity equation

$$\nabla \cdot \left(\rho\vec{v}\right) = 0 \tag{5}$$

Momentum equation

$$\nabla \cdot \left(\rho \vec{v}\vec{v}\right) = \nabla \cdot \left(\mu\left(\nabla\vec{v} + \left(\nabla\vec{v}\right)^T\right)\right) - \nabla \cdot p + \rho\vec{g} + \vec{F} \tag{6}$$

Energy equation for the fluid

$$\nabla \cdot \left[\vec{v}(\rho E)\right] = \nabla \cdot \left(k_f \nabla T\right) \tag{7}$$

where

$$E = \frac{\alpha_{\text{oil}}\rho_{\text{oil}}E_{\text{oil}} + \alpha_{\text{air}}\rho_{\text{air}}E_{\text{air}}}{\alpha_{\text{oil}}\rho_{\text{oil}} + \alpha_{\text{air}}\rho_{\text{air}}} \tag{8}$$

Energy equation for the solid

$$\nabla \cdot (k_s \nabla T) = 0 \tag{9}$$

where $\vec{v}$ is the velocity vector, p is the pressure, $\vec{F}$ is the external force, $\vec{g}$ is the gravity acceleration, T is the temperature, E is the energy, and k_s is the thermal conductivity of the solid structure which varies with different material types.

In order to model fluid turbulence of high-speed swirl inside the bearing, the RNG k-ε model is selected for it is a recognizable model for rotating or swirling flow. The RNG k-ε turbulence model is a model derived from the transient N-S equation using the renormalization group method. Its k equation and ε equation have the similar form with the standard k-ε turbulence model but are more accurate in calculating the flow field with larger velocity gradient and strong rotational flow by increasing an additional term that is more responsive to the rapid curvature of strain and streamlines.

The transport equation of the turbulence kinetic energy k is

$$\frac{\partial(\rho k)}{\partial t} + \frac{\partial(\rho k u_i)}{\partial x_i} = \frac{\partial}{\partial x_i}\left[\alpha_k \mu \frac{\partial k}{\partial x_i}\right] + G_k + \rho\varepsilon \tag{10}$$

The transport equation of the turbulence dissipation rate equation ε is

$$\frac{\partial(\rho\varepsilon)}{\partial t} + \frac{\partial(\rho\varepsilon u_i)}{\partial x_i} = \frac{\partial}{\partial x_i}\left[\alpha_\varepsilon \mu \frac{\partial \varepsilon}{\partial x_i}\right] + \frac{C_{1\varepsilon}\varepsilon}{k} G_k - C_{2\varepsilon}\rho\frac{\varepsilon^2}{k} \tag{11}$$

where G_k is the production term of the turbulent kinetic energy caused by the average velocity gradient. x_i and u_i represent the coordinate directions and the velocity components, respectively. $C_{1\varepsilon}$, $C_{2\varepsilon}$, C_μ, α_k and α_ε are the turbulence model constants.

3.2 Grid meshing

Before the numerical solution of the governing equations, it is important to determine the computational domain and mesh the domain. In order to improve mesh quality and reduce the computer consumption, the computational domain should be simplified and reasonably divided before meshing. **Figure 3** shows the relevant grid schematic diagram of the computational domain, which contains a fluid domain and a solid domain. The fluid domain comprises the internal area of the nozzle, the bearing, and the outlets as shown in **Figure 3(a)**. The flow field

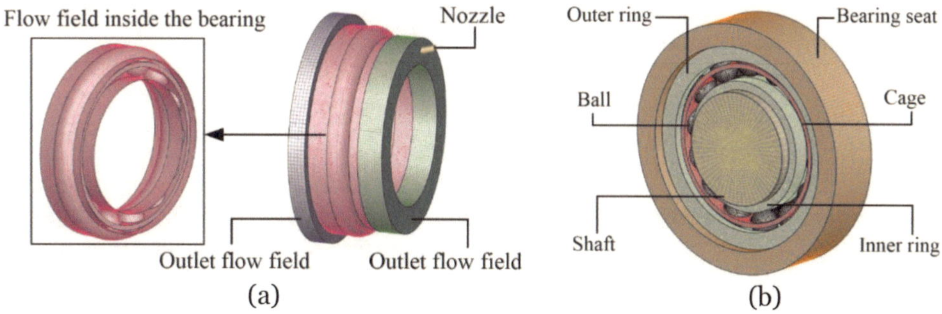

Figure 3.
Three-dimensional bearing heat transfer grid system: (a) mesh of the fluid domain and (b) mesh of the solid domain.

inside the bearing is the space enclosed by the inner surface of the bearing and the outer surface of the raceway, which is responsible for the most part of the interactions between the fluid flow and the solid elements.

Figure 3(b) is the solid domain, which includes the shaft, the rolling elements, the inner ring, the outer ring, the bearing seat, and the cage. The structured hexahedral mesh is used in the areas with regular shapes, such as the inner ring, the outer ring, the bearing seat, and the flow area inside the nozzle, and both of the outlets, etc., in contrast, for areas with relatively complex shapes, such as the flow field inside the bearing, was divided by a tetrahedral unstructured mesh.

The rotation of air-oil flow around the bearing rings is imposed by the rolling elements and the cage, which is rotating at a constant speed, which can be calculated by

$$n_\mathrm{b} = \frac{1}{2} n \left(1 - \frac{D_\mathrm{b} \cos\alpha}{d}\right) \tag{12}$$

where n_b is the rotary speed of the rolling elements, n is the rotary speed of the inner ring, D_b is the ball diameter, α is the bearing contact angle, and d is the pitch diameter of the bearing.

When setting boundary conditions, the nozzle inlet is initialized with a constant mass flow rate and an initial flow temperature. Flow outlet is set at the end faces of the two outlets and is set to be a pressure outlet boundary condition. The ambient environment is set as 0 Pa gauge pressure and 298 K, respectively. The operating pressure is 101,325 Pa. The viscosity of the oil phase changes with the oil temperature, as shown in **Table 3**. The multiple reference frame (MRF) method is used to describe the rotation of the rolling elements and the flow field inside the bearing. The sliding mesh planes are defined to deal with the interferences between the stationary and rotating computational domains. The heat source was applied to the contact area between the rolling elements and the raceway. Based on the experiment condition, the heat source power was calculated according to Harris's method [13]. Regarding the heat transfer between the fluid and solid structures, both the air and oil heat transfer properties are considered. Coupled heat transfer wall condition

Temperature (°C)	20	40	60	80	100	120
Oil viscosity (Pa·s)	0.305	0.146	0.070	0.034	0.016	0.008

Table 3.
Variation of the oil viscosity with the temperature.

was set at the solid-fluid interface. The convection coefficients were calculated by following the energy conservation equation in the computation.

3.3 Solution methods

The ANSYS FLUENT software platform [17] is used to perform the simulations. The solution format is set as a high-order solution mode. For the VOF air-oil two-phase model, the air phase is set to be the main phase. All the governing equations are discretized by finite volume (FV) method with the second-order upwind scheme and solved by the SIMPLEC method.

The residuals and the flux conservation on boundaries, for example, the mass flow rate on inlet and outlet boundaries, are monitored to detect the convergence of the governing equations. The conservation standard is set to be 0.01; that is to say, the computation is considered as convergence, whenever the net mass flow rate between the inlet and outlet boundaries drops to 1% of the mass flow rate on the inlets. Other convergence criteria of residual, such as the volume of fluid function and each velocity component, are all set to be 10^{-5}, except for that of the turbulent kinetic energy and the turbulent kinetic energy dissipation rate which is 10^{-3}.

The criteria are given as follows:

$$\frac{|m_{\text{inlet}} - m_{\text{outlet}}|}{|m_{\text{inlet}}|} < \varepsilon_m \tag{13}$$

$$\frac{|T_{oj+5000} - T_{oj}|}{T_{oj+5000}} < \varepsilon_T \tag{14}$$

where m_{inlet} is the oil mass flow at the inlet, m_{outlet} is the oil mass flow at the outlet, T_O is the outflow temperature, and ε_m and ε_T are the tolerance of oil mass flow and temperature, respectively. The subscript j is the iteration number.

In numerical simulation, the mesh density has a great influence on the accuracy and correctness of the calculation results. Choosing the appropriate number of meshes not only save the workload of the computer but also increase the reliability of the calculation results. Therefore, grid independence verification is an indispensable task in the process of numerical computation. To perform the grid independence verification, a set of numerical calculation is carried out on a computer platform with the flowing configuration: Processor-Intel E5540 × 2@2.53 GHz CPU, RAM-16 GB (4GB × 4), Hard Disk-2 TB, Graphic Card e AMD RADEON HD 6670–2 GB, and Operating system-Window 7 Ultimate 64-bit.

To determine the appropriate number of grids, three different sets of meshes with 135,929, 287,117, and 554,992 cell faces have been tested, and the outlet oil mass flow rate and the average oil volume fraction of the two-phase flow are obtained, as shown in **Table 4**. The numerical calculations show that with the increase of the number of grids, the variations of the calculated outlet oil mass flow rate and the average oil volume fraction are less than 2%. Considering both

	Total number of cell faces	**Outlet mass flow rate (kg/s)**	**Average oil volume fraction (−)**
Mesh 1	135,929	0.04399	0.0107
Mesh 2	287,117	0.04456	0.0103
Mesh 3	554,992	0.04420	0.0106

Table 4.
Technical data of the test apparatus.

calculation time and accuracy, the mesh with 135,929 cell faces is selected as a suitable number of grids.

4. Investigations of the air-oil flow distribution

To better understand the interaction between the oil and the bearing components, a high-speed camera is used to photograph the flow field inside the bearing with jet cooling. The test rig is mainly composed of a test bearing, a motorized spindle, a high-speed camera, a parallel light, and a gear pump. The gear pump and the motorized spindle can accurately control the oil flow rate and the inner ring speed, respectively. The variation of the flow field under different operating conditions can be captured by the high-speed camera.

4.1 Visualization of the air-oil flow field

The visualized flow field with the corresponding simulations inside the bearing at three different speeds is shown in **Figure 4**. The cooling oil is injected in a rate of 0.15 l/min from a 0.5 mm nozzle that leads to a jet velocity around 12 m/s. The bearing contrarotates at speeds of 500, 2000, and 4000 r/min, respectively. It is seen that all the rolling elements are evenly covered with the cooling oil, indicating

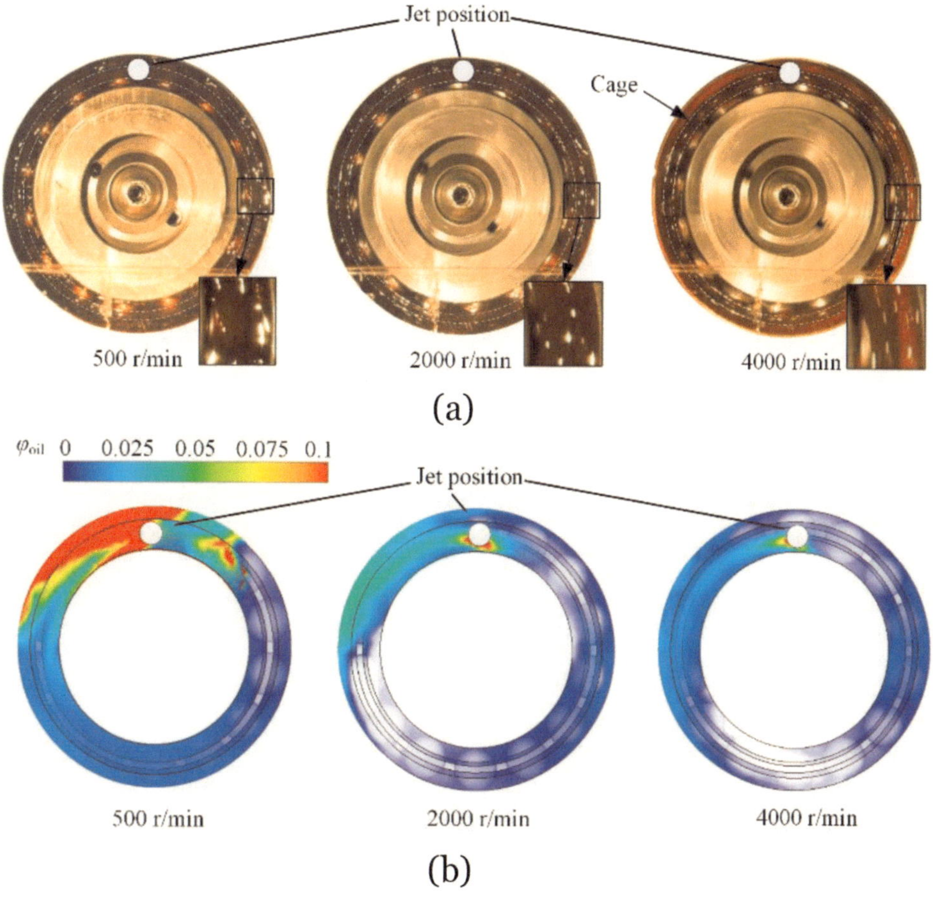

Figure 4.
Cooling oil distribution at different speeds: (a) measured results and (b) simulated results.

a well-distributed oil film at the speed of 500 r/min. When the bearing speed rotates at 2000 r/min, some of the rolling elements become apparent, indicating a decreased film thickness at a higher speed. Moreover, there are no noticeable bubbles observed inside the cooling oil. As the rotating speed reaches 4000 r/min, rolling elements are seen to be uncovered at certain spots, which indicate an insufficient oil supplement at this condition. Microbubbles, which are evident in the air entrainment, are also observed in this condition.

The contours of oil volume fraction in the simulation results well resemble the visualized flow field at each speed, which again confirms that the oil volume fraction inside the bearing decreases with the increase of bearing rotating speed. Thus, the heat transfer characteristics should be treated as a two-phase flow, especially at higher speeds, since the heat capacities of the oil and air are different.

Figure 5 represents the visualized and simulated flow field inside the bearing with respect to different flow rates. The bearing again rotates counterclockwise, and the speed is set to be 4000 r/min constantly. With nozzle diameter of 1.5 mm, the test flow rates are 0.3, 0.5, and 0.9 l/min, respectively. The general observation is that, in all tested flow rates conditions, the thickness of the film increases along the outward radial normal direction, which is evidenced by the phenomena that the cooling oil distributes more preferably around the outer ring due to the centrifugal effect.

A further general observation is that the oil film thickness near the outer ring gradually reduces along the direction of rotation under all conditions. For example,

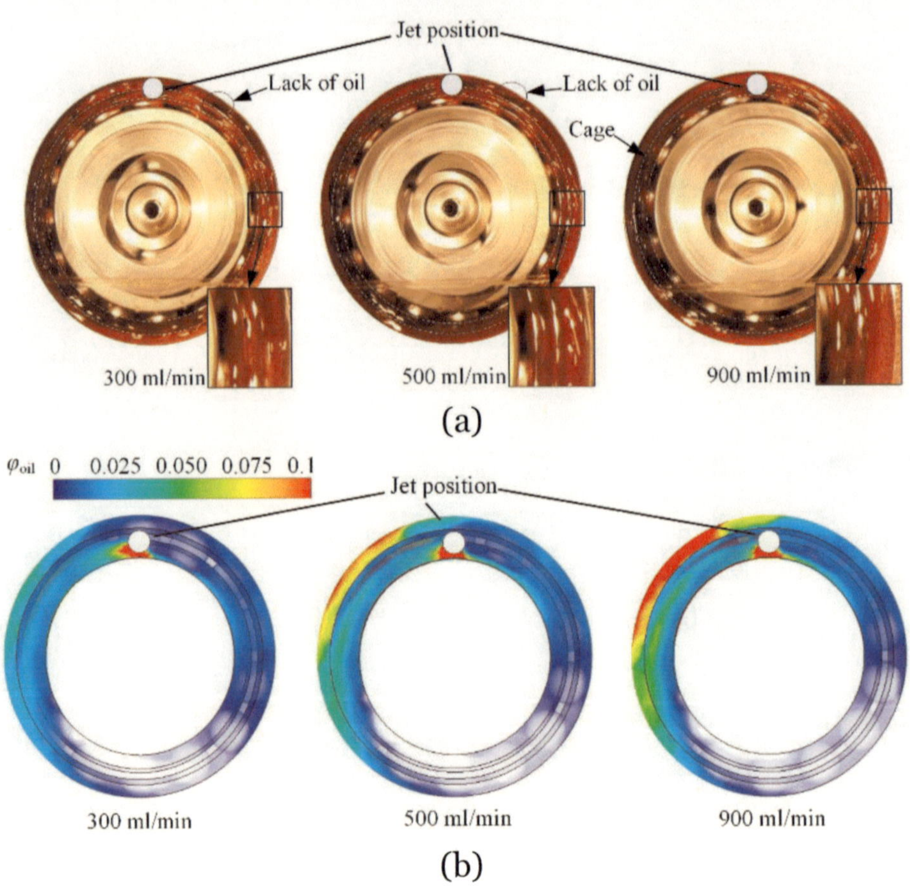

Figure 5.
Cooling oil distribution at different flow rates: (a) measured results and (b) simulated results.

in the 0.3 l/min case, the oil volume fraction becomes lower along the direction of rotation, and the lowest oil volume fraction appears at the upstream of the jet position, leading to a lack of oil. For the 0.5 l/min case, the oil lack area shrinks at the same observation positon, while there is no sign of insufficient oil supplement under the 0.9 l/min condition. This observation indicates that large flow rate is beneficial in homogenization of the cooling oil distribution inside the bearing. Nevertheless, please be noted that the power efficacy could be damaged if an unduly large flow rate of oil is used since the churning loss can be quite high.

Figure 6 shows oil volume fraction simulations inside the same bearing at high-speed conditions. The cooling oil distributions at different speeds show similar trends, as shown in **Figure 6(a)**. The oil volume fraction reaches its peak value at a certain downstream vicinity of the jet position and gradually reduces along the bearing rotation direction from then on. However, the peak oil volume fraction emerges at random position. The area of the peak oil volume fraction also changes under different speeds, as shown in **Figure 6(b)**. It can be seen that the highest oil

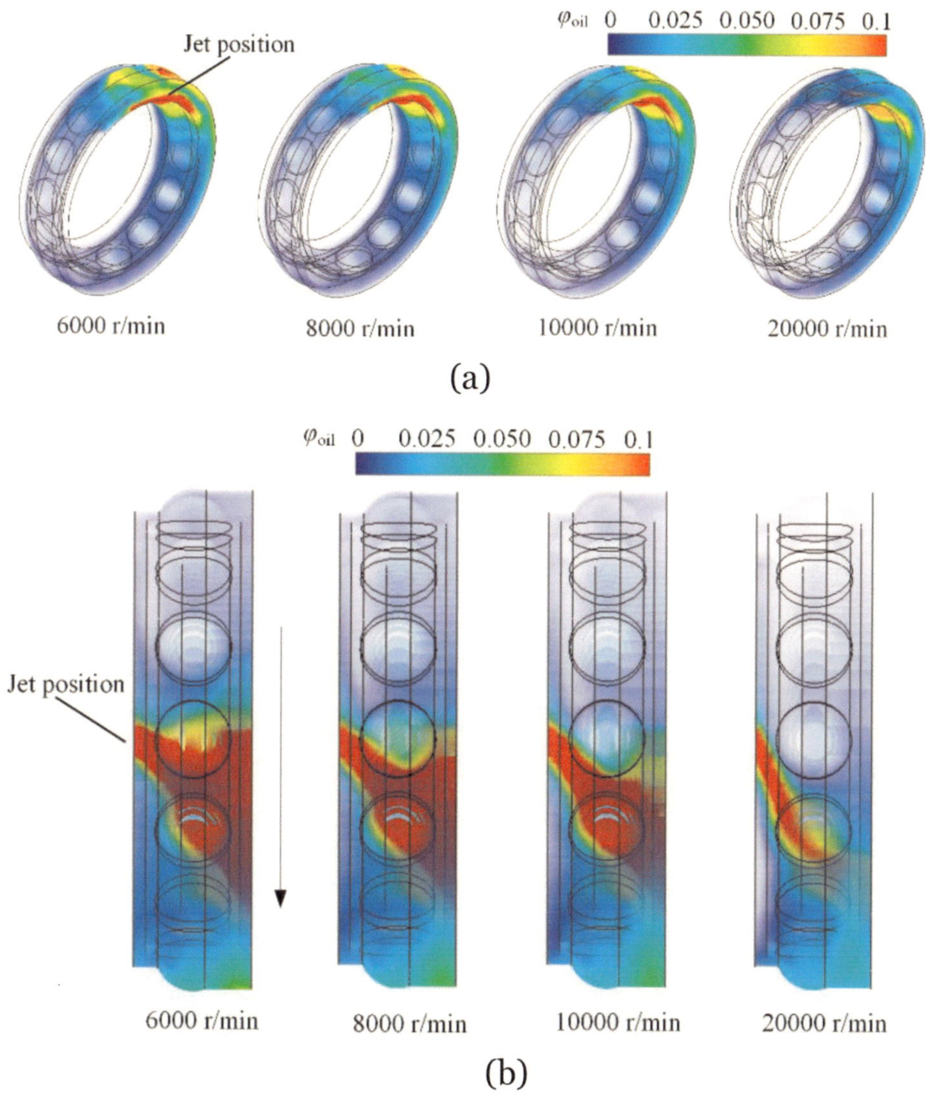

Figure 6.
Cooling oil flow field at higher speeds: (a) simulated oil flow field and (b) oil flow at the jet position.

volume fraction position moves along the rotation direction. This is because that the circumferential angular velocity of the oil increases rapidly at a higher speed. The oil is driven by the roller and cage. Further, part of the oil passes through the bearing directly at a lower speed. This reduces the utilization efficiency of the cooling oil.

4.2 Parametric studies on the air-oil flow distribution

Figure 7 shows the calculated air-oil distribution with different nozzle numbers. The speed of the inner ring is 10,000 r/min, and the oil flow rate is 3.0 l/min. The oil volume fraction contours from three to eight nozzles are given in **Figure 7(a)**. It is seen that the oil volume fraction rises with the increase of the nozzle number. **Figure 7(b)** presents the oil volume fraction distribution around the circumference. The parametric results of the oil volume fraction indicate the effect of the different nozzle numbers. It seems that the uniformity of the oil volume fraction inside the bearing rises with the increase of the nozzle number. The oil volume fraction also

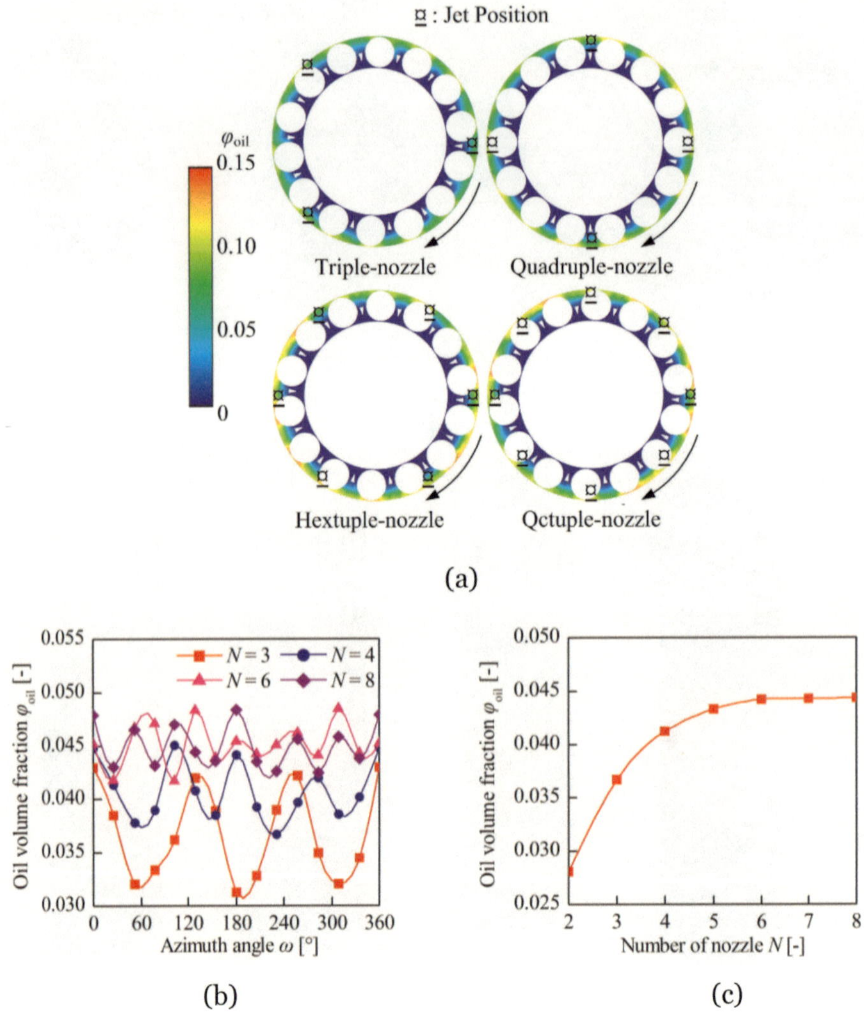

Figure 7.
Air-oil distribution with different numbers of nozzles: (a) oil volume fraction contours in the center cross section; (b) oil volume fraction distributions; and (c) average oil volume fraction around the circumference vs. the number of nozzle.

increases with the increase of the nozzle number. However, when the nozzle number is larger than 6, the oil volume fraction improvement becomes unapparent, as shown in **Figure 7(c)**. Further, when the nozzle number exceeds 4, the variation of the oil volume fraction and the uniformity of the oil volume fraction inside the bearing are less than 10%, as shown in **Table 5**. The multiple-nozzle jet requires a more complex mechanism than the single-nozzle jet.

Figure 8 shows the correlations between the air-oil distribution and jet velocity. The rolling bearing with a single-nozzle jet configuration is simulated; the oil volume fraction is obtained under a revolution speed of 10,000 r/min, and the oil flow rate is 3.0 l/min. The oil volume fraction is shown staying around both the inner

Nozzle numbers	1	2	3	4	6	8
Uniformity of the oil volume fraction	1.6029	0.6864	0.3189	0.2830	0.2624	0.2580
Rate of change	—	57.2%	53.5%	11.3%	7.3%	1.7%

Table 5.
Uniformity of the oil volume fraction with different nozzle numbers.

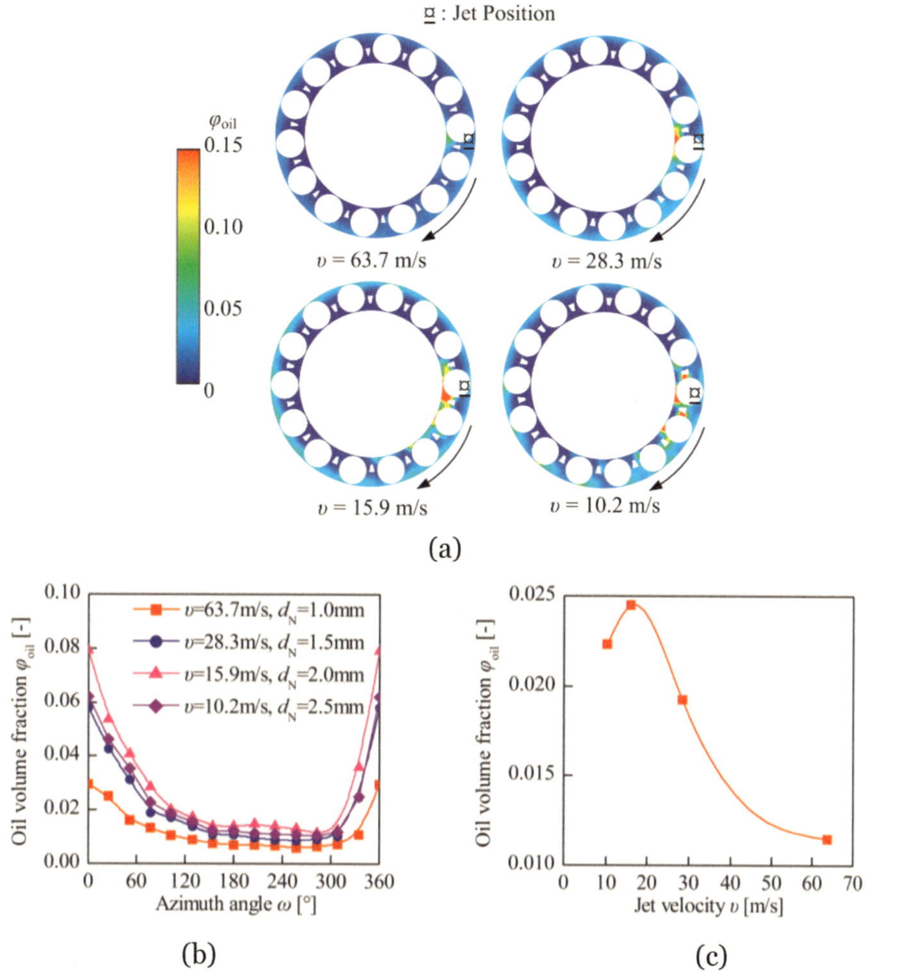

Figure 8.
Air-oil distribution with different jet velocities: (a) oil volume fraction contours in the center cross section; (b) oil volume fraction distributions; and (c) average oil volume fraction around the circumference vs. the jet velocity.

ring and the outer ring which becomes much lower with the increase of the jet velocity, as shown in **Figure 8(a)**. However, the oil volume fraction becomes more uniform with higher jet velocity. The oil volume fraction distribution around the circumference decreases with the higher jet velocity, as shown in **Figure 8(b)**. However, the jet velocity has little effect on the tendency of the oil volume fraction distributions around the circumference. Further, there is a jet velocity at which the average oil volume fraction achieves the largest value. The calculated result of the jet velocity is between 15 and 20 m/s in the given operation conditions, as shown in **Figure 8(c)**. The detailed relationship between the jet velocity and the oil volume fraction still needs more effort to investigate.

Figure 9 shows the simulated air-oil distribution with different oil flow rates under a constant speed of 10,000 r/min. As indicated in **Figure 9(a)**, the oil volume

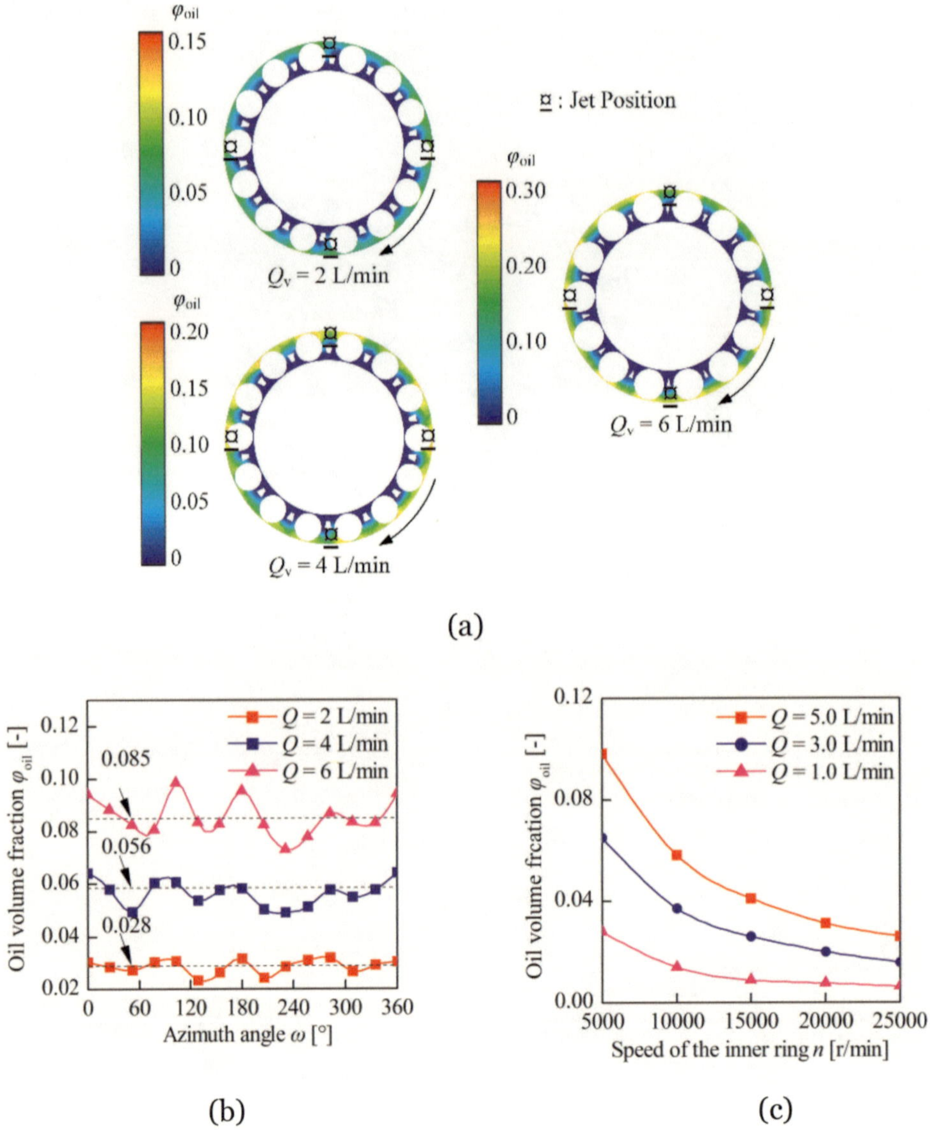

Figure 9.
Air-oil distribution with different oil flow rates: (a) oil volume fraction contours in the center cross section; (b) oil volume fraction distributions and (c) average oil volume fraction around the circumference vs. the operation speed.

fraction inside the bearing is increased as the oil flow rate increases. **Figure 9(b)** indicates that the oil volume fraction distribution under different oil flow rate condition shows similar trends; however, the amplitude of the oil volume fraction rises with the increase of the oil flow rate. Moreover, the oil volume fraction distributed more evenly at locations far from the jet position. The average oil volume fraction increases with the decrease of the speed, and the increase tendency becomes faster at higher speed, as shown in **Figure 9(c)**. Large flow rate is beneficial in homogenization of the cooling oil distribution inside the bearing and therefore the heat dissipation of the bearing. Nevertheless, the power efficacy could also be damaged due to the resulting unduly churning loss.

5. Investigations of the temperature distribution

Tested apparatuses for the temperature distribution of the ball bearing have been built up as shown in **Figure 2**. There are three temperature sensors attached to the outer ring of the test bearing and one more temperature sensor on the inner ring. A data acquisition system is also employed to collect and transmit the data to a computer for further analysis.

5.1 Temperature variations of the bearing

Figure 10 shows simulated outer ring temperature with the consideration of the nonuniform air-oil two-phase distribution under a bearing speed at 10,000 r/min. The oil flow rate is 3.0 l/min, which is equivalent to a jet speed about 10 m/s. The bearing is imposed with an axial load of 5.0 kN in the test. **Table 6** shows the comparison between the test measurements and simulation results. It is shown that the relative tolerance between the tests and the simulations is smaller than 5%,

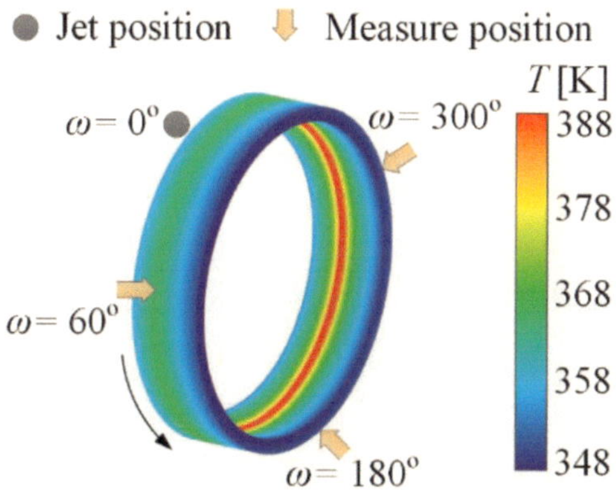

Figure 10.
The simulated temperature of the outer ring.

Location angle [°]	Simulated value [K]	Measured value [K]	Error (%)
60	365.58	373.6	2.19
180	368.06	376.1	2.18
300	369.23	378.4	2.48

Table 6.
Measured and simulated outer ring temperatures.

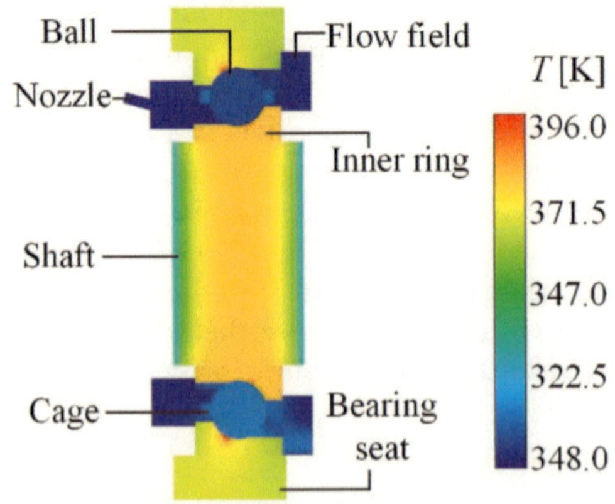

Figure 11.
Temperature distribution of the fluid-solid coupling heat transfer area.

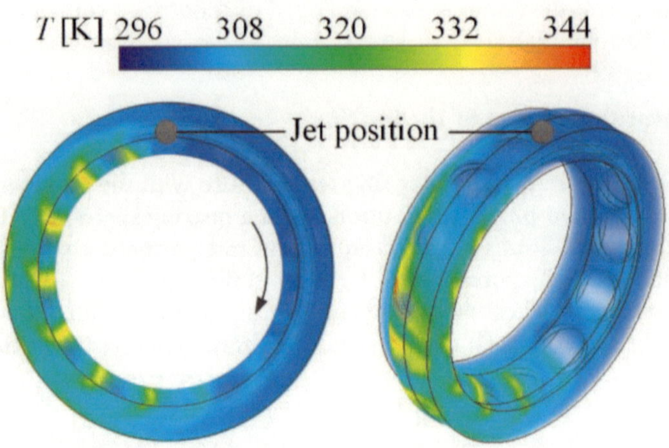

Figure 12.
Temperature distribution of the oil flow field inside the bearing.

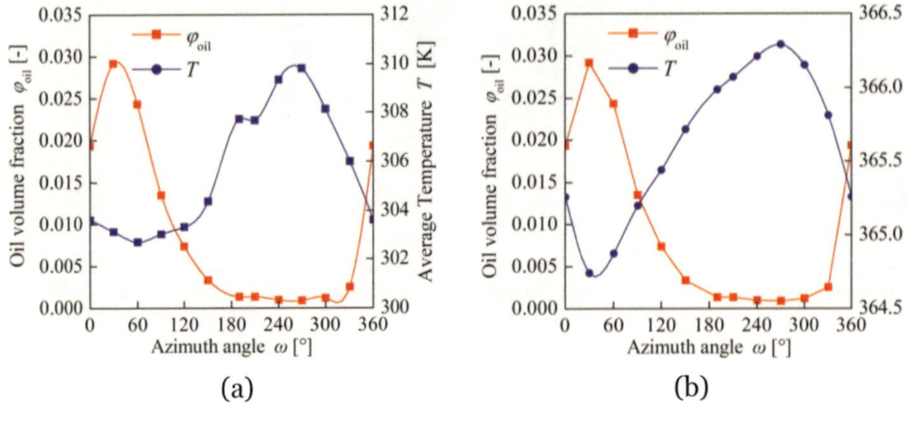

Figure 13.
Simulation of average circumferential temperature and oil volume fraction distributions: (a) average oil temperature and (b) average outer ring temperature.

which indicates that the simulated results are in good agreements with the experimental results and sufficient in preforming quantitative analyses, since the research focuses are the overall heat transfer effects of the air-oil flow inside the bearing.

Figure 11 shows the simulation result of radial temperature distribution of both the fluid and solid domains inside the rolling bearing. The temperature distribution along the radial direction is inhomogeneous. The inner ring shows a relatively high temperature compared to that of the rolling elements. It is reasonable since the distribution of cooling oil around the inner ring is less, as most of the cooling oil is forced to leave the inner ring for the outer ones due to the strong centrifugal effect under high rotating speed. The accumulation of cooling oil near the outer ring surface leads to a better heat transfer effect that contributes to the lower temperature of the outer ring. Owing to the large heat-generating rate between roller and raceway, the temperature in these contact regions is of the highest level.

The 3D display of the simulated temperature distribution shown in **Figure 12** confirms the nonuniformity of the temperature distribution inside the bearing. As shown in **Figure 13**, the temperature distribution of the flow field has a close

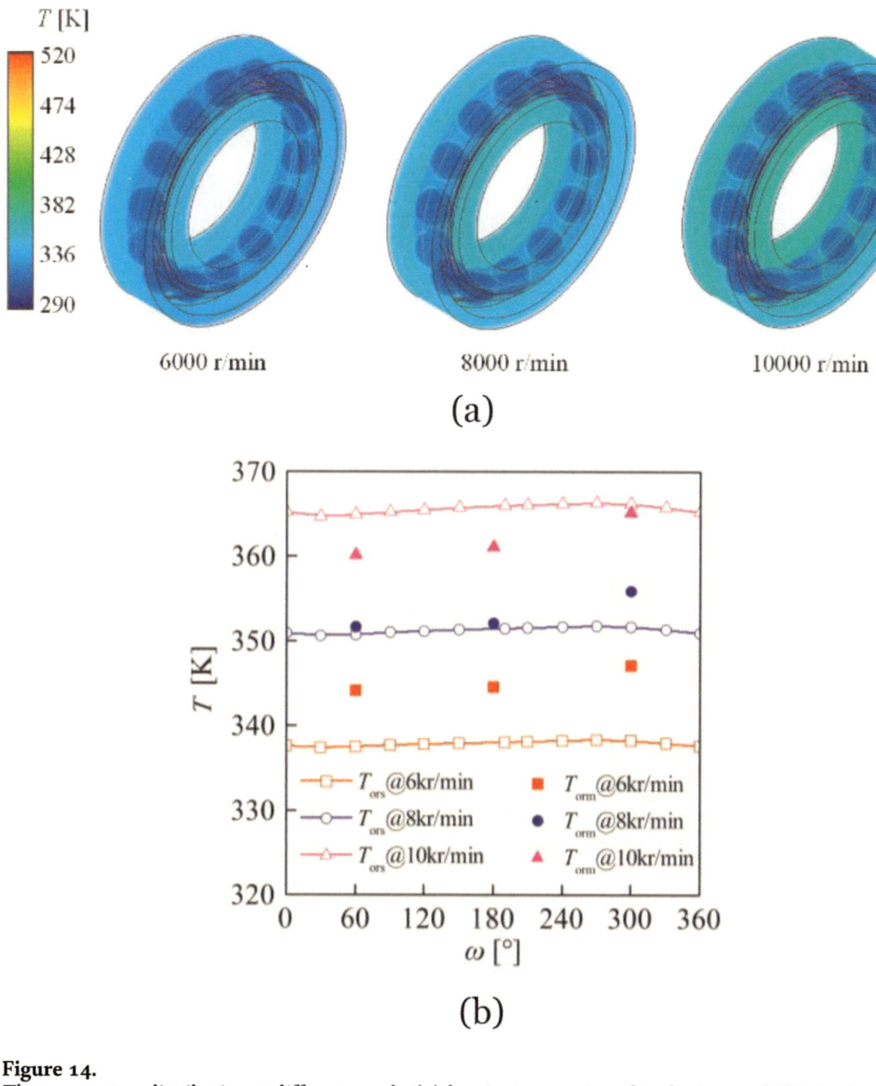

Figure 14.
The temperature distribution at different speeds: (a) bearing temperature distributions and (b) simulated and measured outer ring temperatures around the circumferential direction.

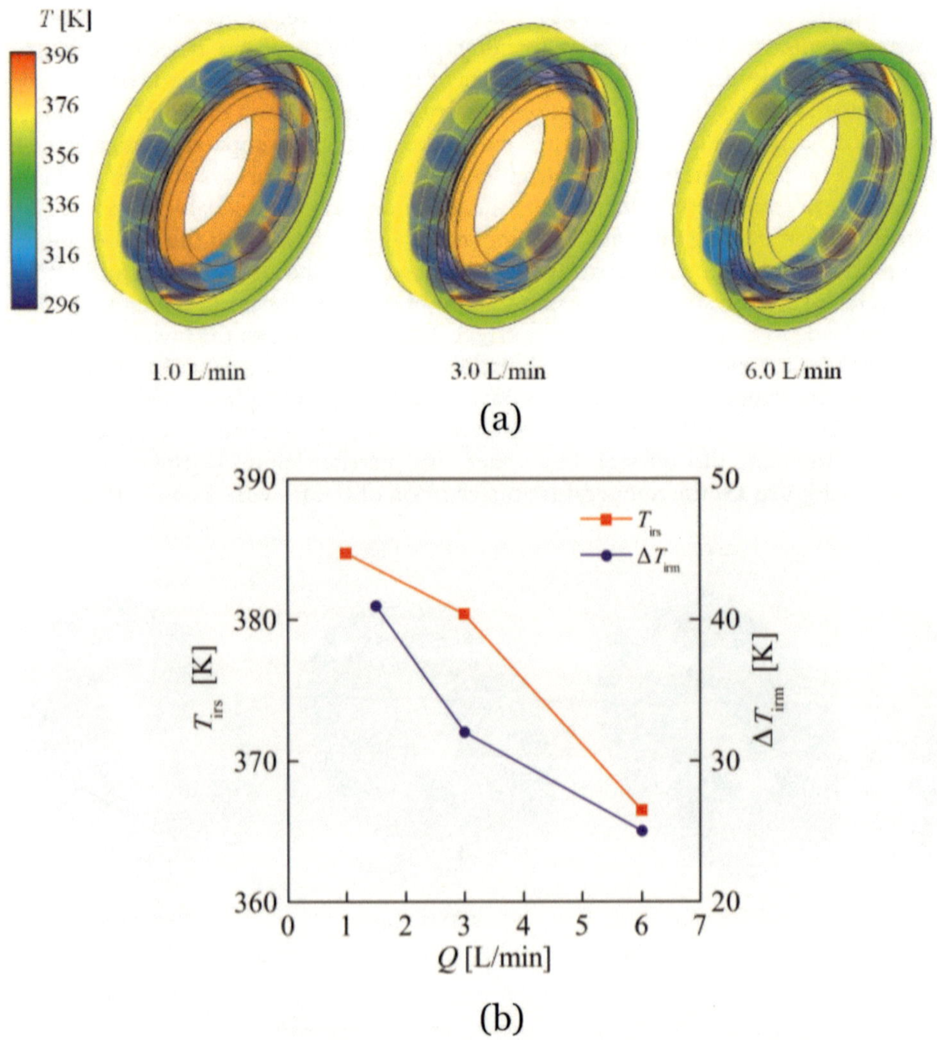

Figure 15.
The temperature distributions at different oil flow rates: (a) Bearing temperature distributions and (b) simulated inner ring temperature and measured inner ring temperature variations.

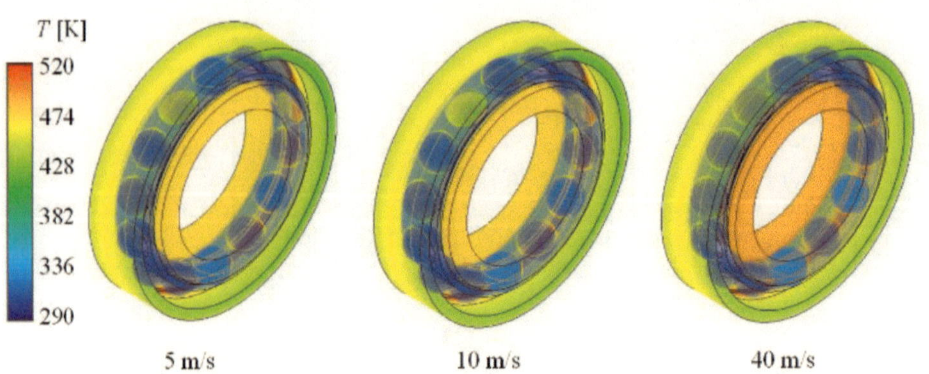

Figure 16.
The temperature distributions at different oil-jet speeds.

association with the distribution of oil and gas. It is known that the heat transfer capacity of oil is considerably higher than that of air. That explains the reason that the lower temperature always achieves at positions with higher oil volume fractions, such as the positions that are close to the jet position. In general, the temperature near the nozzle is the lowest and increases gradually along the direction of rotation. It drops on the other side of the nozzle. This is because the low-temperature oil is discharged from the nozzle, and the low-temperature oil is constant. More oil supplement is the key factor in achieving better heat convection and lower temperatures.

5.2 Parametric studies on the temperature distribution

Figure 14 shows the temperature distribution of the ball bearing at different speeds. The operating speeds of the bearing are 6000, 8000, and 10,000 r/min, respectively. The flow rate is 3.0 L/min and the oil injection speed is 10 m/s. It seems that the bearing temperature varies obviously at different speeds. With the increase of the speed, the bearing temperature rises, especially at a higher speed. Different from the average circumferential temperature distribution in the internal ring, the circumferential temperature difference exists in the outer ring. Further, the average temperature of the internal ring is higher than that of the outer ring.

Figure 15 shows the temperature distribution of the ball bearing under different oil flow rates. The oil flow rates are 1, 3, and 6 L/min with a speed of 10,000 r/min. The internal ring rotating direction is clockwise. The result indicates that the oil volume fraction of the flow field increases gradually. The convection heat transfer coefficient of the bearing boundary will also increase correspondingly. The heat dissipation condition becomes better. It is because that the amount of the oil entering the flow field in a unit time increases. At a constant speed, the stirring ability of the roller and cage changes little. Thus, the internal flow field of the bearing is distributed with more oil. The average temperatures of the rings and the flow field decrease, as shown in **Figure 15(b)**.

Figure 16 shows the temperature distribution of the ball bearing at different jet speeds. The jet speeds are 5, 10, and 40 m/s with an operating speed of 10,000 r/min. The lubrication flow rate is 3.0 L/min. With the increase of the jet speed, the oil volume fraction of the flow field inside the bearing reduces gradually. The temperature of the bearing also rises. The variation of the outer ring temperature is not apparent compared with the inner ring. It is because that the convective heat

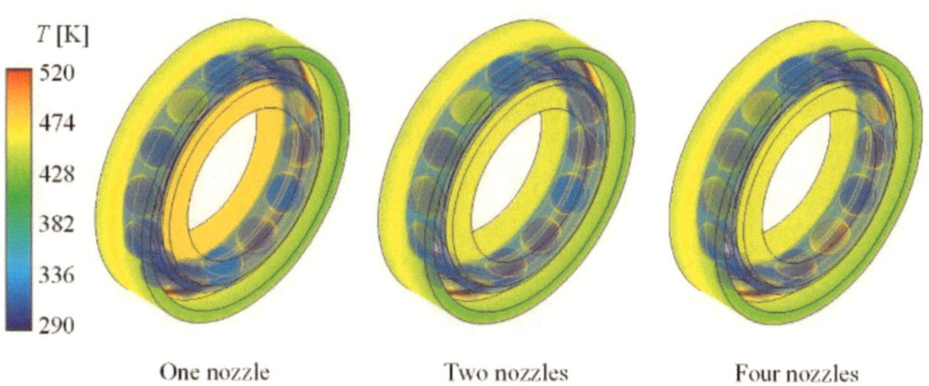

Figure 17.
The temperature distributions with different nozzle numbers.

Speed (r/min)	Temperature sensor					
	Outer ring temperature (°C)					
	One nozzle			Two nozzles		
	No. 1	No. 2	No. 3	No. 1	No. 2	No. 3
6000	71.0	71.4	74.0	69.0	68.3	70.3
10,000	87.2	88.1	92.0	84.0	83.7	86.7

Table 7.
The measured temperatures with different nozzle numbers and speeds.

transfer capacity of the bearing boundary decreases, especially around the inner ring. The heat dissipation condition becomes worse with a higher jet speed.

Figure 17 shows the temperature distribution of the ball bearing under different nozzle numbers. The nozzle numbers are 1, 2, and 4 with an operating speed of 10,000 r/min. The oil-jet speed is 10 m/s and the oil flow rate is 3.0 L/min. The bearing temperature is decreased with a larger nozzle number. The heat convection coefficient inside the bearing increases. The experiments of the bearing with one nozzle and two nozzles have been carried out.

Table 7 presents the experimental results. The measured temperatures were presented. The axis load is 5 kN and the radial load is 10 kN. It can be seen that the increase of the nozzle number can improve the heat dissipation effect under a constant flow rate. The experimental results are consistent with the calculated results.

6. Conclusions

A three-dimensional CFD analysis using the VOF multiphase model has been carried out to study heat transfer and fluid flow behavior in a jet cooling rolling bearing. The nonuniform air-oil distribution and temperature distribution of the two-phase flow inside the bearing had been studied. The results suggest the following:

1. The jet cooling oil inside the ball bearing is not pure oil phase, especially at the high-speed operation. The microbubble is mixed into the oil. The air-oil two-phase flow should be considered in the jet cooling bearing heat transfer analysis.

2. The VOF multiphase model and the coupled boundary conditions can be used to simulate the heat transfer phenomenon of a jet cooling rolling bearing. The influences of the different operation conditions on the bearing component temperatures can be forecasted. The results can be used to guide the design of the lubricating mechanism.

3. The oil volume fraction reaches the highest value at the position close to the nozzle. The highest oil volume fraction position changes with different speeds and moves along the rotation direction.

4. The full wall oil film appears inside the jet cooling bearing at the low-speed stage. It breaks with the increase of the rotating speed. The oil-gas distribution is fairly uniform, and the oil volume fraction is low during the high speed.

5. With the increase of the rotating speed, the bearing temperature increases. With higher flow rate, larger nozzle number, and lower jet speed, the average temperature of the bearing system decreases. The outer ring temperature changes much smaller than that of the inner ring with the variation of the flow rate, nozzle number, and jet speed.

The flow field and thermal behavior analyses give an opportunity for the advanced precision cooling mechanism design of the high-speed rolling bearing. Work on these topics is underway in the National Key Laboratory of Vehicular Transmission at the Beijing Institute of Technology.

Conflict of interest

The authors declared that they have no conflicts of interest with other parties that can unsuitably influence the manuscript entitled "Flow and Heat Transfer in Jet Cooling Rolling Bearing."

Author details

Wei Wu*, Jibin Hu and Shihua Yuan
National Key Laboratory of Vehicular Transmission, Beijing Institute of Technology, Beijing, P.R. China

*Address all correspondence to: wuweijing@bit.edu.cn

References

[1] Gloeckner P. The influence of the raceway curvature ratio on power loss and temperature of a high-speed jet engine ball bearing. Tribology Transactions. 2013;**56**(1):27-32

[2] Jiang S, Mao H. Investigation of the high speed rolling bearing temperature rise with oil-air lubrication. Journal of Tribology-Transactions of the ASME. 2011;**133**(2):1-9

[3] Pinel SI, Signer HR, Zaretsky EV. Comparison between oil-mist and oil-jet lubrication of high-speed, small-bore, angular-contact ball bearings. Tribology Transactions. 2001;**44**(3):327-338

[4] Aidarinis J, Missirlis D, Yakinthos K, Goulas A. CFD modeling and LDA measurements for the air-flow in an aero engine front bearing chamber. Journal of Engineering for Gas Turbines and Power-Transactions of The ASME. 2011;**133**(8):1-8

[5] Hu JB, Wu W, Wu MX, Yuan SH. Numerical investigation of the air-oil two-phase flow inside an oil-jet lubricated ball bearing. International Journal of Heat and Mass Transfer. 2014;**68**:85-93

[6] Wu W, Hu C, Hu J, Yuan S. Jet cooling for rolling bearings: Flow visualization and temperature distribution. Applied Thermal Engineering. Jul. 2016;**105**(SI):217-224. DOI: DOI. 10.1016/j.applthermaleng. 2016.05.147

[7] Hager CH, Doll GL, Evans RD, Shiller PJ. Minimum quantity lubrication of M50/M50 and M50/Si3N4 tribological interfaces. Wear. 2011;**271**(9–10): 1761-1771

[8] Zaretsky EV, Signer H, Bamberger EN. Operating limitations of high-speed jet-lubricated ball bearings. Journal of Lubrication Technology. 1976;**98**(1): 32-39

[9] Oh IS, Kim D, Hong SW, Kim K. Three-dimensional air flow patterns within a rotating ball bearing. Advanced Science Letters. 2013;**19**(8):2180-2183

[10] Yann M, Christophe C, Fabrice V. Numerical investigations on drag coefficient of balls in rolling element bearing. Tribology Transactions. 2014; **57**:778-785

[11] Zhou H, Luo G, Chen G, Wang F. Analysis of the nonlinear dynamic response of a rotor supported on ball bearings with floating-ring squeeze film dampers. Mechanism and Machine Theory. 2013;**59**:65-77

[12] Gloeckner P, Ebert F. Micro-sliding in high-speed aircraft engine ball bearings. Tribology Transactions. 2010; **53**(3):369-375

[13] Harris TA, Barnsby RM, Kotzalas MN. A method to calculate frictional effects in oil-lubricated ball bearings. Tribology Transactions. 2001;**44**(4): 704-708

[14] Harris TA, Kotzalas MN. Rolling Bearing Analysis. 5th ed. New York: Taylor & Francis; 2006

[15] Takabi J, Khonsari MM. Experimental testing and thermal analysis of ball bearings. Tribology International. 2013;**60**:93-103

[16] Hirt C, Nichols B. Volume of fluid (VOF) method for the dynamics of free boundaries. Journal of Computer Physic. 1981;**39**(1):201-225

[17] ANSYS Inc. ANSYS FLUENT 15.0 Tutorial Guide. In: ANSYS Inc. 2015

Chapter 12

Two-Phase Flow Modeling of Cryogenic Loading Operations

Ekaterina Ponizovskaya-Devine, Dmitry Luchinsky, Michael Foygel, Vasil Hafiychuk, Michae Khasin, Jared Sass and Barbara Brown

Abstract

We consider problem of modeling and controlling two-phase cryogenic flows during ground loading operations. We introduce homogeneous moving front and separated two-phase flow solvers that are capable of fast and accurate online predictions of flow dynamics during chilldown and transfer under nominal conditions and in the presence of faults. Concise sets of cryogenic correlations are proposed in each case. We present results of application of proposed solvers to the analysis of chilldown in large-scale experimental cryogenic transfer line build in Kennedy Space Center. We discuss optimization of parameters of cryogenic models obtained using general inferential framework and an application of the solvers to the fault detection and evaluation based on D-matrix approach. It is shown that solver's predictions are in good agreement with experimental data obtained for liquid nitrogen flow in nominal regime and in the presence of faults.

Keywords: cryogenic flow, chilldown, moving front, two-phase flow, optimization, flow boiling, correlations, heat transfer

1. Introduction

Deep space exploration requires development of autonomous management system [1] that can recognize, predict, and control two-phase cryogenic flows online without human interaction.

This long-standing problem is of great complexity because flowing fluids are often far away from thermal equilibrium (e.g., during chilldown), and their dynamics involves heat and mass transfer [2–4]. A sketch of the two-phase flow is shown in the **Figure 1**. Within the accepted approximation [3, 4], the pressure p in the system is common. The temperature of the wall T_w, vapor T_g, and liquid T_l can all be different. Gas and liquid velocities can have different values and directions. The heat and mass exchange and the drag take place both at the wall and at the interface. The shape of the interface can be complex [5].

During the past decades, a number of efficient algorithms have been developed for analysis of multiphase flows [3, 4, 6–9]. However, the problem still presents a major challenge to the scientific community due to the lack of a general agreement regarding the fundamental physical models describing two-phase flows [10] and relatively poor knowledge of heat and mass transfer correlations in boiling cryogenic flows [11–15].

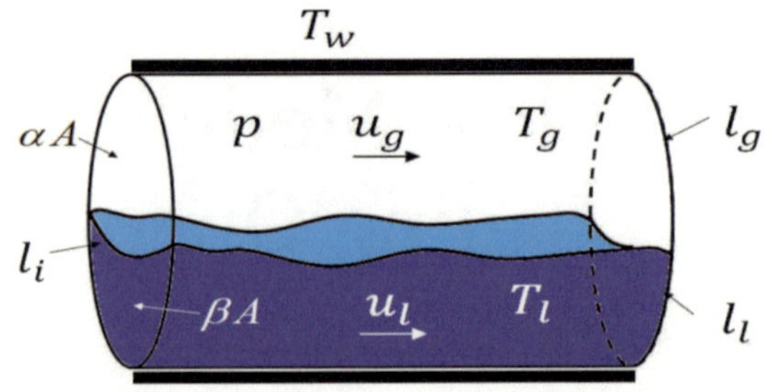

Figure 1.
Sketch of two-phase flow. α is the void fraction, A is the cross-section of the flow, l_g and l_l are dry and wetted perimeters, and l_i is interface perimeter. Other parameters are explained in the text.

To address these issues critical issues, NASA has implemented an impressive program of research [1, 16–18] that resulted in emergence of space-based fluid management technologies; see, e.g. [14, 15, 19–24] and references therein. Specifically, two-phase separated flow models were developed for some of the flow regimes [25, 26]. A number of optimization techniques have become commercially available for analysis of the model parameters and data correlations [27].

Despite these efforts small-time steps and instabilities [26] or implicitness of numerical scheme [28, 29] impose substantial limitations on the speed of the solution and possibility of online application. And an extensive research is currently under way to improve accuracy of predictions of transient cryogenic flows [15, 19, 24, 27].

Some of the grand challenges of this research include development of fast algorithms, choice of cryogenic correlations, and applications of the algorithms to the full-scale practical systems.

In this paper, we report on the development of fast two-fluid models suitable for online analysis of cryogenic flows and discuss applications of these models to the large-scale experimental prototypes of cryogenic transfer lines for ground launch systems.

The paper is organized as follows. In the next section, we describe the moving front model and algorithm of integration of homogeneous cryogenic flow. In Section 3, we discuss the separated two-phase flow model. In Section 4 we briefly discuss applications of these models to the inference of cryogenic correlations and fault detection in cryogenic transfer line. Finally, in the Conclusions, we summarize the obtained results and discuss directions of future work.

2. Homogeneous moving front model

2.1 Model equations

One of the simplest models that can recognize and predict cryogenic two-phase fluid dynamics online in nominal and off-nominal (i.e., in the presence of faults) flow regimes is moving front homogeneous model [30, 31]. The model was developed for the integrated health management system for cryogenic loading operations at the Kennedy Space Center (KSC). It is similar to the model [32] that describes transient phenomena in evaporators and condensers in refrigeration. It accounts for the heat exchange with the pipe walls and for the moving liquid front in chilldown regime. It reproduces with good accuracy the time evolution of the pressure, temperature, and flow rate in spatially distributed cryogenic systems in the nominal loading regime. It can detect and isolate major faults of cryogenic loading operation in the real time. It works much faster than commercially available full-scale fluid dynamics solvers and can be used for preliminary analysis of the loading protocols on the ground and under microgravity conditions [33].

Within this model the cryogenic transfer line is divided into a set of control volumes with constant cross-section shown in **Figure 2**. The long pipes can consist of several control volumes. Following [30] we will assume that within each control volume flow is one-dimensional along the tube's axis in z-direction. The change in pipe diameter considered only at the control volume boundaries. We further neglect the viscous dissipation and spatial variations of the pressure. The model is based on the conservation equations for the mass, momentum, and specific energy e in one dimension:

$$\begin{aligned} &\rho_{,t} + (\rho u)_{,x} = 0 \\ &(\rho u)_{,t} + (u(\rho u))_{,x} + p_x + \frac{1}{A}\tau_w l_w + \rho g \sin\theta = 0 \\ &(\rho e)_{,t} + (\rho u H)_{,x} + \frac{1}{A}\dot{q}_w l_w = 0 \end{aligned} \tag{1}$$

Here H is enthalpy, ρ is density, and u is velocity of the flow. τ_w is the friction losses, l_w is the pipe length, g is the gravity, θ is the angle of the pipe with horizontal plane, $\dot{q}_w$ is the heat transfer from the pipe wall to the fluid mixture, and A is the cross-section area of the pipe.

The wall temperature is described by the energy conservation equation:

$$c_w \rho_w A_w (T_w)_{,t} = h\pi D(T - T_w) + h_{amb}\pi D(T_{amd} - T_w) + \kappa A_w (T)_{,xx} \tag{2}$$

where A_w is the cross-sectional area of the pipe wall with diameter D, which is in thermal contact with the ambient fluid of temperature T_{amb} and with the fluid of temperature T flowing in the pipe. c_w is the specific heat, ρ_w is density, and κ is thermal conductivity of the wall material. h is the heat transfer coefficient at the inner wall surface, and h_{amb} is the heat transfer coefficient at the ambient side.

Two-phase flow is considered to be homogeneous with void fraction of the vapor equal α and $(1-\alpha)$ for the liquid. Each volume can contain pure liquid with $\alpha = 0$, pure gas with $\alpha = 1$, and two-phase flow, and there are two volumes that contain the moving front from liquid to two-phase and from two-phase to gas. The local void fraction and the vapor mass fraction (fluid quality x) in the two-phase region are

$$\alpha = \frac{A_v}{A} = \frac{A_v}{A_v + A_l} \quad \text{and} \quad x = \frac{M_v}{M_v + M_l} = \left(1 + \frac{(1-\alpha)\rho_v}{\alpha/\rho_l S}\right)^{-1} \tag{3}$$

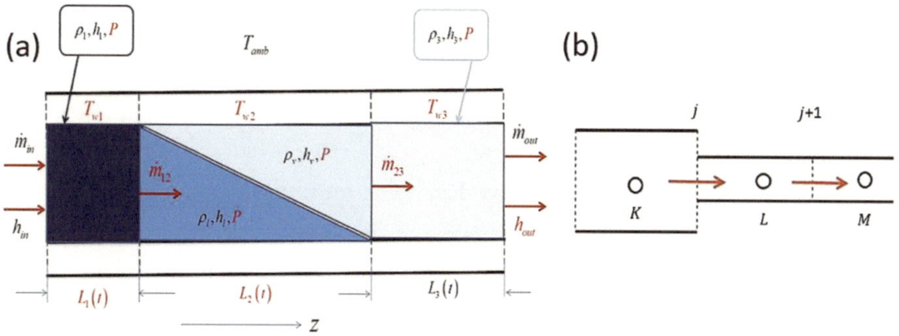

Figure 2.
(a) Three main control volumes of the model. Subindexes 1, 2, and 3 refer to (1) subcooled liquid region; (2) saturated two-phase region; and (3) superheated vapor region. (b) The convention for the nodes and junction notations: K, L and M refer to the centers of control volumes and index j enumerates junctions between control volumes.

where $S = u_v/u_l$ is the slip ratio. It can be shown that for the steady flow without slip ($S = 1$), the local density and specific enthalpy of the mixture are [34]:

$$\rho = \alpha\rho_v + (1-\alpha)\rho_l; \quad H = xH_v + (1-x)H_l \tag{4}$$

The momentum equation is reduced to a set of quasi-static algebraic equations assuming that the flow is relatively slow and there are no shock waves:

$$(A\rho u^2)_{,x} + \frac{1}{A}(\tau_w l_w)_{tp} = p_{,x} - \rho g sin(\theta) \tag{5}$$

The model equations are closed by adding equation of state $\rho_{g,l} = \rho_{g,l}(p, H_{v,l})$ in the form of tables and by providing a set of cryogenic correlations discussed below.

There is a multitude of data on void fraction-quality correlations for different flow patterns in horizontal and upward inclined pipes [30]. Here we use correlation of the form [30]

$$\alpha(x,p) = \left(1 + a(p)\left(\tfrac{1}{x} - 1\right)\right)^{-1} \tag{6}$$

where the factor a is

$$a = \left(\frac{\rho_v}{\rho_l}\right)^{n1} \left(\frac{\mu_l}{\mu_v}\right)^{n2} \tag{7}$$

and μ is the viscosity of the fluid. Coefficients n_1 and n_2 can have different values, e.g., according to Thome [35] $n1 = 0.89, n2 = 0.18$, while it is suggested to use $n1 = 0.67, n2 = 0$ in [34].

The estimated average void fraction α can be used to evaluate the heat transfer coefficient from the tube wall to the TP mixture in the TP region ($n = 2$) as $h_2 = \alpha h_v + (1-\alpha)h_l$, where $h_{v,l} = \left(\frac{\kappa}{D_h} Nu\right)_{v,l}$, Nu is Nusselt number, and Pr is Prandtl number.

For the fully developed laminar flow, $Nu = 3.66$. For the turbulent flow ($\mathrm{Re} > 2300$), one can use the modified Dittus-Boelter approximation with the Nusselt number calculated separately for each of the vapor and liquid components. For example, according to [34]

$$Nu = 0.0214\left(\mathrm{Re}^{4/5} - 100\right)\mathrm{Pr}^{2/5}\left(1 + (D_h/L_2)^{2/3}\right) \tag{8}$$

for $0.5 < \mathrm{Pr} < 1.5$ and

$$Nu = 0.012\left(\mathrm{Re}^{0.87} - 280\right)\mathrm{Pr}^{2/5}\left(1 + (D_h/L_2)^{2/3}\right) \tag{9}$$

for $1.5 < \mathrm{Pr} < 500$. Here the effective Reynolds number for each of the components, $\mathrm{Re}_{v,l} = \left(\frac{\dot{m}D_h}{\mu A}\right)$, is estimated by means of the average hydraulic diameters $D_{h,v} = \alpha D$ and $D_{h,l} = (1-\alpha)D$.

For the forced convection, the heat transfer coefficient [9] at the environmental side is

$$h_{amb} = \frac{2\kappa}{\pi D} Nu_{amb}, \qquad Nu_{amb} = Nu_{lam} + Nu_{turb}, \tag{10}$$

where laminar and turbulent contributions to the Nusselt number

$$Nu_{lam} = 0.0664\text{Re}^{1/2}\text{Pr}^{1/3} \quad and \quad Nu_{turb} = \frac{0.037\text{Re}^{0.8}\text{Pr}}{1 + 2.443\text{Re}^{-0.1}\left(\text{Pr}^{2/3} - 1\right)} \tag{11}$$

are calculated for the ambient fluid.

The specific form of the linearized conservation equations depends on the type of the flow. Using Taylor expansion for density ρ in terms of pressure p and enthalpy H, we can rewrite the mass and energy conservation equations in terms of pressure p and enthalpy H or pressure p and void fraction α. In the case of two-phase control volume, the equations are (with $\beta = 1 - \alpha$):

$$\begin{aligned}
&\left(\alpha\frac{\partial\rho_v}{\partial p} + \beta\frac{\partial\rho_l}{\partial p}\right)\frac{dp}{dt} + (\rho_v - \rho_l)\frac{d\alpha}{dt} = \frac{\dot{m}_L - \dot{m}_R}{LA} \\
&\left(\alpha\frac{\partial(\rho H)_v}{\partial p} + \beta\frac{\partial(\rho H)_l}{\partial p} - 1\right)\frac{dp}{dt} + (\rho_v H_v - \rho_l H_l)\frac{d\alpha}{dt} = \frac{\dot{m}_L H_L - \dot{m}_R H_R}{LA} + \frac{4h(T_w - T)}{D}
\end{aligned} \tag{12}$$

and in the case of single-phase control volume (e.g., vapor), we have

$$\begin{aligned}
&\frac{\partial\rho}{\partial p}\frac{dp}{dt} + \frac{\partial\rho}{\partial H}\frac{dH}{dt} = \frac{\dot{m}_L - \dot{m}_R}{LA} \\
&\left(\frac{\partial(\rho H)}{\partial p} - 1\right)\frac{dp}{dt} + \left(\frac{\partial\rho H}{\partial H}\right)\frac{dH}{dt} = \frac{\dot{m}_L H_L - \dot{m}_R H_R}{LA} + \frac{4h(T_w - T)}{D}
\end{aligned} \tag{13}$$

Here subindexes R and L refer to the right and left boundaries; ρ_v and ρ_l are the saturated density for vapor and liquid in Eq. (12), while in Eq. (13) for the single-phase, ρ is the density of vapor.

The case when the saturated two-phase flow and gas flow coexist in one control volume integration of the conservation equations that accounts for the moving front position L_2 results in the following set of equations:

$$\begin{aligned}
&\left(\alpha\frac{\partial\rho_v}{\partial P} + \beta\frac{\partial\rho_l}{\partial p}\right)\frac{dp}{dt} + (\rho_v - \rho_l)\left(\frac{d\alpha}{dt} - \beta\frac{dln(L_2)}{dt}\right) = \frac{\dot{m}_L - \dot{m}_R}{L_2 A} \\
&\left(\alpha\frac{\partial\rho_v H_v}{\partial p} + \beta\frac{\partial\rho_l H_l}{\partial p} - 1\right)\frac{dp}{dt} + (\rho_v H_v - \rho_l H_l)\left(\frac{d\alpha}{dt} - \beta\frac{dln(L_2)}{dt}\right) \\
&\quad = \frac{\dot{m}_L H_L - \dot{m}_R H_v}{L_2 A} + \frac{4h}{D}(T_w - T) \\
&\frac{\partial\rho}{\partial p}\frac{dp}{dt} + \frac{\partial\rho}{\partial H}\frac{dH}{dt} + (\rho - \rho_v)\frac{dln(L_3)}{dt}) = \frac{\dot{m}_L - \dot{m}_R}{(L_3)A} \\
&\left(\frac{\partial\rho H}{\partial p} - 1\right)\frac{dp}{dt} + \frac{\partial\rho H}{\partial H}\frac{dH}{dt} + (\rho H - \rho_v H_v)\frac{dln(L_3)}{dt} = \frac{\dot{m}_L H_l - \dot{m}_R H_R}{(L_3)A} + \frac{4h}{D}(T_w - T)
\end{aligned} \tag{14}$$

Here $L_3 = L - L_2$ and L is the length of the control volume; see **Figure 2**. These equations are solved simultaneously to find pressure, enthalpy, and density in each region and the position of the moving front [30]. The wall temperature is calculated for each region according to Eq. (2).

The mass fraction x is used to calculate the enthalpy flux at the boundaries, and it is assumed to change linearly with coordinate in the phase transition region. The new enthalpy for two-phase flow depends on x, and the new density is defined by the void fraction α in Eq. (4), while the relation between x and α is defined by Eq. (6).

The model was applied to consider two-phase flow of liquid nitrogen in the KSC experimental cryogenic transfer line. It takes into consideration compressibility of the vapor and condensation-evaporation effects in the flowing cryogenic fluid through the initially hot transfer line. Both the transient and steady-state regimes of flow can be described by accounting for the viscous and thermal interaction between the two-phase fluid and the pipe wall. Similarly, the fluid dynamics in the storage and vehicle tanks includes evaporation at the liquid surface and the heat exchange at the tanks walls as described in [36].

2.2 Algorithm

The conservation equations (1) are solved for the set of control volumes that form a staggered grid. The solution proceeds by iterating the following steps. For each volume we calculate pressure P, position of the front L, void fraction α, vapor (v) and liquid (l) enthalpies $H_{v(l)}$, and two wall temperatures before and after the front T_{w1} and T_{w2}. These variables are calculated at the center of control volume.

Next the mass $\dot{m}$ and enthalpy fluxes $H\dot{m}$ are calculated at the boundaries of each control volume; see **Figure 2**, using the following equation:

$$\frac{\dot{m}_j^2}{A_n^2}\left(\frac{1}{\rho_n}\left(f_n\frac{L_n}{D_n}+\sum K_n\right)+\left(\frac{x^2}{\gamma\rho_g}+\frac{(1-x)^2}{(1-\gamma)\rho_l}\right)_n\right)=-\Delta p_j-\Delta z_j\rho_j g sin\theta \qquad (15)$$

The subindex n for frictional factor f and minor losses K refers to the control volume on the left- or right-hand side of the junction. The stability of this approximation was enforced by the donor-like formulation of the frictional losses for the mass fluxes [35].

For each control volume k, the continuity condition is added as follows: $\dot{m}_{k,in}=\dot{m}_{k-1,out}$ for the serial connections of the $k-1$, k, and $\dot{m}_{k-1,out}=\dot{m}_{k1,in}+\dot{m}_{k2,in}$ if the k-th control volume consists of two parallel control volumes k_1 and k_2.

Updated values of the thermodynamic variables, enthalpy, and mass fluxes are used to calculate new parameters of friction and heat transfer correlations.

The algorithm can be briefly summarized as follows: (i) calculate variables P, H, L, α, and T_w at the center of the control volume using mass and heat fluxes from the previous step, (ii) update location of the moving fronts, (iii) calculate new mass and heat fluxes at the boundaries of each control volume, and (iv) update friction and heat transfer correlations and return to the first step.

The moving front model allows to account for multiple faults in cryogenic transfer lines including, e.g., valve clogging, valve stuck open/closed, and heat and mass leaks [37, 38]. However, the homogeneous moving front model has some important limitations. It does not include shock waves and cannot distinguish nonhomogeneous ($u_g \neq u_l$) and non-equilibrium ($T_g \neq T_l$) flows such as e.g. counter flows and stratified flows. To overcome these limitations we developed two-phase separated solver for cryogenic flows discussed below.

3. Two-phase flow model

Having in mind fast online applications of the solver we have tested a number of algorithms for the separated two-phase flow [30, 31, 39–44]. It was shown that the nearly implicit algorithm (NIA), similar to one developed in [7], can be applied successfully for online predictions of nonhomogeneous and non-equilibrium flows. Below we describe briefly the NIA and the corresponding model equations. The details can be found in [39, 40, 42–45].

In the nearly implicit algorithm, the six-equation model consists of a set of conservation laws for the mass, momentum, and energy of the gas [6, 7, 39]

$$\begin{aligned}
&\left(A\alpha\rho_g\right)_{,t} + \left(A\alpha\rho_g u_g\right)_{,x} = A\Gamma_g \\
&\left(A\alpha\rho_g u_g\right)_{,t} + \left(A\alpha\rho_g u_g^2\right)_{,x} + A\alpha p_{,x} = A\Gamma_g u_{ig} - A\alpha g\rho_g z_{,x} - \tau_{gw} l_{wg} - \tau_{gi} l_i \\
&\left(A\alpha\rho_g E_g\right)_{,t} + \left(A\alpha\rho_g E_g u_g\right)_{,x} = A\Gamma_g H_g - Ap\alpha_{,t} - \left(pA\alpha u_g\right)_{,x} + \dot{q}_{gw} l_{wg} + \dot{q}_{gi} l_i
\end{aligned} \tag{16}$$

and liquid phases

$$\begin{aligned}
&(A\beta\rho_l)_{,t} + (A\beta\rho_l u_l)_{,x} = -A\Gamma_g \\
&(A\beta\rho_l u_l)_{,t} + \left(A\beta\rho_l u_l^2\right)_{,x} + A\beta p_{,x} = -A\beta g\rho_l z_{,x} - \tau_{lw} l_{wl} - \tau_{li} l_i - A\Gamma_g u_{il} \\
&(A\beta E_l\rho_l)_{,t} + (A\beta E_l\rho_l u_l)_{,x} = -Ap\beta_{,t} - (pA\beta u_l)_{,x} + \dot{q}_{lw} l_{wl} + \dot{q}_{li} l_i - A\Gamma_g H_l.
\end{aligned} \tag{17}$$

Variables in Eqs. (16) and (17) are defined following convention of Eq. (1); the details are provided in the nomenclature. Importantly, velocities and temperatures of the gas and liquid phases are now independent variables; the mass flux and the drag are now defined both at the wall and at the interface.

The equations above are coupled to the equation for the wall temperature T_w:

$$\rho_w c_w d_w \frac{\partial T_w}{\partial t} = h_{wg}\left(T_g - T_w\right) + h_{wl}(T_l - T_w) + h_{amb}(T_{amb} - T_w). \tag{18}$$

In the model (16), (17) is non-hyperbolic [46]. It does not have a complete set of real eigenvalues and does not represent a well-posed system of equations [39, 47]. It also lacks positivity and possesses instabilities due to phase appearance/disappearance process [8, 48].

Despite these difficulties it was successfully employed to predict two-phase flows of boiling water in large-scale system [6, 7]. In our developments we were following the guidelines of this research.

3.1 Algorithm

The algorithm originates from [49, 50] all-speed implicit continuous-fluid Eulerian algorithm. In our work we closely follow recent modifications [6, 7] that enhance stability and eliminate material CFL restrictions using two-step nearly implicit schemes. Within this approach Eqs. (16) and (17) are expanded and solved [39, 42] for velocities and pressure in the first step, and the unexpanded conservation equations are solved for densities and energies in the second step of the algorithm.

The near implicitness of the algorithm is determined by the fact that nearly all terms in the numerical scheme are taken at the new time step. Yet the algorithm is very efficient because linearized form of equations is used, which allows to use block tridiagonal solver for velocity in the first step and to reduce the second step to the solution of four independent tridiagonal matrices.

The stability of the code was further enhanced by using upwinding and staggered grid methods as well as ad hoc smoothing and multiple time step control techniques [39]. The non-hyperbolicity was suppressed using so-called virtual mass term [7]. The time step was controlled to keep all the thermodynamic variables within the predetermined limits, to ensure that the change of these variables at any given time step does not exceed 25% of their values obtained at the previous time step and to enforce mass conservation in each control volume and in the system as whole. In addition, smooth transition between phases [39, 51] was used to adjust temperature, velocity, and density.

The models (15)–(17) have to be completed with the equations of state and the constitutive relations. The equations of state were implemented using tables for each phase. The constitutive relations are briefly summarized below.

3.2 Constitutive relations

Following the algorithm outlined above, we calculate new velocities, pressure, densities, and energies for both phases using Eqs. (15)–(17). Once these variables are calculated, we have to determine the flow pattern and boiling regime in the system and update drag, heat, and mass fluxes using appropriate correlations. First, the flow patterns have to be recognized by the code. There are numerous flow pattern correlations, but their validation for cryogenic flows remains an open problem [11, 12, 45]. Here we considered only a few flow regimes following earlier work by Wojtan et al. [52] including stratified-Wavy-to-stratified transition, stratified-Wavy-to-annular-intermittent transition, and dry-out transition. We approximated the location of these boundaries by low-dimensional polynomials and used polynomial coefficients as fitting parameters. The details of the transformation can be found in [45].

Next, the heat transfer and pressure losses for a given flow pattern have to be calculated. There are literally hundreds of heat transfer correlations proposed in the literature [53]. A concise subset of these correlations briefly discussed below was selected as a result of extensive numerical analysis and validation using two experimental setups; see Section 4 and Refs. [44, 45, 54].

3.2.1 Heat and mass transfer

As was mentioned above, the total mass transfer in the two-phase model is the sum $\Gamma_g = \Gamma_{wg} + \Gamma_{ig}$ of the mass transfer at the wall and at the interface:

$$\Gamma_{wg} = \frac{\dot{q}_{wl}}{H_g^* - H_l^*}; \qquad \Gamma_{ig} = \frac{\dot{q}_{li} + \dot{q}_{gi}}{H_g^* - H_l^*}; \tag{19}$$

where $H_g^* - H_l^*$ is the difference between saturated gas and liquid enthalpies $H_{g,s} - H_l$ for positive flux $\Gamma > 0$ while it is $H_g - H_{l,s}$ for negative flux.

The heat transfer coefficients h for the gas (subscript g) and liquid (l) at the wall (w) and at the interface (i) are defined by the following relations:

$$\begin{aligned} \dot{q}_{wg} &= h_{wg}(T_w - T_g);\ \dot{q}_{ig} = h_{ig}(T_{l,s} - T_g); \\ \dot{q}_{wl} &= h_{wl}(T_w - T_l);\ \dot{q}_{il} = h_{il}(T_{l,s} - T_l). \end{aligned} \tag{20}$$

For relatively low mass fluxes considered in this work (<1 metric ton per square meter per second), the heat transfer correlations can be obtained as the mass flux and void fraction dependent corrections to the pool boiling correlations [55–59]. Accordingly, the following pool boiling mechanisms of the heat transfer were modeled: convective heat transfer, nucleate, transition, and film boiling.

Convective heat transfer coefficient h_{cb} was calculated for four flow regimes [6, 60]. The value of $(D_h\, h_{cb}/\kappa)$ was taken to be $\sim$4.36 in the forced laminar regime, while in forced turbulent regime, it was calculated as $0.023 \cdot \mathrm{Re}^{0.8}\mathrm{Pr}^{0.4}$ [53]. The coefficient $(D_h\, h_{cb}/\kappa)$ for natural convection was calculated as $0.1(Gr \cdot \mathrm{Pr})^{1/3}$ for laminar and as $0.59(Gr \cdot \mathrm{Pr})^{1/4}$ for turbulent flows [61]. Here

$$\mathrm{Pr} = \frac{\mu C_p}{\kappa} \quad \text{and} \quad Gr = \frac{\rho^2 g \beta_T \left(T_w - T_{l(g)}\right) D^3}{\mu^2} \tag{21}$$

are Prandtl and Grashof numbers, respectively, β_T is the coefficient of thermal expansion, and D_h is the hydraulic diameter. The maximum value of h_{cb} calculated for four regimes is taken as the value for the convective heat transfer.

The next critical temperature that defines the shape of the boiling curve and influences the chilldown corresponds to the onset of nucleation boiling T_{onb}.

The onset of nucleate boiling correlations is based on the analysis of the balance between mechanical and thermodynamical equilibrium [3]. The corresponding T_{onb} and heat flux $\dot{q}_{onb}$ are [62–64].

$$T_{onb} = T_s + F\left(1 + \sqrt{1 + \frac{2\Delta T_{sub}}{F}}\right), \quad \dot{q}_{onb} = \frac{B}{\mathrm{Pr}^2}\, \Delta T_{sat}^2 = h_{cb}(T_{onb} - T_l) \tag{22}$$

where $B = \frac{\rho_g h_{lg} \kappa_l}{8\sigma T_s}$, $F = \frac{h_{cb} \mathrm{Pr}_l^2}{2B}$, $\Delta T_{sat} = T_{onb} - T_s$ is the wall superheat, and $\Delta T_{sub} = T_s - T_l$ is liquid subcooling temperature.

When the wall superheat exceeds $\Delta T_{sat} = T_{onb} - T_s$, the nucleation boiling begins, and the heat flux to the wall may increase by more than an order of magnitude. This increase continues until the heat flux approaches its critical value $\dot{q}_{chf}$.

The critical heat flux $\dot{q}_{chf}$ and the critical wall superheat T_{chf} correlations are crucial for predicting chilldown and dry-out phenomena in cryogenic flows. Yet the corresponding experimental data remains sparse, and values of $\dot{q}_{chf}$ and T_{chf} are often estimated using mechanistic models [3, 15, 56, 65].

For the pool boiling value of $\dot{q}_{chf,0}$, we used Zuber [58] correlation

$$\dot{q}_{chf,0} = \frac{\pi}{24} h_{lg} \rho_g \left(\frac{\sigma g \left(\rho_l - \rho_g\right)}{\rho_g^2}\right)^{1/4} \left(\frac{\rho_l}{\rho_l + \rho_g}\right)^{1/2}. \tag{23}$$

In boiling flows further corrections have to be introduced to take into account the dependence of the heat flux on the void fraction, velocity, and subcooling of the flow [55, 56].

$$\dot{q}_{chf} = \dot{q}_{chf,0}(\alpha_{cr} - \alpha)\left(1 + a_1\left(\frac{\rho_l c_l \Delta T_{sub}}{\rho_g h_{lg}}\right) + a_2 \mathrm{Re}_l + a_3 \sqrt{\frac{\mathrm{Re}_l \rho_l c_l \Delta T_{sub}}{\rho_g h_{lg}}}\right), \tag{24}$$

where α_{cr} is the critical value of the void fraction and a_i are constants, e.g., $a_1 = 0.0144$, $a_2 = 10^{-6}$, $a_3 = 0.5 \times 10^{-3}$ [57], and $\alpha_{cr} = 0.96$ [55] for water.

In practice, we often used a simpler expression; see for the details [55, 66].

The temperature T_{chf} for the critical heat flux was estimated in this work using the approach proposed by Theler and Freis [67].

$$T_{chf} = \frac{T_s}{1 - \frac{T_s R_g}{h_{lg}} \log\left(2k_g + 1\right)}, \tag{25}$$

where k_g is the isoentropic expansion factor that for ideal diatomic gases is 7/2 and R_g is the specific gas constant. T_{chf} was further corrected for boiling flows using Iloeje-type correlations [66] similar to the one applied ΔT_{mfb}, see below.

When wall superheat exceeds $\Delta T_{chf} = T_{chf} - T_s$, the transition boiling begins, and the heat flux to the wall decreases sharply as a function of the wall temperature until the latter reaches minimum film boiling temperature T_{mfb}.

The minimum film boiling regime signifies a complete separation of the fluid flow from the wall by the vapor film. The corresponding value of the wall superheat $\Delta T_{mfb} = T_{mfb} - T_s$ was estimated by Berenson as [68, 69].

$$\Delta T_{mfb,0} = 0.127 \frac{\rho_g h_{lg}}{\kappa_g} \times \left[\frac{g\left(\rho_l - \rho_g\right)}{\rho_l + \rho_g}\right]^{2/3} \left[\frac{\sigma}{g\left(\rho_l - \rho_g\right)}\right]^{1/2} \left[\frac{\mu_g}{\left(\rho_l - \rho_g\right)}\right]^{1/3} \tag{26}$$

Iloeje [55, 66] has corrected Berenson equation to take into account the dependence of the ΔT_{mfb} on the quality and mass flux of the boiling flows in the form

$$\Delta T_{mfb} = c_1 \Delta T_{mfb,0}\left(1 - c_2 \chi_e^{c_3}\right)\left(1 + c_4 G^{c_5}\right), \tag{27}$$

where χ_e is the equilibrium quality, G is liquid mass flux, and a_i are constants, e.g., $a_1 = 0.0144$, $a_2 = 10^{-6}$, $a_3 = 0.5 \times 10^{-3}$ [57], and $\alpha_{cr} = 0.96$ [55] for water.

In this work the heat flux to the wall in the film boiling regime was taken in the form of Bromley correlations:

$$h_{fb} = c_1 \cdot \left[g \rho_g \kappa_g^2 \left(\rho_l - \rho_g\right) \frac{\tilde{h}_{lg} c_{pg}}{D\left(T_w - T_{spt}\right) \mathrm{Pr}_g}\right]^{0.25} \left(1 - c_2 \chi_e^{c_3}\right)\left(1 + c_4 G^{c_5}\right) \tag{28}$$

Typical values of the parameters used in simulations are the following: (i) $c_1 = 2.0$, (ii) $c_2 = 1.04$, (iii) $c_3 = 2.0$, (iv) $c_4 = 0.2$, and (v) $c_5 = 0.1$.

The minimum film boiling heat flux can now be defined as

$$\dot{q}_{mfb} = h_{fb} \Delta T_{mfb}. \tag{29}$$

In the region of single-phase gas flow, the heat transfer coefficient is given by h_{cb} discussed above with appropriately modified parameters. Transition to the single-phase heat transfer is initiated when dry-out transition is detected.

Interpolation: The transitions between various flow boiling regimes are characterized by a number of critical points including (i) onset of nucleate boiling, (ii) critical heat flux, (iii) minimum film boiling, and (iv) onset of dry-out. The values of the heat flux between the first three critical points were interpolated as follows.

In the regime where the wall superheat is increasing from ΔT_{onb} to ΔT_{chf}, the heat flux was defined using the following interpolation [6]:

$$\dot{q}_{nb} = y^n \dot{q}_{onb} + (1 - y^n) \dot{q}_{chf}, \tag{30}$$

where n is constant, y is defined as $\left(T_w - T_{onb}\right)/\left(T_{chf} - T_{onb}\right)$, while $\dot{q}_{chf}$ and $\dot{q}_{onb}$ are given by Eqs. (24) and (22), respectively.

Similarly, for the regime where the wall superheat is increasing from ΔT_{chf} to ΔT_{mfb}, the transition boiling heat flux was interpolated as [6].

$$\dot{q}_{tb} = f_{tb} \cdot \dot{q}_{chf} + (1 - f_{tb})\dot{q}_{mfb}, \tag{31}$$

where $f_{tb} = \left(\frac{T_w - T_{mfb}}{T_{chf} - T_{mfb}}\right)^2$, while T_{chf}, T_{mfb}, and $\dot{q}_{mfb}$ are given by Eqs. (25), (27), and (29), respectively.

Using parameterization of the heat transfer correlations described in Section 3.2, we could obtain smooth transition from the pool boiling correlations to the flow boiling regime as the mass flux and void fraction of the flow increase [44].

3.2.2 Pressure drop

The constitutive relations are completed by providing pressure drop correlations as follows. For the single-phase flow, the wall drag was calculated using standard relations $\tau_{wl} = 0.5 f_{wl} \rho_l u_l^2$ and $\tau_{wg} = 0.5 f_{wg} \rho_g u_g^2$ with the friction factors $f_{wl(g)}$ for turbulent and laminar flow given by Churchill approximation [70].

The two-phase pressure drop $\left(\frac{dp}{dz}\right)_{2\phi}$ is given by the Lockhart-Martinelli correlations [71]. The pressure losses are partitioned between the phases as follows [7]:

$$\tau_{wg} l_{wg} = \alpha_g \left(\frac{dp}{dz}\right)_{2\phi} \left(\frac{1}{\alpha_g + \alpha_l Z^2}\right), \qquad \tau_{wl} l_{wl} = \alpha_l \left(\frac{dp}{dz}\right)_{2\phi} \left(\frac{Z^2}{\alpha_g + \alpha_l Z^2}\right) \tag{32}$$

where Z^2 is given by

$$Z^2 = \left(f_{wl} \mathrm{Re}_l \rho_l u_l^2 \frac{\alpha_{wl}}{\alpha_l}\right) \Big/ \left(f_{wg} \mathrm{Re}_g \rho_g u_g^2 \frac{\alpha_{wg}}{\alpha_g}\right), \tag{33}$$

Friction factors $f_{wg(l)}$ are the same as in above, while coefficients α_{wl} and α_{wg} depend on the flow pattern [7].

The interface drag is given by $\tau_{ig} = -\tau_{il} = \frac{1}{2} C_D \rho_g |u_g - u_l| (u_g - u_l)$, where interfacial drag coefficient C_D depends on the flow pattern [6].

We note that the functional form of the correlations adopted in this work is not unique and a number of alternative presentations can be used [2, 6, 45, 60].

4. Applications

The algorithms discussed above were used to develop a number of applications related to the management of cryogenic flows including, e.g., fault detection [72], diagnostics [38], evaluation [37], parameter inference, and optimization [44, 73].

To develop these applications and to validate our approach, we used experimental data obtained for cryogenic chilldown in two systems: (i) horizontal transfer line at National Bureau of Standards (currently NIST) [74] and (ii) large-scale experimental transfer line build at KSC [18].

The first system [74] is a 61 m long vacuum jacketed line with internal diameter of the copper pipe ~3/4 inches. Four measurement stations were located along the pipe as shown in **Figure 3**. In this case the cold front continuously propagates through the initially hot pipe, and chilldown is achieved at approximately 30, 80,

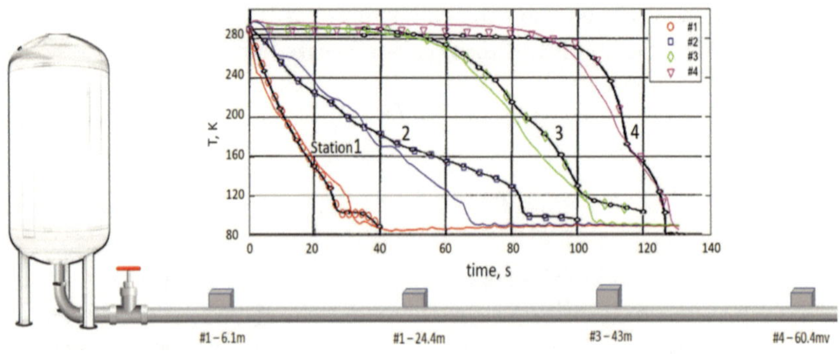

Figure 3.
Sketch of the cryogenic transfer line built at NIST. It includes storage tank (ST) and horizontal transfer line with four measurement stations. The model predictions for the temperature at four stations are compared with experimental data in the inset [44]. The experimental data are shown by solid black lines and open symbols. The model predictions are shown by colored solid lines.

100, and 130 s at each station. The model predictions for the fluid temperature are compared with the experimental data in the inset of the **Figure 3**. These results were obtained for subcooled flow and tank pressure 4.2 atm. A detailed comparison for various chilldown regimes is given in [44].

A more complex cryogenic test-bed (CTB) system with multiple control and bleed valves was built at KSC; see **Figure 4**. It consists of a 6000 gallon storage tank (ST) connected to a 2000-gallon vehicle tank (VT), control (CV) and bleed (BV) valves, and pump that control the conditions of cryogen flow. The total length of the transfer line is about 45 m. The diameter of the stainless steel pipe varies from 0.1524 to 0.0254 m. A set of the temperature (TT) and pressure (PT) sensors allows for accurate detection of the flow conditions and gradual chilldown of the system. The liquid motion through the transfer line is driven by an elevated pressure in the storage tank.

We note that both algorithms can be used for online fault detection and control of the flow. The homogeneous algorithm is about 10 times faster as compared to the NIA and can integrate 2 h of real-time loading in the CTB in less than 1 s. However, the NIA provides deeper insight into mechanism of two-phase flow and can distinguish a number of important non-equilibrium regimes including, e.g., counter flow and vertical stratification. We therefore proposed to use for online applications both algorithms in parallel to take advantage of speed of homogeneous model and greater

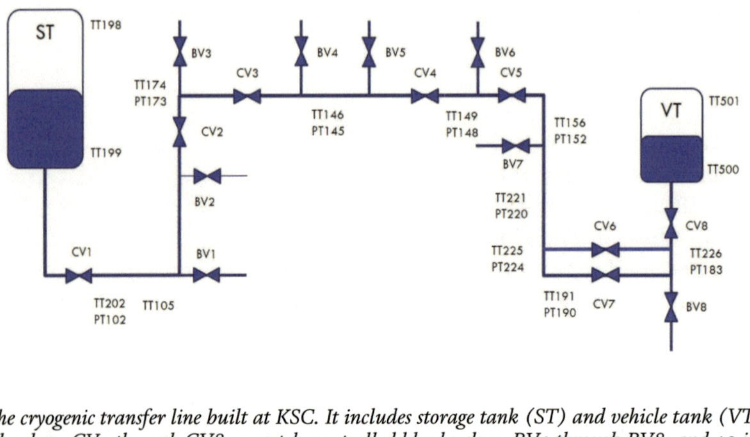

Figure 4.
Sketch of the cryogenic transfer line built at KSC. It includes storage tank (ST) and vehicle tank (VT); the in-line control valves, CV1 through CV8; remotely controlled bleed valves, BV1 through BV8; and 10 in-line temperature sensors (TT) and 9 in-line pressure sensors (PT).

precision of the two-phase flow model. Below we discuss some of the applications of these algorithms.

The validation and convergence of the algorithms using first horizontal cryogenic transfer line build in NIST were considered in details in our earlier work [44, 45, 54]. Here we briefly discuss some of the applications of the algorithms to the analysis of the flow in the CTB system.

4.1 Parameter inference and optimization

Practical application of the algorithms to management of specific large-scale cryogenic system involves adjustment of parameters of cryogenic flow boiling and system component correlations. In addition, parameter inference and optimization are required for continuous learning of multiple faults in the system (e.g., [45, 54] clogging, valves stuck open/closed, heat and mass leaks) and development of the recovery strategy. Parameter inference is important because not only functional form of the correlations is not unique, but also the values of numerical "constants" in these correlations should be considered as fitting parameters [44].

To address this problem, we developed inferential framework [44] that allows for simultaneous estimation and optimization of a large number of parameters of cryogenic models.

The application of this framework to the analysis of cryogenic correlations in NIST system was discussed in details in our earlier work [44]. Here we provide a brief example of parameter optimization during chilldown by applying two-phase flow model to CTB system [45, 54]; see **Figure 4**. In the first example we use direct search for preliminary estimation of some of the model parameters. The search is performed simultaneously for six parameters, which are scales for flow rate of input valve CV1, in-line valve CV2, dump valve BV3, dump valve BV2, dump valve BV2, and minimum film boiling heat transfer. The cost function is the sum of the square distances between experimental data and model predictions for four temperature sensors (TT202, TT105, TT162, and TT174) and four pressure sensors (PT102, PT104, PT161, PT173). It can be seen from **Figure 5** that model predictions nearly converge to the sensor's readings after ~100 iterations.

To optimize and infer the full set of nearly 100 parameters of the KSC system, we used a number of steps. In the first step, we performed extensive sensitivity analysis to find parameters of the model that can significantly modify the dynamics of the system. At this step the number of optimization parameters can be reduced by the factor of ~2 by selecting only significant parameters for further optimization. Next, we inferred the obtained subset of sensitive parameters of the heat transfer

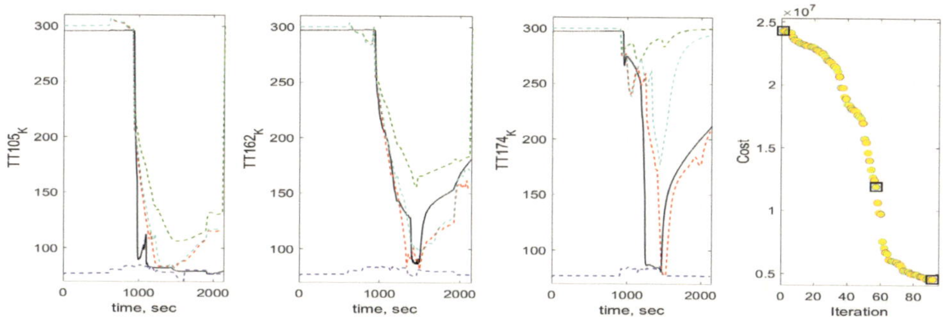

Figure 5.
Convergence of the direct search algorithm for the model predictions (dashed lines) toward experimental data (solid black lines) measured for three temperature sensors: TT105, TT162, and TT174. The cost values for three dashed lines are highlighted by squares in the right panel.

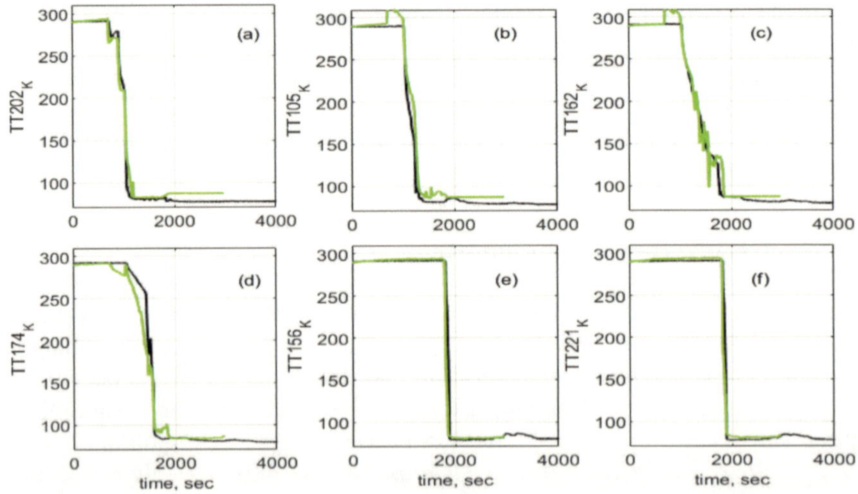

Figure 6.
Separated model predictions of the fluid temperature (green) during complete sequence of the CTB chilldown as compared to the experimental data (black lines) for the following temperature sensors: (a) TT202, (b) TT105, (c) TT162, (d) TT174, (e) TT156, (f) TT221.

and pressure correlations using NIST system [44] that does not have system components other than input valve. Next, we inferred parameters of the components of more complex CTB system using heat transfer and pressure correlations obtained at the previous step. The optimization itself was performed in two steps. First, we used direct search to roughly estimate parameter values close to global minimum, see example above and **Figure 5**. Then we refined the results using global stochastic optimization methods. The details of the approach are provided in [44, 54].

The comparison of the model predictions with the temperature sensor measurements obtained within this inferential framework is shown in **Figure 6**. It can be seen from the figure that by using this approach, we were able to accurately fit all the sensor's measurements of the CTB system. We note that similar results were obtained using homogeneous model but for a fewer sensor's measurements.

4.2 Fault detection

One of the major applications of the algorithms in the cryogenic flow management is detection and evaluation of the system faults [37, 38]. Our approach to the fault management is based on the ability of the models to accurately predict fluid dynamics in the system, which was discussed above. To efficiently manage the faults, we developed [37, 38] approach based on the application of D-matrix, which is a causal representation of the relationship between faults and tests with 1 representing the relationship that the test can detect corresponding failure in the component and 0, otherwise.

Within this approach each fault is represented by a binary array of sensor readings, and each distinguishable failure mode has its own "signature" binary array. To account for strong variability of the fluid dynamics during chilldown, the original approach was extended [38] by introducing time-dependent D-matrix. The system failure modes were defined (as binary arrays) in several time windows corresponding to the different stages of chilldown.

The extended approach was verified by performing numerical analysis of 7 faults using 12 sensors readings [37, 38]. Here we provide an example of fault detection using this method. In this example the fault (valve BV3 is stuck open) is injected at

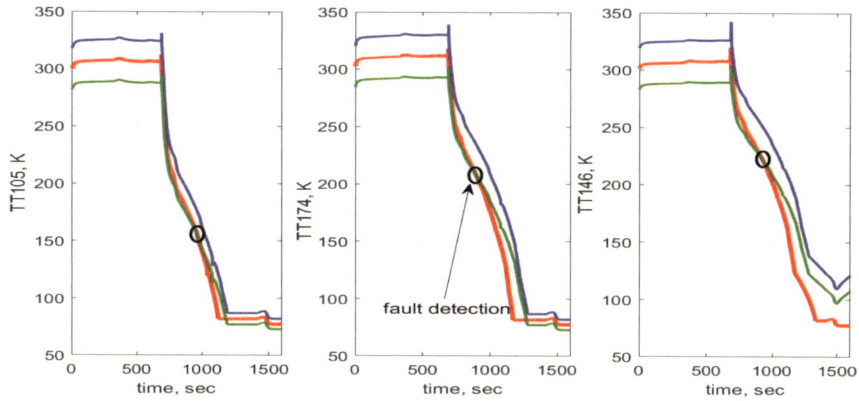

Figure 7.
Fault detection in the transfer line when the valve BV3 is stuck opened. Top (red) and bottom (green) lines indicate margins of the nominal regime. Middle (blue) line correspond sensor readings during loading operation. The fault (BV3 stuck open) is injected at 300 s. Fault detection (crossing the margins) is shown by open circles.

$t = 300$ s. The system is simulated over 1600 s using homogeneous moving front model. At each time step, the model checks if the sensor readings stay within predefined maximum and minimum threshold limits corresponding to the nominal regime. Once the sensor reading crosses the thresholds, the fault is detected as shown in **Figure 7**. The failure mode is determined by comparing the binary array of the fault with the set of signature arrays.

4.3 Fault evaluation and model predictions

To successfully apply the fault detection and evaluation algorithm in real cryogenic systems, we need to validate the predictions of the models in the presence of faults. The validation was performed by conducting experiment with injected valve stuck open/closed faults in the CTB. In this experiment three chilldown regimes are explored, in which the opening of valve CV2 had values 15% (stuck closed), 25% (nominal), and 40% (stuck open). All other controls remain at the fixed values corresponding to the nominal flow considered in Section 4.1. During simulations the model parameters are kept constant at the values found during optimization

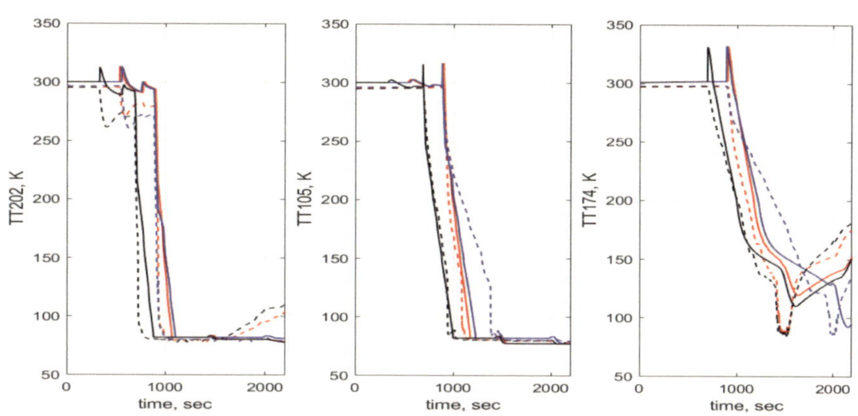

Figure 8.
Comparison of the model predictions for the temperature time-series data (colored dashed lines) with the experimental results (colored solid lines) for the different openings of the valve cv112: (i) 15% blue line, (ii) 25% black line (nominal), and (iii) 40% red line.

described above. Only the value of the model parameter corresponding to the valve cv112 opening is modified according to experimental data.

The results of the comparison of the homogeneous model predictions of the unseen experimental data are shown in **Figure 8**. It can be seen from the figure that the model predicts reasonably well the main trends of the fault including the shift in the chilldown initiation and the change in the slopes of the time traces of temperature. Similar results were obtained using two-phase flow model [54].

We note, however, that the accuracy of the predictions remains limited for both models. To improve the accuracy, we proposed to embed the optimization algorithms discussed above within the machine learning framework. Such an approach would allow for semiautomated continuous learning of the model parameters using multiple databases and specifically databases obtained during fault modeling.

5. Conclusion

To summarize, we developed fast and reliable homogeneous moving front and separated two-fluid solvers for cryogenic loading operations. We proposed concise sets of cryogenic flow boiling correlations capable of reproducing a wide range of experimental time-series data obtained for two cryogenic transfer lines using both solvers.

Both solvers offer similar performance. Homogeneous solver is ~10 times faster and can integrate 2 h of real-time loading in the CTB in less than 1 s. The two-phase flow algorithm provides access to important correlation parameters of two-phase flow, describes more accurately sensor readings, and can distinguish a number of non-equilibrium regimes unavailable in homogeneous flow including counterflow and vertical stratification.

We validated the performance of the models using horizontal transfer line at National Bureau of Standards [74] and large-scale experimental transfer line built at KSC [18]. We concluded that both solvers can be used for online control of cryogenic loading operations, and by applying two models in parallel, it is possible to resolve the trade-off issue between speed and accuracy of each solver.

We demonstrated an example of applications of these solvers to the analysis of cryogenic flow in the CTB transfer line including inference of the parameters of cryogenic correlations and chilldown modeling. We discussed an approach to the fault detection in cryogenic loading system-based D-matrix approach and demonstrated a capability of the solvers to resolve faults observed experimentally in the CTB system during chilldown.

We demonstrated that solvers allow for efficient optimization of the model parameters and discussed an extended approach to the inference of cryogenic correlations developed in our earlier work [44, 54]. We proposed that optimization algorithms should be included into machine learning framework for efficient offline learning of cryogenic correlations in the future work.

We emphasize that the machine learning approach will most likely underlie autonomous control and fault management of two-phase flows in the future space missions.

Acknowledgements

This work was supported by the Advanced Exploration Systems and Game Changing Development programs at NASA HQ.

Nomenclature

A	pipe cross-section
D	pipe diameter
D_h	hydraulic pipe diameter
E	specific energy
Gr	Grashof number
H	specific enthalpy
K_n	minor losses
N_u	Nusselt number
Pr	Prandtl number
Re	Reynolds number
T	temperature
χ	mass fraction
$\dot{m}$	mass flux
$\dot{q}$	heat flux
τ_w	friction losses
c	specific heat
c_p	specific heat for constant pressure
g	gravity
h	heat transfer coefficient
h_{lg}	latent heat of evaporation
p	pressure
u	fluid velocity

Greek symbols

α	void fraction
β	liquid void fraction
Γ	mass flow rate
k	thermal conductivity
μ	viscosity
ρ	density
σ	surface tension
τ	shear stress
h	heat transfer coefficient

Subscripts

e	equilibrium
g	gas
l	liquid
s	saturated
sub	subcooled
w	wall

Author details

Ekaterina Ponizovskaya-Devine[1], Dmitry Luchinsky[2*], Michael Foygel[3], Vasil Hafiychuk[2], Michae Khasin[2], Jared Sass[4] and Barbara Brown[4]

1 University of California—Davis, Davis, CA, USA

2 SGT, Inc., Ames Research Center, CA, USA

3 South Dakota School of Mines and Technology, SD, USA

4 Kennedy Space Center, FL, USA

*Address all correspondence to: dmitry.g.luchinsky@nasa.gov

References

[1] Chato DJ. Cryogenic fluid transfer for exploration. Cryogenics. 2008;**48** (5–6):206-209

[2] Collier JG, Thome JR. Convective Boiling and Condensation. Oxford, Clarendon Press; 1994

[3] Ghiaasiaan SM. Two-phase flow, boiling, and condensation. In: Conventional and Miniature Systems. Cambridge: Cambridge University Press; 2007

[4] Ishii M, Hibiki T. Thermo-Fluid Dynamics of Two-Phase Flow. Bücher: Springer; 2010

[5] Two-Phase Flow Regimes. 2010. Available from: https://www.sabic.com/en/products/specialties/ultem-resins/ultem-resin

[6] TRACE V5.0 Theory Manual - Volume 1: Field Equations, Solution Methods, and Physical Models. U. S. Nuclear Regulatory Commission, Washington: Nuclear Regulatory Commission; 2008. ML120060218

[7] RELAP5-3D Code Manual Volume I: Code Structure, System Models and Solution Methods, Revision 4.2, Idaho National Laboratory, Idaho Falls; 2014. INL-EXT-98-00834-V1

[8] Nourgaliev R. Solution Algorithms for Averaged Equations. Idaho National Laboratory, Idaho Falls; 2012. INL/EXT-12-27187

[9] Cheng L, Ribatski G, Wojtan L, Thome JR. New flow boiling heat transfer model and flow pattern map for carbon dioxide evaporating inside horizontal tubes. International Journal of Heat and Mass Transfer. 2006;**49** (21–22):4082-4094

[10] Prosperetti A, Tryggvason G. Computational Methods for Multiphase Flow. Cambridge: Cambridge University Press; 2007

[11] Van Dresar NT, Siegwarth JD, Hasan MM. Convective heat transfer coefficients for near-horizontal two-phase flow of nitrogen and hydrogen at low mass and heat flux. Cryogenics. 2001;**41**(11–12):805-811

[12] Jackson JK. Cryogenic Two-Phase Flow During Chilldown: Flow Transition and Nucleate Boiling Heat Transfer. Gainesville: University of Florida; 2006

[13] Wang S, Wen J, Li Y, Yang H, Li Y, Tu J. Numerical prediction for subcooled boiling flow of liquid nitrogen in a vertical tube with MUSIG model. Chinese Journal of Chemical Engineering. 2013;**21**(11):1195-1205

[14] Darr SR, Hu H, Shaeffer R, Chung J, Hartwig JW, Majumdar AK. Numerical Simulation of the Liquid Nitrogen Chilldown of a Vertical Tube. AIAA SciTech. Kissimmee, FL: American Institute of Aeronautics and Astronautics; 2015

[15] Konishi C, Mudawar I. Review of flow boiling and critical heat flux in microgravity. International Journal of Heat and Mass Transfer. 2015;**80**: 469-493

[16] Notardonato W. Active control of cryogenic propellants in space. Cryogenics. 2012;**52**(4–6):236-242

[17] Kim J. Review of reduced gravity boiling heat transfer: US research. Journal of The Japan Society of Microgravity Application. 2003;**20**(4): 264-271

[18] Johnson R, Notardonato W, Currin K, Orozco-Smith E. Integrated ground operations demonstration units testing

plans and status. In: AIAA SPACE 2012 Conference & Exposition. SPACE Conferences & Exposition. American Institute of Aeronautics and Astronautics; 2012

[19] Hartwig JW, Vera J. Numerical modeling of the transient chilldown process of a cryogenic propellant transfer line. In: 53rd AIAA Aerospace Sciences Meeting. AIAA SciTech. American Institute of Aeronautics and Astronautics; 2015

[20] Kawaji M. Boiling heat transfer during quenching under microgravity. In: Fluid Dynamics Conference. Fluid Dynamics and Co-located Conferences. American Institute of Aeronautics and Astronautics; 1996

[21] Majumdar A, Steadman T. Numerical modeling of thermofluid transients during chilldown of cryogenic transfer lines. In: 33rd International Conference on Environmental Systems, Vol. 3; Vancouver, B.C.: Society of Automotive Engineers; 2003

[22] Majumdar AK, Ravindran SS. Fast, nonlinear network flow solvers for fluid and thermal transient analysis. International Journal of Numerical Methods for Heat and Fluid Flow. 2010; **20**(6–7):617-637

[23] Chung JN, Yuan K. Recent progress on experimental research of cryogenic transport line chilldown process. Frontiers in Heat and Mass Transfer. 2015;**6**:1-17

[24] Darr S, Dong J, Glikin N, Hartwig J, Majumdar A, Leclair A, et al. The effect of reduced gravity on cryogenic nitrogen boiling and pipe chilldown. NPJ Microgravity. 2016;**2**:16033

[25] Hedayatpour A, Antar BN, Kawaji M. Cool-down of a vertical line with liquid-nitrogen. Journal of Thermophysics and Heat Transfer. 1993;**7**(3):426-434

[26] Yuan K, Ji Y, Chung JN. Numerical modeling of cryogenic chilldown process in terrestrial gravity and microgravity. International Journal of Heat and Fluid Flow. 2009;**30**(1):44-53

[27] Cullimore BA. Optimization and automated data correlation in the nasa standard thermal/fluid system analyzer. In: Proceedings 33rd Intersociety Engineering Conference on Energy Conversion; 1998

[28] Majumdar A, Ravindran SS. Numerical prediction of conjugate heat transfer in fluid network. Journal of Propulsion and Power. 2011;**27**(3): 620-630

[29] SINDA/FLUINT, General Purpose Thermal/Fluid Network Analyzer, Version 5.8. Boulder, Colorado: C&R Technologies, Inc; 2015

[30] Hafiychuk V, Foygel M, Ponizovskaya-Devine E, Smelyanskiy V, Watson MD, Brown B, et al. Moving-boundary model of cryogenic fuel loading, I: Two-phase flow in a pipe. Journal of Thermophysics and Heat Transfer. 2015;**29**:533-544

[31] Hafiychuk V, Foygel M, Ponizovskaya-Devine E, Smelyanskiy V, Watson MD, Brown B, et al. Moving-boundary model of cryogenic fuel loading, II: Theory versus experiments. Journal of Thermophysics and Heat Transfer. 2015;**29**:545-550

[32] Grald EW, MacArthur JW. A moving-boundary formulation for modeling time-dependent two-phase flow. International Journal of Heat and Fluid Flow. 1992;**13**:266-272

[33] Doherty MP, Gaby JD, Salerno LJ, Sutherlin SG. Cryogenic Fluid Management Technology for Moon and Mars Missions. AIAA SPACE 2009 Conference & Exposition, Pasadena, California; 2009

[34] Schmidt FW, Henderson RE, Wolgemuth CH. Introduction to Thermal Sciences; New York: Wiley, John \& Sons, Incorporated; 1993

[35] Thome JR, The heat transfer engineering data book III, Wieland-Werke AG, Edition 3, Ulm; 2016

[36] Osipov VV, Hafiychuk H, Ponizovskaya-Devine E, Khasin M, Smelyanskiy V. Risk Assessment and Scaling for the SLS LOx ET. NASA, ARC; 2012. NASA/TP—2012–216485

[37] Ponizovskaya-Devine E, Hafiychuk VV, Luchinsky DG, Khasin M, Perotti J, Brown B. Fault diagnostics and evaluation in cryogenic loading system using optimization algorithm. In: Annual Conference of the Prognostics and Health Management Society; 2015

[38] Kodali A, Ponizovskaya-Devine E, Robinson P, Luchinsky D, Bajwa A, Khasin M, et al. D-matrix based fault diagnostics for cryogenic loading systems. In: Annual Conference of the Prognostics and Health Management Society; 2015

[39] Luchinsky DG, Smelyanskiy VN, Brown B. Physics based model for cryogenic chilldown and loading. Part IV: Code structure. 2014. NASA, ARC. NASA/TP-2014-218399

[40] Luchinsky DG, Smelyanskiy VN, Brown B. Physics based model for cryogenic chilldown and loading. Part II: Verification and validation. 2014. NASA, ARC. NASA/TP-2014

[41] Ponizovskaya-Devine E, Luchinsky DG, Khasin M, Perotti J, Sass J, Brown B. Towards physics based autonomous control of the cryogenic propellant loading system. In: 51st AIAA/SAE/ASEE Joint Propulsion Conference. Propulsion and Energy Forum. American Institute of Aeronautics and Astronautics; 2015

[42] Luchinskiy DG, Ponizovskaya-Devine E, Hafiychuk V, Kashani A, Khasin M, Timucin D, et al. Hierarchy of two-phase flow models for autonomous control of cryogenic loading operation. IOP Conference Series: Materials Science and Engineering. 2015;**101**(1):012069

[43] Luchinsky D, Ponizovskaya-Devine E, Khasin M, Kodali A, Perotti J, Sass J, et al. Two-phase flow modelling of the cryogenic propellant loading system. In: 51st AIAA/SAE/ASEE Joint Propulsion Conference. Propulsion and Energy Forum. American Institute of Aeronautics and Astronautics; 2015. p. 4214

[44] Luchinsky DG, Khasin M, Timucin D, Sass J, Brown B. Inferential framework for two-fluid model of cryogenic chilldown. International Journal of Heat and Mass Transfer. 2017; **114**:796-808

[45] Luchinsky DG, Khasin M, Timucin D, Sass J, Johnson RG, Perotti J, et al. Physics based model for cryogenic chilldown and loading. Part III: Correlations. 2016. NASA, ARC. NASA/TP-2016

[46] Dinh TN, Nourgaliev RR, Theofanous TG. Understanding of the Ill-posed two-fluid model. In: Proceedings of the Tenth International Topical Meeting on Nuclear Reactor Thermal Hydraulics. KNS; 2003. p. 1CD-ROM

[47] Liou MS, Nguyen L, Chang CH, Sushchikh S, Nourgaliev R, Theofanous T. Hyperbolicity, discontinuities, and numerics of two-fluid models. In: Deconinck H, Dick E, editors. Springer, Berlin: Computational Fluid Dynamics 2006. 2009. p. 625

[48] Cordier F, Degond P, Kumbaro A. Phase appearance or disappearance in two-phase flows. Journal of Scientific Computing. 2014;**58**(1):115-148

[49] Harlow FH, Amsden AA. A numerical fluid dynamics calculation method for all flow speeds. Journal of Computational Physics. 1971;**8**(2): 197-213

[50] Liles DR, Reed WH. A semi-implicit method for two-phase fluid dynamics. Journal of Computational Physics. 1978; **26**(3):390-407

[51] Liou MS, Chang CH, Nguyen L, Theofanous TG. How to solve compressible multifluid equations: A simple, robust, and accurate method. AIAA Journal. 2008;**46**(9):2345-2356

[52] Wojtan L, Ursenbacher T, Thome JR. Investigation of flow boiling in horizontal tubes: Part I—A new diabetic two-phase flow pattern map. International Journal of Heat and Mass Transfer. 2005;**48**(14):2955-2969

[53] Nellis G, Klein S. Heat Transfer. Cambridge: Cambridge University Press; 2009

[54] Luchinsky DG, Khasin M, Timucin D, Sass J, Johnson RG, Perotti J, et al. Physics based model for cryogenic chilldown and loading. Part V: Optimization techniques. 2016. NASA, ARC. NASA/TP-2016

[55] Franchello G. Development of a Heat Transfer Package Applicable to a Large Variety of Fluids. Luxembourg: European Commission; 1993. EUR 14985 EN

[56] Seader JD, Miller WS, Kalvinskas LA. Boiling Heat Transfer for Cryogenics. NASA, Washington: NASA CR-243; 1965

[57] Griffith P. The Correlation of Nucleate Boiling Burnout Data. Massachusetts Institute of Technology; 1957. NS-035-267

[58] Zuber N, Tribus M. Further Remarks on the Stability of Boiling. Oak Ridge, AECU-3631: Technical Information Service Extension; 1958. 68 pages

[59] Kutateladze SS. Critical heat flux to flowing, wetting, subcooled liquids. Energetika. 1959;**2**:229-239

[60] RELAP5-3D Code Manual Volume IV: Models and Correlations, Revision 4.2, Idaho National Laboratory, Idaho Falls; 2014. INL-EXT-98-00834-V4

[61] Holman JP. Heat Transfer. Mechanical Engineering Series. Boston: McGraw-Hill; 1989

[62] Sato T, Matsumura H. On the conditions of incipient subcooled-boiling with forced convection. Bulletin of JSME. 1964;7(26):392-398

[63] Frost W, Dzakowic GS. An Extension of the Method for Predicting Incipient Boiling on Commercially Finished Surfaces. New York: ASME; 1967. p. 67-HT-61

[64] Huang L. Evaluation of onset of nucleate boiling models. In: ECI International Conference on Boiling Heat Transfer, Vol. 40. INUS; 2009. p. 40079216

[65] Tong LS, Tang YS. Boiling Heat Transfer and Two-Phase Flow. Series in Chemical and Mechanical Engineering. New York: Taylor & Francis; 1997

[66] Iloeje OC, Plummer DN, Rohsenow WM, Griffith P. Effects of mass flux, flow quality, thermal and surface properties of materials on rewet of dispersed flow film boiling. Journal of Heat Transfer. 1982;**104**(2):304-308. DOI: 10.1115/1.3245088

[67] Theler G, Freis D. Theoretical critical heat flux prediction based on non-equilibrium thermodynamics considerations of the subcooled boiling phenomenon. Mecánica Computacional.

2011;**XXX**:1713-1732. Möller O, Signorelli JW, Storti MA, editors

[68] Carbajo JJ. A study on the rewetting temperature. Nuclear Engineering and Design. 1985;**84**(1):21-52

[69] Berenson PJ. Film-boiling heat transfer from a horizontal surface. Journal of Heat Transfer. 1961;**83**: 351-358

[70] Churchill SW. Friction-factor equation spans all fluid-flow regimes. Chemical Engineering. 1977;**84**(24): 91-92

[71] Chisholm D. A theoretical basis for the Lockhart-Martinelli correlation for two-phase flow. International Journal of Heat and Mass Transfer. 1967;**10**(12): 1767-1778

[72] Kashani A, Ponizhovskaya E, Luchinsky D, Smelyanskiy V, Sass J, Brown B, et al. Physics based model for online fault detection in autonomous cryogenic loading system. AIP Conference Proceedings. 2014;**1573**(1): 1305-1310

[73] Kashani A, Luchinskiy DG, Ponizovskaya-Devine E, Khasin M, Timucin D, Sass J, et al. Optimization of cryogenic chilldown and loading operation using SINDA/FLUINT. IOP Conference Series: Materials Science and Engineering. 2015;**101**(1):012115

[74] Brennan JA, Brentari EG, Smith RV, Steward WG. Cooldown of Cryogenic Transfer Lines. Boulder, Colorado: National Bureau of Standards; 1966. NBS Report 9264

Chapter 13

Numerical Solution to Two-Dimensional Freezing and Subsequent Defrosting of Logs

Nencho Deliiski and Natalia Tumbarkova

Abstract

Two-dimensional mutually connected mathematical models have been created, solved, and verified for the transient non-linear heat conduction in logs during their freezing and subsequent defrosting. The models reflect the influence of the internal sources of latent heat of both the free and bound water on the logs' freezing process and also the impact of the temperature on the fiber saturation point of wood species, with whose participation the current values of the thermo-physical characteristics in each separate volume point of the subjected to freezing and subsequent defrosting logs are computed. The chapter presents solutions of the models with explicit form of the finite-difference method and their validation towards own experimental studies. Results from experimental and simulative investigation of 2D non-stationary temperature distribution in the longitudinal section of beech and pine logs with a diameter of 0.24 m and length of 0.48 m during their many hours freezing in a freezer and subsequent defrosting at room temperature are presented, visualized, and analyzed.

Keywords: logs, modeling, heat conduction, latent heat, freezing, defrosting, free water, bound water

1. Introduction

It is known that the duration of the thermal treatment of the frozen logs in the winter aimed at their plasticizing for the production of veneer and the very high energy consumption needed for this treatment depend on the degree of the logs' icing [1–10]. For example, for the defrosting and plasticizing of beech and oak logs with an initial temperature of $-10°C$ and moisture content of 0.6 $kg{\cdot}kg^{-1}$ approximately 68 and 81 $kWh{\cdot}m^{-3}$ thermal energy respectively [10] is needed.

In the specialized literature there are limited reports about the temperature distribution in subjected to defrosting frozen logs [8, 11–22] and there is very little information about research of the temperature distribution in logs during their freezing given by the authors only [23–25]. That is why the modeling and the multi-parameter study of the mutually connected freezing and defrosting processes of logs are of considerable scientific and practical interest.

Considerable contribution to the calculation of the non-stationary distribution of the temperature in frozen and non-frozen logs and to the duration of their heating at conductive boundary conditions has been made by Steinhagen [11, 12] and later one-dimensional and two-dimensional models have been created and solved [13–17]. The thermal energy, which is needed for the melting of the ice, which has been formed from the freezing of the bound water in the wood, has not been taken into account in the models of cited references.

The models assume that the fiber saturation point is identical for all wood species (i.e., $u_{fsp} = 0.3\ kg{\cdot}kg^{-1} = const$). However, it is known that there are significant differences between the fiber saturation points of the different wood species [1–8]. The indicated complications and incompleteness in these models have been overcome in the suggested by Deliiski [9, 18, 19] 2-dimensional mathematical model of the transient non-linear heat conduction in frozen logs during their heating and defrosting.

For different engineering calculations it is needed to be able to determine the icing degree of the wood materials depending on the temperature of the influencing on them gas or liquid medium and on the duration of their staying in this medium. Such calculations are carried out using mathematical models, which describe adequately the complex processes of the freezing of both the free and bound water in the wood.

The computer solutions of these models give the non-stationary distribution of the temperature in the materials during their cooling below temperatures, at which a freezing of the whole amount of the free water and a freezing of respective, depending on the temperature, part of the bound water in the wood occurs [24, 25].

This chapter presents the creation, numerical solving and validation of two mutually connected two-dimensional mathematical models of the transient non-linear heat conduction in logs during their freezing and subsequent defrosting at convective boundary conditions.

The model of the freezing process takes into account for the first time the impact of the internal sources of latent heat of both the free and bound water on the temperature distribution. The both models reflect the impact of the temperature on the fiber saturation point of each wood species, with whose participation the current values of the thermo-physical characteristics in each separate volume point of the subjected to freezing and subsequent defrosting logs are computed.

The chapter also presents and visualizes the results from experimental and simulative investigation of the 2D non-stationary temperature distribution in the longitudinal section of beech and poplar logs with a diameter of 0.24 m, length of 0.48 m, and different moisture content during many hours their freezing in a freezer and subsequent defrosting at curvilinear changing temperature of the processing air medium.

2. Mechanism of the temperature distribution in logs during their freezing and subsequent defrosting

2.1 Mathematical model of the 2D temperature distribution in logs subjected to freezing

When the length of the logs, L, is larger than their diameter, D, by not more than 3–4 times, for the calculation of the change in the temperature in the logs' longitudinal sections (i.e., along the coordinates r and z of these sections) during their freezing in air medium the following 2D mathematical model can be used [24]:

$$c_{\text{we-fr}} \cdot \rho_{\text{w}} \frac{\partial T(r,z,\tau)}{\partial \tau} = \lambda_{\text{wr}} \left[\frac{\partial^2 T(r,z,\tau)}{\partial r^2} + \frac{1}{r} \cdot \frac{\partial T(r,z,\tau)}{\partial r} \right] + \frac{\partial \lambda_{\text{wr}}}{\partial T} \left[\frac{\partial T(r,z,\tau)}{\partial r} \right]^2 + \lambda_{\text{wp}} \frac{\partial^2 T(r,z,\tau)}{\partial z^2} + \frac{\partial \lambda_{\text{wp}}}{\partial T} \left[\frac{\partial T(r,z,\tau)}{\partial z} \right]^2 + q_{\text{v}} \qquad (1)$$

with an initial condition

$$T(r,z,0) = T_{w0} \tag{2}$$

and boundary conditions for convective heat transfer:

- along the radial coordinate r on the logs' frontal surface during the freezing process (refer to **Figure 3**):

$$\frac{\partial T(r,0,\tau)}{\partial r} = -\frac{\alpha_{\text{wp-fr}}(r,0,\tau)}{\lambda_{\text{wp}}(r,0,\tau)}[T(r,0,\tau) - T_{\text{m-fr}}(\tau)], \tag{3}$$

- along the longitudinal coordinate z on the logs' cylindrical surface during the freezing:

$$\frac{\partial T(0,z,\tau)}{\partial z} = -\frac{\alpha_{\text{wr-fr}}(0,z,\tau)}{\lambda_{\text{wr}}(0,z,\tau)}[T(0,z,\tau) - T_{\text{m-fr}}(\tau)]. \tag{4}$$

Eqs. (1)–(4) represent a common form of a mathematical model of the logs' freezing process, i.e., of the 2D temperature distribution in logs subjected to freezing.

2.2 Mathematical model of the 2D temperature distribution in logs subjected to defrosting

In cases when the length of the logs, L, is larger than their diameter, D, by not more than 3–4 times, for the calculation of the change in T in the longitudinal sections of the logs during their defrosting in air processing medium the following 2D mathematical model can be used [8, 9]:

$$\begin{aligned} c_{\text{we-dfr}} \cdot \rho_{\text{w}} \frac{\partial T(r,z,\tau)}{\partial \tau} &= \lambda_{\text{wr}} \left[\frac{\partial^2 T(r,z,\tau)}{\partial r^2} + \frac{1}{r} \cdot \frac{\partial T(r,z,\tau)}{\partial r}\right] + \frac{\partial \lambda_{\text{wr}}}{\partial T}\left[\frac{\partial T(r,z,\tau)}{\partial r}\right]^2 \\ &+ \lambda_{\text{wp}} \frac{\partial^2 T(r,z,\tau)}{\partial z^2} + \frac{\partial \lambda_{\text{wp}}}{\partial T}\left[\frac{\partial T(r,z,\tau)}{\partial z}\right]^2 \end{aligned} \tag{5}$$

with an initial condition

$$T(r,z,0) = T(r,z,\tau_{\text{fr}}) \tag{6}$$

and boundary conditions for convective heat transfer:

- along the radial coordinate r on the logs' frontal surface during the defrosting process:

$$\frac{\partial T(r,0,\tau)}{\partial r} = -\frac{\alpha_{\text{wp-dfr}}(r,0,\tau)}{\lambda_{\text{wp}}(r,0,\tau)}[T(r,0,\tau) - T_{\text{m-dfr}}(\tau)], \tag{7}$$

- along the longitudinal coordinate z on the logs' cylindrical surface during the defrosting process:

$$\frac{\partial T(0,z,\tau)}{\partial z} = -\frac{\alpha_{\text{wr-dfr}}(0,z,\tau)}{\lambda_{\text{wr}}(0,z,\tau)}[T(0,z,\tau) - T_{\text{m-dfr}}(\tau)]. \tag{8}$$

Eqs. (5)–(8) represent a mathematical model of the logs' defrosting process, i.e., of the 2D temperature distribution in logs subjected to defrosting immediately after their freezing.

2.3 Mathematical description of the thermo-physical characteristics of logs

On **Figures 1** and **2** the temperature ranges are presented, at which the processes of the logs' freezing and subsequent defrosting respectively are carried out when $u > u_{fsp}$.

There thermo-physical characteristics of the logs and of both the frozen free and bound water in them have also been shown. The information on these characteristics is very important for the solving of the mutually connected mathematical models given above.

During the first range of the logs' freezing process from T_{w0} to T_{fr-fw} only a cooling of the logs with fully liquid water in them occurs (**Figure 1**). During the second range from T_{fr-fw} to T_{fr-bwm} a further cooling of the logs occurs until reaching of the state needed for starting of the crystallization of the free water. During that range also the phase transition of this water into ice is carried out. The second range is absent when u is less than u_{fsp}. During the third range from T_{fr-bwm} to $T_{w-fre-avg}$ a further cooling of the logs is carried out until reaching of the state needed for starting of the crystallization of the bound water. During this range also the phase transition of the bound water into ice is performed.

During the first range of the logs' defrosting process from $T_{w-fre-avg}$ to $T_{dfr-bwm}$ a heating of the frozen logs is carried out until reaching of the state needed for starting and realization of the gradually melting of the frozen bound water in them

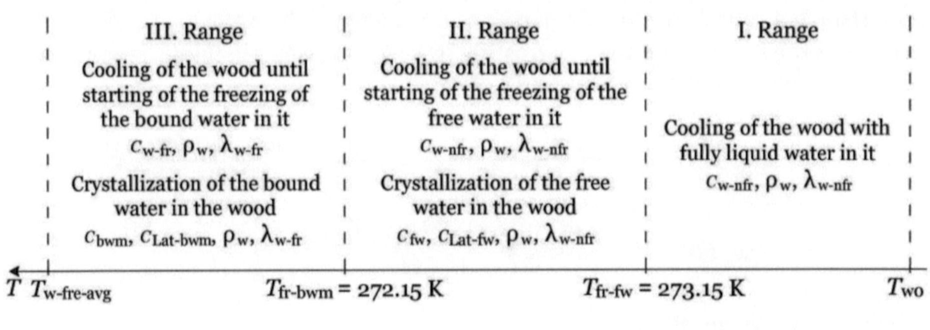

Figure 1.
Temperature ranges of the logs' freezing process at $u > u_{fsp}$ and thermo-physical characteristics of the wood and of the frozen free and bound water in it.

Figure 2.
Temperature ranges of the logs' defrosting process at $u > u_{fsp}$ and thermo-physical characteristics of the wood and of the frozen free and bound water in it.

(**Figure 2**). During the second range from $T_{\text{dfr-bwm}}$ to $T_{\text{dfr-fw}}$ a further heating of the logs occurs until reaching of the state needed for starting and realization of the melting of the frozen free water in them. During the third range from $T_{\text{dfr-fw}}$ to $T_{\text{w-dfre-avg}}$ a further heating of the logs with fully liquid water in them occurs.

The effective specific heat capacities of the logs during the pointed 3 ranges of the freezing process (see **Figure 1**) above the hygroscopic range, $c_{\text{we-fr}}$, are equal to the below:

$$\text{I. range}: c_{\text{we-fr1}} = c_{\text{w-nfr}}, \tag{9}$$

$$\text{II. range}: c_{\text{we-fr2}} = c_{\text{w-nfr}} + c_{\text{fw}} - c_{\text{Lat-fw}}, \tag{10}$$

$$\text{III. range}: c_{\text{we-fr3}} = c_{\text{w-fr}} + c_{\text{bwm}} - c_{\text{Lat-bwm}}. \tag{11}$$

The effective specific heat capacities of the logs during the pointed 3 ranges of the defrosting process (see **Figure 2**) above the hygroscopic range, $c_{\text{we-dfr}}$, are equal to the below:

$$\text{I. range}: c_{we\text{-dfr1}} = c_{\text{w-fr}} + c_{\text{bwm}}, \tag{12}$$

$$\text{II. range}: c_{\text{we-dfr2}} = c_{\text{w-nfr}} + c_{\text{fw}}, \tag{13}$$

$$\text{III. Range}: c_{\text{we-dfr3}} = c_{\text{w-nfr}}. \tag{14}$$

Mathematical descriptions of the specific heat capacities $c_{\text{w-nfr}}$, $c_{\text{w-fr}}$, c_{fw}, c_{bwm}, and also of the thermal conductivities of non-frozen, $\lambda_{\text{w-nfr}}$, and frozen wood, $\lambda_{\text{w-fr}}$, have been suggested by the first co-author earlier [9, 10, 19] using the experimentally determined in the dissertations by Kanter [26] and Chudinov [2] data for their change as a function of t and u. These relations are used in both the European [5–10] and the American specialized literature [11–17] when calculating various processes of the wood thermal treatment.

Mathematical descriptions of the specific heat capacities, which are formed by the release of the latent heat of both the free and bound water during their crystallization in the wood, $c_{\text{Lat-fw}}$ and $c_{\text{Lat-bwm}}$ respectively (refer to Eqs. (10) and (11)) have been given by the authors in [23]. Mathematical descriptions of the internal heat sources separately for the free and bound water, $q_{\text{v-fw}}$ and $q_{\text{v-bw}}$ respectively, which participate in the right-hand part of Eq. (1), have been also given.

Our study has shown that for the calculation (in $\text{W}\cdot\text{m}^{-2}\cdot\text{K}^{-1}$) of the radial and longitudinal transfer coefficients of the logs, which participate in the boundary conditions of the model the following equations are most suitable [27]:

- in the radial direction on the cylindrical surface of the horizontally situated logs:

$$\alpha_{\text{wr-fr}} = 2.56[T(0,z,\tau) - T_{\text{m-fr}}(\tau)]^{E_{\text{fr}}}, \tag{15}$$

$$\alpha_{\text{wr-dfr}} = 2.56[T(0,z,\tau) - T_{\text{m-fr}}(\tau)]^{E_{\text{dfr}}} \tag{16}$$

- in the longitudinal direction on the frontal surface of the logs:

$$\alpha_{\text{wp-fr}} = 1.123[T(r,0,\tau) - T_{\text{m-fr}}(\tau)]^{E_{\text{fr}}}, \tag{17}$$

$$\alpha_{\text{wp-dfr}} = 1.123[T(r,0,\tau) - T_{\text{m-fr}}(\tau)]^{E_{\text{dfr}}}, \tag{18}$$

where E_{fr} and E_{dfr} are exponents, whose values are determined during the solving and verification of the models through minimization of the root square mean error (*RSME*) between the calculated by the models and experimentally obtained results about the change of the temperature fields in subjected to freezing and subsequent defrosting logs.

2.4 Transformation of the models to a form suitable for programming

Due to the correct cylindrical shape of the logs, for the solution of the above presented two mutually connected mathematical models an explicit form of the finite-difference method has been used. That form allows for the exclusion of any simplifications of the models and also of the necessity for any linearization of the mathematically described variables in them [8, 9, 18, 19].

The large calculation resources of the contemporary computers eliminate the inconvenience, which creates the limitation for the value of the step along the time coordinate $\Delta\tau$ by using the explicit form [9].

According to the idea of the finite-difference method, the temperature, which is a uninterrupted function of space and time, is presented using a grid vector, and the derivatives $\frac{\partial T}{\partial r}$, $\frac{\partial T}{\partial z}$ and $\frac{\partial T}{\partial \tau}$ are approximated using the built computational mesh along the spatial and time coordinates through their finite-difference (discrete) analogues.

2.4.1 Transformation of the equations of thermo-conductivity in the models

The transformation of Eqs. (1) and (5) of the models in their suitable for programming discrete analogues has been realized using the presented on **Figure 3** coordinate system. That system shows the positioning of the knots of the calculation mesh, in which the non-stationary distribution of the temperature in the longitudinal section of subjected to freezing and subsequent defrosting log has been calculated. The calculation mesh has been built on ¼ of the longitudinal section of the log due to the circumstance that this ¼ is mirror symmetrical towards the remaining 3/4 of the same section.

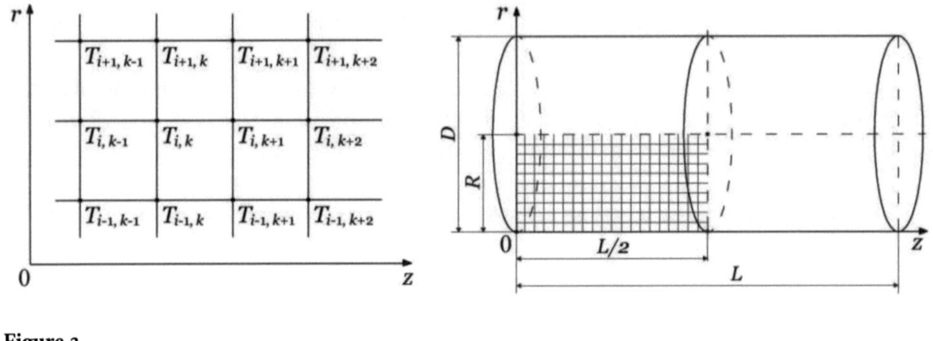

Figure 3.
Calculation mesh for the solution of the models (left) and positioning of the knots of 2D calculation mesh on 1/4 of longitudinal section of a log subjected to freezing and subsequent defrosting (right).

Taking into consideration the relationships [9, 19]

$$\lambda_{\mathrm{w}} = \lambda_{\mathrm{w0r,\,p}} \cdot \gamma \cdot [1 + \beta \cdot (T - 273.15)], \tag{19}$$

$$\lambda_{\mathrm{w0r}} = K_{\mathrm{wr}} \cdot \upsilon \cdot \left[0.165 + (1.39 + 3.8u) \cdot \left(3.3 \cdot 10^{-7}\rho_{\mathrm{b}}^{2} + 1.015 \cdot 10^{-3}\rho_{\mathrm{b}}\right)\right], \tag{20}$$

$$\lambda_{\mathrm{w0p}} = K_{\mathrm{wp}} \cdot \upsilon \cdot \left[0.165 + (1.39 + 3.8u) \cdot \left(3.3 \cdot 10^{-7}\rho_{\mathrm{b}}^{2} + 1.015 \cdot 10^{-3}\rho_{\mathrm{b}}\right)\right], \tag{21}$$

and using the coefficient

$$K_{\mathrm{wp/wr}} = \frac{K_{\mathrm{wp}}}{K_{\mathrm{wr}}}, \tag{22}$$

after applying the explicit form of the finite-differences method to Eq. (1), it is transformed into the following system of equations:

$$T_{i,k}^{n+1} = T_{i,k}^{n} + \frac{\lambda_{\mathrm{w0r}} \cdot \gamma \cdot \Delta\tau}{c_{\text{we-fr1, 2, 3}}^{n} \cdot \rho_{\mathrm{w}} \cdot \Delta r^{2}} \cdot \left\{ \begin{array}{l} \left[1 + \beta \cdot \left(T_{i,k}^{n} - 273.15\right)\right] \cdot \left[\begin{array}{l} T_{i-1,k}^{n} + T_{i+1,k}^{n} + K_{\mathrm{p/r}}\left(T_{i,k-1}^{n} + T_{i,k+1}^{n}\right) - \\ -\left(2 + 2K_{\mathrm{p/r}}\right)T_{i,k}^{n} + \dfrac{1}{i-1}\left(T_{i-1,k}^{n} - T_{i,k}^{n}\right) \end{array}\right] \\ + \beta \cdot \left[\left(T_{i-1,k}^{n} - T_{i,k}^{n}\right)^{2} + K_{\mathrm{wp/wr}}\left(T_{i,k-1}^{n} - T_{i,k}^{n}\right)^{2}\right] \end{array} \right\}, \tag{23}$$

where λ_{wor} and λ_{wop} are the wood thermal conductivities at t = 0°C in radial and longitudinal direction respectively, W·m^{-1}·K^{-1}; γ, β, and ν—coefficients for the determination of λ_{w}, λ_{wor} and λ_{wop}. Equation for calculation of γ, β, and ν for non-frozen and frozen wood are given in [9, 19]; K_{wr} and K_{wp}—empirically determined coefficients, which take into account the influence of the radial and longitudinal anatomical directions on the wood thermal conductivity, -; $\Delta\tau$—interval between time levels, i.e., step along the time coordinate, with which the model is solving, s; n—current number of the step $\Delta\tau$ along the time coordinate: n = 0, 1, 2, 3,... $\tau_{\mathrm{fr}}/\Delta\tau$; τ_{fr}—duration of the freezing process of logs, s; $\Delta r = \Delta z$—step along the coordinates r and z, by which the model is being solved, m; i—current number of the knots of the calculation mesh along the coordinate r: $1 \le i \le 1 + (R/\Delta r)$; k—current number of the knots along the coordinate z: $1 \le k \le 1 + [(L/2)/\Delta r]$; R—log's radius, m; L—log's length, m.

The effective specific heat capacities of the log during the pointed above three ranges of its freezing process (see **Figure 1**), $c_{\text{we-fr1}}$, $c_{\text{we-fr2}}$, and $c_{\text{we-fr3}}$, which are unitedly represented as $c_{\text{we-fr1, 2, 3}}$ in Eq. (23), are computed according to Eqs. (9)–(11) separately for each knot of the calculation mesh.

The density of the log's wood above the hygroscopic range, ρ_{w}, which participates in Eq. (23), is calculated (in kg·m^{-3}) according to the following widely accepted in the literature equation [1–17, 28]:

$$\rho_{\mathrm{w}} = \rho_{\mathrm{b}} \cdot (1 + u), \tag{24}$$

where ρ_b is the basic density of the wood, based on dry mass divided to green volume, kg·m^{-3}; u—wood moisture content, kg·kg^{-1}.

After applying the explicit form of the finite-differences method to Eq. (5), it is transformed into the following system of equations, which is similar to Eq. (23):

$$T_{i,k}^{n+1} = T_{i,k}^{n} + \frac{\lambda_{\text{w0r}} \cdot \gamma \cdot \Delta\tau}{c_{\text{we-dfr1, 2, 3}}^{n} \cdot \rho_{\text{w}} \cdot \Delta r^2} \cdot \left\{ \begin{array}{l} \left[1+\beta \cdot \left(T_{i,k}^{n} - 273.15\right)\right] \cdot \left[\begin{array}{l} T_{i-1,k}^{n} + T_{i+1,k}^{n} + K_{\text{p/r}}\left(T_{i,k-1}^{n} + T_{i,k+1}^{n}\right) - \\ -\left(2+2K_{\text{p/r}}\right)T_{i,k}^{n} + \dfrac{1}{i-1}\left(T_{i-1,k}^{n} - T_{i,k}^{n}\right) \end{array} \right] \\ + \beta \cdot \left[\left(T_{i-1,k}^{n} - T_{i,k}^{n}\right)^2 + K_{\text{wp/wr}}\left(T_{i,k-1}^{n} - T_{i,k}^{n}\right)^2\right] \end{array} \right\}, \tag{25}$$

where the effective specific heat capacities of the log during the pointed above three ranges of its defrosting process (see **Figure 2**), $c_{\text{we-dfr1}}$, $c_{\text{we-dfr2}}$, and $c_{\text{we-dfr3}}$, which are unitedly represented as $c_{\text{we-dfr1, 2, 3}}$in Eq. (25), are computed according to Eqs. (12)–(14) separately for each knot of the calculation mesh.

The current number of the step $\Delta\tau$ along the time coordinate, n, in Eq. (25) is equal to: $n = \tau_{\text{fr}}/\Delta\tau$, $(\tau_{\text{fr}}/\Delta\tau) + 1$, $(\tau_{\text{fr}}/\Delta\tau) + 2$, $(\tau_{\text{fr}}/\Delta\tau) + 3$, ..., $(\tau_{\text{fr}}+\tau_{\text{dfr}})/\Delta\tau$, where τ_{dfr} is the duration of the log's defrosting process, s.

2.4.2 Transformation of the equations of initial conditions in the models

The initial condition (2) of the model of logs' freezing process obtains the following discrete finite-difference form:

$$T_{i,k}^{0} = T_{\text{w0}}. \tag{26}$$

The initial condition (6) of the model of logs' defrosting process obtains the following discrete finite-difference form:

$$T_{i,k}^{0} = T_{i,k}^{n=\tau_{\text{fr}}/\Delta\tau}. \tag{27}$$

where τ_{fr} is the duration of the freezing process of logs, which precedes the beginning of their subsequent defrosting, s.

2.4.3 Transformation of the equations of boundary conditions in the models along the radial coordinate

The discrete finite difference analogue of the left-hand part of Eq. (3), which is suitable for programming in FORTRAN, has the following form [8, 9]:

$$\frac{\partial T(r,0,\tau)}{\partial r} \approx \frac{T_{i,1}^{n+1} - T_{i,2}^{n}}{\Delta r}. \tag{28}$$

The discrete analogue of the right-hand part of Eq. (3) has the following form:

$$-\frac{\alpha_{\text{wp-fr}}(r,0,\tau)}{\lambda_{\text{wp}}(r,0,\tau)}\left[T(r,0,\tau) - T_{\text{m-fr}}(\tau)\right] \approx -\frac{\alpha_{\text{wp-fr}}^{n}}{\lambda_{\text{w0p}} \cdot \gamma \cdot \left[1+\beta \cdot \left(T_{i,1}^{n} - 273.15\right)\right]}\left(T_{i,1}^{n+1} - T_{\text{m-fr}}^{n+1}\right), \tag{29}$$

where according to Eq. (17)

$$\alpha^n_{\text{wp-fr}} = 1.123\left[T^n_{i,1} - T^n_{\text{m-fr}}\right]^{E_{\text{fr}}}. \tag{30}$$

After alignment of Eq. (28) with Eq. (29), it is obtained that

$$\frac{T^{n+1}_{i,1} - T^n_{i,2}}{\Delta r} \approx -\frac{\alpha^n_{\text{wp-fr}}}{\lambda_{\text{w0p}} \cdot \gamma \cdot \left[1 + \beta \cdot \left(T^n_{i,1} - 273.15\right)\right]}\left(T^{n+1}_{i,1} - T^{n+1}_{\text{m-fr}}\right). \tag{31}$$

During the solving of the model it is needed to determine the temperature on the log's frontal surface, $T^{n+1}_{i,1}$, for each next step $n+1$ along the time coordinate. Using Eq. (31), $T^{n+1}_{i,1}$ is equal to

$$T^{n+1}_{i,1} \approx T^n_{i,2} - \frac{\Delta r \cdot \alpha^n_{\text{wp-fr}} \cdot T^{n+1}_{i,1}}{\lambda_{\text{0p}} \cdot \gamma \cdot \left[1 + \beta \cdot \left(T^n_{i,1} - 273.15\right)\right]} + \frac{\Delta r \cdot \alpha^n_{\text{wp-fr}} \cdot T^{n+1}_{\text{m-fr}}}{\lambda_{\text{w0p}} \cdot \gamma \cdot \left[1 + \beta \cdot \left(T^n_{i,1} - 273.15\right)\right]}. \tag{32}$$

After designation of

$$G^n_{i,1\text{-fr}} = \frac{\Delta r \cdot \alpha^n_{\text{wp-fr}}}{\lambda_{\text{w0p}} \cdot \gamma \cdot \left[1 + \beta \cdot \left(T^n_{i,1} - 273.15\right)\right]}, \tag{33}$$

the boundary condition (3) of the logs' freezing process obtains the following final form, suitable for programming:

$$T^{n+1}_{i,1} = \frac{T^n_{i,2} + G^n_{i,1\text{-fr}} \cdot T^{n+1}_{\text{m-fr}}}{1 + G^n_{i,1\text{-fr}}}. \tag{34}$$

Analogously, the boundary condition (7) of the logs' defrosting process obtains the following final form, suitable for programming:

$$T^{n+1}_{i,1} = \frac{T^n_{i,2} + G^n_{i,1\text{-dfr}} \cdot T^{n+1}_{\text{m-dfr}}}{1 + G^n_{i,1\text{-dfr}}}. \tag{35}$$

The variable $G^n_{i,1\text{-dfr}}$ in Eq. (35) is equal to

$$G^n_{i,1\text{-dfr}} = \frac{\Delta r \cdot \alpha^n_{\text{wp-dfr}}}{\lambda_{\text{w0p}} \cdot \gamma \cdot \left[1 + \beta \cdot \left(T^n_{i,1} - 273.15\right)\right]}, \tag{36}$$

where according to Eq. (18)

$$\alpha^n_{\text{wp-dfr}} = 1.123\left[T^n_{i,1} - T^n_{\text{m-fr}}\right]^{E_{\text{dfr}}}. \tag{37}$$

2.4.4 Transformation of the equations of boundary conditions in the models along the longitudinal coordinate

The discrete finite difference analogue of the left-hand part of Eq. (4) at $\Delta z = \Delta r$ has the following suitable for programming form [8, 9]:

$$\frac{\partial T(0, z, \tau)}{\partial z} \approx \frac{T^{n+1}_{1,k} - T^n_{1,k}}{\Delta r}. \tag{38}$$

The discrete analogue of the right-hand part of Eq. (4) has the following form:

$$-\frac{\alpha_{\text{wr-fr}}(0,z,\tau)}{\lambda_{\text{wr}}(0,z,\tau)}[T(0,z,\tau)-T_{\text{m-fr}}(\tau)] \approx -\frac{\alpha^{n}_{\text{wr-fr}}}{\lambda_{\text{w0r}}\cdot\gamma\cdot\left[1+\beta\cdot\left(T^{n}_{1,k}-273.15\right)\right]}\left(T^{n+1}_{1,k}-T^{n+1}_{\text{m-fr}}\right). \tag{39}$$

where according to Eq. (15)

$$\alpha^{n}_{\text{wr-fr}} = 2.56\left[T^{n}_{1,k}-T^{n}_{\text{m-fr}}\right]^{E_{\text{fr}}}. \tag{40}$$

After alignment of Eq. (38) with Eq. (35), it is obtained that

$$\frac{T^{n+1}_{1,k}-T^{n}_{2,k}}{\Delta r} \approx -\frac{\alpha^{n}_{\text{wr-fr}}}{\lambda_{0\text{r}}\cdot\gamma\cdot\left[1+\beta\cdot\left(T^{n}_{1,k}-273.15\right)\right]}\left(T^{n+1}_{1,k}-T^{n+1}_{\text{m-fr}}\right). \tag{41}$$

Using Eq. (41), the temperature $T^{n+1}_{1,k}$ is equal to

$$T^{n+1}_{1,k} \approx T^{n}_{2,k} - \frac{\Delta r\cdot\alpha^{n}_{\text{wr-fr}}\cdot T^{n+1}_{1,k}}{\lambda_{\text{w0r}}\cdot\gamma\cdot\left[1+\beta\cdot\left(T^{n}_{1,k}-273.15\right)\right]} + \frac{\Delta r\cdot\alpha^{n}_{\text{wr-fr}}\cdot T^{n+1}_{\text{m-fr}}}{\lambda_{\text{w0r}}\cdot\gamma\cdot\left[1+\beta\cdot\left(T^{n}_{1,k}-273.15\right)\right]}. \tag{42}$$

After designation of

$$G^{n}_{1,k-\text{fr}} = \frac{\Delta r\cdot\alpha^{n}_{\text{wr-fr}}}{\lambda_{\text{w0r}}\cdot\gamma\cdot\left[1+\beta\cdot\left(T^{n}_{1,k}-273.15\right)\right]}, \tag{43}$$

the boundary condition (4) of the logs' freezing process obtains the following final form, suitable for programming:

$$T^{n+1}_{1,k} = \frac{T^{n}_{2,k}+G^{n}_{1,k-\text{fr}}\cdot T^{n+1}_{\text{m-fr}}}{1+G^{n}_{1,k-\text{fr}}}. \tag{44}$$

Analogously, the boundary condition (8) of the logs' defrosting process obtains the following final form, suitable for programming in FORTRAN:

$$T^{n+1}_{1,k} = \frac{T^{n}_{2,k}+G^{n}_{1,k-\text{dfr}}\cdot T^{n+1}_{\text{m-fr}}}{1+G^{n}_{1,k-\text{dfr}}}. \tag{45}$$

The variable $G^{n}_{1,k-\text{dfr}}$ in Eq. (45) is equal to

$$G^{n}_{1,k-\text{dfr}} = \frac{\Delta r\cdot\alpha^{n}_{\text{wr-dfr}}}{\lambda_{\text{w0r}}\cdot\gamma\cdot\left[1+\beta\cdot\left(T^{n}_{1,k}-273.15\right)\right]}, \tag{46}$$

where according to Eq. (16)

$$\alpha^{n}_{\text{wr-dfr}} = 2.56\left[T^{n}_{1,k}-T^{n}_{\text{m-dfr}}\right]^{E_{\text{dfr}}}. \tag{47}$$

3. Experimental research of the logs' freezing and defrosting

3.1 Experimental research of the 2D temperature distribution in beech and poplar logs during their freezing and subsequent defrosting

For the validation of the suggested above mathematical models, it is necessary to have experimentally obtained data about the 2D temperature distribution in logs during their freezing and subsequent defrosting. That is why we carried out such experiments.

The logs subjected to freezing and subsequent defrosting in our experimental research were with a diameter of 240 mm, length of 480 mm, and $u > u_{fsp}$. This means, that the logs contained the maximum possible amount of bound water. They were produced from the sap-wood of a freshly felled beech (*Fagus sylvatica* L.) and poplar (*Populus nigra* L.) trunks.

Before the experiments, four holes with diameters of 6 mm and different lengths were drilled in each log parallel to its axis until reaching of the characteristic points of the log [23].

The coordinates of the characteristic points of the logs are given on **Figure 4**. These coordinates of the points allow for the determination of the 2D temperature distribution in logs during their freezing and subsequent defrosting.

For the freezing of the logs according to the suggested methodology by the authors [23], a horizontal freezer was used with adjustable temperature range from −1 to −30°C.

Sensors Pt100 with long metal casings were positioned in the drilled four holes of the logs. The automatic measurement and record of t_m, φ_m, and t in the characteristic points of the logs during the experiments was realized by Data Logger type HygroLog NT3 produced by the Swiss firm ROTRONIC AG (http:/www.rotronic.com). The Data Logger has software HW4 for graphical presentation of the experimentally obtained data. After reaching of about −28°C in the log's center during the freezing, the freezer was switch off. Then its lid was opened and a defrosting of the log at room temperature was carried out.

In **Figures 5** and **6** the change in the temperature of the processing air medium, t_m, and in its humidity, φ_m, and also in the temperature in 4 characteristic points of a beech and of a poplar log respectively during their separately 50 and 70 h is presented. The record of all data was made automatically by Data Logger with

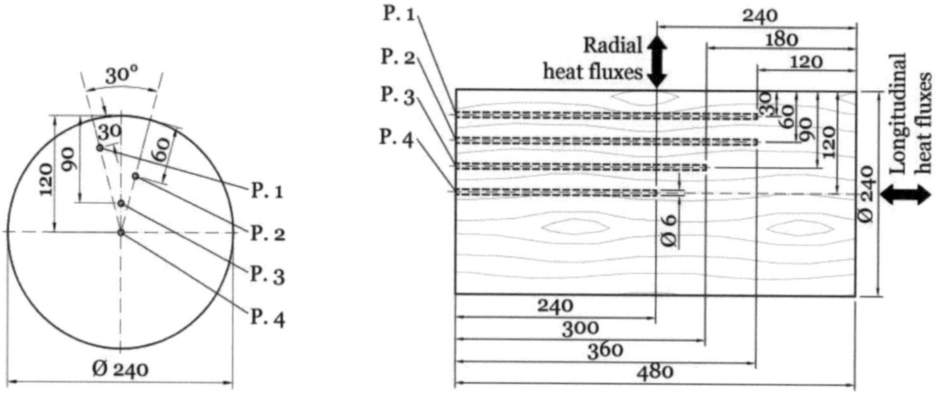

Figure 4.
Radial (left) and longitudinal (right) coordinates of four characteristic points for measurement of the temperature in logs subjected to freezing and subsequent defrosting.

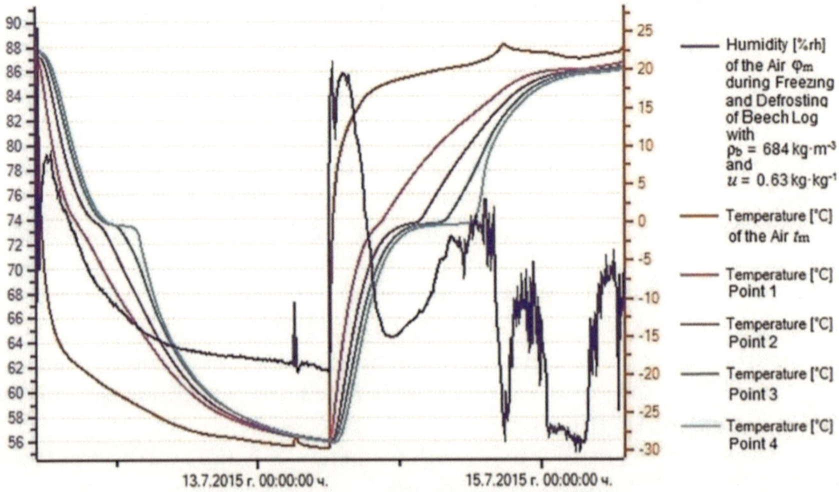

Figure 5.
Experimentally determined change in t_m, φ_m, *and t in four points of the studied beech log during its 50 h freezing and 50 h subsequent defrosting.*

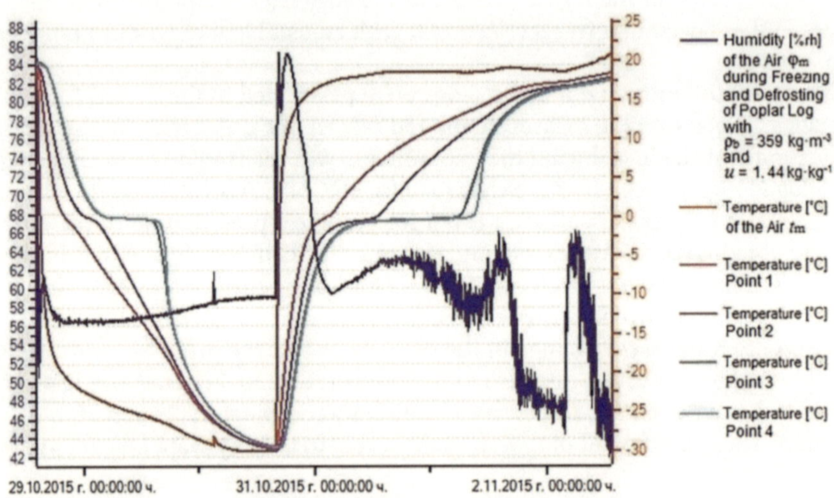

Figure 6.
Experimentally determined change in t_m, φ_m, *and t in four points of the studied poplar log during its 50 h freezing and 70 h subsequent defrosting.*

intervals of 15 min. The left coordinate axis on the figures is graduated at % of φ_m, and the left one is graduated at °C of t.

The initial temperature, t_{w0}, basic density, ρ_b, moisture content, u, duration of the freezing, τ_{fr}, and duration of the subsequent defrosting, τ_{dfr}, of the logs during the experiments were as follow:

- for the beech log: t_{w0} = 22.4°C, ρ_b = 684 kg·kg^{-1}, u = 0.63 kg·kg^{-1}, τ_{fr} = 50 h, and τ_{dfr} = 50 h;

- for the poplar log: t_{w0} = 19.8°C, ρ_b = 359 kg·kg^{-1}, u = 1.44 kg·kg^{-1}, τ_{fr} = 50 h, and τ_{dfr} = 70 h.

3.2 Mathematical description of the air medium temperature during logs' freezing and subsequent defrosting

The change in the shown on **Figures 5** and **6** freezing and defrosting air medium temperatures, T_{m-fr} and T_{m-dfr}, with very high accuracy (correlation 0.98 and Root Square Mean Error (RSME), $\sigma < 1.0°C$) has been approximated with the help of the software package Table Curve 2D (http://www.sigmaplot.co.uk/products/tablecurve2d/tablecurve2d.php) by the following equations:

- during the freezing of the beech log:

$$T_{m-fr} = \frac{a_{fr} + c_{fr}\tau^{0.5}}{1 + b_{fr}\tau^{0.5}}, \tag{48}$$

whose coefficients are equal to:

a_{fr} = 294.3352069, b_{fr} = 0.010648218, c_{fr} = 2.468350514;

- during the freezing of the poplar log:

$$T_{m-fr} = \frac{a_{fr} + c_{fr}\tau + e_{fr}\tau^2 + g_{fr}\tau^3}{1 + b_{fr}\tau + d_{fr}\tau^2 + f_{fr}\tau^3 + h_{fr}\tau^4}, \tag{49}$$

whose coefficients are equal to: a_{fr} = 301.8210985, b_{fr} = 0.0004515197, c_{fr} = 0.111484207, $d_{fr} = -6.6585073 \cdot 10^{-9}$, $e_{fr} = -1.6653 \cdot 10^{-6}$, $f_{fr} = 2.52712 \cdot 10^{-14}$, $g_{fr} = 6.46801 \cdot 10^{-12}$, $h_{fr} = 2.94924 \cdot 10^{-21}$;

- during the defrosting of both the beech and poplar logs:

$$T_{m-dfr} = \frac{a_{dfr} + c_{dfr}\tau^{0.5}}{1 + b_{dfr}\tau^{0.5}}, \tag{50}$$

whose coefficients are equal to as follows:
a_{dfr} = 297.1420433, $b_{dfr} = -0.00237763$, $c_{dfr} = -0.70526837$ for the beech log;
a_{dfr} = 296.3637194, $b_{dfr} = -0.00236425$, $c_{dfr} = -0.69281743$ for the poplar log.

Eqs. (48) and (49) were used for the solving of Eqs. (3) and (7) and Eq. (50) was used for the solving of Eqs. (4) and (8) of the model.

4. Numerical solution of the mathematical models of the logs' freezing and subsequent defrosting processes

For the numerical solution of the mutually connected mathematical models, a software package was prepared in Visual FORTRAN Professional developed by Microsoft. Using the package, computations were carried out for the determination of the 2D non-stationary change of t in the characteristic points of ¼ of the longitudinal sections of the studied logs, whose experimentally determined temperature fields are presented on **Figures 3** and **4**.

The models have been solved with the help of explicit schemes of the finite difference method in a way, analogous to the one used and described in [8–10, 18, 19]. For the computation of the temperature distribution in ¼ of the longitudinal section of the logs, which is mirror symmetrical towards the remaining 3/4 of the same section, the models were solved with step $\Delta r = \Delta z = 0.006$ m along the coordinates r and z and with the same initial and boundary conditions, as they were during the experimental research.

During the solving of the models, mathematical descriptions of the thermo-physical characteristics of beech wood with $u_{\text{fsp}}^{293.15} = 0.31\ \text{kg·kg}^{-1}$, $K_{\text{wr}} = 1.35$, and $K_{\text{wp}} = 2.40$, and also of poplar sapwood with $u_{\text{fsp}}^{293.15} = 0.35\ \text{kg·kg}^{-1}$, $K_{\text{wr}} = 1.48$, and $K_{\text{wp}} = 2.88$ [8, 10].

The mathematical models of the logs' freezing and subsequent defrosting processes have been solved with different values of the exponents E_{fr} and E_{dfr} in Eqs. (15)–(18). The calculated by the models change of the temperature in the four characteristic points of the longitudinal logs' sections with each of the used values of E_{fr} and E_{dfr} during the freezing and defrosting has been compared mathematically with the corresponding one experimentally determined change of t in the same points with an interval of 15 min. The aim of this comparison was to find that the values of E_{fr} and E_{dfr}, which ensures the best qualitative and quantitative compliance between the calculated and experimentally determined temperature fields in the logs' longitudinal sections.

As a criterion of the best compliance between the compared values of the temperature total for the four characteristic points the minimum value of RSME, σ_{avg}, has been used.

The average value of RSME, σ_{avg}, has been calculated according to the following equation:

$$\sigma_{\text{avg}} = \sqrt{\frac{\sum_{n=1}^{N}\sum_{p=1}^{P}\left(t_{p,n}^{\text{calc}} - t_{p,n}^{\text{exp}}\right)^2}{P\cdot(N-1)}}, \tag{51}$$

where $t_{p,n}^{\text{calc}}$ and $t_{p,n}^{\text{exp}}$ are the calculated and experimentally established temperatures in the characteristic points; p—number of the characteristic points in the log's longitudinal section: $p = 1, 2, 3, 4$, i.e., $P = 4$ is inputted into Eq. (51); n—number of the moments of the freezing and subsequent defrosting processes: $n = 1, 2, 3,,.....,$ $N = (\tau_{\text{fr}}+\tau_{\text{dfr}})/(150\Delta\tau)$ because of the circumstance that the comparison of the calculated values of t with experimentally determined values in the same points has been made with an interval of 15 min = 900 s = $150\Delta\tau$.

For the determination of σ_{avg} software program in the calculation environment of MS Excel has been prepared. At $\tau_{\text{fr}}+\tau_{\text{dfr}} = 100$ h for the beech log and at $\tau_{\text{fr}}+\tau_{\text{dfr}} = 120$ h for the poplar log, RSME has been calculated with the help of the program simultaneously for total $(N + 1)\cdot P = 1604$ temperature–time points for the beech log and for total 1924 such points for the poplar log during their freezing and subsequent defrosting. It was determined that the minimum values of RSME overall for the studied 4 characteristic points are equal to $\sigma_{\text{avg}} = 1.29$°C for the beech log and to $\sigma_{\text{avg}} = 1.50$°C for the poplar log.

The minimum values of σ_{avg} were obtained with the following values of the exponents in Eqs. (15)–(18):

- for the beech log: $E_{\text{fr}} = 0.52$ and $E_{\text{dfr}} = 0.32$;
- for the poplar log: $E_{\text{fr}} = 0.43$ and $E_{\text{dfr}} = 0.22$.

Figures 7 and **8** presents the calculated change in $t_{\text{m-fr}}$ and $t_{\text{m-dfr}}$, which are represented unitedly as t_{m}, and also in logs' surface temperature t_s and t of 4 characteristic points of the studied beech and poplar logs respectively.

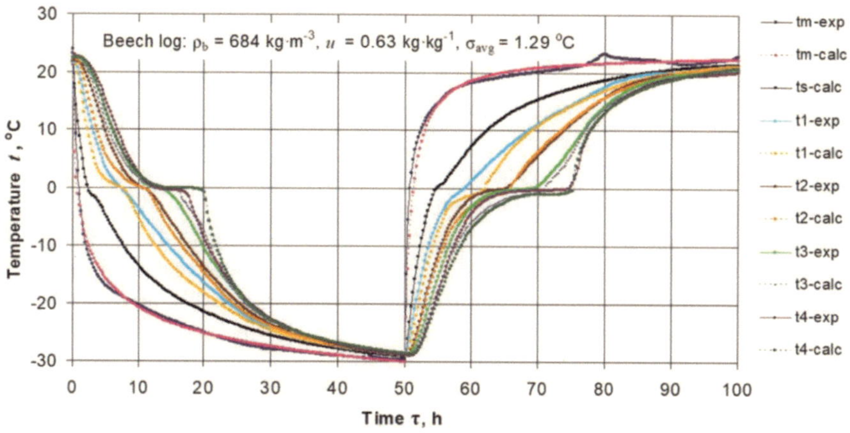

Figure 7.
Experimentally determined and calculated change in t_m, φ_m, and t in four points of the beech log during its 50 h freezing and 50 h subsequent defrosting.

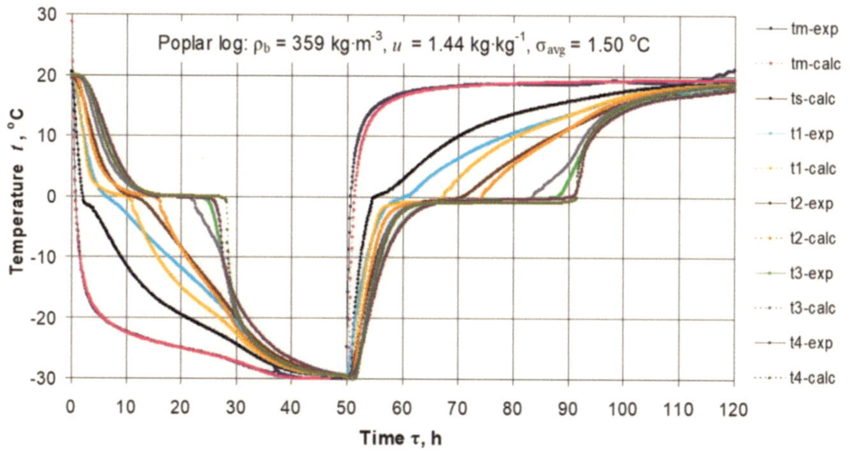

Figure 8.
Experimentally determined and calculated change in t_m, φ_m, and t in four points of the poplar log during its 50 h freezing and 70 h subsequent defrosting.

The comparison to each other of the analogical curves on **Figures 5** and 7, and also on **Figures 6** and **8** shows good conformity between the calculated and experimentally determined changes in the very complicated temperature fields of the studied logs during their freezing and subsequent defrosting.

During our wide simulations with the mathematical models, we observed good compliance between computed and experimentally established temperature fields of logs various wood species with different moisture content.

The overall RSME for the studied 4 characteristic points in the logs does not exceed 5% of the temperature ranges between the initial and the end temperatures of the logs subjected to freezing or subjected to subsequent defrosting.

5. Conclusions

This chapter describes the creation, solution, and validation of two mutually connected 2D non-linear mathematical models for the transient heat conduction in subjected to freezing and subsequent defrosting logs with any $u \geq u_{fsp}$. The model of the freezing process takes into account the impact of the internal sources of latent heat of both the free and bound water on the temperature distribution. The both models reflect the impact of the temperature on the fiber saturation point of each wood species, with whose participation the current values of the thermo-physical characteristics in each separate volume point of the subjected to freezing and subsequent defrosting logs are computed.

The mechanism of the temperature distribution in the longitudinal section of the logs during their freezing and subsequent defrosting has been mathematically described by 2D equations of heat conduction. Boundary conditions for convective heat transfer have been implemented in the models. For the transformation of both models into discrete analogues, which are suitable for programming, an explicit form of the finite-difference method has been used.

For the numerical solution of the discrete analogues of the models a software package has been prepared using the programming language FORTRAN, which has been input in the calculation environment of Visual Fortran Professional.

A validation of the models towards own experimentally determined 2D temperature distribution in beech and poplar logs with a diameter of 0.24 m, length of 0.48 m during their separate 50 h freezing in a freezer and many hours subsequent defrosting at room temperature has been carried out. The influence of the curvilinear changing temperature of the air medium in the freezer until reaching of approximately −30°C and also of the air processing medium during the defrosting of logs has been investigated. The following minimum values of the average RSME total for the temperature change in four characteristic points in each of the logs have been obtained:

- σ_{avg} = 1.29°C for the beech log with t_{wo} = 22.4°C, ρ_b = 684 $kg{\cdot}m^{-3}$, and u = 0.63 $kg{\cdot}kg^{-1}$;

- σ_{avg} = 1.50°C for the poplar log with t_{wo} = 19.8°C, ρ_b = 359 $kg{\cdot}m^{-3}$, and u = 1.44 $kg{\cdot}kg^{-1}$.

During our experimental research it has been determined that in situated on the logs' inner layers characteristic points the specific practically horizontal sections of retention of the temperature for many hours in the range from 0 to −1°C arise, while in these layers a complete freezing of the free water occurs. Analogous retention of the temperature in the range from −1 to 0°C arises during the logs' defrosting. The further the point is distanced from the logs' surfaces and the larger the amount of the free water in the wood is, that much more these sections with temperature retention are extended. Our simulations show that this phenomenon of the freezing and defrosting processes has been correctly reflected in the models (see **Figures** 7 and 8).

Good adequacy and precision of the models towards the results from wide own experimental studies allow for the carrying out of various calculations with the models, which are connected to the non-stationary temperature distribution in logs from different wood species during their freezing and subsequent defrosting.

The validation of the models with curvilinear change in the temperature of both the freezing and defrosting air mediums will allow us in the future to solve the models with curvilinear changing of the climate temperature [29] over many winter days and nights. The solution will allow for the calculation of the temperature distribution, icing degrees from both the free and bound water, and also different energy characteristics of logs for each desired moment of the their freezing and subsequent defrosting.

The mutually connected models of the freezing and defrosting processes can be applied for the development of scientifically based and energy saving optimized regimes for thermal treatment of frozen logs and also in the software for controllers used for model predictive automatic control [21, 22, 30] of this treatment.

Nomenclature

c	specific heat capacity, $J{\cdot}kg^{-1}{\cdot}K^{-1}$
D	diameter, m
E	exponent, –
L	latent heat, $J{\cdot}kg^{-1}$, or length, m
q	internal heat source, $W{\cdot}m^{-3}$
R	radius, m
r	radial coordinate: $0 \le r \le R$, m
T	temperature, K
t	temperature, °C
u	moisture content, $kg{\cdot}kg^{-1}$ = %/100
z	longitudinal coordinate: $0 \le z \le L/2$, m
α	heat transfer coefficients between log's surfaces and the surrounding air medium, $W{\cdot}m^{-2}{\cdot}K^{-1}$
λ	thermal conductivity, $W{\cdot}m^{-1}{\cdot}K^{-1}$
φ	relative humidity, %
ρ	density, $kg{\cdot}m^{-3}$
σ	root square mean error (RSME), °C
τ	time, s
Δr	step along the coordinates r and z for solving of the models, m
$\Delta\tau$	step along the time coordinate for solving of the models, s
Subscripts	
avg	average (for relative icing degree or for root square mean error)
b	basic (for wood density, based on dry mass divided to green volume)
bw	bound water
bwm	maximum possible amount of the bound water in the wood
cr	crystallization
dfre	end of defrosting
fr	freezing
fre	end of freezing
fsp	fiber saturation point
fw	free water
i	knot of the calculation mesh in the direction along the logs' radius: i = 1, 2, 3,…, 21.
k	knot of the calculation mesh in longitudinal direction of the logs: k = 1, 2, 3, …, 41.
Lat	latent heat
m	medium (for the air during the logs' freezing and defrosting)
p	parallel to the wood fibers
r	radial direction
v	volume
w	wood
we	wood effective (for specific heat capacity)
w-fr	wood with frozen water in it
w-nfr	wood with fully liquid water in it

0	initial or at 0°C
wop	parallel to the wood fibers at °C
wor	radial direction of wood at °C

Superscripts

n	current number of the step $\Delta\tau$ along the time coordinate: $n = 0, 1, 2, \ldots$
272.15	at 272.15 K, i.e. at −1°C
293.15	at 293.15 K, i.e. at 20°C

Author details

Nencho Deliiski* and Natalia Tumbarkova
University of Forestry, Sofia, Bulgaria

*Address all correspondence to: deliiski@netbg.com

References

[1] Vorreiter L. Holztechnologisches Handbuch. Vienna: Fromm; 1949. 2080p

[2] Chudinov BS. Theoretical research of thermo physical properties and thermal treatment of wood [Thesis for DSc.]. Krasnojarsk, USSR: SibLTI; 1966 (in Russian)

[3] Kollmann FF. Technologie Des Holzes and Holzwerkstoffe. I. Bd. 2 Auflage ed. Berlin/Göttingen, Heidelberg: Springer; 1951. 2233p

[4] Kollmann FF, WAC Jr. Principles of Wood Science and Technology. I. Solid Wood. Berlin, Heidelberg, New York: Springer-Verlag; 1984. 592p

[5] Shubin GS. Drying and Thermal Treatment of Wood. Moscow: Lesnaya Promyshlennost, USSR; 1990. 337p (in Russian)

[6] Požgaj A, Chovanec D, Kurjatko S, Babiak M. Structure and Properties of Wood. 2nd ed. Bratislava: Priroda a.s; 1997. 485p (in Slovak)

[7] Trebula P, Klement I. Drying and Hydro-Thermal Treatment of Wood. Slovakia: Technical University in Zvolen; 2002. 449p (in Slovak)

[8] Deliiski N, Dzurenda L. Modelling of the Thermal Processes in the Technologies for Wood Thermal Treatment. Slovakia: Technical University in Zvolen; 2010. 224p (in Russian)

[9] Deliiski N. Transient heat conduction in capillary porous bodies. In: Ahsan A, editor. Convection and Conduction Heat Transfer. Rijeka, Croatia: InTech Publishing House; 2011. pp. 149-176. DOI: 10.5772/21424

[10] Deliiski N. Modelling of the Energy Needed for Heating of Capillary Porous Bodies in Frozen and Non-frozen States. Saarbrücken, Germany: Lambert Academic Publishing, Scholars' Press; 2013. 116p. http://www.scholars-press.com//system/covergenerator/build/1060

[11] Steinhagen HP. Computerized finite-difference method to calculate transient heat conduction with thawing. Wood and Fiber Science. 1986;**18**(3): 460-467

[12] Steinhagen HP. Heat transfer computation for a long, frozen log heated in agitated water or steam—A practical recipe. Holz als Roh- und Werkstoff. 1991;**49**(7–8):287-290. DOI: 10.1007/BF02663790

[13] Steinhagen HP, Lee HW. Enthalpy method to compute radial heating and thawing of logs. Wood and Fiber Science. 1988;**20**(4):415-421

[14] Steinhagen HP, Lee HW, Loehnertz SP. LOGHEAT: A computer program of determining log heating times for frozen and non-frozen logs. Forest Products Journal. 1987;**37**(11/12):60-64

[15] Khattabi A, Steinhagen HP. Numerical solution to two-dimensional heating of logs. Holz als Roh- und Werkstoff. 1992;**50**(7–8):308-312. DOI: 10.1007/BF02615359

[16] Khattabi A, Steinhagen HP. Analysis of transient non-linear heat conduction in wood using finite-difference solutions. Holz als Roh- und Werkstoff. 1993;**51**(4):272-278. DOI: 10.1007/BF02629373

[17] Khattabi A, Steinhagen HP. Update of "numerical solution to two-dimensional heating of logs". Holz als Roh- und Werkstoff. 1995;**53**(1):93-94. DOI: 10.1007/BF02716399

[18] Deliiski N. Modelling and automatic control of heat energy consumption

required for thermal treatment of logs. Drvna Industrija. 2004;**55**(4):181-199

[19] Deliiski N. Computation of the 2-dimensional transient temperature distribution and heat energy consumption of frozen and non-frozen logs. Wood Research. 2009;**54**(3):67-78

[20] Deliiski N, Dzurenda L, Tumbarkova N, Angelski D. Computation of the temperature conductivity of frozen wood during its defrosting. Drvna Industrija. 2015; **66**(2):87-96. DOI: 10.5552/drind.2015.1351

[21] Hadjiski M, Deliiski N. Cost oriented suboptimal control of the thermal treatment of wood materials. IFAC-PapersOnLine. 2015;**48**(24):54-59. DOI: 10.1016/j.ifacol.2015.12.056

[22] Hadjiski M, Deliiski N. Advanced control of the wood thermal treatment processing. Cybernetics and Information Technologies, Bulgarian Academy of Sciences. 2016;**16**(2): 179-197. DOI: 10.1515/cait-2016-0029

[23] Deliiski N, Tumbarkova N. A methodology for experimental research of the freezing process of logs. Acta Silvatica et Lignaria Hungarica. 2016; **12**(2):145-156. DOI: 10.1515/aslh-2016-0013

[24] Deliiski N, Tumbarkova N. An approach and an algorithm for computation of the unsteady icing degrees of logs subjected to freezing. Acta Facultatis Xilologiae Zvolen. 2017; **59**(2):91-104

[25] Deliiski N, Tumbarkova N. An approach for computation the heat sources in logs subjected to freezing. Acta Silvatica et Lignaria Hungarica. 2018;**14**(1):35-49. http://aslh.nyme.hu/index.php?id=29435&L=0

[26] Kanter KR. Investigation of the thermal properties of wood [Thesis]. Moscow, USSR: MLTI; 1955 (in Russian)

[27] Telegin AS, Shvidkiy BS, Yaroshenko UG. Heat- and Mass Transfer. Moscow: Akademkniga; 2002. 456p (in Russian)

[28] Hrčka R. Model in free water in wood. Wood Research. 2017;**62**(6): 831-837

[29] Deliiski N. Thermische Frequenzkennlinien von wetterbeanspruchten Holzbalken. Holz als Roh- und Werkstoff. 1988;**20**(2): 59-65. DOI: 10.1007/BF02612530

[30] Hadjiski M, Deliiski N, Grancharova A. Spatiotemporal parameter estimation of thermal treatment process via initial condition reconstruction using neural networks. In: Hadjiski M, Atanasov KT, editors. Intuitionistic Fuzziness and Other Intelligent Theories and their Applications. Cham, Switzerland: Springer International Publishing AG; 2019. pp. 51-80. DOI: 10.1007/978-3-319-78931-6

Section 6

Thermodynamics and Thermophysical Properties

Chapter 14

Minimal Dissipation Processes in Irreversible Thermodynamics and Their Applications

Margarita Anatolyevna Zaeva and Anatoly Mikhailovich Tsirlin

Abstract

It is known that the maximum efficiency of conversion of thermal energy into mechanical work or separation work is achieved in reversible processes. If the intensity of the target flux is set, the processes in the thermodynamic system are irreversible. In this case, the role of reversible processes is played by the processes of minimal dissipation. The review presents the derivation of conditions for minimum dissipation in general form and their specification for heat and mass transfer processes with arbitrary dynamics. It is shown how these conditions follow the solution of problems on the optimal organization of two-flux and multiflux heat exchange. The algorithm for the synthesis of heat exchange systems with given water equivalents and the phase state of the flows is described. The form of the region of realizability of systems using thermal energy and the problem of choosing the order of separation of multicomponent mixtures with the minimum specific heat consumption are considered. It is shown that the efficiency of the rectification processes in the marginal productivity mode monotonously depends on the reversible efficiency, which makes it possible to ignore irreversible factors for choosing the order of separation in this mode.

Keywords: entropy production, conditions of minimal dissipation, optimal heat transfer, multithreaded heat exchange system, rectification, separation of multicomponent mixtures, boundary of the realizability of thermal machines

1. Problems and methodology of finite-time thermodynamics

Applied thermodynamics originates from the work of Sadi Carnot in 1824 [1]. One of the problems of thermodynamics is the study of problems on the limiting possibilities of thermodynamic systems. For a long time, these tasks boiled down to finding the maximum efficiency of heat and refrigeration machines, separation systems, and various chemical processes. The solution of these problems led to the fact that the maximum efficiency value was determined in the case when the process under study was reversible. Reversibility will include processes in which the coefficients of heat and mass transfer are arbitrarily large or the fluxes of energy and matter in the system under study are arbitrarily small. With the development of nuclear energy, a new task was set—to obtain such a cycle of a heat engine that would correspond to its maximum power with certain fixed exchange ratios with sources. This task is due to the fact that the capital expenditures for the construction of nuclear power facilities are high with a relatively low cost of fuel spent. Variants of solving the problem of optimization thermodynamics were proposed in [2, 3].

Further development of finite-time thermodynamics was stimulated by a great deal of work of very many investigators. Here, we list names of just a few first researchers: R.S. Berry, B. Andresen, K.H. Hoffmann, P. Salamon, L.I. Rozonoer, and some others (see [1–32]).

Typical problems of optimization thermodynamics include the following: processes with minimal irreversibility; determination of the limiting possibilities of heat engines, cold cycles, and heat pumps (maximum power, maximum efficiency, many realizable modes); and analysis of the processes of separation of mixtures.

The general approach to solving problems is as follows. It is assumed that the whole system is divided into subsystems. In each subsystem, at any time moment, the deviations of the intensive variables from their average values over the volume are negligible. Consequently, the change of these variables (temperatures, pressures, etc.) occurs only at the boundaries of the subsystems, which means that the system as a whole is in a nonequilibrium state. This assumption makes it possible to use the equation of state in the description of individual subsystems, which are valid under equilibrium conditions, and ordinary differential equations can be used to describe the dynamics of the subsystems. The solution of extremal problems in this case is performed by methods of the optimal control theory for lumped parameter systems.

To study the limiting possibilities of thermodynamic systems, it is first necessary to make balance relations for matter, energy, and entropy. Moreover, the balance ratio for entropy includes dissipation σ, that is, the production of entropy. It characterizes the irreversibility of processes in the system. If all processes are reversible, then the dissipation is zero. If the processes are irreversible, then dissipation takes positive values. Dissipation depends on the dynamics of the processes. The set of realizability of the system in the parameter space of input and output streams is determined by the nonnegativity of dissipation. Reversible processes lie on the boundary of this set.

When a minimum possible dissipation is found as a function of flux intensities, then the inequality $\sigma \geq \sigma_{\text{min}}$ holds in an arbitrary real system; this contracts the region of realization. In this formulation, the set obtained incorporates the effects of process dynamics, the magnitude of fluxes, and the system's extent (due to the presence of heat and mass transfer coefficients).

In any real system, it is possible to narrow the realizability region if we find the minimum possible dissipation value as a function of the flux intensity ($\sigma \geq \sigma_{min}$). This will take into account the dynamics, the flux intensity, and even the size of the installation through the coefficients of heat and mass transfer.

Then, from the balance equations, it is necessary to derive the connection between the system performance indicators and dissipation σ. Performance indicators usually monotonically deteriorate with the increase of σ. The best values of efficiency indicators are achieved in a reversible process, which allows using them similarly to the Carnot efficiency indicators.

Next, it is necessary to solve the problem of the organization of processes in such a way that, with the given constraints, the dissipation as a function of the flux intensities is minimal. This is the most difficult step in analyzing the capabilities of thermodynamic systems.

Consider the process of studying the limiting possibilities in more detail, and begin with thermodynamic balances. Thermodynamic balances show the relationship between the fluxes (matter, energy, and entropy) that the system exchanges with the environment and the changes in these values in the system [19]. Let us summarize all the fluxes, considering incoming fluxes as positive and outcoming fluxes as negative. Fluxes can be convective and diffusive. Convective fluxes are forced into the system and removed from it. The diffusive flux depends on the differences between the intensive variables of the system at the point where it enters and the intensive variables of the environment.

The energy balance shows the rate of change in the energy of a system, which is determined by the flux of energy that enters or is removed along with the convective fluxes of matter, the change in energy due to the diffusional exchange of matter, the currents of conductively transmitted heat, and the power of the work done. Material balance shows the change in the number of moles of substances in the system. Entropy balance shows the change in the entropy of the system, which occurs due to the influx of entropy together with the incoming substances, the influx or removal of heat, and the production of entropy due to the irreversibility of exchange processes.

If the system operates cyclically, the balances can be recorded on average for the equipment working cycle. In this case, the total change in energy, amount of matter, and entropy per cycle is zero, since the state of the system at the start and the end of the cycle is the same. Balances are transformed into a system of relations of averages over cycle-averaged components.

The equations of thermodynamic balances show the relationship between process efficiency indicators, external fluxes, and the structure of the system. The increase in entropy σ causes an increase in the entropy of output fluxes. At the same time, either the temperature of the fluxes at the outlet decreases, or the outlet flux of heat increases at a constant temperature. This leads to a reduction in the work of separation, the mechanical work produced by the system.

Consider the operation of a thermal machine that converts the heat received from a hot source with temperature T_+ into work. The working fluid gives a part of the energy to a cold source with a temperature T_-. The working fluid changes its state cyclically. As an indicator of efficiency, we will consider the thermal efficiency ($\eta = p/q_+$)—the ratio of the work produced to the amount of heat collected from a hot source.

Let us denote the average intensity of the heat flux taken from the hot source q_+ and that given to the cold source, q_-. For the generated power p, we write the equations of energy balances:

$$q_+ - q_- - p = 0, \tag{1}$$

and

$$\frac{q_+}{T_+} - \frac{q_-}{T_-} + \sigma = 0. \tag{2}$$

Since the state of the working fluid either does not change in time (for steam and gas turbines) or changes cyclically (for steam engines), then there are zeros in the right parts of the equations.

Thermal efficiency $\eta = p/q_+$ follows from Eq. (1) in the form

$$\eta = \frac{p}{q_+} = 1 - \frac{q_-}{q_+}. \tag{3}$$

Taking into account the fact that the Eq. (2) implies

$$\frac{q_-}{q_+} = \frac{T_-}{T_+} + \sigma\frac{T_-}{q_+}. \tag{4}$$

Therefore,

$$\eta = \left(1 - \frac{T_-}{T_+}\right) - \sigma\frac{T_-}{q_+} = \frac{1 - T_-/T_+}{1 + \sigma T_-/p}. \tag{5}$$

Thermal efficiency η is equal to Carnot efficiency when any irreversible phenomena are absent in the system.

The growth of σ leads to a growth entropy of output streams; under other equal conditions, this growth reduces the flux temperature at the outlet or at a fixed temperature increases the waste heat flux. And in this and in another case, this leads to a decrease in the mechanical work produced by the system or the work of separation. Energy efficiency of thermodynamic system, characterized by the relation of useful work, produced in it, to the energy costs, reaches a maximum in the invertible processes, when $\sigma = 0$.

2. Processes with a minimal dissipation

It is known that the maximum efficiency of conversion of thermal energy into mechanical work or separation work is achieved in reversible processes. If the intensity of the target flux is set, the processes in the thermodynamic system are irreversible. In this case, the role of reversible processes is played by the processes of minimal dissipation, so it is necessary to determine conditions under which thermodynamic processes exhibit minimal dissipation for a prescribed average intensity (prescribed averaged value of driving forces).

2.1 The minimal dissipation's conditions

Consider two systems interacting with each other. Intensive variables for the ith system will be denoted by u_i and extensive variables by x_i. In general, these are vector variables. When systems are in contact, the difference between u_1 and u_2 leads to the appearance of flux $J(u_1, u_2)$. Function J is continuous, is differentiable, and has the following properties:

$$\frac{\partial J_j}{\partial u_{1j}} > 0, \qquad \frac{\partial J_j}{\partial u_{2j}} < 0, \qquad J(u_1, u_2) = 0, \quad \text{at} \quad u_2 = u_1 \tag{6}$$

for scalar u_1 and u_2.

The difference between vectors u_1 and u_2 (of the same sign as flux J_j) leads to appearance of driving forces X_j. Each force defines exclusively u_{1j} and u_{2j}, satisfying conditions analogous to those in Eq. (6). Entropy production σ, which characterizes the process irreversibility, is equal to the scalar product of the flux vector and the driving force vector, and average value of entropy is described by the formula

$$\overline{\sigma} = \frac{1}{L}\int_0^L \sum_{j=1}^m J_j(u_1, u_2) X_j(u_{1j}, u_{2j}) dl, \tag{7}$$

where the independent variable l has the interpretation of a space-time or a contacting area measure. The integrand of this functional is defined non-negatively.

We shall assume that in our algorithm (at least) one intensive variable appears, by definition $u_2(l)$, which may assume its values from within a certain manifold V. Yet, because of the variability of extensive variables of the first subsystem $\left(\frac{dY_{1j}}{dl} = -J_j(u_1, u_2)\right)$, the second variable changes in accordance with the formula

$$\frac{du_{1j}}{dl} = \varphi_j(u_1, u_2), \quad u_1(0) = u_{10}, \quad j = 1, \ldots, m. \tag{8}$$

Average values of all or some selected fluxes are prescribed:

$$\frac{1}{L}\int_0^L J_j(u_1, u_2)dl = \overline{J_j}, \quad j = 1, \ldots, k_1, \quad k_1 \leq m. \tag{9}$$

Further on, we consider only the case of a scalar flux. The problem for vector fluxes and its solution is considered with details in [32, 33].

The scalar flux problem involves minimizing of the integral

$$\overline{\sigma} = \frac{1}{L}\int_0^L J(u_1, u_2)X(u_1, u_2)dl \to \min_{u_2 \in V} \tag{10}$$

subject to constraining conditions:

$$\frac{du_1}{dl} = \varphi(u_1, u_2), \quad u_1(0) = u_{10}, \tag{11}$$

$$\frac{1}{L}\int_0^L J(u_1, u_2)dl = \bar{J}. \tag{12}$$

The problem (10)–(12) simplifies in an important case when the rate of change of variable u_1 is proportional to the flux:

$$\varphi(u_1, u_2) = c(u_1)J(u_1, u_2). \tag{13}$$

In this case the condition of minimal dissipation assumes the form

$$J^2(u_1, u_2) = \lambda_2\left(\frac{\partial J(u_1, u_2)}{\partial u_2} : \frac{\partial X(u_1, u_2)}{\partial u_2}\right), \tag{14}$$

whereas the condition of prescribed flux intensity can be written as

$$\int_{u_{10}}^{u_{1L}} \frac{du_1}{c(u_1)} = \bar{J} \cdot L. \tag{15}$$

The value of u_{1L} is determined regardless of the optimal solution $u_2^*(u_1)$.

If the flux is proportional to the driving force with constant coefficient α, then the minimum entropy production equals

$$\overline{\sigma} = \frac{\bar{J}^2}{\alpha}. \tag{16}$$

2.2 Minimal dissipation's conditions of selected processes

Consider the conditions for the minimum dissipation of heat exchange. Let us take the temperature of the body being heated as the controlling intense variable. The driving force in the minimum dissipation problem is

$$X(T_1, T_2) = \left(\frac{1}{T_2} - \frac{1}{T_1}\right), \tag{17}$$

whereas the heat flux is $q(T_1, T_2)$. In the majority of cases, we may assume the energy balance in the form

$$\frac{dT_1}{dl} = -\frac{1}{c_1(T_1)} q(T_1, T_2), \quad T_1(0) = T_{10}, \tag{18}$$

where $c_1(T_1)$ is the heat capacity of the hot source.

If the process takes place in time, then the parameter l has the meaning of time, and the parameter L—the duration of the process. If a pipe heat exchanger is considered, in which the hot flux temperature changes from section to section, the value of c is the water equivalent of the flux, and L is the length of the heat exchanger.

In agreement with conditions (14), (15) describing the minimum dissipation subject a prescribed average intensity of heat flux $\overline{q}$, we can obtain a condition of minimum dissipation for an arbitrary law of heat transfer:

$$q^2(T_1, T_2) : \frac{\partial q}{\partial T_2} T_2^2 = -\lambda_2 = \text{const}, \tag{19}$$

$$\int_{T_{1L}}^{T_{10}} c_1(T_1) dT_1 = \overline{q} \cdot L, \qquad \int_{T_{1L}}^{T_{10}} \frac{c_1(T_1) dT_1}{q(T_1, T_2)} = L. \tag{20}$$

The first of these conditions determines $T_2^*(T_1, \lambda_2)$, second—T_{1L}, and third—constant λ_2.

For the Newtonian law of heat transfer

$$q = \alpha(T_1 - T_2) \tag{21}$$

with a constant heat capacity (water equivalent) c, we obtain from conditions (18)–(20)

$$\alpha^2(T_1 - T_2)^2 = -\lambda_2(-\alpha)T_2^2 \Rightarrow \alpha\left(\frac{T_1}{T_2} - 1\right)^2 = \lambda_2. \tag{22}$$

Therefore, for an arbitrary l of an optimal process, the ratio $\frac{T_1}{T_2}$ should be constant. This constant equals

$$\frac{T_1}{T_2} = 1 + \sqrt{\frac{\lambda_2}{\alpha}}. \tag{23}$$

As it follows from Eq. (20), $T_{1L} = T_{10} - \overline{q}L/c$. Finally, the condition (20) leads to the following equality:

$$\frac{\sqrt{\frac{\lambda_2}{\alpha}}}{1+\sqrt{\frac{\lambda_2}{\alpha}}} = -\frac{c}{\alpha L}\ln\left(1-\frac{\bar{q}L}{cT_{10}}\right). \tag{24}$$

Substituting Eqs. (23) and (24) into the expression

$$\sigma = \frac{c}{L}\int_{T_{1L}}^{T_{10}}\left(\frac{1}{T_2(T_1)}-\frac{1}{T_1}\right)dT_1, \tag{25}$$

minimal entropy production is obtained in the form

$$\sigma_{min} = \frac{c^2\ln^2\left(1-\frac{\bar{q}L}{cT_{10}}\right)}{\left[\alpha L + c\ln\left(1-\frac{\bar{q}L}{cT_{10}}\right)\right]L}. \tag{26}$$

Table 1 presents analogous conditions of minimal dissipation for some well-known processes and corresponding expressions for minimal entropy production.

As shown in [34], the conditions of minimal dissipation make it significantly easier to estimate the limiting possibilities of thermodynamic systems. In a system with multithreaded heat exchange [35], the total heat load q and the total heat transfer coefficient $\bar{\alpha}$ are fixed. At the input of the system, k heating fluxes with temperatures T_{i0} and water equivalents W_i come in. It is necessary to choose the parameters of heat fluxes, the structure of the system, and the distribution of heat transfer coefficients.

The conditions under which the minimum possible production of the entropy of the $\bar{\sigma}^*$ trait is reached are also defined in [34]: *(1) At each point of contact of the heating and heated streams, the minimum dissipation conditions must be satisfied. (2) Temperatures of heating fluxes at the outlet of the system should be equal to each other, as well as the temperature of the heated fluxes at the outlet. (3) Heating fluxes, in*

Process	**Conditions of minimal dissipation and entropy production**
Heat transfer $q = \alpha(T_2 - T_1)$	$\frac{T_1(l)}{T_2(l)} = 1 - \frac{\beta}{\alpha L}; \sigma_{\min} = \frac{\beta^2}{\alpha L - \beta}$ $\beta = W\ln\left(1-\frac{\bar{q}}{WT_1(0)}\right)$
Vector flux, linearly depending on driving forces $J = LX$	$X = \text{const.} J = \bar{J}; \sigma_{\min} = \bar{J}^T L^{-1}\bar{J}$
One-sided isothermal mass transfer $g(c_1, c_2) = k(c_1(l) - c_2(l))$	$c_2(l) = c_1(l) + \frac{m}{2} - \sqrt{c_1(l)m + \frac{m}{4}}$ $\int_{c_1(L)}^{c_1(0)} \frac{GdC_1}{k\left(1-c_1^2\right)\sqrt{c_1 m + \frac{m}{4}} - \frac{m}{2}} = L;$ $\sigma_{\min} = \int_{c_1(L)}^{c_1(0)} \frac{RG}{(1-c_1)^2}\ln\frac{c_1 dc_1}{c_1 + \frac{m}{2} - \sqrt{c_1 m + \frac{m^2}{4}}}$
Two-sided isothermal equimolar mass transfer	$\frac{\partial g}{\partial c_1} / \frac{\partial g}{\partial c_2} = m\frac{c_2(l)(1-c_2(l))}{c_1(l)(1-c_1(l))}$ $\frac{dc_1}{dl} = -\frac{g(c_1, c_2)}{G_1}$ $c_1(0) = C_{10}, c_1(L) = c_{1L};$ $\sigma_{min} = R\int_0^L g(c_1, c_2)\ln\left[\frac{c_1(1-c_2)}{c_2(1-c_1)}\right]dl$

Table 1.
Conditions of minimal dissipation in thermodynamic processes.

which the inlet temperature is less than the calculated temperature $\overline{T}$, do not participate in heat exchange.

Computational relations for Newtonian heat transfer are

$$\left.\begin{aligned}
&\overline{T}=\frac{\sum_{i=1}^{k}T_{i0}W_i-\overline{q}}{\sum_{i=1}^{k}W_i},\\
&q^*(T_{i0})=W_i(T_{i0}-\overline{T}),\\
&\alpha^*(T_{i0})=\frac{\overline{\alpha}W_i(\ln T_{i0}-\ln\overline{T})}{\sum_{i=1}^{k}W_i(\ln T_{i0}-\ln\overline{T})},\\
&m=1-\frac{1}{\overline{\alpha}}\sum_{i=1}^{k}W_i(\ln T_{i0}-\ln\overline{T}),\\
&\overline{\sigma}^*=\overline{\alpha}\frac{(1-m)^2}{m},\\
&\alpha^*(T_{i0})=q^*(T_{i0})=W_i=0,\quad T_{i0}\le\overline{T}.
\end{aligned}\right\}\tag{27}$$

The system in which the entropy production calculated with parameters of all fluxes

$$\sigma=\sum_{\nu}W_{\nu}\ln\frac{T_{\nu}^{out}}{T_{\nu}^{in}}\tag{28}$$

is lower than a certain value cannot exist in reality.

Analogous relations can easily be obtained in the case when the inlet parameters of heated fluxes are prescribed.

3. Synthesis of heat exchange systems

In [36] the problem of the limiting possibilities of the heat exchange system ("ideal" heat exchange) was considered. The minimum possible entropy production σ^* was found in the system with the given values of water equivalents and input temperatures of hot or cold fluxes and given the total heat load and the total heat transfer coefficient. It is shown that for the case when the heat flux is proportional to the temperature difference (Newtonian dynamics), this irreversibility limit can be reached if at each point of contact the ratio of the absolute temperatures of the fluxes is the same, and their temperatures at the outlet of the system are the same for all fluxes whose input temperatures are fixed (hot or cold).

Conditions of ideal heat exchange impose very strict requirements on the characteristics of the system:

—Each double-flux cell must be a counter-flux heat exchanger.

—The ratio of the water equivalents of the hot and cold flux in it should be equal to the ratio in degrees Kelvin of the temperature of the cold flux at the outlet of the heat exchange cell to the temperature of the hot flux at its inlet—conditions of thermodynamic consistency.

—This ratio and its corresponding minimum possible entropy production at fixed temperatures and water equivalents of hot fluxes are related to their inlet temperatures T_i^0, the water equivalents W_i, and the total heat transfer coefficient K as:

$$\left.\begin{aligned} m &= 1 - \frac{1}{K}\sum_{i=1}^{n} W_i\left(\ln T_i^0 - \ln \overline{T}_+\right), \\ \sigma^* &= K\frac{(1-m)^2}{m}. \end{aligned}\right\} \tag{29}$$

—The temperature of the hot streams at the outlet should be the same and, as it follows from the conditions of the energy balance, is equal to:

$$\overline{T}_+ = \frac{\sum_{i=1}^{k} T_{i0} W_i - \overline{q}}{\sum_{i=1}^{k} W_i}, \tag{30}$$

—Hot fluxes with initial temperatures less than $\overline{T}_+$ do not participate in the heat exchange system.

If a part of the hot fluxes condenses in the process of heat transfer, then in the expression for m (Eq. (24)), the water equivalent of the corresponding term tends to infinity. Assign the index k to the condensing fluxes and find the limit.

$W_k\left(\ln T_k^0 - \ln \overline{T}_+\right) = W_k\left(\ln T_k^0 - \ln\left(T_k^0 - \frac{q_k}{W_k}\right)\right)$ when W_k tends to infinity. Using L'Hospital's rule to disclose the uncertainties, we find that

$$\lim_{W_k\to\infty} W_k\left(\ln T_k^0 - \ln\left(T_k^0 - \frac{q_k}{W_k}\right)\right) = \frac{q_k}{T_k^0} = \frac{g_k r_k}{T_{bk}}. \tag{31}$$

Here, it is taken into account that the temperature T_k^0 is equal to the condensation temperature and the thermal load is the product of the flux rate of the latent heat of vaporization.

Thus, the expression for m in the presence of condensing fluxes will be rewritten in the form:

$$m = 1 - \frac{1}{K}\left(\sum_{i\neq k} W_i\left(\ln T_i^0 - \ln \overline{T}_+\right) + \sum_k \frac{g_k r_k}{T_{bk}}\right). \tag{32}$$

In a multithreaded system integrated with the technological process, the values of water equivalents of both hot and cold fluxes are set, and often their outlet temperatures are set. Therefore, the performance of the ideal heat exchange system cannot be achieved. It is natural to set the task of synthesis of the heat exchange system of the minimum irreversibility at more rigid restrictions on characteristics of streams. The conditions of ideal heat transfer can only serve as a "guiding star" like Carnot's efficiency for thermal machines, and the value of the ratio σ^* to the real production of entropy in the designed system is an indicator of its thermodynamic perfection.

Next, we propose the calculated relations for the bottom estimate of the minimum dissipation in the system with the above restrictions and the synthesis of a hypothetical system in which such an estimate is implemented.

Consider a multithreaded heat exchange system containing a set of hot (index i) and cold fluxes (index j), with given water equivalents W_i, W_j. For each of the cold (heated) fluxes, its inlet and outlet temperatures are set to T_j^0 and $\overline{T}_j > T_j^0$.

For hot (cooled) fluxes, except for water equivalents, their temperatures at the inlet to the heat exchanger T_i^0 are set. If some flux in the system changes its phase state, then for it except water equivalents, the flux rate g_i, g_j and heat of vaporization (condensation) r_i, r_j are fixed. The ambient temperature will be denoted as T_0.

Under these conditions, the thermal load of the system is equal to the total energy required for heating all cold fluxes and is determined by the equality:

$$\overline{q} = \sum_j q_j = \sum_j W_j\left(\overline{T}_j - T_j^0\right). \tag{33}$$

The difference in the conditions imposed on the hot and cold fluxes is due to the fact that for cold fluxes leaving the system with a temperature less than a predetermined one, heating is required, i.e., additional energy costs, and for hot ones, if their outlet temperature is greater than a predetermined one, cooling is required, which is much easier.

Entropy production is the difference between the total entropy of outgoing fluxes and the total entropy of incoming fluxes. Initially, we assume that all fluxes enter and leave the system in the same phase state, the pressure change in the system is small, and the heat capacity is constant. Then, the change in the entropy of each flux is the product of its water equivalent by the logarithm of the ratio of its inlet and outlet temperatures in degrees Kelvin [37]. So, it follows from the conditions of the thermodynamic entropy balance that:

$$\sigma = \sigma_+ + \sigma_- = \sum_i W_i\left(\ln \overline{T}_i - \ln T_i^0\right) + \sum_j W_j\left(\ln \overline{T}_j - \ln T_j^0\right). \tag{34}$$

The first of these terms is negative, the second is positive, and their sum is always greater than $\sigma^* > 0$.

Note that all variables determining the value of the entropy growth of cold fluxes are given by the conditions of the problem, so that the minimum entropy production corresponds to the minimum at a given thermal load of the first summand by temperatures $\overline{T}_i$.

The formal statement will take the form:

$$\sigma_+ = \sum_i W_i\left(\ln \overline{T}_i - \ln T_i^0\right) \to \min / \sum_i W_i\left(T_i^0 - \overline{T}_i\right) = \overline{q} = \sum_j W_j\left(\overline{T}_j - T_j^0\right). \tag{35}$$

The Lagrangian of this problem

$$L = \sum_i W_i\left(\ln \overline{T}_i - \ln T_i^0\right) - \lambda \sum_i W_i\left(\overline{T}_i - T_i^0\right). \tag{36}$$

The conditions of its stationarity in $\overline{T}_i$ lead to equality:

$$\overline{T}_i = \frac{1}{\lambda}. \tag{37}$$

Thus, *in any water equivalents and input temperatures of hot streams, minimum dissipation corresponds to such an organization of heat transfer for which the temperatures of hot streams at the exit are the same.*

In general, coolant fluxes at the system inlet can have different phase states: vapor, liquid, or vapor-liquid mixture. The same states can be at the output of the stream.

—If the flux does not change its phase state, but changes only the temperature, then we assume that its temperature at the input to the cell T_k^0, the water equivalent

W_k, and for cold fluxes the temperature at the output $\overline{T}_k$ are known. The temperature of the hot streams at the output of the $\overline{T}_+$ system is selectable (see Eq. (30)).

—If the cold flux changes its phase state so that at the inlet it is a liquid at boiling point and at the outlet it is saturated with steam (let us define it as "evaporating"), the weight flux rate g_j, the boiling point T_{bj}, and the heat of vaporization r_j are given. The same is true for hot "condensing" streams. They have a saturated steam state at the inlet and a liquid state at the boiling point at the outlet.

Thus, the first step in the synthesis algorithm of heat exchange systems is the preparation of initial data, in which actual fluxes and their characteristics are converted into calculated fluxes. They can be of two types: those that do not change their phase state (heated and cooled) and those that change it at the boiling point (evaporating and condensing). End-to-end fluxes are not included in the calculation. To calculate the total heat load production, use the following expression:

$$\sum_{jh} W_{jh}\left(\overline{T}_{jh} - T^0_{jh}\right) + \sum_{bj} g_{bj} r_{be} = \overline{q}. \tag{38}$$

Minimum dissipation implies fulfillment of the "counterflow principle": the cold streams with higher temperatures must be in contact with the hot flux with a higher temperature. The latter requirement, as well as the equality of temperatures of hot streams at the outlet, corresponds to the conditions of the ideal heat transfer [36].

As the hot fluxes move from one contact cell to the next, their temperature changes due to the recoil of the heat flux. At the output of the system, the heat flux given by them is $\overline{q}$, and the temperature is $\overline{T}_+$. Let us denote by q the given heat load in some intermediate state of hot fluxes. As the hot streams cool down, it changes from zero to $\overline{q}$.

In this case, we assume that when the hot flux with the highest input temperature (first) is cooled to a temperature of T^0_2, the first and second fluxes are combined, so that their water equivalents are summed. A similar union occurs with the third flux, etc., until the temperature of the equivalent hot flux drops to the previously calculated formula (30) $\overline{T}_+$. If a condensing flux is at a certain temperature T_{bi} in the number of hot fluxes, the temperature of the equivalent flux is constant and equal to T_{bi} until the equivalent flux transfers the heat of condensation $q_{bi} = g_{bi} r_i$. The dependence of the temperature of the hot flux equivalent on the given heat load $T_+(q)$, we will call the contact temperature of the hot fluxes.

Cold fluxes are ordered by their outlet temperature, so that $j = 1$ corresponds to the flux with the highest output temperature. For cold fluxes, the value of q is equal to the current required heat load, i.e., the heat they need to obtain to satisfy the conditions imposed on their temperature and output state. The greater the q, the lower the cold flux temperature corresponding to this value. In this case, we assume that when the temperature of the first flux decreases, two events are possible:

1. Its temperature will drop to the set temperature at the output of the second stream.

2. Its temperature will drop to its initial temperature.

In the first case, the first cold flux is calculated combined with the second. In the second case, it is excluded from the system and transferred to the heating of the second stream. This procedure continues until an equivalent cold flux reaches the lowest cold flux temperature at the system inlet. The number of threads included in the equivalent cold flux is changed by adding fluxes with lower temperatures at the outlet and due to the exclusion from streams with the highest temperatures at the entrance. But each value of q corresponds to the value of $T_-(q)$ of the contact temperature of cold fluxes.

The dependencies of the current contact temperatures can be calculated from energy balance conditions similar to the expression (25). For equivalent hot flux:

$$T_+(q) = \frac{\sum_{i=1}^{S_+(T_+)} W_i T_i^0 - q}{\sum_{i=1}^{S_+(T_+)} W_i}, \tag{39}$$

where $S_+(T_+)$ is the set of indices of hot fluxes for which the inlet temperature is greater than the current contact temperature ($T_i^0 > T_+(q)$).

Similarly, for the contact temperature of the equivalent cold flux, we have:

$$T_-(q) = \frac{\sum_{j\ni S_-(T_-)} W_j \overline{T}_j - q}{\sum_{j\ni S_-(T_-)} W_j}, \tag{40}$$

where $S_-(T_-)$ is the set of indices of cold fluxes for which the contact temperature T_- satisfies the inequality $\overline{T}_j > T_-(q) > T_j^0$.

The curves of the current contact temperatures decrease monotonically with the growth of q, with $T_+(q) > T_-(q)$. On each of these curves, the points (nodes) are selected, in which either the composition of the fluxes entering the equivalent flux changes or the condensation/evaporation process takes place. In the latter case, horizontal sections appear on the curves. On the curve $T_-(q)$, there can be vertical jumps if the flux temperature $T_j^0 > \overline{T}_{j-1}$.

The interval δq_ν from one of the nodes on any of the contact curves to the nearest node on the same or another curve is characterized by the same composition and phase state of the contacting fluxes. We will call it *the homogeneity interval*.

For each such interval of δq_ν, three combinations of contacting fluxes are possible:

1. Both equivalent fluxes change their phase states.

2. The hot equivalent flux is cooled and the cold is heated.

3. One of the fluxes changes its phase state, and the other is cooled or heated.

Contact temperature curves provide all the data necessary to calculate the heat transfer coefficient of the cell in which the contact is made:

—Water equivalents of W_+, W_- is equal to the sum of the equivalents of water fluxes which are part of the equivalent contacting fluxes.

—Temperatures of equivalent fluxes at the inlet and outlet of the interval of homogeneity is known.

—The thermal load of such a computational cell is δq_ν.

Depending on which of these contact combinations is implemented, it is possible to select the type of cell hydrodynamics and find K_ν. Finding the heat transfer coefficient K_ν for each ν interval of homogeneity and summing these coefficients over all intervals, we obtain the total coefficient K, which can be achieved by organizing the countercurrent heat exchange of equivalent fluxes. In turn, knowledge of the heat load $\overline{q}$ and the total heat transfer coefficient allows us to calculate the minimum possible entropy production σ^* by the formula (29) and estimate the degree of thermodynamic perfection of the constructed system as $\eta = \frac{\sigma^*}{\sigma^0}$, where σ^0 is total entropy production in the system.

4. The region of realizability of systems

An irreversible factor affecting machine power or pump performance is finite heat transfer coefficients α_i between heat sources and the working fluid. We substitute the minimum possible entropy production into Eq. (5) and obtain the condition determining the maximum power:

$$p = \eta_k q_+ - \sigma_{min}(q_+)T_-. \tag{41}$$

As $\sigma_{min}(q_+)$ increases faster than q_+, there exist a maximum of power in the region of realizability.

To minimize the production of entropy, it is necessary that with each contact of the working medium with the sources the conditions of minimum dissipation, which depend on the dynamics of heat transfer, are met. For a source of infinite capacity and the temperature of the working fluid in contact with, it should be constant. For Newtonian dynamics, the ratio of working fluid temperature and sources should have been constant. So, if the temperature of the source changes due to the final capacity, then the temperature of the working fluid should change, remaining proportional to the temperature of the source.

For sources of infinite capacity, the optimal cycle of a heat machine with maximum power for any heat transfer dynamics should consist of two isotherms and two adiabats, and it turned out that the efficiency corresponding to the maximum power (it is called the Novikov-Curzon-Ahlborn, η_{nca}) is only a function of the Carnot efficiency:

$$\eta_{nca} = 1 - \sqrt{1 - \eta_C}. \tag{42}$$

The maximum difference between η_C and η_{nca} is achieved when the ratio of absolute temperature of the hot and cold sources is 0.25, when η_C=0.75.

For power that is less than the maximum possible, the maximum efficiency of the heat machine is equal to

$$\eta_{\max}(p) = 1 - \frac{1}{2T_+}\left[T_+ + T_- - \frac{p}{\alpha} - \sqrt{(T_+ - T_-)^2 + \left(\frac{p}{\alpha}\right)^2 - 2\frac{p}{\alpha}(T_+ + T_-)}\right]. \tag{43}$$

In this case, α is expressed as

$$\alpha = \frac{\alpha_+ \alpha_-}{\alpha_+ + \alpha_-}. \tag{44}$$

As $p \to 0$, the efficiency $\eta_{\max}$ tends to Carnot efficiency and as

$$p \to p_{\max} = \alpha\left(\sqrt{T_+} - \sqrt{T_-}\right)^2. \tag{45}$$

Corresponding thermal efficiency approaches the efficiency value obtained by Novikov, Curzon, and Ahlborn (42).

The nature of the set of realization modes is shown in **Figure 1**.

Similar results can be obtained for the heat pumps. Since the flux of costs is mechanical energy, the set of realizable modes has the form of a convex upward and unbounded parabola.

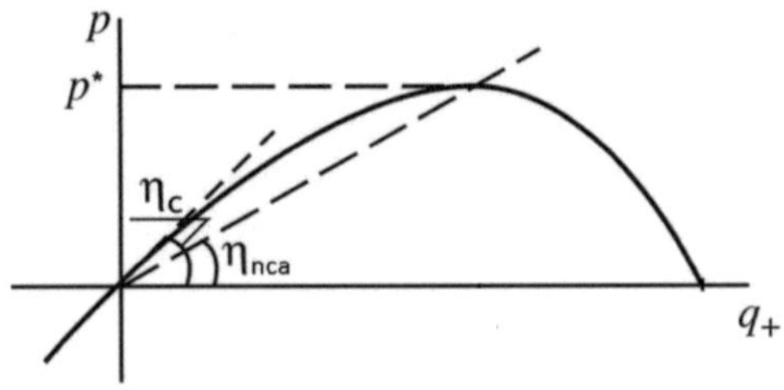

Figure 1.
Power of thermal machine as a function of driving heat flux.

5. Rectification processes

In the separation process, energy is spending on getting the work of separation. The work of separation can be obtained as an increase in the free energy of the streams leaving the system compared to the energy of the mixture flux at the system inlet. The energy expended can be thermal or mechanical. In systems of separation with thermal energy, the set of realizable modes coincides in the form with heat engines. In this case, the rectification processes will be the most important and energy-intensive. In the section below, the process of thermal separation of a two-component mixture is considered, and considerations which allow one to proceed to the determination of the order of separation of multicomponent mixtures are obtained.

Let the following parameters be defined for a mixture of two components: q_i, T_i, s_i, p_i, h_i, x_i, μ_i—molar consumption, temperature, molar entropy, pressure, enthalpy, concentration of key component, and its chemical potential in ith stream. Assume the index $i = 0$ for the separated stream, index $i = 1$ for the stream of the enriched key component (for which $x_1 > x_0$), and index $i = 2$ for the stream cleared of the key component ($x_2 < x_0$). Heat flux q_++, brought (coming) from a source of temperature T_+, is supplied to the separation system, whereas heat flux q_- is rejected to a source of temperature T_-. Equations of thermodynamic balances are of the form:

$$\begin{aligned} &g_0 = g_1 + g_2, \ g_0 x_0 - g_1 x_1 - g_2 x_2 = 0, \\ &q_+ - q_- + g_0 h_0 - g_1 h_1 - g_2 h_2 = 0, \\ &\frac{q_+}{T_+} - \frac{q_-}{T_-} + g_0 s_0 - g_1 s_1 - g_2 s_2 + \sigma = 0. \end{aligned} \tag{46}$$

The ratio of target mass flux g_1 and heat flux q_+ may be accepted as the thermal efficiency of the separation process:

$$\eta = \frac{g_1}{q_+}. \tag{47}$$

Using material balances of Eq. (46), we shall express g_2 in terms of g_1 and introduce coefficient $a = (x_1 - x_0)/(x_0 - x_2)$. Then, the second flux satisfies $g_2 = a g_1$. Eq. (46) assumes the forms

$$q_+ - q_- + g_1(\Delta h_{01} + a \Delta h_{02}) = 0, \tag{48}$$

$$\frac{q_+}{T_+} - \frac{q_-}{T_-} + g_1(\Delta s_{01} + a \Delta s_{02}) + \sigma = 0. \tag{49}$$

Here, $\Delta s_{01} = s_0 - s_1$, $\Delta s_{02} = s_0 - s_2$ are the entropy increases, whereas $\Delta h_{01} = h_0 - h_1$, $\Delta h_{02} = h_0 - h_2$ are enthalpy increases in corresponding streams. It is mandatory that the concentrations of the key component in streams are prescribed.

We transform Eq. (48) to the form $q_- = q_+ + g_1(\Delta h_{01} + a\Delta h_{02})$ and substitute the expression obtained into Eq. (49). The frontier of the realization set is characterized by the following equation:

$$g_1 = \frac{1}{F}\left(1 - \frac{T_-}{T_+}\right)q_+ - \sigma\frac{T_-}{F}. \tag{50}$$

Here, $F = T_-(\Delta s_{01} + a\Delta s_{02}) - \Delta h_{01} - ah_{02}$. The increases of enthalpy and entropy contained in F have the forms

$$\Delta h_{0i} = C_p(T_0 - T_i), \quad i = 1, 2, \tag{51}$$

$$\Delta s_{0i} = C_{p0}\ln T_0 - C_{pi}\ln T_i - R\ln\frac{P_0}{P_i} + \Delta s_{\text{mix}\,0} - \Delta s_{\text{mix}\,i}, \quad i = 1, 2.$$

The entropy of mixing per one mole of mixture is:

$$\Delta s_{\text{mix}\,i} = R[(1 - x_i)\ln(1 - x_i) + x_i\ln x_i], \quad i = \overline{0, 2}. \tag{52}$$

Note that the ratio T-/F depends on reversible factors only. In the reversible process, the entropy production σ is zero, and the thermal efficiency reaches a maximum equal to the multiplier at the heat flux in Eq. (42).

As a productivity you can take any of the streams, even the stream of a separated mixture, because with given compositions of the streams they are proportional.

A reversible estimate of the thermal efficiency of the separation process and the shape of the border of the realizability region can be clarified by finding the minimum possible for a given productivity and dynamics of heat and mass transfer value σ and its dependence on the coefficients of dynamics and heat flux.

If the dynamics of heat transfer can be approximated by the Fourier law and the mass transfer flux is proportional to the difference of chemical potentials, then the minimum dissipation is proportional to the square of the cost of heat. The boundary of the set of realizable modes in this case has a parabolic form

$$g = bq - aq^2. \tag{53}$$

Then, the efficiency of a separation column in the maximum productivity mode is equal to one half of the reversible efficiency:

$$\eta^* = 0,5\eta^0 = 0,5b. \tag{54}$$

Qualitative expressions linking characteristic coefficients a and b with parameters of the separation column were obtained [38]:

$$b = \frac{T_B - T_D}{T_B A_G} = \frac{\eta_k}{A_G}, \quad a = \left[\frac{1}{\beta_B T_B T_+} + \frac{1}{\beta_D T_D T_-} + \frac{2(x_D - x_B)}{kr^2}\right]\frac{T_D}{A_G}. \tag{55}$$

Here, A_G is the molar reversible work of mixture separation, equal to the difference between molar free energy of streams leaving the column and the free energy of raw stream, T_D, T_B are the temperatures in the condenser and the kettle of the column, r is the molar evaporation heat, β_D, β_B are the coefficients of heat exchange in the condenser and the kettle, and k is the effective coefficient of mass transfer for column height.

The coefficients a and b can be found not only by Eq. (55) but also by the results of measurements on the current column. This allows us to solve many problems associated with finding a set of realizability, including the problem of choosing the order of separation. It is important that the efficiency in maximum performance mode depends only on the reversible efficiency. The following condition, sufficient for to be independent of a, is valid.

The sufficiency condition for independence of η^* of a [34]:

If the partial derivative $\frac{\partial\sigma(a,b,q)}{\partial q}$ *depends continuously on some scalar function* $z(a,q)$ *and the ratio* $\frac{\sigma(a,b,q)}{q}$ *is a function of z, then the thermal efficiency in the maximum productivity mode is defined exclusively in terms of variables b, characterizing the reversible process*. The condition is satisfied for thermal machines and for binary rectification.

In **Figure 2** shows an example of the boundaries of realizable sets in cases where $\sigma((a,b,q))$ does not depend on a.

With decreasing dynamic coefficients, the entropy production increases. The set of realizable modes is compressed, while the maximum performance points with a corresponding heat flux remain on a straight line with a slope of η^*.

5.1 Order of separation: rule of temperature multipliers

We arrange the substances according to the property γ used for separation (boiling point in the rectification processes). We normalize γ so that it is in the range from 0 to 1. The order of separation of substances may be direct or reverse. In the case of a direct order, the stream with the components $\gamma < \gamma 1$ is first separated, and then, the stream with large values of γ is divided into two sub-streams so that the first one has $\gamma 1 \leq \gamma < \gamma 2$ and the second one $\gamma \geq \gamma 2$. In the case of reverse order, the stream with the components $\gamma > \gamma 2$ is first separated, and then, the stream with lower values of γ is divided into two sub-streams so that the first one has $\gamma 1 \leq \gamma < \gamma 2$ and the second one $\gamma \leq \gamma 1$.

Let A denote the work of separation of the mixture. The A includes the factor RT_0, where R is the universal gas constant and T_0 is the ambient temperature. Then, the entropy of the separated fluxes will decrease in proportion to the work of separation A. In a reversible case, a decrease in the entropy of the material flux should be compensated by an increase in the entropy of the heat flux. If T_+ and T_-

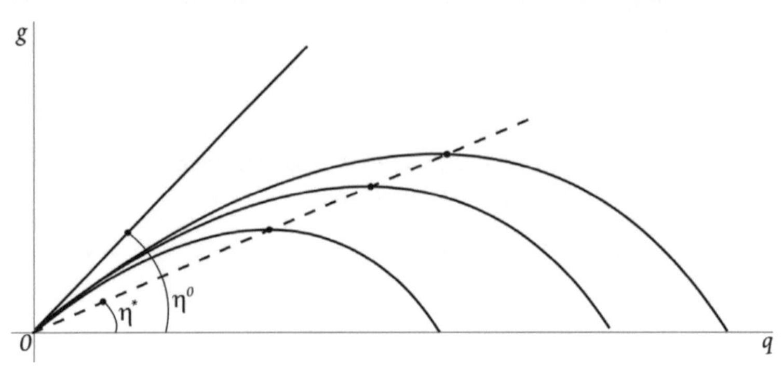

Figure 2.
Nature of change of the realization frontier with the irreversibility increase.

are the temperatures of heat supply and removal, then the increase in entropy of the heat flux can be expressed as $\delta s_q = q(1/T_- - 1/T+)$. The total cost of heat depends on the order of separation, while the total work does not depend on the order of separation. At each separation stage, the heat flux is proportional to the separation work at this stage, with the proportionality multiplier $K_T = \frac{T_+T_-}{T_+-T_-}$. Temperature factors are determined by the choice of the separation boundary. Denote by K_{T1} and K_{T2} the temperature factors corresponding to the separation boundary $\gamma 1$ and $\gamma 2$, respectively, and by A_{21} and A_{22} the work of separation of the mixture in the second stage in the direct and inverse order of separation. The total cost of heat per mole of the input mixture in the direct and reverse order of separation can be written as:

$$q_1 = K_{T1}(A_0 - A_{21}) + K_{T2}A_{21}, \quad q_2 = K_{T2}(A_0 - A_{22}) + K_{T1}A_{22}. \tag{56}$$

To determine the separation order, it is necessary to calculate the difference:

$$\Delta q = (K_{T1} - K_{T2})[A_0 - A_{21} - A_{22}]. \tag{57}$$

If the result of the calculation Eq. (57) is negative, then it is reasonable to choose a direct separation order. If the result is positive—a reverse order.

In the case of a multistage system, this rule applies to each of two successive stages. It is easy to see that the expression in square brackets in Eq. (57) is non-negative. From here follows the rule of temperature multipliers (see [39]): The separation boundaries must be chosen so that the temperature multipliers do not decrease from stage to stage. In the case when the separation efficiency in the maximum performance mode depends only on the reversible efficiency, the rule of temperature multipliers is also valid. It is important that the information that is needed to calculate temperature factors is much more accessible and accurate than the information on the dynamics of the processes in the column.

6. Conclusions

This chapter discusses the problems of optimization of thermodynamics and methods of analysis of systems and describes the types of thermodynamic balances, the relationship between the performance of the process, and the production of entropy. Also, it is shown that in the absence of irreversibility, the thermal efficiency is equal to the Carnot efficiency.

The conditions are found under which the thermodynamic processes at a given average intensity have minimal dissipation, expressions for determining the minimum dissipation and entropy with the Newtonian heat transfer law are obtained, and expressions for the cases of vector flux, one-sided isothermal, and two-sided equimolar mass transfer are given.

The synthesis algorithm makes it possible to build heat exchange systems with minimal irreversibility, in which restrictions on water equivalents, temperatures, and phase states of the flows are fulfilled, which imply combining the fluxes into two equivalent ones. The nature of the set of realizable modes of heat engines and pumps is described. It is shown that the efficiency corresponding to the maximum power mode does not depend on heat transfer coefficients, but is only a function of the Carnot efficiency.

Separation processes are considered, and estimates of the thermal efficiency of the separation process and the shape of the realizable area boundary are obtained for them. It is shown that the efficiency in the mode of maximum performance depends only on the reversible efficiency. The rule of temperature multipliers is described, which allows to determine the separation order in multistage systems.

Author details

Margarita Anatolyevna Zaeva[1*†] and Anatoly Mikhailovich Tsirlin[2]

1 Institute of Cyber Intelligence Systems, National Research Nuclear University MEPhI (Moscow Engineering Physics Institute), Moscow, Russia

2 A. Aylamazyan Program Systems Institute of Russian Academy of Sciences, Russia

*Address all correspondence to: mazayeva@mephi.ru

†Ad memoriam Anatoly V. Zaev

References

[1] Carnot S. Reflections on the motive power of fire and on machines fitted to develop that power. In: Thurston RH, editor. Ecole Polytechnique. No. 55. Chez Bachelier, Libraire, Quai des Augustins: A Paris; 1824

[2] Novikov II. The efficiency of atomic power stations. Atomnaya Energiya. 1957;**3**(11):409; English translation in Journal of Nuclear Energy. Part B. 1958; 7(2):25-128

[3] Reitlinger HB. Sur l'utilisation de la chaleur dans les machines a feu. Liege: Vaillant-Carmanne; 1929

[4] Andresen B, Salamon P, Berry RS. Thermodynamics in finite time: Extremals for imperfect heat engines. The Journal of Chemical Physics. 1977; **66**(4):1571-1577

[5] Andresen B. Finite-Time Thermodynamics. Copenhagen: University of Copenhagen; 1983

[6] Andresen B. Finite time thermodynamics. Current trends in finite time thermodynamics. Angewandte Chemie, International Edition. 2011;**50**:2690-2704

[7] Bejan A. Entropy generation minimization: The new thermodynamics of finite size devices and finite time process. Journal of Applied Physics. 1996;**79**: 1191-1218

[8] Berry RS, Kazakov VA, Sieniutycz S, Szwast Z, Tsirlin AM. Thermodynamic Optimization of Finite Time Processes. Chichester: Wiley; 1999

[9] Chen J, Yan Z, Lin G, Andresen B. On the Curzon-Ahlborn efficiency and its connection with the efficiencies of real heat engines. Energy Conversion and Management. 2001;**42**:173-181

[10] Chen L, Wu C, Sun F. Finite time thermodynamic optimization or entropy generation minimization of energy systems. Journal of Non-Equilibrium Thermodynamics. 1999;**24**:327-359

[11] Curzon FL, Ahlburn B. Efficiency of a Carnot engine at maximum power output. American Journal of Physics. 1975;**43**:22-24

[12] Hoffman KH, Watowich SJ, Berry RS. Optimal paths for thermodynamic systems: the ideal Diesel cycle. Journal of Applied Physics. 1985;**58**(6): 2125-2134

[13] Linezky SB, Rodnjansky LE, Tsirlin AM. Optimal cycles of chillers and heat pumps. Izvestiya An SSSR—Energetika i Transport. 1985;**5**(6):42-49

[14] Mironova VA, Amelkin SA, Tsirlin AM. Mathematical Methods of Thermodynamics with a Finite Time. M: Ximija; 2000

[15] Moloshnikov BE, Tsirlin AM. The thermodynamically optimal concentration profiles in the problems of isothermal irreversible mass transfer. Theoretical Foundations of Chemical Engineering. 1990;(24):129-137

[16] Mironova V, Tsirlin A, Kazakov V, Berry RS. Finite-time thermodynamics: Exergy and optimization of time-constrained processes. Journal of Applied Physics. 1994;(76):629-636

[17] Ondrechen MJ, Andresen B, Mozurkewich M, Berry RS. Maximum work from a finite reservoir by sequential Carnot cycles. The American Journal of Physiology. 1981;**49**:681

[18] Ondrechen MJ, Berry RS, Andresen B. Thermodynamics in finite time: A chemically driven engine. The Journal of Chemical Physics. 1980;**72**(9):5118-5124

[19] Ondrechen MJ, Berry RS, Andresen B. Thermodynamics in finite time: Processes with temperature-dependent chemical reactions. The Journal of Chemical Physics. 1980;**73**(11): 5838-5843

[20] Orlov VA, Pozonoer LI. Estimates of the efficiency of the thermodynamic processes of controlled substances based on the energy balance equations and entropy. In: X-Union Conference on Control. M.: Nauka; 1986

[21] Rozonoer LI, Tsirlin AM. Optimal control of thermodynamic systems. Automation and Remote Control. 1983; **44**(1):70-79; **44**(2):88-101; **44**(3):50-64

[22] Salamon P, Nitzan A. Finite time optimizations of a Newton's law Carnot cycle. The Journal of Chemical Physics. 1981;**74**(6):3546-3560

[23] Salamon P, Band YB, Kafri O. Maximum power from a cycling working fluid. Journal of Applied Physics. 1982;**53**(1):197-202

[24] Salamon P, Hoffman KH, Schubert S, Berry RS, Andresen B. What conditions make minimum entropy production equivalent to maximum power production? Journal of Non-Equilibrium Thermodynamics. 2001;**26** (1):73-84

[25] Salamon P, Nitzan A, Andresen B, Berry RS. Minimum entropy production and the optimization of heat engines. Physical Review A. 1980;**21**:2115-2129

[26] Salamon P, Nulton JD, Siragusa G, Andresen TR, Limon A. Principles of control thermodynamics. The International Journal of Energy Research. 2001;**26**(3):307-319

[27] Salamon P. Physics versus engineering of finite-time thermodynamic models and optimizations. In: Bejan A, Mamut E, editors. Thermodynamic Optimization of Complex Energy Systems. Dordrecht, The Netherlands: Kluwer Academic Publishers; 1999. pp. 421-424

[28] Salamon P, Hoffman K-H, Tsirlin AM. Optimal control in a quantum cooling problem. Applied Mathematics Letters. 2012;(25):1263-1266

[29] Salamon P, Hoffman K-H, Rezek K-H Y. Maximum work in minimum time from a conservative quantum system. Chemical Physics. 2009;(11): 1027-1032

[30] Sieniutycz S, Jezowski J. Energy Optimization in Process Systems and Fuel Cells. Oxford: Elsevier; 2013

[31] Sieniutycz S. Thermodynamic Approaches in Engineering Systems. Oxford: Elsevier; 2016

[32] Spirkl W, Ries H. Optimal finite-time endoreversible processes Physical Review E. 1995;**52**(4):3455-3459

[33] Tsirlin AM, Mironova VA, Amelkin SA. Processes minimal dissipation. Theoretical Foundations of Chemical Engineering. 1997;**31**(6):649-658

[34] Sieniutycz S, Tsirlin A. Finding limiting possibilities of thermodynamic systems by optimization. Philosophical Transactions of the Royal Society A: Mathematical, Physical and Engineering Sciences. 2017;**375**(2088): 20160219

[35] Tsirlin AM, Ackremenkov AA. The optimal organization of heating and cooling systems. Theoretical Foundations of Chemical Engineering. 2012;**46**(1):109-114

[36] Tsirlin AM. Ideal heat exchange systems. Journal of Engineering Physics and Thermophysics. 2017;**90**(5)

[37] Kondepudi D, Prigogine I. Modern Thermodynamics. From Heat Engines

to Dissipative Structures. John Wiley & Sons; 1998

[38] Tsirlin AM, Sukin IA. Finite-time thermodynamics: The maximal productivity of binary distillation and selection of optimal separation sequence for an ideal ternary mixture. Journal of Non-Equilibrium Thermodynamics. 2014;**39**(1):13

[39] Tsirlin AM, Balunov AI, Sukin IA. Estimates of energy consumption and selection of optimal distillation sequence for multicomponent distillation. Theoretical Foundations of Chemical Engineering.2016;**50**(3):250-259

Chapter 15

Thermal Properties on Metals at Cryogenic Temperatures

Cemil Koyunoğlu

Abstract

The thermophysical and some other properties of solids are of great importance for the use in the chemical, military, and even aerospace industries and for the design of efficient cryogenic equipment. Considering the heat loads, cooling, thermal fluctuations, or stresses or cryogenic fluids in boilers, the thermophysical properties should be considered. There is a considerable literature on the mechanical and structural properties of solids at cryogenic temperatures, but unfortunately there is not enough literature available for thermophysical properties. This chapter is recommended to close this gap. This chapter basically states: thermophysical properties of metals at cryogenic temperatures, specific heats, and thermal conductivity.

Keywords: cryogenic temperature, specific heat, thermal conductivity, advanced aviation technology

1. Thermophysical properties of metals at cryogenic temperatures

The mechanical and fabrication properties of solids are of paramount importance for the design of efficient cryogenic equipment for the chemical, military, and aerospace industries. The thermophysical properties are also important whenever heat loads, refrigeration, thermal fluctuations, or stresses or boiling cryogenic fluids must be considered. There is a relatively large amount of information available on the mechanical and structural properties of solids at cryogenic temperatures, but unfortunately the opposite is true for thermophysical properties.

The solid materials for which there are complete thermophysical data are only a minor fraction of the total number that are important and commercially available. This dearth of information will undoubtedly continue in the future, in spite of increased interest, activity, and financial support for cryogenic research, because of the considerable difficulty and expense of the actual experimental measurements. It is, therefore, of the utmost importance that a design or materials engineer must be able to estimate accurate values for new or untested materials. Unlike many mechanical or fabrication properties, the thermophysical properties (except thermoelectricity) for a given material may often be predicted from theoretical or semiempirical knowledge combined with data on similar materials.

This presentation stresses the basic phenomena and the fundamental concepts and assumptions. Each property is discussed with emphasis on temperature dependences and contributions to the total observed effect.

Also noted are the interrelations between the various properties and the value of those interrelations as aids in prediction. Besides the usual references to articles and books, information is given on useful compilations and sources of specialized up-to-date bibliographies and data collections.

Because of the breadth of subject and limitations of space, it is impractical to give in this chapter either a complete review of the data on various thermophysical properties of solids or a detailed explanation of the underlying assumptions and theories. To make the following discussions more manageable, the solid materials are limited to metals, and the thermophysical properties are limited to specific heat and thermal conductivity. As this is a review article, the subjects will not be confined to our own experimental programs and results.

2. Specific heats

The specific heats of metals and alloys are fitted very well by existing theories, at least as far as engineering data are concerned. Therefore, it is common either to tabulate experimental values or to represent them on a common, reduced temperature graph. **Figure 1** shows a graph, but without specific metals or temperatures represented. The essential experimental problem is to ascertain the characteristic temperature, Θ. Conversely, given the values of the characteristic temperatures, values for specific heat are easily obtained. The specific heat at constant volume of a metal may be defined as $C_v = (\frac{\partial U}{\partial T})v$, where U is the total internal energy of the metal or alloy system and T is the absolute temperature. It is important to determine the theoretical problem of energy transformation. The first of these is that the energy and indirectly the specific heat are slowly changing temperature functions as shown in **Figure 1**. For example, the energy of an ionic lattice structure of a metal and electrons in its free or conductive behavior can be given. Secondly, it can be said that a particular type of internal energy will change significantly only in a limited temperature range. The transformation energies for phase changes and magnetic order are the examples given for the second type. The effect of the specific heat for these processes can only be observed in the same limited temperature range as the energy change. Although the limited temperature range or abnormal specific temperatures are of

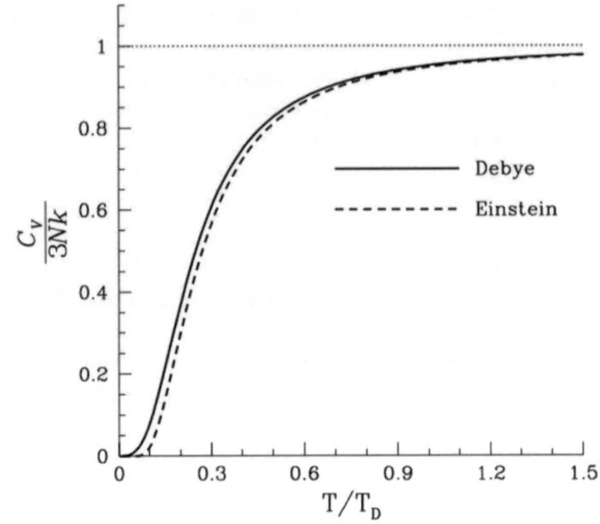

Figure 1.
Typical internal energy and specific heat curve for metals [5].

great importance for academic research in physics and chemistry, it can often be said that they are not of great importance in commercial materials [1–4].

The lattice specific heat is much larger than the electronic specific heat at most temperatures. It will be discussed first. Einstein's representation of the ionic lattice as a system of independent oscillators led to the equation $C_v = 3RE(\Theta E/T)$, where Θ is a characteristic temperature, R is the gas constant, and E is the Einstein function as defined by $E(\Theta/T) = (\Theta/T)^2 e^{\Theta/T}/(e^{\Theta/T}-1)^2$. This gives a good fit and above room temperatures, approximating the earlier observed Dulong-Petit universal value for heat capacity at high temperatures. It does not fit well at low temperatures, however. The lattice of ions is known to interact. When this is taken into consideration and some simplifying approximations are made for the distributions of energies, the Debye theory is obtained. This theory is almost too good; Its predictive ability was so successful that theoretical refinements were not considered seriously for many years [1, 2]. The Debye specific heat (for the lattice) is $C_v = 3RD(\Theta D/T)$, where the Debye function, D, is defined as

$$D\frac{\Theta}{T} = 3\left(\frac{T}{\Theta}\right)^3 \int_0^{\Theta/T} \frac{x^4 e^x}{(e^x - 1)^2} dx$$

The Einstein and Debye specific heat and energy curves are shown in **Figure 1**.

The main characteristics of the Debye curve are easily seen. At low temperatures the specific heat varies as T^3; at high temperatures it is approximately constant. The theory was developed for isotropic, homogeneous porous metals; how does it apply to alloys?

In near room temperature, the specific heat of an alloy is obtained quite well by the Kapp-Joule rule: the total specific heat is a linear combination of the specific heats of the constituents, each weighted according to its relative abundance. At low temperatures, one can either combine additively the actual specific heats of the constituents or take a weighted average of the characteristic temperatures. Either of the procedures will give approximately correct results [6–8].

The electronic specific heat is small compared to the lattice contribution at high temperatures, but it is linear in its temperature variation. Since the lattice term decreases as T^3, the electronic term will become significant only at the lowest temperatures. It is usually not significant for engineering applications [5, 9].

3. Thermal conductivity

The thermal conductivities of several solids, values for metals, and their alloys are represented in **Figures 2–4**. Even within those restrictions, there is considerable variation. Is it possible to make order out of this variety; Is it possible to be able to predict reasonably well the thermal conductivity for new or untested materials? It is, if one utilizes knowledge of the fundamental phenomena [10, 11].

Two parallel mechanisms connected at low temperatures are primarily responsible for transporting heat from a metal. First one is the thermal energy transmission with the transmission electrons movement, and the most important is the transmission of electronic heat. The second is the thermal energy transport by the lattice thermal conduction, the movement of conductive electrons. The mesh is the thermal conduction; the thermally induced interactive lattice ions are carried by directional cooperative quantitative vibration (phonons). These are the same phonons which are responsible for the specific warming and thermal expansion observed in metals. Mesh thermal conductivity is insignificantly compared with electronic thermal conductivity, for pure metals and dilute alloys. For alloys

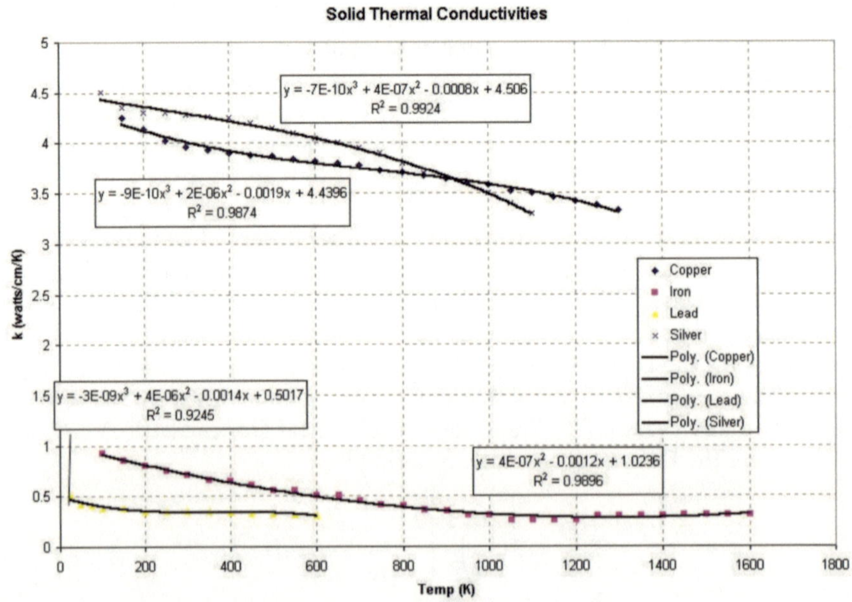

Figure 2.
Thermal conductivity of some solids [12].

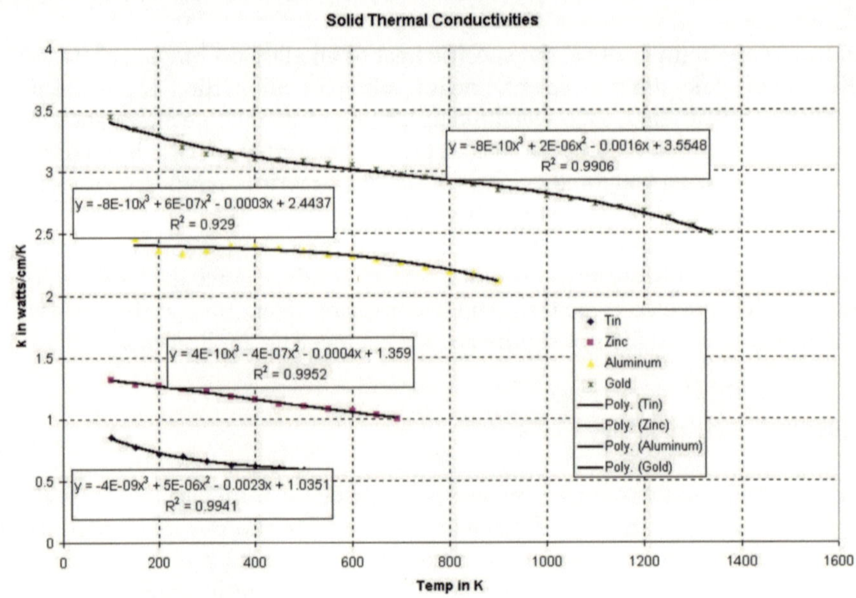

Figure 3.
Thermal conductivity of some solids [12].

containing several or more percent additives, decreasing electronic thermal conductivity ensures that the lattice additives are still small, but still small compared to the electronic additive. For most metals and alloys, total conductivity means K_g (g refers to Gitter, German word for cage), $K = K_e + K_g$. This equation is the same as the two conductors used for total conductivity in parallel in electrical circuit theory. Both conductive mechanisms, K_e and K_g, are limited to various scattering processes that limit electronic conductivity in the above statement [12–14]. A sample of the analog is presented in **Figure 5**.

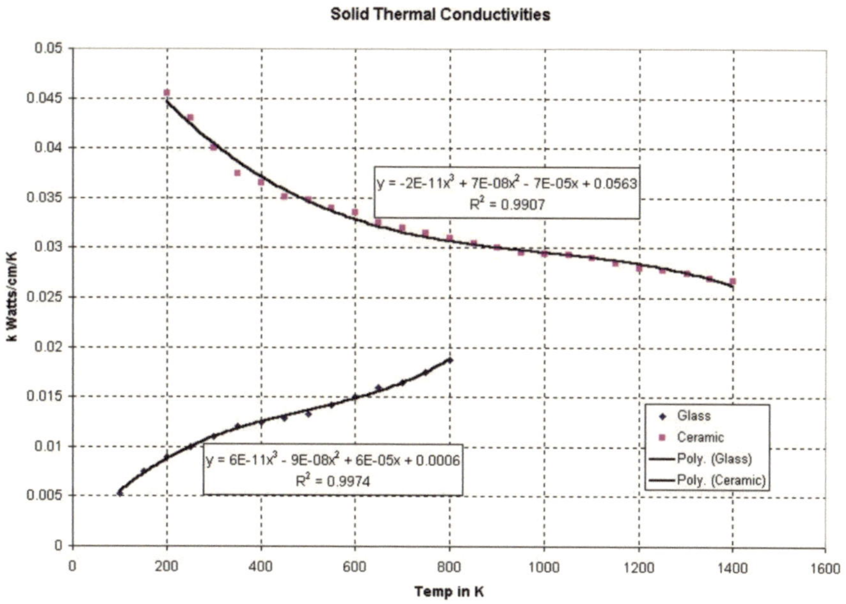

Figure 4.
Thermal conductivity of some solids [12].

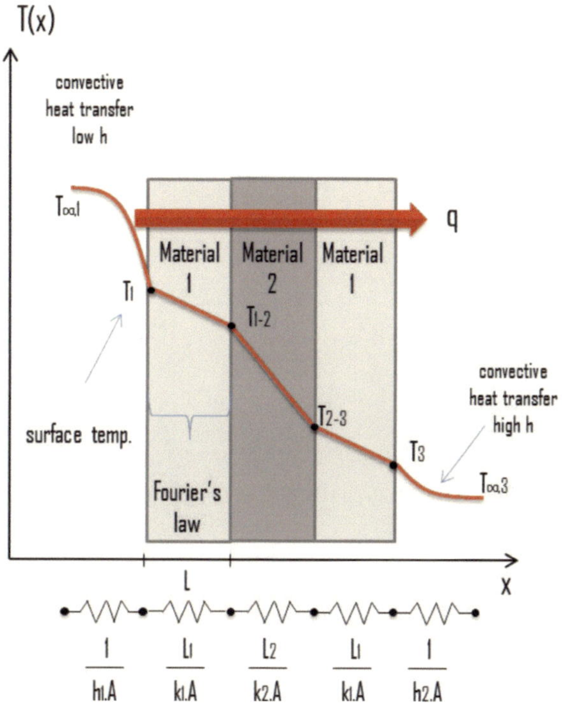

Figure 5.
Thermal conduction's sample electrical analog [15].

The first of these is the diffusion of the electrons that are represented as a characteristic feature for a given metal to the electron-phonon resistance (W_L). This spreading is most important at higher intermediate temperatures (about 40–80 K) and at higher temperatures. The second process is the conductive electron propagation by defects (both lattice defects and impurity atoms) as represented by electron-defect resistance W_0. This scattering is more important at low temperatures. K_e, which is the total electronic thermal conductivity equivalent, also gives the total electronic heat resistance. The resistances W_L and W_D are treated as the default of the sum of W_{Lo}, that is, plus a small deviation term, $1/K_e$. And $1/K_e = W_e = W_L + W_o + W_{Lo}$. This equation is similar in the electrical circuit theory to the equation used for the total resistance of the resistors in the series. $W_{Lo} = \alpha W_L W_\sigma/(\beta W_L + (W_o)$. Here, α, β, and γ are constants and can be determined experimentally. Although theoretically meaningful, these terms are only numerically important for very pure metals [12–14].

When the interaction term W_{Lo} is negligible, the Matthiessen's electrical resistance rule for thermal equivalent is roughly as accurate as $W_e = W_L + W_o$. A graph for this relationship and its equivalent for conductivity is given in **Figure 6**. The estimation of total electronic thermal resistance can be done by two separate components [12–14].

Both theoretical and empirical researches are defined by different expressions for the magnitude of electron-phonon and electron-defect resistance and temperature dependencies: $W_L = AT^n$ ($n = 2$–3, $T < 40$ K); $W_L = a$ is constant (near room temperature); $W_o = B/T$ (at all temperatures). The term constant A in the term electron-phonon resistance refers to the specific characteristics (or characteristic temperature) of a given metal and will not change for the small chemical additive addition or physical defects and the specific sample for residual electrical resistance. At all temperatures above 40 K, often labeled $W_\infty = B/T$, the electron-phonon resistivity approaches a constant value. The constant A in the electron-phonon resistivity term is related to a given metal intrinsic properties (including the characteristic temperature, Θ) and will not change for chemical impurities for minor additions or physical imperfections; B in the electron-defect resistivity term is related to the given imperfections amount and the specific specimen's residual electrical resistivity. Above 40 K, the electron-phonon resistivity approaches a constant value, often labeled W_∞.

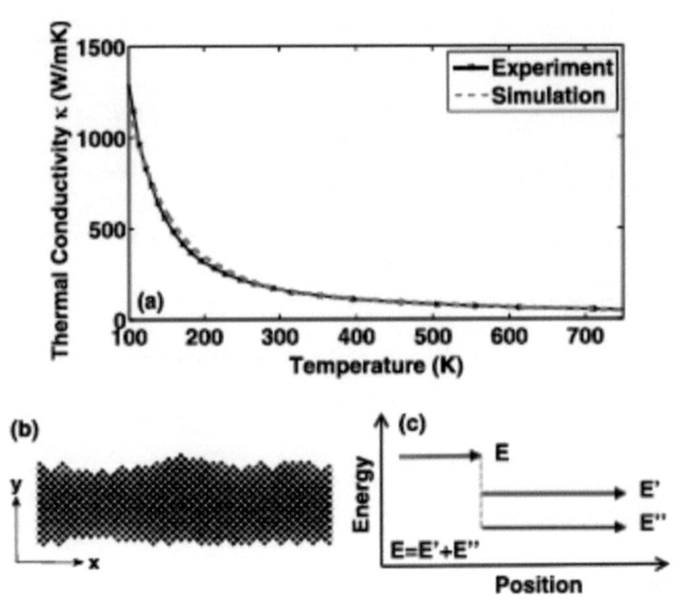

Figure 6.
A sample graph explaining Matthiessen's rule for electrical resistivity (a) thermal conductivity of bulk Si, (b) Si nanowire's surface roughness and (c) anharmonic phonon-phonon scattering process [15].

In a detailed analysis, the term electron-phonon will show that W_L is unchanged, that is, it is really specific to aluminum. **Figures 3** and **4** clearly illustrate adding more impurities effects to a given metal, thus increasing W_0. At low temperatures, when the curves are parallel, the high impurity alloy is low. When the curves approach each other, the differences are reduced by approximately 1/T, at high temperatures.

The curve shapes for the aluminum given in **Figure 3** are specific to pure metals: electronic conductivity is predominant and mesh conductivity is insignificant. For a metal sample with physical defects or without chemical impurities, the electron-phonon scattering component caused by the thermally induced ionic cage W_L directs its dependence on the temperature of the conductivity. As the temperature is lowered, the resistivity decreases in approximate proportion to T^2; the conductivity rises equivalently. Superimposed on this decreasing ideal of electron-phonon resistivity is the electron-defect resistivity which increases as the temperature is decreased. The maximum heat conduction resistance at the minimum temperature in the two temperature-spreading mechanisms is considered to be approximately the same. However, at high temperatures, electron-phonon scatter is dominant; at low temperatures, heat loss from electron defect is dominant. On the other hand, the conductivity-limiting defects at low temperatures can be quite diverse: chemical impurities, residues, gaps, inter-atoms, dislocations, particle boundaries, outer surface boundaries, etc. The waste heat dissipation resistance, except for the final defect for almost every situation, has the same temperature dependency B/T as previously mentioned. At very low temperatures, boundary scattering is difficult to investigate and can only be observed in extremely pure metals [12–14].

Today, it is not possible to accurately estimate the thermal conductivity of pure metals on the basis of chemical or physical properties because every kind of chemical impurity is specific to that substance. It also depends on the magnetic interactions of matter and the differences between the ionic mass, ion volume, and the host and impurity in the valance electrons. Much work has been done on the specific effects of impurities in electrical resistance, but little has been done about thermal conductivity. Sometimes a chemical analysis in this regard can hardly be helpful, given that a given impurity is much more effective as a scattering agent in the form of a solid solution, rather than leaving the test boundaries or residues. This decomposition effect is very pronounced in copper. Similar interpretive difficulties arise for physical defects in test items [15, 16].

Thermal conductivity changes are caused by relatively physical defects or small chemical impurities in contrast to losses in thermal expansion or specific temperatures. Successful estimation of thermal conductivity depends on an ingenious analysis of possible propagation mechanisms of about 10–20% and obtaining experimental results on a very similar metal or alloy. As will be shown later, the low temperature electrical resistance data are also very valuable for predicting thermal conductivity [15, 16].

In many low-conductivity alloys, the lattice structure of the material's thermal conductivity can be measured in K_o, and separate propagation components can be observed for this. For most alloys, there are three main processes that limit the lattice structure conductivity. The first operation is the conductive electrons scattering, as represented by phonon-electron resistance. This is electron equivalent scattered by phonons. The second process is the W_P distribution from the phonon-scattering resistance, resulting from construction defects. At low temperatures, the first two processes limit the lattice structure's conductivity of the material; the third condition is the limitation at higher temperatures [17, 18].

The phonon-loss and phonon-electron resistors have the same temperature dependence. Thus, for a sample metal or alloy, two scattering mechanisms cannot be clearly distinguished. For annealed samples, the two resistances are approximately the same size; for non-annealed samples, loss resistance is heavier than phonon-electron resistance. The conductivity of the metal lattice structure below the temperature of about 40 K can be represented by the following formula:

$1/K_g = W_g = W_e + W_d + W_p = ((E + D)\ T^{-2} + PT)$. Due to the effect of the phonon-electron and phonon-loss resistors, T^{-2} dependence is expected to be dominant at low temperatures and negligible at high temperatures. The maximum conductivity in the cage structure conductivity is generally in the range of 50–100 K for most alloys. However, above these temperatures, the lattice structure's heat conduction cannot be readily separated from the electronic conductivity experimentally because the electronic conductivity has a much larger magnitude [19, 20].

For the aluminum alloys of **Figure 4**, there are special shapes of curves; for example, electronic conductivity is dominant, but the conductivity of the metal lattice structure can be observed. In addition, the effects of impurities are obvious. For aluminum alloys, the lattice conduction may be about the total's 10%; for some iron or nickel alloys, it may be much higher [21, 22].

Due to the Wiedemann-Franz-Lorenz law, $\rho = LW_eT$, where L is the Lorenz ratio, it is assumed to be a fundamental constant given by the Sommerfeld value $L = (\pi 2/3)(K/e)^2 = 2.44 \times 10^{-8}$ [watt-ohm/(deg K)^2], whether there is no direct information on the metal thermal conductivity; after that data on the electrical resistivity may be related to the electronic thermal resistivity [23, 24].

The separation and scattering of components for electrical conductivity is almost the same as described above on electronic thermal conductivity. Total electrical resistance, ρ, is considered to be about two separate resistances, electron-phonon or total resistances, ρ_L and electron, or residual defective resistance, ρ_o. Alternatively, Matthiessen is estimated to be close to the principle of electrical resistance expressed by $\rho = \rho_L + \rho_o$ [25, 26].

The above-described internal system and residual terms have temperature statements of approximately $\rho_L = \alpha Tn$ ($n = 4$–5, $T < 40$ K); $\rho_L = \alpha' T$ (300 K close to T); $\rho_o = \beta$ (constant). For a general metal, a separation graph of the electrical resistance components is shown in **Figure 6** [25, 26]. The experimental results for a typical series of alloys are given in **Figure 7**.

For the aluminums and aluminum alloys, calculated Lorenz ratios are given in **Figures 2, 3**, and **7**. The numbers calculated and shown in **Figure 8** express the ratios of the electrical resistivities. At the lowest temperatures, the extrapolated values Lo, however, should in electronic terms only represent ratios since the lattice contribution to the total thermal conductivity is greatly reduced [23]. A graph for Lorenz ratios for a general metal is given in **Figure 8**.

The ratios of Lorenz for high-conductivity materials are extrapolated to roughly Sommerfeld at 0 K, but at lower temperatures, that falls considerably below this value. However, the low-conductivity alloy behavior is different: the conductivity value is higher than the values between about 10 and 60 K, while the conductivity value is lower than 60 K value. In the case of where the ratios are above the Sommerfeld value, the temperature ranges in which the metal lattice structure conductivity is important are shown in [24, 25].

In metal structure, the Lorenz ratio should be constant in the conduction electrons' elastic distribution. This is almost the case when the thermal vibration at the high temperatures causing the large electron-phonon scatter is maximum, and the residual term expressed in the above equation is dominant in the electrical resistance at low temperatures. It can be said that the Lorenz ratios are significantly reduced from Sommerfeld in the case of medium temperatures where there is a large thermal vibration amount causing large electron-phonon scattering and in low temperatures where the term is dominant in the elasticity and the thermal

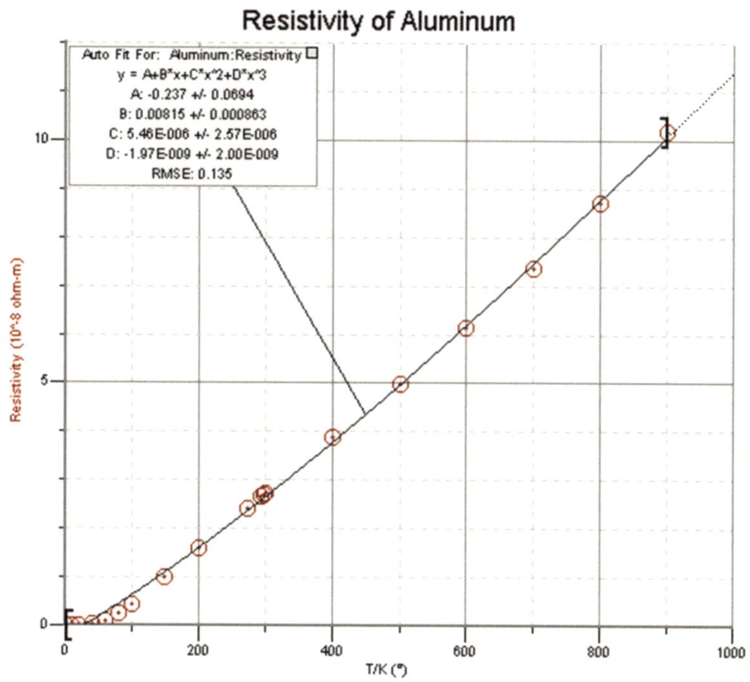

Figure 7.
Thermal resistivity of aluminum [23].

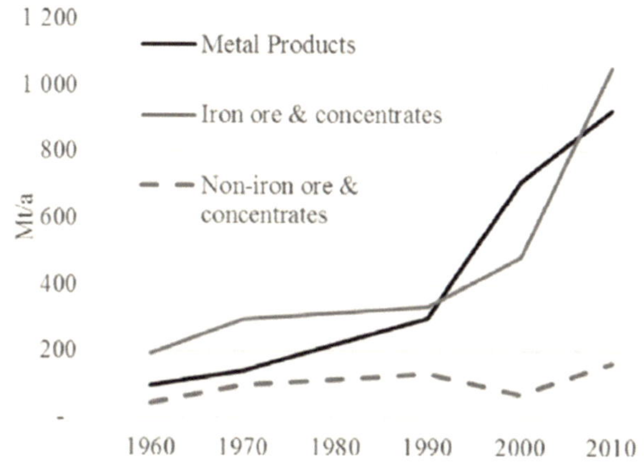

Figure 8.
Lorenz curve of waste rock and metal extraction primary y-axis, by area x-axis in 2010 [24].

conductivity of the metal cage structure is insignificant. For situations where a significant lattice thermal conductivity amount is present, the increase in the Lorenz ratio can only be said to be above the value of the electronic term in the thermal conductivity formula [24, 25].

Up to now, on the commercial alloys' Lorenz ratio, very little research has been reported. Whenever Lorenz ratios for the general class of materials and electrical resistivities for a special material are available, reasonable predictions for thermal conductivity can be obtained [23, 26].

It has been shown that at the present time, it is not possible to accurately predict thermal conductivities for metals and alloys from the fundamentals. It is possible to make adequate predictions, however, if there are data on the thermal or electrical resistivities of similar materials and if one uses proper interpolation formulas and a knowledge of the effects of minor changes in the chemical impurities or physical imperfections. It is imperative, of course, that good compendiums of experimental data exist. Fortunately, they do [23, 26].

Nomenclature

C_v	Specific heat at constant volume, J/(kg·°C)
U	Total internal energy, kg/ms^2
T	Absolute temperature, K
Θ	Characteristic temperature, K
R	Gas constant, J/mol·K
E	Einstein function
D	Debye function
x	length, m
K	Total conductivity, S/m
W	Electron-phonon resistivity
A	Intrinsic properties
B	Physical imperfections
P	Point imperfections
ρ	Total electrical resistivity, (Ω m)
L	Lorentz ratio, (watt-ohm/(K)^2)
e	electron charge, Coulomb

Author details

Cemil Koyunoğlu
Energy Systems Engineering Department, Yalova University, Yalova, Turkey

*Address all correspondence to: cemil.koyunoglu@yalova.edu.tr

References

[1] Mizutani U. Introduction to the Electron Theory of Metals. Cambridge University Press; 2001

[2] Collings EW. Applied Superconductivity, Metallurgy, and Physics of Titanium Alloys: Fundamentals Alloy Superconductors: Their Metallurgical, Physical, and Magnetic-Mixed-State Properties. US: Springer; 1986

[3] Testing ASf, Materials, Testing ASf, Sciences MDoM. Behavior of Materials at Cryogenic Temperatures. American Society for Testing and Materials; 1966

[4] Winterbottom Y, Laboratory BN. Proceedings: National Technical Information Service; 1972

[5] Ashcroft NW, Wilkins JW. Effect of Electron-Phonon Interaction on Electronic Specific Heat of Simple Metals. Defense Technical Information Center; 1964

[6] Godbee HW, Roberts JT, Commission USAE. Survey on the Measurement of Thermal Conductivity of Solids Produced by Evaporation and Calcination of Synthetic Fuel Reprocessing Solutions. Oak Ridge National Laboratory; 1959

[7] Brenden BB, Newkirk HW, Operation HAP, Works H, Commission USAE, Research USE, et al. A Method of Determining the Thermal Conductivity of Incandescent Solids. U.S. Atomic Energy Commission; 1959

[8] Childs GE, Ericks LJ, Powell RL, Division IfBSC. Thermal Conductivity of Solids at Room Temperature and below: A Review and Compilation of the Literature. National Bureau of Standards. [for sale by the Supt. of Docs., U.S. Govt. Print. Off.]; 1973

[9] Ahmet T. The Specific Heat of Matter At Low Temperatures. World Scientific Publishing Company; 2003

[10] Barron RF, Nellis GF. Cryogenic Heat Transfer. CRC Press; 2017

[11] Madelung O, Klemens PG, Neuer G, White GK, Sundqvist B, Uher C. Thermal Conductivity of Pure Metals and Alloys/Wärmeleitfähigkeit von Reinen Metallen und Legierungen. Berlin Heidelberg: Springer; 1991

[12] Tritt TM. Thermal Conductivity: Theory, Properties, and Applications. US: Springer; 2006

[13] Hust JG, Sparks LL. Lorenz Ratios of Technically Important Metals and Alloys. U.S: National Bureau of Standards; 1973

[14] Touloukian YS. Thermal Conductivity: Metallic Elements and Alloys. IFI/Plenum; 1970

[15] Bartlit JR, Bell KJ, Engineers AIoC. Advances in Cryogenic Heat Transfer. American Institute of Chemical Engineers; 1968

[16] Engineers AIoC. Chemical Engineering Progress Symposium Series. American Institute of Chemical Engineers; 1968

[17] Tiryakioğlu M, International ASM. Advances in the Metallurgy of Aluminum Alloys: Proceedings from Materials Solutions Conference 2001, the James T. Staley Honorary Symposium on Aluminum Alloys, 5-8 November 2001, Indianapolis, Indiana. ASM International; 2001

[18] Acton QA. Light Metals: Advances in Research and Application. 2011th ed. ScholarlyEditions; 2012

[19] Iwasa Y. Case Studies in Superconducting Magnets: Design and Operational Issues. US: Springer; 2006

[20] Ju YS, Goodson KE. Microscale Heat Conduction in Integrated Circuits and their Constituent Films. US: Springer; 1999

[21] Brechna H. Superconducting magnet systems. Springer; 1973

[22] Reed RP, Clark AF, ASf M. Materials at Low Temperatures. American Society for Metals; 1983

[23] Hust JG, Sparks LL. Lorenz Ratios of Technically Important Metals and Alloys. U.S. Government Printing Office; 1973

[24] Bauer E, Sigrist M. Non-Centrosymmetric Superconductors: Introduction and Overview. Berlin Heidelberg: Springer; 2012

[25] Klemens P. Thermal Conductivity 14. US: Springer; 2013

[26] Timmerhaus K. Advances in Cryogenic Engineering. US: Springer; 2012

Section 7

Mass Transfer Processes

Chapter 16

Comparison and Analysis of Diffusion Models for the Fe_2B Layers Formed on the AISI 12L14 Steel by Using Powder-Pack Technique

Martín Ortiz Domínguez

Abstract

Boriding is a thermochemical surface treatment, a diffusion process similar to carburizing and nitriding in that boron is diffused into a metal base. An indispensable tool to choose the suitable process parameters for obtaining boride layer of an adequate thickness is the modeling of the boriding kinetics. Moreover, the simulation of the growth kinetics of boride layers has gained great interest in the recent years. In this chapter, the AISI 12L14 steel was pack-borided in the temperature range of 1123–1273 K for treatment times between 2 and 8 h. A parabolic law for the kinetics of growth of Fe_2B layers formed on the surface of AISI 12L14 steel was deducted. Two diffusion models were proposed for estimating the boron diffusion coefficients through the Fe_2B layers. The measurements of the thickness (Fe_2B), for different temperature of boriding, were used for calculations. As a result, the boron activation energy for the AISI 12L14 steel was estimated as 165.0 kJ/mol. In addition, to extend the validity of the present models, two additional boriding conditions were done. The Fe_2B layers grown on AISI 12L14 steel were characterized by use of the following experimental techniques: X-ray diffraction, scanning electron microscopy and energy dispersive X-ray spectroscopy.

Keywords: diffusion model, activation energy, parabolic growth law, diffusion coefficient, growth kinetics

1. Introduction

Surface hardening of steel can be achieved, mainly through two procedures: modifying the chemical composition of the surface by diffusion of some chemical element (carbon, nitrogen, boron, sulfur, etc.) in which case it is known as thermochemical treatment (**Table 1**) or modifying only the microstructure of the surface by thermal treatment, then known as surface treatment. The current technological demands highlight the need to have metallic materials with high performance under critical service conditions, consequently, the increase in the wear resistance, preserving its ductility and the toughness of the core.

According to **Table 1**, there are three methods of surface hardening:

IntechOpen

- Diffusion process that modifies the chemical composition where a component in a solid mixture can diffuse through another at a speed is measurable, if there is a suitable concentration gradient and the temperature is high enough. The effects of diffusion in solids are very important in metallurgy (increase surface hardness) as well: The continuous flux of carbon, nitrogen, boron, and so on, can form a hard coating, where the mass transfer is described by Fick's laws.

- Applied energy process, the interesting about these processes is that it is not necessary to incorporate any element to the substrate. For example, tempering is a heat treatment in which steel is heated up to austenization temperatures and subsequently it is cooled rapidly, with in order to obtain a transformation that provides a structure martensitic hard and resistant. Surface tempering is generally used to components that need a hard surface and a substrate with a high value of fracture toughness.

- Coating and surface hardness, the coating covers the surface of the substrate, obtained after the deposition process, substrates considerably increase the physical characteristics of hardness and corrosion resistance, maintaining the original morphological characteristics (roughness and brilliance) unchanged, making the functional and decorative coating at the same time.

The current technological requirements highlight the need to have metallic materials with specific characteristics, for increasingly critical service conditions. For example, the metal dies used in the metallurgical processes of cold working and hot metals need a high toughness and surface hardness, especially at high temperature. Surface hardening of steel can be achieved, basically, by two processes: modifying the chemical composition of the surface by diffusion of some chemical elements (carbon, nitrogen, sulfur, boron, aluminum, zinc, chromium, and so on). Only boriding process for surface hardening is briefly reviewed in this chapter, boriding is a thermochemical treatment in which boron atoms are diffused into the surface of a workpiece and form borides with the base metal. Apart from constructional materials, which meet these high demands, processes have been developed which have a positive effect on the tribological applications including abrasive, adhesive, fatigue and corrosion wear of the component surface [1–3]. Boride layers are of particular benefit when the components have to withstand abrasive wear. The fundamental advantage of the borided layers (FeB and Fe_2B) is that they can reach high hardness near the surface (1800 $HV_{0.1}$ and 2000 $HV_{0.1}$), maintained at high temperatures [4–8]. In this chapter, the growth kinetics of single phase layer (Fe_2B) on the ferrous substrate was studied during the iron powder-pack boriding (steady state and non-steady state). The parabolic growth law for the borided layers was

Diffusion methods	Applied energy methods	Coating and surface modification
Carburizing	Flame hardening	Hard chromium planting
Nitriding	Induction hardening	Electroless nickel plating
Carbonitriding	Laser beam hardening	Thermal spraying
Boriding	Electron beam hardening	Weld hardfacing
Thermal diffusion process		Chemical vapor deposition
		Physical vapor deposition
		Ion implantation
		Laser surface processing

Table 1.
Engineering methods for surface hardening of steels.

mathematically estimated. Likewise, a mass balance equation was proposed at the Fe_2B/substrate (AISI 12L14) interface. Moreover, the boron diffusion coefficients (D_{Fe_2B}) in the Fe_2B layers were determined considering two mathematical models for mass transfer. The Fe_2B layers formed on the alloy surface is controlled by the diffusion of boron atoms, and the presence of the Fe_2B layers was checked by the XRD technique. Finally, the distribution of the alloy elements in AISI 12L14 borided steel was verified by chemical microanalysis technique (EDS) used in conjunction with SEM.

1.1 The diffusion models

One of the most important parameters that characterizes the Fe_2B layers is the thickness, since the properties of the coating depend on it, such as: resistance to wear, fatigue, hardness, and dynamic loads, as well as to a large extent determining the grip with the substrate. Having an expression that allow estimating the layer thickness during the boriding process, facilitates the appropriate selection of the technological parameters, in order to guarantee the desired properties. The layer thickness exhibits a time dependence such that:

$$\text{layer thickness v} \approx t^{1/2}, \tag{1}$$

1.1.1 Derivation of the parabolic growth law

In diffusion processes, parabolic kinetics occurs when the mass gain on a sample is proportional to the square root of time. In general, parabolic kinetics indicates that diffusion of reactants (such as boron) through a growing layer is rate-determining. If the diffusion of B atoms is rate-determining, the layer rate is proportional to the flux through the substrate:

$$\frac{dx}{dt} \approx J_{Fe_2B}(x,t). \tag{2}$$

El flux, $J_{Fe_2B}(x,t)$, can be written as

$$J_{Fe_2B}(x,t) = C_{Fe_2B}(x,t)(dx/dt), \tag{3}$$

where $C_{Fe_2B}(x,t)$ is the boron concentration profile in mol/m^3 and is the velocity dx/dt of Fe_2B layer in m/s, $J_{Fe_2B}(x,t)$ giving units of mol/m^2 s. The velocity of a particle is proportional to the force, F, on the particle:

$$dx/dt = B_{Fe_2B}F, \tag{4}$$

where B_{Fe_2B} is the mobility of the boron. Writing the chemical potential as μ_{Fe_2B}, this force is written as

$$F = -\partial\mu_{Fe_2B}/\partial x, \tag{5}$$

for a Fe_2B layer with thickness x. Combining Eq. (4) and (5) yields

$$J_{Fe_2B}(x,t) = -C_{Fe_2B}(x,t)B_{Fe_2B}\partial\mu_{Fe_2B}/\partial x, \tag{6}$$

from the relationship

$$\mu_{Fe_2B} = \mu^o_{Fe_2B} + k_B T \ln a_{Fe_2B}, \tag{7}$$

where k_B is the Boltzmann's constant, we can write

$$\frac{\partial \mu_{Fe_2B}}{\partial x} = \frac{\partial\left(\mu^o_{Fe_2B} + k_B T \ln a_{Fe_2B}\right)}{\partial x} = k_B T \frac{\partial \ln a_{Fe_2B}}{\partial x}. \tag{8}$$

In an ideal system, the concentration, $C_{Fe_2B}(x,t)$, is equivalent to activity, $a_{Fe_2B}(x,t)$. Substituting the Eq. (8) into Eq. (6), we get

$$J_{Fe_2B}(x,t) = -C_{Fe_2B}(x,t) B_{Fe_2B} k_B T \frac{\partial \ln C_{Fe_2B}(x,t)}{\partial x} = -B_{Fe_2B} k_B T \frac{\partial C_{Fe_2B}(x,t)}{\partial x}. \tag{9}$$

As shown in Eq. (2),

$$\frac{dx}{dt} = (\text{constant}) J_{Fe_2B}(x,t), \tag{10}$$

so that a combination of Eqs. (2) and (9) gives

$$\frac{dx}{dt} = -(\text{constant}) B_{Fe_2B} k_B T \frac{\partial C_{Fe_2B}(x,t)}{\partial x}. \tag{11}$$

If we assume that the potential is fixed at each boundary of the Fe_2B layer, we can replace $\partial C_{Fe_2B}(x,t)/\partial x$ in Eq. (11) with the slope ($= \Delta C_{Fe_2B}/x$). We then introduce the parabolic growth constant k_{Fe_2B}, and set:

$$k_{Fe_2B} = -(\text{constant}) B_{Fe_2B} k_B T \Delta C_{Fe_2B}. \tag{12}$$

Combining Eqs. (11) and (12) then gives

$$\frac{dx}{dt} = \frac{k_{Fe_2B}}{x}. \tag{13}$$

Eq. (13) can be rewritten as

$$x dx = k_{Fe_2B} dt. \tag{14}$$

Upon integration of Eq. (14),

$$\int_{x=0}^{x=\text{v}} x dx = k_{Fe_2B} \int_{t=0}^{t=t} dt. \tag{15}$$

We arrive at the parabolic growth law:

$$\text{v}^2 = 2 k_{Fe_2B} t. \tag{16}$$

1.1.2 Steady state diffusion model

Steady state means that there will not be any change in the composition profile with time. A linear boron concentration profile is considered along the depth of the Fe_2B layer as depicted in **Figure 1**. The $f(x)$ represents to the boron distribution in the substrate before the nucleation of iron boride layers on AISI 12L14 steel. $t_0^{Fe_2B}$ is the boride incubation time indispensable to form the Fe_2B phase. Moreover, $C_{up}^{Fe_2B}$

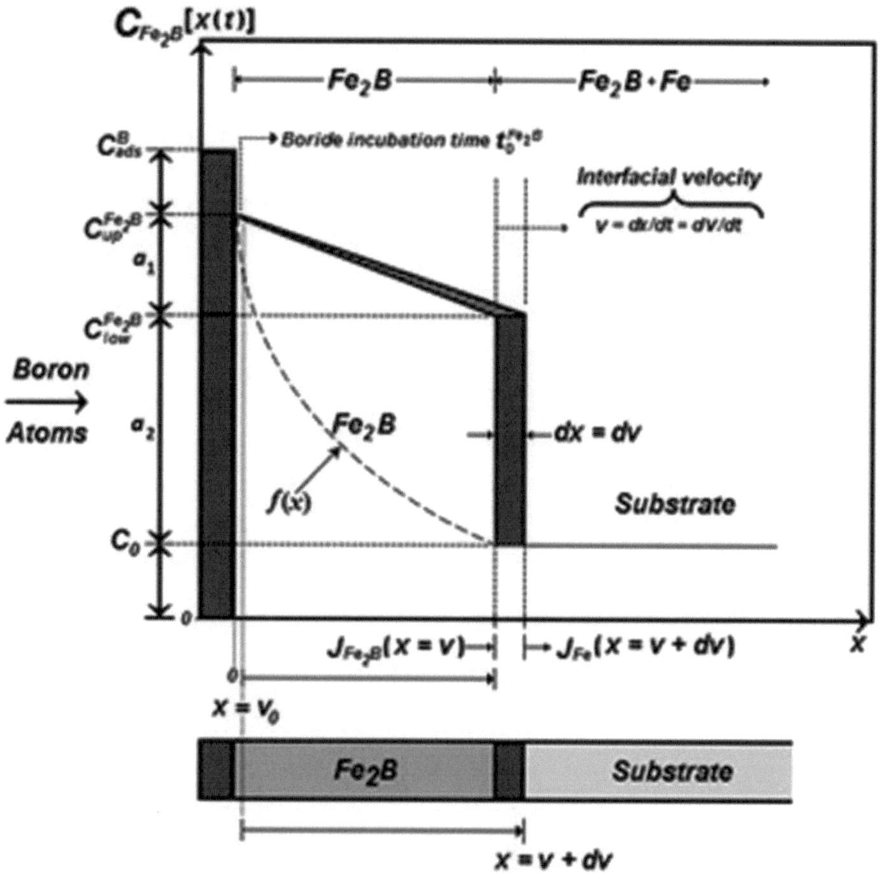

Figure 1.
A schematic linear concentration profile of boron through the Fe_2B layer is used to describe the steady state diffusion model.

represents the boron concentration on the surface in Fe_2B layer ($= 60 \times 10^3$ molm^{-3}), $C_{low}^{Fe_2B}$ represents the boron concentration at the Fe_2B/substrate interface ($= 59.8 \times 10^3$ molm^{-3}) and $x(t = t) = \mathrm{v}$ is the layer thickness of the boride layer (m) [9, 10].

The term C_{ads}^{B} is the effective adsorbed boron concentration during the boriding process [11]. From **Figure 1**, $a_1 = C_{up}^{Fe_2B} - C_{low}^{Fe_2B}$ defines the homogeneity range of the Fe_2B layer, $a_2 = C_{low}^{Fe_2B} - C_0$ represents the range of miscibility and C_0 is the boron concentration in the substrate (AISI 12L14) assumed as null [10, 12, 13]. During the establishment of the steady-state diffusion model, a linear concentration-profile of boron along the Fe_2B layer is considered. Likewise, the assumptions proposed by Campos-Silva et al. [8], are taken account.

v_0 is the first boride layer formed on the surface of the substrate (ASI 12 L14) during the boride incubation time [14], its thickness is very small in magnitude compared to the thickness of the boride layer (v). Moreover, regarded the mass balance equation at the growth interface (Fe_2B/substrate), which is described as follows [15–18]:

$$\left(\frac{C_{up}^{Fe_2B} + C_{low}^{Fe_2B} - 2C_0}{2}\right)\frac{dx(t)}{dt}\bigg|_{x(t)=\mathrm{v}} = -D_{Fe_2B}\frac{\partial C_{Fe_2B}[x(t=t), t=t]}{\partial x}\bigg|_{x(t)=\mathrm{v}}. \quad (17)$$

When the concentration field is independent of time and D_{Fe_2B} is independent of $C_{Fe_2B}(x,t)$, Fick's second law is reduced to Laplace's equation,

$$\nabla^2 C_{Fe_2B}[x(t)] = \frac{d^2 C_{Fe_2B}[x(t)]}{dx^2} = 0. \tag{18}$$

By solving Eq. (18), and applying the boundary conditions proposed in **Figure 1**, the distribution of boron concentration in Fe_2B is expressed as:

$$C_{Fe_2B}[x(t)] = \frac{C_{low}^{Fe_2B} - C_{up}^{Fe_2B}}{\mathrm{v}} x + C_{up}^{Fe_2B}. \tag{19}$$

By substituting the derivative of Eq. (19) with respect of the distance $x(t)$ into Eq. (17), we have

$$\left(\frac{C_{up}^{Fe_2B} + C_{low}^{Fe_2B} - 2C_0}{2} \right) \frac{d\mathrm{v}}{dt} = D_{Fe_2B} \frac{C_{up}^{Fe_2B} - C_{low}^{Fe_2B}}{\mathrm{v}}, \tag{20}$$

for $0 \le x \le \mathrm{v}$.

By substituting Eq. (16) into Eq. (20)

$$D_{Fe_2B} = \frac{1}{2} \left(\frac{C_{up}^{Fe_2B} + C_{low}^{Fe_2B} - 2C_0}{C_{up}^{Fe_2B} - C_{low}^{Fe_2B}} \right) k_{Fe_2B} \tag{21}$$

1.1.3 Non-steady state diffusion model in one dimension

The general diffusion equation for one-dimensional analysis under non-steady state condition is defined by Fick's second law. The growth of single phase layer (Fe_2B) with one diffusing element (boron) is observed as illustrated in **Figure 2**.

The $f(x,t)$ function represents the boron distribution in the ferritic matrix before the nucleation of Fe_2B phase as a function of time. Likewise, for analysis, the kinetic model is imposing the same restrictions as in the previous model, except the last one, it is replaced by:

- The concentration-profile of boron is the solution of the Fick's second law and depends on initial and boundary conditions through the Fe_2B zone.

The mass balance equation at the (Fe_2B/substrate) interface can be formulated by Eq. (22) as follows:

$$\left(\frac{C_{up}^{Fe_2B} + C_{low}^{Fe_2B} - 2C_0}{2} \right) \frac{dx(t)}{dt}\bigg|_{x(t)=\mathrm{v}} = - D_{Fe_2B} \frac{\partial C_{Fe_2B}[x(t=t), t=t]}{\partial x}\bigg|_{x(t)=\mathrm{v}}. \tag{22}$$

Fick's second law, isotropic one-dimensional diffusion, D_{Fe_2B} independent of concentration:

$$\frac{\partial C_{Fe_2B}[x(t),t]}{\partial t} = D_{Fe_2B} \frac{\partial^2 C_{Fe_2B}[x(t),t]}{\partial x^2}. \tag{23}$$

By solving Eq. (23), and applying the boundary conditions proposed in **Figure 2**, the boron concentration profile in Fe_2B is expressed by Eq. (24), if the boron diffusion coefficient (D_{Fe_2B}) in Fe_2B is constant for a particular temperature:

$$C_{Fe_2B}[x(t),t] = C_{up}^{Fe_2B} + \frac{C_{low}^{Fe_2B} - C_{up}^{Fe_2B}}{erf\left(\frac{v}{2\sqrt{D_{Fe_2B}t}}\right)} erf\left(\frac{x}{2\sqrt{D_{Fe_2B}t}}\right). \tag{24}$$

By substituting Eq. (24) into Eq. (22), Eq. (25) is obtained:

$$\left(\frac{C_{up}^{Fe_2B} + C_{low}^{Fe_2B} - 2C_0}{2}\right)\frac{dv}{dt} = \sqrt{\frac{D_{Fe_2B}}{\pi t}} \frac{C_{up}^{Fe_2B} - C_{low}^{Fe_2B}}{erf\left(\frac{v}{2\sqrt{D_{Fe_2B}t}}\right)} \exp\left(-\frac{v^2}{4D_{Fe_2B}t}\right), \tag{25}$$

for $0 \le x \le v$.

Substituting the expression of the parabolic growth law obtained from Eq. (16) ($v = \sqrt{2k_{Fe_2B}t}$) into Eq. (25), we have

$$\left(\frac{C_{up}^{Fe_2B} + C_{low}^{Fe_2B} - 2C_0}{4}\right)(2k_{Fe_2B})^{1/2} = \sqrt{\frac{D_{Fe_2B}}{\pi}} \frac{C_{up}^{Fe_2B} - C_{low}^{Fe_2B}}{erf\left(\sqrt{\frac{k_{Fe_2B}}{2D_{Fe_2B}}}\right)} \exp\left(-\frac{k_{Fe_2B}}{2D_{Fe_2B}}\right). \tag{26}$$

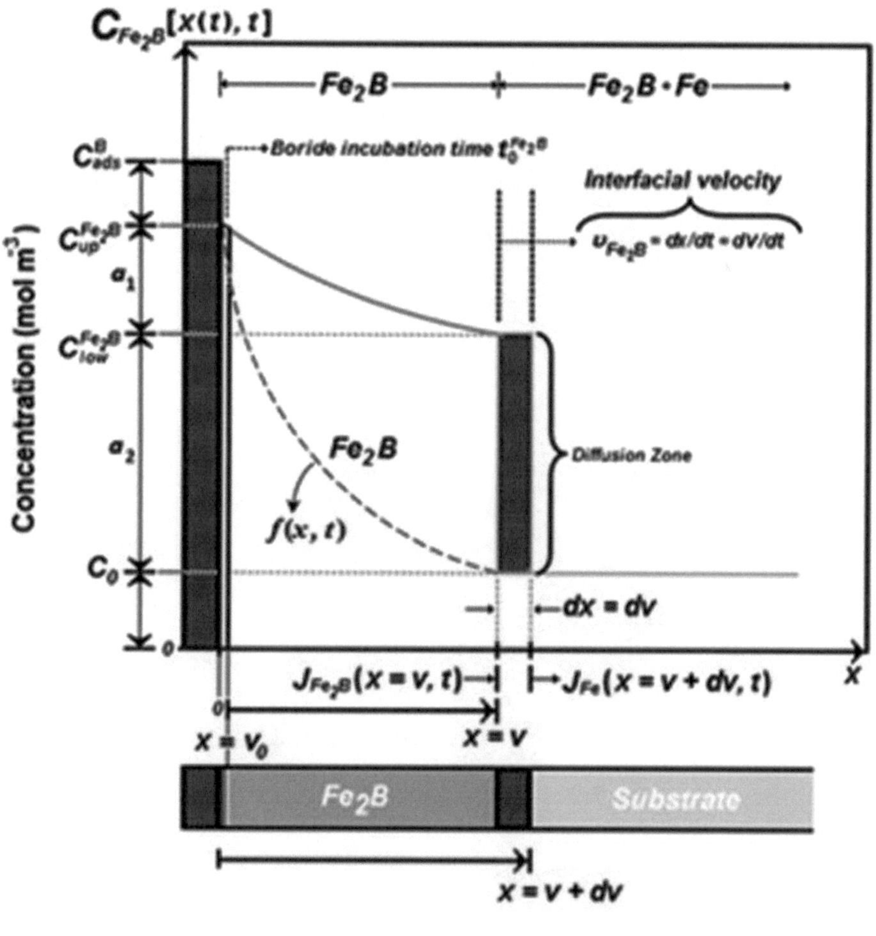

Figure 2.
A schematic non-linear concentration profile of boron through the Fe_2B layer is used to describe the non-steady state diffusion model in on dimension.

The diffusion coefficient (D_{Fe_2B}) can be estimated numerically by the Newton–Raphson method. It is assumed that expressions $C_{up}^{Fe_2B}$, $C_{low}^{Fe_2B}$, and C_0, do not depend significantly on temperature (in the considered temperature range) [10].

1.2 Materials and methods

1.2.1 Powder pack boriding process

AISI 12L14 steel was used for investigation. It had a nominal chemical composition of 0.10–0.15% C, 0.040–0.090% P, 0.15–0.35% Pb, 0.80–1.20% Mn, 0.25–0.35% S, 0.10% Si. The typical applications are: brake hose ends, pulleys, disc brake pistons, wheel nuts and inserts, control linkages, gear box components (case hardened), domestic garbage bin axles, concrete anchors, padlock shackles, hydraulic fittings, vice jaws (case hardened). The samples were sectioned into cubes with dimensions of 10 mm × 10 mm × 10 mm. Prior to the boriding process, the samples were polished with SiC sandpaper up 2500 grade, ultrasonically cleaned in an alcohol solution and deionized water for 15 min at room temperature, and dried and stored under clean-room conditions. The mean hardness was 237 HV. The samples were embedded in a closed cylindrical case (AISI 316L) as shown in **Figure 3**, using Ekabor 2 as a boron-rich agent.

The powder-pack boriding process was performed in a conventional furnace under a pure argon atmosphere. It is important to note that oxygen-bearing compounds adversely affect the boriding process [1]. The thermochemical treatment

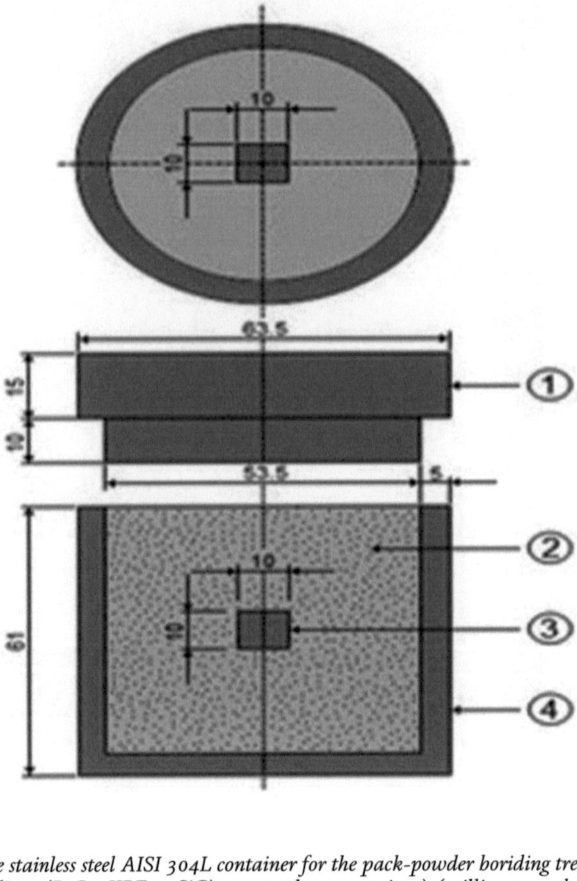

Figure 3.
Schematic view of the stainless steel AISI 304L container for the pack-powder boriding treatment (1: lid; 2: powder boriding medium (B_4C + KBF_4 + SiC); 3: sample; 4: container) (millimeter scale).

was performed at temperatures of 1123, 1173, 1223, and 1273 K with 2, 4, 6 and 8 h of exposure time. When the boriding process was concluded, the steel container was removed from the heating furnace and placed in a room temperature.

1.2.2 Characterization of boride layers

The borided samples were prepared metallographically for their characterization using GX51 Olympus equipment. As a result of preliminary experiments it was estimated that boriding started at approximately $t_0^{Fe_2B} = 29.55$ min after transferring the sample to the furnace; after that, the so-called boride incubation time sets in. The borided and etched samples were cross-sectioned, for microstructural investigations, to be observed by scanning electron microscope. The equipment used was the Quanta 3D FEG-FEI JSM7800-JOEL. **Figure 4** shows the cross-sections of boride layers formed on the surfaces of AISI 12L14 steel at different exposure times (2, 4, 6 and 8 h) and for 1173 K of boriding temperature.

The resultant microstructure of Fe_2B layers appears to be very dense and homogenous, exhibiting a sawtooth morphology where the boride needles with different lengths penetrate into the substrate [19, 20]. These elements tend to concentrate in the tips of boride layers, reducing the boron flux in this zone. The Fe_2B crystals preferably grow along the crystallographic direction [0 0 1], because it is the easiest path for the diffusion of boron in the body-centered tetragonal lattice of the Fe_2B phase [19].

It is seen that the thickness of Fe_2B layer increased with an increase of the boriding temperature (**Figure 4**) since the boriding kinetics is influenced by the

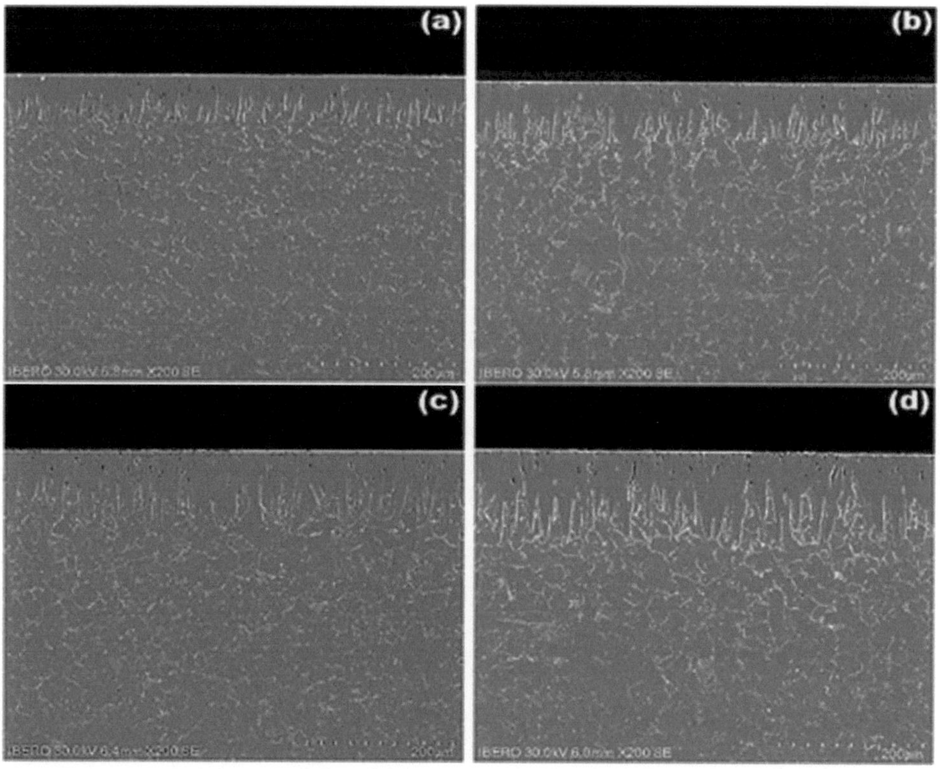

Figure 4.
SEM micrographs of the cross-sections of AISI 12L14 steel samples borided at 1173 K during different exposure times: (a) 2, (b) 4, (c) 6, and (d) 8 h.

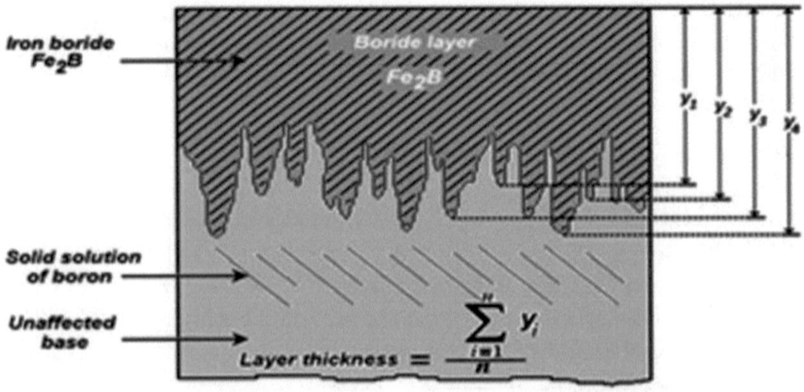

Figure 5.
Schematic diagram illustrating the procedure for estimation of boride layer thickness in AISI 12L14 steel.

treatment time. To estimate the boride layer thickness, 50 measurements were made from the surface to the long boride teeth in different sections, as shown in **Figure 5**; the boride layer thickness was measured using specialized *software* [20–22].

The identification of phases was carried out on the top surface of borided sample by an X-ray diffraction (XRD) equipment (Equinox 2000) using CoK_{α} radiation of 0.179 nm wavelength. In addition, the elemental distribution of the transition elements within the cross-section of boride layer was determined by electron dispersive spectroscopy (EDS) equipment (Quanta 3D FEG-FEI JSM7800-JOEL) from the surface.

1.3 Results and discussions

1.3.1 SEM observations and EDS analysis

The metallography of coating/substrate formed in AISI 12L14 borided steel at different exposure times (2, 4, 6 and 8 h) and for 1173 K of boriding temperature are shown in **Figure 4**. The EDS analysis obtained by SEM is shown in **Figure 6(a)** and **(b)**.

The results show in **Figure 6(a)** that the sulfur can be dissolve in the Fe_2B phase, in fact, the atomic radiuses of S (= 0.088 nm) is smaller than that of Fe (= 0.156 nm), and it can then be expected that S dissolved on the Fe sublattice of the borides. In **Figure 6(b)**, the resulting EDS analyses spectrums revealed that the manganese, carbon and silicon do not dissolve significantly over the Fe_2B phase and they do not diffuse through the boride layer, being displaced to the diffusion zone, and forms together with boron, solid solutions [10, 23, 24]. On boriding carbon is driven ahead of the boride layer and, together with boron, it forms borocementite, $Fe_3(B, C)$ as a separate layer between Fe_2B and the matrix with about 4 mass% B corresponding to $Fe_3(B_{0.67}C_{0.33})$ [10]. Thus, part of the boron supplied is used for the formation of borocementite. Likewise, silicon forming together with boron, solid solutions like silicoborides ($FeSi_{0.4}B_{0.6}$ and Fe_5SiB_2) [24].

1.3.2 X-ray diffraction analysis

Figure 7 shows the XRD pattern recorded on the surface of borided AISI 12L14 steel at a temperatures of: 1123 K for a treatment time of 2 h, and 1273 K for a treatment time of 8 h. The patterns of X-ray diffraction (see **Figure 7**) show the

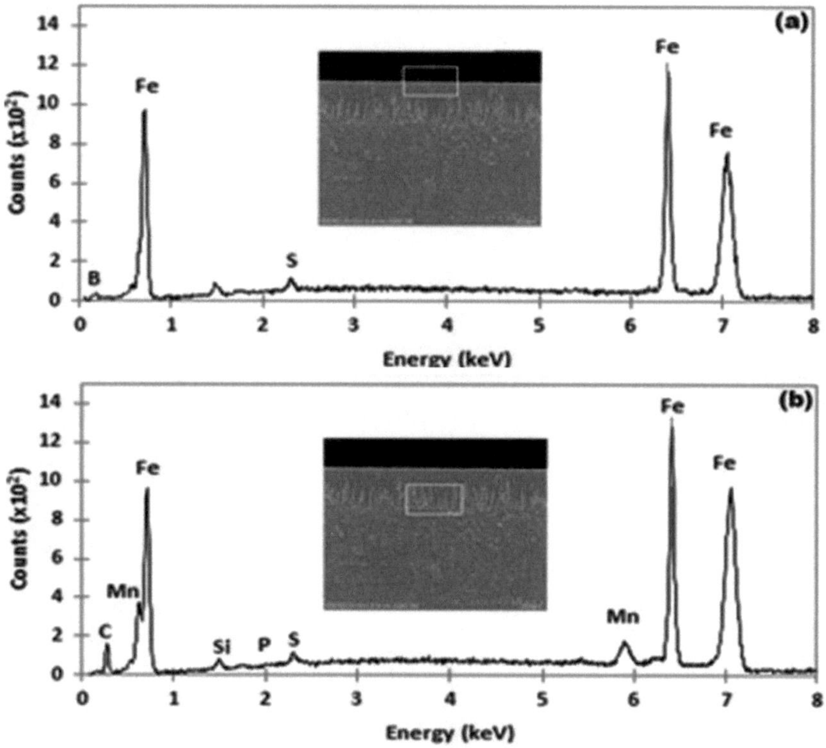

Figure 6.
A SEM micrographs of the cross-sections of the borided AISI 12L14 steel micrograph image of microstructure of the AISI 12L14 boride layer obtained at 1173 K with exposure time of 8 h, (a) and (b) EDS spectrum of borided sample.

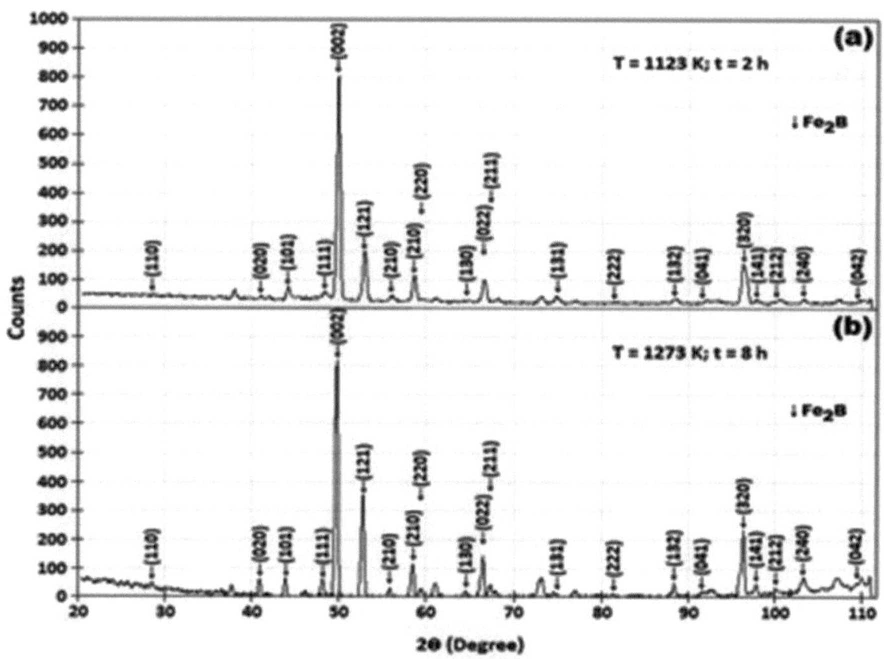

Figure 7.
XRD patterns obtained at the surface of borided AISI 12L14 steels for two boriding conditions: (a) 1123 K for 2 h and (b) 1273 K for 8 h.

presence of Fe_2B phase which is well compacted. Likewise, the patters show that there is a preferential orientation in the crystallographic plane (0 0 2) whose strength increases as the depth of the analysis increases. In a study by Martini et al. [18], the growth of the iron borides (Fe_2B) near at the Fe_2B/substrate interface only shows the diffraction peak of Fe_2B in the crystallographic plane (002).

1.3.3 Estimation of boron activation energy with steady state model

The growth kinetics of Fe_2B layers formed on the AISI 12L14 steel was used to estimate the boron diffusion coefficient through the Fe_2B layers by applying the suggested steady state diffusion model. In **Figure 8** is plotted the time dependence of the squared value of Fe_2B layer thickness for different temperatures.

In **Figure 8**, the square of boride layer thicknesses were plotted vs. the treatment time, the slopes of each of the straight lines provide the values of the parabolic growth constants ($= 2k_{Fe_2B}$). In addition, to determinate the boride incubation time, was necessary extrapolating the straight lines to a null boride layer thickness (see **Figure 8**). **Table 2** provides the estimated value of growth constants in Fe_2B at each temperature. The results, which are summarized in **Table 2**, reflect a diffusion-controlled growth of the boride layers.

In **Table 2**, the boron diffusion coefficient in the Fe_2B layers (D_{Fe_2B}) was estimated for each boriding temperature. So, an Arrhenius equation relating the boron diffusion coefficient to the boriding temperature can be adopted.

As a consequence, the boron activation energy (Q_{Fe_2B}) and pre-exponential factor (D_0) can be calculated from the slopes and intercepts of the straight line shown in coordinate system: lnD_{Fe_2B} as a function of reciprocal boriding temperature, it is presented in **Figure 9**. The boron diffusion coefficient through Fe_2B layers was deducted by steady state diffusion model as:

$$D_{Fe_2B} = 2.444 \times 10^{-3} \exp\left(-165.0329\, kJmol^{-1}/RT\right) \quad \left[m^2s^{-1}\right]. \tag{27}$$

where: $R = 8.3144621$ $\left[Jmol^{-1}K^{-1}\right]$ and T absolute temperature [K]. From the Eq. (27), the pre-exponential factor $\left(D_0 = 2.444 \times 10^{-3}\, m^2/s\right)$ and the activation

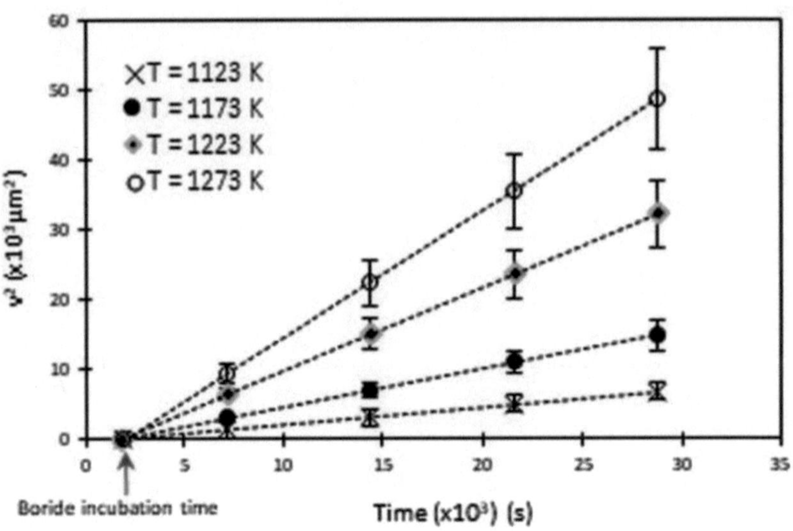

Figure 8.
Averaged squares of the Fe_2B layers (v^2) vs. boriding time (t) at different temperature.

Temperature (K)	Type of layer	Growth constants $2k_{Fe_2B}$ ($m^2\,s^{-1}$)	$D_{Fe_2B} = \frac{1}{2}\left(\frac{C_{up}^{Fe_2B}+C_{low}^{Fe_2B}-2C_0}{C_{up}^{Fe_2B}-C_{low}^{Fe_2B}}\right)k_{Fe_2B}$ Eq. (21) ($m^2\,s^{-1}$)
1123	Fe_2B	2.91×10^{-13}	5.13×10^{-11}
1173		6.12×10^{-13}	1.08×10^{-10}
1223		1.291×10^{-12}	2.27×10^{-10}
1273		2.29×10^{-12}	4.04×10^{-10}

Table 2.
The growth constants and boron diffusion coefficients as a function of boriding temperature.

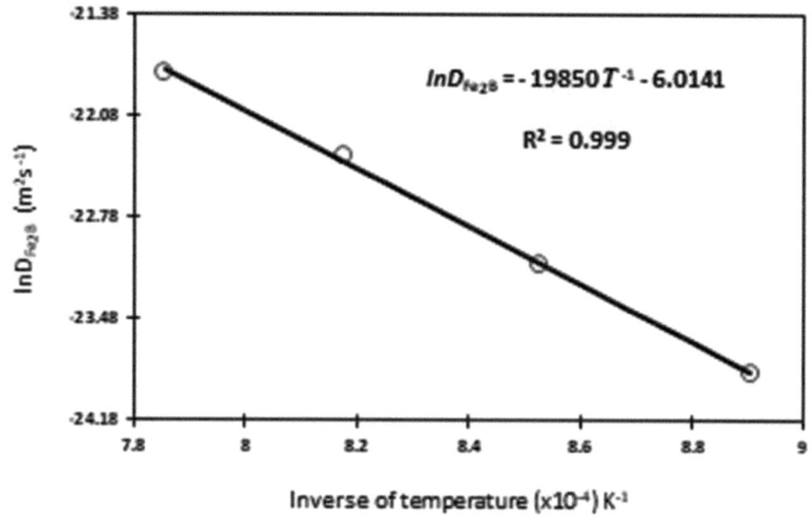

Figure 9.
Arrhenius relationship for boron diffusion coefficient (D_{Fe_2B}) through the Fe_2B layer obtained with the steady state diffusion model.

energy ($Q_{Fe_2B} = 165.0329\ kJmol^{-1}$) values are affected by the contact surface between the boriding medium and the substrate, as well as the chemical composition of the substrate [9, 10, 23–28].

1.3.4 Estimation of boron activation energy with non-steady state diffusion model

In **Table 2** provides the growth constants ($2k_{Fe_2B}$) at each temperature, and in **Table 3** provides the boron diffusion coefficients (D_{Fe_2B}), they were estimated numerically by the Newton-Raphson method from the Eq. (26).

The boron activation energy ($Q_{Fe_2B} = 164.999\ \mathrm{kJ\,mol^{-1}}$) and pre-exponential factor ($D_0 = 2.072 \times 10^{-3}\ \mathrm{m^2/s}$) can be calculated from the slopes and intercepts of the straight line shown in coordinate system: $\ln D_{Fe_2B}$ as a function of reciprocal boriding temperature, it is presented in **Figure 10**, in the same way as above.

The boron diffusion coefficient through Fe_2B layers was deducted by non-steady state diffusion model as:

$$D_{Fe_2B} = 2.072 \times 10^{-3} \exp\left(-164.999\ kJmol^{-1}/RT\right)\ \left[\mathrm{m^2 s^{-1}}\right]. \qquad (28)$$

Temperature (K)	Type of layer	$\left(\frac{C_{up}^{Fe_2B}+C_{low}^{Fe_2B}-2C_0}{4}\right)(2k_{Fe_2B})^{1/2}=\sqrt{\frac{D_{Fe_2B}}{\pi}}\frac{C_{up}^{Fe_2B}-C_{low}^{Fe_2B}}{erf\left(\sqrt{\frac{k_{Fe_2B}}{2D_{Fe_2B}}}\right)}\exp\left(-\frac{k_{Fe_2B}}{2D_{Fe_2B}}\right)$ **Eq. (26) ($m^2\,s^{-1}$)**
1123	Fe_2B	4.362×10^{-11}
1173		9.174×10^{-11}
1223		1.933×10^{-10}
1273		3.433×10^{-10}

Table 3.
The boron diffusion coefficients (D_{Fe_2B}) as a function of boriding temperature.

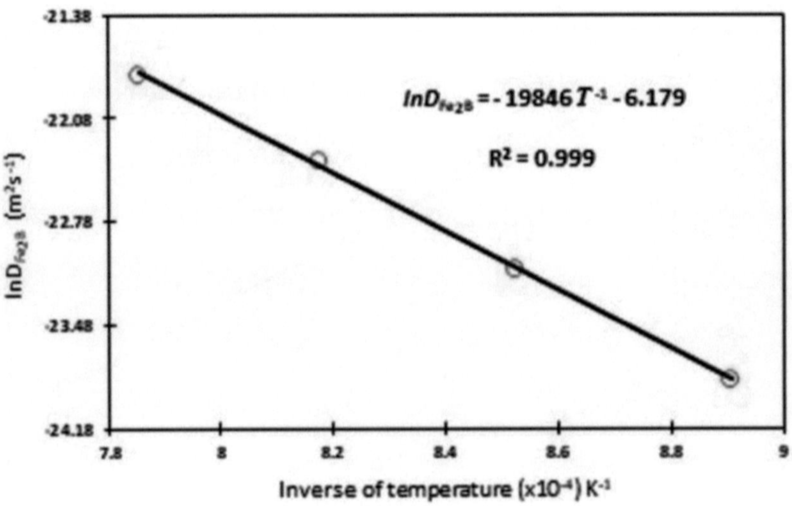

Figure 10.
Arrhenius relationship for boron diffusion coefficient (D_{Fe_2B}) through the Fe_2B layer obtained with the non-steady state diffusion model in on dimension.

1.3.5 The two diffusion models

In this section we want to illustrate the differences between the two diffusion models have been used to describe the growth kinetics of boride layers. It is noticed that the estimated values of boron activation energy $\left(Q_{Fe_2B}=165.0\ kJmol^{-1}\right)$ for AISI 12L14 steel by steady state (see Eq. (27)) and non-steady state (see Eq. (28)), is exactly the same value for both diffusion models. Likewise, the estimated values of pre-exponential factor by steady state $\left(D_0=2.444\times10^{-3}\ m^2/s\right)$ and non-steady state $\left(D_0=2.072\times10^{-3}\ m^2/s\right)$, there is a small variation. To find out how this similarity is possible in the diffusion coefficients obtained by two different models, we first focus our attention on Eq. (23). The error function $erf\left(\sqrt{k_{Fe_2B}/2D_{Fe_2B}}\right)$ is a monotonically increasing odd function of $\sqrt{k_{Fe_2B}/2D_{Fe_2B}}$. Its Maclaurin series (for small $\sqrt{k_{Fe_2B}/2D_{Fe_2B}}$) is given by [29]:

$$erf\left(\sqrt{k_{Fe_2B}/2D_{Fe_2B}}\right)=\frac{2}{\sqrt{\pi}}\left(\sqrt{k_{Fe_2B}/2D_{Fe_2B}}-\frac{\left(\sqrt{k_{Fe_2B}/2D_{Fe_2B}}\right)^3}{3\cdot1!}+\frac{\left(\sqrt{k_{Fe_2B}/2D_{Fe_2B}}\right)^5}{5\cdot2!}-\cdots\right). \tag{29}$$

According to the numerical value of the $\sqrt{k_{Fe_2B}/2D_{Fe_2B}}$, Eq. (29) can be rewritten as:

$$erf\left(\sqrt{k_{Fe_2B}/2D_{Fe_2B}}\right) = 2\sqrt{k_{Fe_2B}/2D_{Fe_2B}\pi}. \tag{30}$$

Similarly for the real exponential function $\exp\left(-k_{Fe_2B}/2D_{Fe_2B}\right) : \mathbb{R} \rightarrow \mathbb{R}$ can be characterized in a variety of equivalent ways. Most commonly, it is defined by the following power series [30]:

$$exp\left(-k_{Fe_2B}/2D_{Fe_2B}\right) = 1 - k_{Fe_2B}/2D_{Fe_2B} + \left(-k_{Fe_2B}/2D_{Fe_2B}\right)^2/2! + \left(-k_{Fe_2B}/2D_{Fe_2B}\right)^3/3! + \cdots. \tag{31}$$

Thus, Eq. (31) can be written as:

$$exp\left(-k_{Fe_2B}/2D_{Fe_2B}\right) = 1. \tag{32}$$

By substituting the Eqs. (30) and (32) into Eq. (26), we have

$$\left(\frac{C_{\text{up}}^{\text{Fe}_2\text{B}} + C_{\text{low}}^{\text{Fe}_2\text{B}} - 2C_0}{4}\right)\left(2k_{Fe_2B}\right)^{1/2} = \sqrt{\frac{D_{Fe_2B}}{\pi}}\frac{C_{\text{up}}^{\text{Fe}_2\text{B}} - C_{\text{low}}^{\text{Fe}_2\text{B}}}{2\sqrt{\frac{k_{Fe_2B}}{2D_{Fe_2B}\pi}}}(1), \tag{33}$$

$$D_{Fe_2B} = \frac{1}{2}\left(\frac{C_{\text{up}}^{\text{Fe}_2\text{B}} + C_{\text{low}}^{\text{Fe}_2\text{B}} - 2C_0}{C_{\text{up}}^{\text{Fe}_2\text{B}} - C_{\text{low}}^{\text{Fe}_2\text{B}}}\right)k_{Fe_2B}. \tag{34}$$

The result obtained by Eq. (34) is the same as that obtained in Eq. (21) estimated by steady state diffusion model. The result from the Eq. (21) would appear to imply that the non-steady state diffusion model is superior to the steady state diffusion model and so should always be used. However, in many interesting cases the models are equivalent.

1.4 Fe_2B layer's thicknesses

The estimation of the Fe_2B layers' thicknesses can be determined through the Eqs. (35) and (36).

- Steady state diffusion model

$$\text{v} = \sqrt{\left(3.33 \times 10^{-3}\right)\left(2.444 \times 10^{-3}\right)\exp\left(-\frac{165.0329\ \text{kJ mol}^{-1}}{RT}\right)t} \quad [\text{m}], \tag{35}$$

- Non-steady state diffusion model

$$\text{v} = \sqrt{\left(3.33 \times 10^{-3}\right)\left(2.072 \times 10^{-3}\right)\exp\left(-\frac{164.999\ \text{kJ mol}^{-1}}{RT}\right)t} \quad [\text{m}], \tag{36}$$

Hence, Eqs. (35) and (36) can be used to estimate the optimum boride layer thicknesses for different borided ferrous materials.

2. Conclusions

The following conclusions can be drawn from the present study:

- Two simple kinetic models were proposed for estimating the boron diffusion coefficient in Fe_2B (steady state and non-steady state).
- A value of activation energy for AISI 12L14 steel was estimated as 165.0 kJ mol^{-1}.
- Two useful equations were derived for predicting the Fe_2B layer thickness as a function of boriding parameters (time and temperature).

Finally, these diffusion models are in general not identical, but are equivalent models, and this fact can be used as a tool to optimize the boriding parameters to produce boride layers with sufficient thicknesses that meet the requirements during service life.

Acknowledgements

The work described in this paper was supported by a grant of National Council of Science and Technology (CONACyT) and PRODEP México.

Conflict of interest

The author declares that there is no conflict of interests regarding the publication of this paper.

Nomenclature

v	boride layer thickness (m).
v_0	is a thin layer with a thickness of ≈ 5 nm that formed during the nucleation stage.
k_{Fe_2B}	rate constant in the Fe_2B phase (m^2/s).
t_v	effective growth time of the Fe_2B layer (s).
t	treatment time (s).
$t_0^{Fe_2B}$	boride incubation time (s).
$C_{up}^{Fe_2B}$	upper limit of boron content in Fe_2B (= 60×10^3 mol/m^3).
$C_{low}^{Fe_2B}$	lower limit of boron content in Fe_2B (= 59.8×10^3 mol/m^3).
C_0	terminal solubility of the interstitial solute (≈ 0 mol/m^3).
$C_{Fe_2B}[x(t), t]$	boron concentration profile (≈0 mol/m^3).
D_{Fe_2B}	boron diffusion coefficient (m^2/s).

Author details

Martín Ortiz Domínguez
Autonomous University of Hidalgo State-Higher School of Ciudad Sahagún,
Ciudad Sahagún, Hidalgo, México

*Address all correspondence to: martin_ortiz@uaeh.edu.mx

References

[1] Wahl G. Boronizing, a method for the production of hard surfaces for extreme wear. Durferrit-Technical Information. Reprint from VDI-Z117. Germany; 1975. pp. 785-789

[2] Graf von Matuschka A. Boronizing. 1st ed. Munich: Carl Hanser Verlag; 1980

[3] Davis JR. Surface Hardening of Steels Understanding the Basics. USA: ASM International; 2002. pp. 213-223

[4] Singhal SC. A hard diffusion boride coating for ferrous materials. Thin Solid Films. 1977;**45**:321-329

[5] Genel K, Ozbek I, Bindal C. Kinetics of boriding of AISI W1 steel. Materials Science and Engineering. 2003;**A347**: 311-314. DOI: 10.1016/S0921-5093(02) 00607-X

[6] Yapar U, Arısoy CF, Basman G, Yesilcubuk SA, Sesen MK. Influence of boronizing on mechanical properties of En-C35e steel. Key Engineering Materials. 2004;**264-268**:633-636. DOI: 10.4028/www.scientific.net/ KEM.264-268.629

[7] Fichtl W. Boronizing and its practical application. Revue Internationale des Hautes Températures. 1980;**17**:33-43. DOI: 10.1016/0261-3069(81)90034-0

[8] Campos-Silva I, Ortiz-Domínguez M, López-Perrusquia N, Meneses-Amador A, Escobar-Galindo R, Martínez-Trinidad J. Characterization of AISI 4140 borided steels. Applied Surface Science. 2010;**256**:2372-2379. DOI: 10.1016/j.apsusc.2009.10.070

[9] Kulka M, Makuch N, Pertek A, Maldzinski L. Simulation of the growth kinetics of boride layers formed on Fe during gas boriding in H_2-BCl_3 atmosphere. Journal of Solid State Chemistry. 2013;**199**:196-203. DOI: 10.1016/j.jssc.2012.12.029

[10] Brakman CM, Gommers AWJ, Mittemeijer EJ. Boriding Fe and Fe—C, Fe—Cr, and Fe—Ni alloys; boride-layer growth kinetics. Journal of Materials Research. 1989;**4**:1354-1370. DOI: 10.1557/JMR.1989.1354

[11] Yu LG, Chen XJ, Khor KA, Sundararajan G. FeB/Fe_2B phase transformation during SPS pack-boriding: Boride layer growth kinetics. Acta Materialia. 2005;**53**:2361-2368. DOI: 10.1016/j.actamat.2005.01.043

[12] Massalski TB. Binary Alloy Phase Diagrams. Materials Park, Ohio, USA: ASM International; 1990. p. 280. DOI: 10.1002/adma.19910031215

[13] Okamoto H. B—Fe (boron-iron). Journal of Phase Equilibria and Diffusion. 2004;**25**:297-298. DOI: 10.1007/s11669-004-0128-3

[14] Dybkov VI. Reaction Diffusion and Solid State Chemical Kinetics. Switzerland-UK-USA: Trans Tech Publications; 2010. p. 7. DOI: 10.4028/ www.scientific.net/MSFo.67-68

[15] Jost W. Diffusion in Solids, Liquids, Gases. New York: Academic Press Inc; 1960. pp. 69-72. DOI: 10.1002/ ange.19530651912

[16] Shewmon P. Diffusion in Solids. USA: Minerals Metals and Materials Society; 1989. p. 40

[17] Porter DA, Easterling KE. Phase Transformations in Metals and Alloys. London: Chapman and Hall; 1981. p. 105

[18] Martini C, Palombarini G, Carbucicchio M. Mechanism of thermochemical growth of iron borides on iron. Journal of Materials Science.

2004;**39**:933-937. DOI: 10.1023/B:JMSC.0000012924.74578.87

[19] Palombarini G, Carbucicchio M. Growth of boride coatings on iron. Journal of Materials Science Letters. 1987;**6**:415-416. DOI: 10.1007/BF01756781

[20] Ortiz Domínguez M. Contribución de la Modelación Matemática en el Tratamiento Termoquímico de Borurización [thesis]. México: SEPI-ESIME from the Instituto Politécnico Nacional; 2013

[21] Campos-Silva I, Bravo-Bárcenas D, Meneses-Amador A, Ortiz-Dominguez M, Cimenoglu H, Figueroa-López U, et al. Growth kinetics and mechanical properties of boride layers formed at the surface of the ASTM F-75 biomedical alloy. Surface and Coatings Technology. 2013;**237**:402-414. DOI: 10.1016/j.surfcoat.2013.06.083

[22] Campos-Silva I, Ortiz-Domínguez M, Bravo-Bárcenas O, Doñu-Ruiz MA, Bravo-Bárcenas D, Tapia-Quintero C, et al. Formation and kinetics of FeB/Fe_2B layers and diffusion zone at the surface of AISI 316 borided steels. Surface and Coatings Technology. 2010;**205**:403-412. DOI: 10.1016/j.surfcoat.2010.06.068

[23] Eyre TS. Effect of boronising on friction and wear of ferrous metals. Wear. 1975;**34**(3):383-397. DOI: 10.1016/0043-1648(75)90105-2

[24] Dukarevich IS, Mozharov MV, Shigarev AS. Redistribution of elements in boride coatings. Metallovedenie Termicheskaya i Obrabotka Metallov. 1973;(2):164-166. DOI: 10.1007/BF00679753

[25] Elias-Espinosa M, Ortiz-Domínguez M, Keddam M, Flores-Rentería MA, Damián-Mejía O, Zuno-Silva J, et al. Growth kinetics of the Fe_2B layers and adhesion on Armco iron substrate. Journal of Materials Engineering and Performance. 2014;**23**:2943-2952. DOI: 10.1007/s11665-014-1052-2

[26] Ortiz-Domínguez M, Keddam M, Elias-Espinosa M, Damián-Mejía O, Flores-Rentería MA, Arenas-Flores A, et al. Investigation of boriding kinetics of AISI D2 steel. Surface Engineering. 2014;**30**:490-497. DOI: 10.1179/1743294414Y.0000000273

[27] Ortiz-Domínguez M, Flores-Rentería MA, Keddam M, Elias-Espinosa M, Damián-Mejía O, Aldana-González JI, et al. Simulation of growth kinetics of Fe_2B layers formed on gray cast iron during the powder-pack boriding. Materiali in Tehnologije/Materials and Technology. 2014;**48**(6):905-916. Available from: http://mit.imt.si/Revija/izvodi/mit146/ortiz.pdf

[28] Campos-Silva I, Ortiz-Domínguez M, Tapia-Quintero C, Rodriguez-Castro G, Jimenez-Reyes MY, Chavez-Gutierrez E. Kinetics and boron diffusion in the FeB/Fe_2B layers formed at the surface of borided high-alloy steel. Journal of Materials Engineering and Performance. 2012;**21**:1714-1723. DOI: 10.1007/s11665-011-0088-9

[29] Ferraro G. The Rise and Development of the Theory of Series up to the Early 1820s. Denmark: Springer; 2008. pp. 147-149. DOI: 10.1007/978-0-387-73468-2

[30] Rubin W. Real and Complex Analysis. USA: McGraw-Hill; 1987. p. 1. https://59clc.files.wordpress.com/2011/01/real-and-complex-analysis.pdf

Chapter 17

Mean Aspects Controlling Supercritical CO_2 Precipitation Processes

Antonio Montes, Clara Pereyra and Enrique J. Martínez de la Ossa

Abstract

The use of supercritical CO_2 is an excellent alternative in extraction, particle precipitation, impregnation and reaction processes due to its special properties. Solubility of the compound in supercritical CO_2 drives the precipitation process in different ways. In supercritical antisolvent process, mass and heat transfers, phase equilibria, nucleation, and growth of the compound to be precipitated are the main phenomena that should be taken into account. Mass transfer conditions the morphology and particle size of the final product. This transfer could be tuned altering operating conditions. Heat transfer in non-isothermal process influences on mixing step the size of generated microparticles. In rapid expansion of supercritical solution, phenomena as the phase change from supercritical to a CO_2 gas flow, rapid mass transfer and crystallization of the compound, and expansion jet define the morphology and size of the final product. These phenomena a priori could be modulated tuning a large number of operating parameters through the experiments, but the correlations and modeling of these processes are necessary to clarify the relative importance of each one. Moreover, particle agglomeration in the expansion jet and CO_2 condensation are determinant phenomena which should be avoided in order to conserve fine particles in the final product.

Keywords: supercritical, antisolvent, rapid expansion, particle, super saturation

1. Introduction

Nowadays, particle formation technology is continuously finding out how to solve the problems that conventional crystallization finds as residues of organic solvent in final product, thermal and chemical solute degradation, heterogeneous batches, and difficult to control the particle size distribution [1]. These problems are essential and must be overcome in pharmaceutical, cosmetic, and food industries. Controlling thermodynamic and kinetic factors together with mass and heat transfer and nucleation and growth of crystal is the key to carry out successfully the crystallization phenomena. Moreover, these forces interplay between them making it difficult to establish trends about it. Anyway, supercritical fluid technology tends to reduce or remove most of the drawbacks previously cited, so active substance with controlled particle size distribution in the micro- and nanometer range and quite stable is achieved. Solvent power and selectivity of supercritical fluids can be tuned altering the experimental conditions. Moreover, their large diffusivities result in higher mass transfer rates.

There are two main particle formation processes using supercritical fluids: rapid expansion of supercritical solution (RESS) and supercritical antisolvent (SAS) processes. The high or low solubility of the active substance in the supercritical fluid will determinate the choice between one and another process.

Numerous investigations in particle formation with supercritical fluids have been carried out to shed light the parameter to govern the crystallization mechanism using CO_2 as antisolvent (SAS) [2–15] or as solvent (RESS) [16–28].

The SAS process consists of spraying a solution of the solute to be precipitated into the vessel that contains the bulk supercritical fluid. Then, there is a rapid solubilization of supercritical fluid into the solvent causing its volumetric expansion and thus reducing the solvation capacity of the solvent. This fact causes the supersaturation of the liquid phase and the consequent generation of the particles. Thus, mass transfer of the process seems to be a key factor affecting particle morphology [29].

SAS technique has been investigated using solutes in a wide range of industrial fields with applications as polymers [30, 31], explosives [32], pharmaceutical compounds [2, 3], superconductors [33], catalysts [34], coloring matter [35], and functional food [6, 7], among others. On the other hand, the application of RESS technique has been enclosed to pharmaceutical field [19–27].

Once SAS process involves the interaction of several phenomena as thermodynamics, mass transfer, jet hydrodynamics, and nucleation kinetics, it is difficult to isolate one of them as responsible for a particle feature trend [36]. Anyway, the mass transfer which occurs between a droplet of organic solvent and a compressed antisolvent is a crucial step of the process especially below mixture critical point in partially miscible conditions. It is well known that in miscible conditions, above mixture critical point, there is no obvious way to define the interface between the two fluids. However, Dukhin et al. have evidenced the existence of a dynamic interfacial tension to the transient existence of droplets at conditions slightly above the mixture critical point (MCP) [37].

Particle size and morphology are influenced by the antisolvent-solvent mass transfer ruled by diffusion process and by the jet breakup, although it seems that the first phenomenon has often more importance [38, 39] in isothermal mixing.

Heat transfer should also be taken into account in this kind of process due to the temperature can be an effective parameter to modulate the lifetimes of the droplets [40]. Thus, there is another way to change the mixing time without pressure and temperature modifications, if the process occurs in non-isothermal conditions. The mixing time will increase if a colder solution is immersed into a warmer CO_2 and will decrease if a warmer solution is immersed in colder CO_2 [40]. Thus, size of microparticles could be tuned.

On the other hand, RESS process takes advantage of the solubility of the solute to be precipitated into supercritical fluids. In this case the supercritical solution, CO_2 plus solute, is rapidly depressurized to atmospheric pressure through a nozzle, thus causing the precipitation, extremely fast, of the solute.

With regard to RESS process, nucleation, condensation, and coagulation models for crystallization of solute have been explored [41–44]. In order to evaluate these models, it is crucial to predict the thermophysical properties and flow characteristics in the nozzle, where supercritical solution comes from the nozzle inlet to the outlet of the expansion chamber [45].

In the work SAS and RESS processes are described making incidence on the main aspects which should be controlled to get a successful precipitation of fine particles.

2. Supercritical CO_2

A supercritical fluid is defined as whatever substance above its critical temperature and pressure. In **Figure 1** a typical pressure-temperature phase diagram is shown where a singular region is shaded. Supercritical fluids have special properties as solvent power similar to liquids, but diffusivities two orders of magnitude larger than those of typical liquids, resulting in higher mass transfer rates. Moreover, solvent power and selectivity can be tuned with the modification of the density of this fluid. This one can be adjusted continuously by altering the experimental conditions of temperature and pressure.

Singularities in compressibility and viscosity, diminishing difference in liquid and vapor phases, are other exceptional characteristics that are present in this state.

CO_2 is the fluid more used in supercritical conditions due to its low danger and relative low cost. CO_2 is a solvent generally recognized as safe (GRAS), non-flammable, non-toxic, gaseous at room temperature, and easily removed from the process. Moreover, it presents relative mild conditions of its critical point (31.1°C and 71.8 bar) permitting the processing of thermolabile solutes which are primordial in pharmaceutical, cosmetic, and functional food industries.

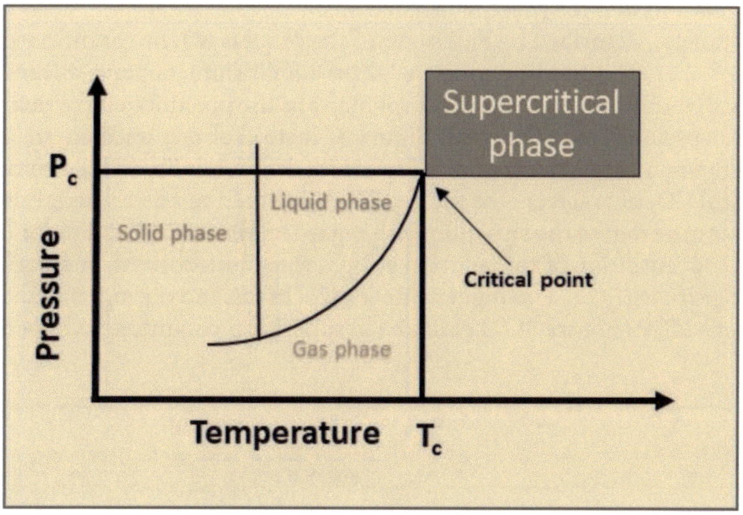

Figure 1.
Pressure-temperature phase diagram.

3. Supercritical antisolvent process

In SAS process an organic solution of the solute is sprayed through a micrometric nozzle to improve the mass transfer between CO_2 and the microdroplets of solution. CO_2 is solubilized into organic solvent and vice versa in lower degree causing a volumetric expansion of the solvent and a super saturation of the solution and the consequent precipitation.

The substances which can be processed by this technique must fulfill two main requirements: (1) It must be not soluble in supercritical CO_2 and (2) must be soluble in an organic solvent that is miscible with supercritical CO_2.

Most of the active ingredients and added value substances are not soluble in supercritical CO_2. So, the main problem is finding out the appropriate solvent to be used in the process. In **Figure 2** a scheme of the process is shown.

In a typical experiment of SAS process, CO_2 is cooled to 5°C (HE1) and posteriorly pumped by a high-pressure pump (P1) to a precipitation vessel (V). CO_2 flow rate is controlled by a Coriolis flowmeter (FM). The stream of CO_2 is heated (HE2) at the required set point. The pressure level is held with the aid of automated back pressure regulator (ABPR). The system works in a semicontinuous way, so until the operating conditions are not reached in this vessel, the organic solution is not pumped to the vessel by a solution pump (P2). This solution is put into the vessel through a nozzle with hundreds of microns to improve the mass transfer. Then, two opposed processes happen in different degrees. CO_2 is solubilized into the solvent, and solvent is evaporated into the bulk CO_2. Anyway, simulations of some systems have shown that absorption of CO_2 into the liquid phase is always faster than solvent evaporation and, consequently, the produced mircodroplets swell as soon as they get into contact with CO_2 [46]. Then, CO_2 is rapidly solubilized into the small droplets of the solution provoking the loss of solvation power of solvent and generating a supersaturation of the solution and the consequent precipitation of solute in the form of micro- or even nanoparticles. CO_2 and solvent are posteriorly separated in a cyclone (CS) venting out the CO_2 for the top and collecting the liquid for the bottom. CO_2 is continuously flowing for a time, called washing time, to ensure no solvent is present into the vessel when depressurization is carried out and could redissolved into the precipitate particles.

Mass transfer, described by Fick's law, of the process will be conditioned by the situation of operating conditions on the phase equilibrium diagram solvent-solute-CO_2, thus affecting on the size and morphology of the precipitated particles. In this sense, a pseudobinary diagram (**Figure 3**) instead of ternary diagram often is considered by authors due to the solute having low solubility in the mixture supercritical CO_2 plus solvent, so the equilibrium would not be altered. Pressure and temperature define the situation with regard to mixture critical point (MCP). But if the concentration of the solution is high, the solute-solvent-interaction could increase significantly MCP to higher values [47]. In this sense programed experiments above MCP (**Figure 3**(2)) could be in subcritical conditions as experiments

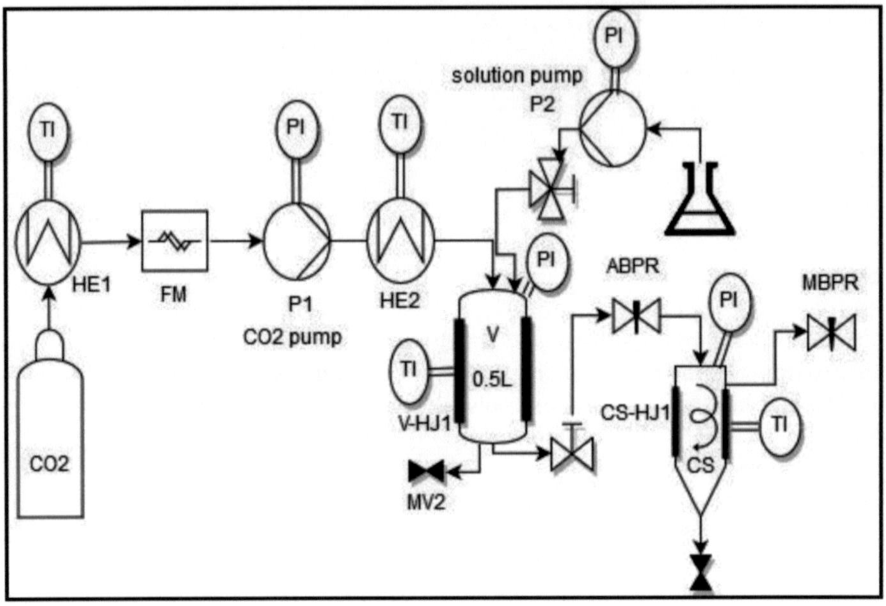

Figure 2.
Scheme of SAS process.

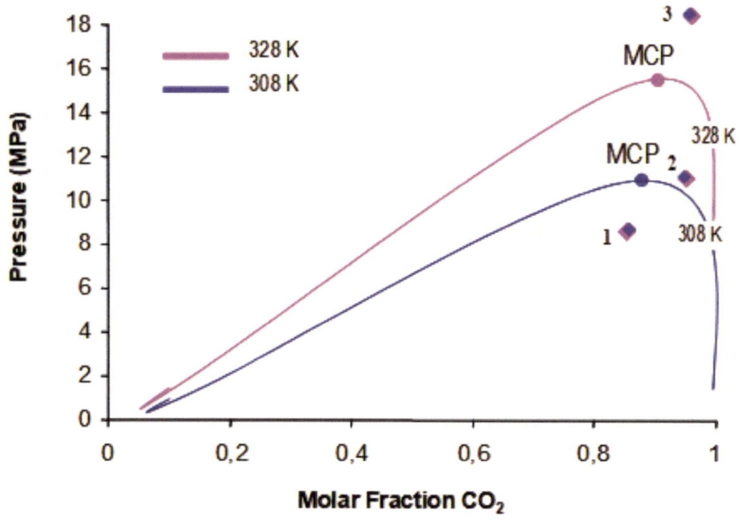

Figure 3.
Phase equilibrium diagram.

designed below mixture critical point CO_2-organic solvent where interfacial tension exists (**Figure 3**(1)). However, far above MCP (**Figure 3**(3)), there will be no interfacial tension in the system solvent-CO_2 when the sprayed solution is in contact with the bulk supercritical CO_2.

To understand this fact, kinetic and hydrodynamic factors should be taken into account. Injected solution has a surface tension as a liquid, but the vanishing of this tension solvent-CO_2 is faster than the breakup jet, so there is no interfacial tension when the solution finds bulk supercritical CO_2 at hard conditions. When supercritical CO_2 is slight above MCP (**Figure 3**(2)), there is a residual surface tension at the moment to be immersed into the bulk CO_2, so mass transfer is lower and microparticles are generated. In this case droplet lifetime is the key to observe difference in morphology and particle size of the precipitates. In this region operating conditions could be tuned to influence mass transfer. As a rule, an increase of pressure leads to a decrease of droplet lifetimes [40], so smaller particles would be precipitated. On the other hand, droplets could swell or shrink depending on the difference in density and diffusivity between the two phases [36] so that droplets shrink when the solvent's mass density is lower than those of the bulk CO_2 and swell when the CO_2 density is lower.

Not only pressure but also temperature modifies sensibly the density, so combination of pressure and temperature is an effective tool to modify the droplet lifetimes [40]. Temperature even has more influence than pressure near the MCP (**Figure 3**(2)). In this sense, the diffusion coefficients are quite sensitive to temperature, and their increase would also accelerate the transfer.

Since an increase of pressure seems to lead to a shorter droplet lifetime as a general rule, an increase of temperature might have opposite effects depending on the pressure range and the chosen solvent. For instance, using acetone or ethanol, the lifetimes will be short at relatively low pressure and larger at higher pressure [40].

On the other hand, heat transfer should be taken into account due to the solvent droplet is often immersed in bulk supercritical CO_2 at a different temperature. In non-isothermal conditions the temperature gradient can be an additional tool to tune the droplet lifetime. Thus, the immersion of colder solvent in warmer CO_2 permits to increase the mixing time that is less interesting for crystal production. In this case low supersaturation ratio can be generated by the injection of colder solution than

CO_2 due to often solubility of compounds increasing with temperature [40]. This fact merged to the less concentration of the solution that leads to larger particles. On the contrary, a warmer solvent immersed in colder CO_2 accelerates the mixing process. Thus, high supersaturation ratio can be achieved, so smaller particle size of the precipitated is generated. An additional advantage is the opportunity to inject solutions more concentrated that come in addition to the crystallization by antisolvent [40].

Mass transfer kinetics are also influenced by hydrodynamic aspect, particularly, the disintegration mode of the liquid jet into the supercritical bulk fluid. There are several main modes of disintegration that affect to mass transfer. In the dripping mode, droplets are formed in the outlet of the nozzle as consequences of lower flow rates and/or higher nozzle diameters. A laminar mode where the jet is continuous at the outlet of the nozzle device and the rupture jet is produced as monodisperse droplets symmetrically or asymmetrically. In the first one called axisymmetric mode, the breakup is produced by axially symmetric disturbances. In the asymmetrical mode, the breakup is caused by disturbances that are symmetrical about a helical axis that starts at the nozzle orifice. Then, atomization mode is achieved when the jet leaves the nozzle smoothly till the zone of highly chaotic rupture is reached. In this case a cone of atomized liquid is formed improving the mass transfer. It is possible to tune the process toward one or another mode acting on the critical atomization velocity, which is the velocity corresponding to the boundary between the asymmetrical and the atomization modes. The critical atomization velocity depends on the liquid solution flow rate and the nozzle device. Liquid solution flow rate depends in turn on viscosity and surface tension of solvent. As a general rule, relative higher flow rate and smaller nozzle diameter are recommended to achieve atomization mode.

Not only the nozzle diameter but also the length or the geometry have been evaluated in order to improve the mass transfer of the process achieving the critical atomization velocity [48, 49]. Even the nozzle relative position to CO_2 inlet modifies the mass transfer due to influences on hydrodynamics and mixing between solution and CO_2. For instance, if CO_2 is injected through the annulus, the fluid that diffuses into the jet does not have organic solvent residues increasing the supersaturation, and smaller particles are generated [50]. Modification of SAS process has been carried out to improve the mixing. In this sense solution-enhanced dispersion by supercritical fluid (SEDS) process uses a coaxial nozzle to introduce the supercritical fluid antisolvent and solution [51, 52].

4. Rapid expansion of supercritical solution process

This technique is appropriated when the active substance is soluble in supercritical fluid. In this case a supercritical solution is formed and then is expanded through a nozzle to ambient pressure provoking the fast supersaturation and the precipitation in form of nano- or microparticles. Nozzle must be thermostatized due to the expansion from supercritical to ambient pressure which comes associated with a dramatic loss of heat and the consequent freezing by Joule effect. It must be taken into account that supercritical solution achieves in the nozzle the speed of sound and expands into the expansion chamber to a supersonic flow provoking an expansion jet with multiple shocks that influence the coagulation process [45].

In **Figure 4**, a general scheme of RESS technique is shown. The procedure is a batch process: on the one hand, the charge of solutes and, on the other hand, the charge of CO_2 and mixing. Solutes which are going to be micronized are weighted and placed into the solubilization chamber (V1) which is posteriorly sealed. Then, CO_2 at 5°C (HE1) of temperature is pumped to V1 after temperature set point is achieved (HE2). CO_2 and solute are held in the vessel during a contact time before

MV2 valve opening. In this point the supercritical solution is sprayed through the thermostatized nozzle to vessel V2 at room conditions. Nano- or microparticles are produced by the drop of pressure once solubility of the solute in supercritical CO_2 is drastically reduced. This fact is associated to a high supersaturation of the solution leading to the formation of fine particles with narrow particle size distribution.

Nucleation, condensation, and coagulation govern the crystallization of solute. During RESS process, properties of the fluid rapidly change from those of a supercritical fluid to those of a gas as the fluid crosses the critical pressure in the nozzle implying mass and heat transfer. Moreover, the flow is accelerated from a static condition to supersonic speeds in the expansion [45].

Solubility of solute in the supercritical CO_2 solution dominates the saturation rate in this process. This rate is the key to know where the supersaturation and the consequent nucleation of the solute will take place. This point will coincide when supercritical fluid crosses the critical pressure toward the gas phase.

Authors have studied the main experimental parameter that influences on particle size and particle size distribution as temperature and pressure of solubilization chamber, pre- and postexpansion pressure and temperature, diameter, geometry and length of nozzle, spray distance, and so on.

Solvating power depends on both temperature and pressure into the solubilization chamber, because the density of the solvent depends on these parameters. Thus, the higher the pressure, the higher the density while temperature is held constant. This increase in density is related to better solvating power of the supercritical fluid and solute concentration in supercritical fluid. Once concentration is increased, the supersaturation and the nucleation rates are increased during the expansion step causing a smaller particle size precipitation [53].

On the other hand, temperature could influence on two competing phenomena. If temperature is increased there is a decrease in the density of CO_2 (decreasing the solvent power of CO_2 and the solute saturation) but a concurrent increase of the solute vapor pressure (increasing the solubility inf CO_2 and the solute saturation). Depending on which phenomena prevail, the trend on particle size will be different, and higher particles or smaller partciles will be produced if the first or second phenomen do respectively.

Corazza et al. pointed out that the supersaturation profile in the free jet region depends on the pre-expansion conditions that could have a remarkable effect on

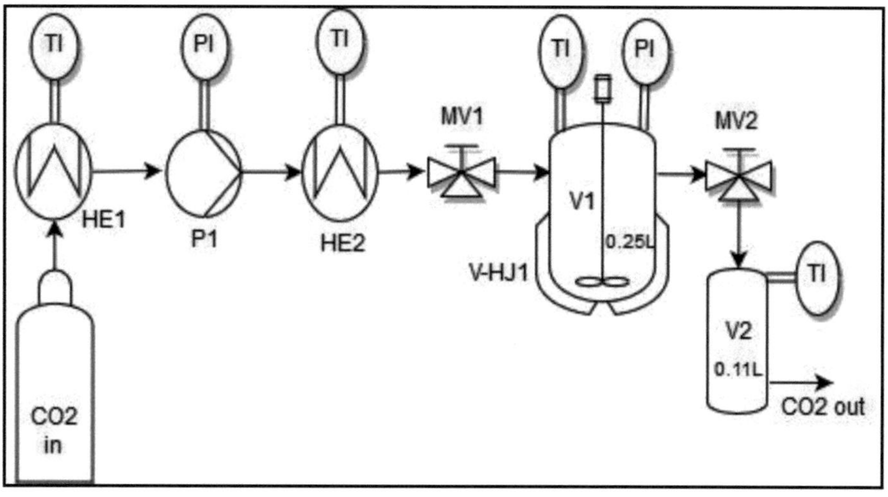

Figure 4.
Scheme of RESS process.

the characteristics of precipitated particles [17]. Higher pre-expansion pressure and lower temperatures are recommended to obtain smaller particle size of benzoic acid, cholesterol, and aspirin precipitated by RESS process [16].

The use of liquid [20] and solid [21] cosolvents to improve the solubility of the solute in supercritical fluid and modify the saturation rate has been associated to RESS process by numerous investigations.

Not only thermodynamic but also hydrodynamic aspects are quite relevant for the success of the process. In this sense, Huang et al. reduced the coalescence and particle size below the micron developing a new clearance nozzle with small exit size in a few microns [18]. Moreover, the spray distance in the expansion unit and the residence time of the precipitated particles are also important factors. Higher spray distance is equal to longer fly time of the particles which allow for their growing and to produce bigger particles [53]. This time is related to the collection distance. Thus, the coagulation of the particles that happens in the expansion region could be minimized using short residence time or short collection distance [53].

Moreover, coagulation phenomena could be drastically reduced modifying the expansion process, so stabilized separated particles were generated. Rapid expansion from supercritical to aqueous solution (RESAS) [54] and rapid expansion of supercritical solutions into liquid solvent (RESOLV) [55] processes appear as alternative to classical RESS process where the air spraying is done. In RESAS the solution is sprayed into surfactant water solution and in RESOLV process into another solvent different to water as ethanol. Thus, particle growth is minimized, and the coagulation of the powder is prevented.

In our facilities nonsteroidal anti-inflammatory drugs (NSAIDs), as ibuprofen and naproxen [22, 23] (**Table 1**) and vanillin [28] (**Table 2**), a solute more soluble in supercritical CO_2, were successfully precipitated by this technique, and parameter influence was investigated. In general, pressure and temperature were the factors which affect strongest as the particle size. Note that naproxen precipitation was aided with methanol as cosolvent due to less solubility of naproxen in supercritical CO_2 than ibuprofen. In both groups of compounds, lower particle size was obtained when pressures were increased. This can be explained due to solvating power of the supercritical fluid which is improved increasing the solute concentration in the supercritical solution, so higher super saturation induces higher nucleation rate during the expansion period provoking fine particle precipitation. In the case of vanillin, the main effects of variables on particle size were found out through a design of experiments

Sample	P (bar)	T (K)	Methanol (%)	Particle size
Naproxen	200	333	5	0.92 ± 1.16
Naproxen	300	333	5	0.22 ± 0.09
Naproxen	200	333	5	0.06 ± 0.01
Naproxen	300	333	5	0.10 ± 0.04
Ibuprofen	100	308	—	1.29 ± 0.38
Ibuprofen	150	308	—	1.26 ± 0.58
Ibuprofen	300	308	—	0.69 ± 0.31
Ibuprofen	100	318	—	1.04 ± 1.48
Ibuprofen	150	318	—	0.67 ± 0.38
Ibuprofen	300	318	—	0.22 ± 0.06

Table 1.
Operating conditions in RESS NSAID assays.

Factors	Low level	High level	Particle size effects
Pressure (bar)	100	300	−6.49
Temperature (K)	313	343	5.29
Contact time (h)	1	2	−1.37
Nozzle diameter (μm)	100	200	2.41

Table 2.
Effects of main operating conditions set in RESS process on particle size of vanillin.

of two levels. It seems particle growth may be dominant over nucleation in these operating conditions and solutes [25]. Higher temperature was recommended in the case of ibuprofen to get smaller particles. The opposite trend happened with vanillin where higher temperature led to higher particle size. Moreover, contact time or nozzle diameter had less impact on particle size recommending smaller nozzle diameter and higher contact time. Anyway, due to the multiple phenomena that interplay in this process, it is difficult sometime to establish general trends.

Modeling of RESS process could permit to shed light to the process. In this sense Moussa et al. simulated particle transport and Brownian coagulation in the expansion chamber by resolving the general dynamic equation. The results showed that postexpansion conditions are an important factor to control particle size of precipitates and demonstrated that particle growth is not completed in the supersonic free jet during the RESS process [56]. In this way Reverchon and Pallado modeled the hydrodynamics of the RESS process and concluded that most of the pressure drop and temperature decrease took place in the postexpansion chamber bringing to light the important role of the process parameters connected to the postexpansion device [57].

Thermophysical flows of RESS process were numerically simulated from the nozzle inlet to the outlet of the expansion chamber using program package for thermophysical properties of fluids (PROPATH) [58] developed at Kyushu University [59] and naphthalene as model molecule. The results indicated that the location at which the pressure crosses the critical pressure is sensitive to the pressure at the nozzle inlet and nucleation starts near that place [45].

Helfgen et al. developed a one-dimensional, steady-state flow field model based on mass, momentum, and energy balance and the Bender equation of state. In the inlet region of the nozzle, a pressure drop, friction, and heat exchange are considered for the flow along the nozzle and a possible condensation in the supersonic free jet. They resolve the general dynamic equation for simultaneous nucleation, condensation, and coagulation. The results showed a decrease of pressure and temperature at the capillary nozzle outlet, which leads to immediate precipitation of the solved substance, and most of the particle growth occurred inside the expansion chamber, but the calculated particle sizes are too high in relation to the measured ones [60]. Further investigations are needed to clarify the process and their dominant forces.

5. Conclusions

Special properties of supercritical CO_2 as solvent power similar to liquids but high diffusivities similar to gases, resulting in higher mass transfer rates. This mass transfer rate is quite important in particle formation process using CO_2 as solvent or antisolvent. In antisolvent processes mass transfer rate will influence on the size and morphology of the particles. Higher mass transfer rate is related to smaller particle size. Anyway, mass transfer of supercritical antisolvent process depends on the one hand on the situation of phase equilibrium diagram of operating conditions and on the other hand on the hydrodynamics of the process. The hydrodynamics of the process is controlled by flow rates of CO_2 and solution and by the nozzle that forms the spray. This spray reduces the boundary between CO_2 and microdroplets of solution, improving the mass transfer of the process. Heat transfer in supercritical antisolvent process influences directly on the mixing step in non-isothermal process. If the solution is colder than bulk CO_2, the mixing time that is less interesting for crystal production will increase. However, in the opposite situation, a warmer solvent immersed in colder CO_2 accelerates the mixing process. Thus, high supersaturation ratio can be achieved, so smaller particles are precipitated.

In rapid expansion of supercritical solution, mass transfer happens in solubilization chamber and in nozzle and expansion chamber. Saturation of the solution which is governed by solubility of solute in supercritical fluid will decide the place where to start the precipitation. This location could be into the nozzle or after in the expansion chamber. Heat transfer happens at the exit of the nozzle due to the drastic change of pressure by Joule effect. Nozzle should be thermostatized to prevent freezing and clogging.

Acknowledgements

We gratefully acknowledge the Spanish Ministry of Science and Technology (Project CTQ2010-19368) for their financial support.

Author details

Antonio Montes*, Clara Pereyra and Enrique J. Martínez de la Ossa
Faculty of Sciences, Department of Chemical Engineering and Food Technology, University of Cádiz, International Excellence Agrifood Campus (CeiA3), Cádiz, Spain

*Address all correspondence to: antonio.montes@uca.es

References

[1] He WZ, Suo QL, Jiang ZH, Shan A, Hong H. Precipitation of ephedrine by SEDS process using a specially designed prefilming atomizer. Journal of Supercritical Fluids. 2004;**31**:101-110. DOI: 10.1016/j.supflu.2004.01.009

[2] Franco P, Reverchon E, De Marco I. PVP/ketoprofen coprecipitation using supercritical antisolvent process. Powder Technology. 2018;**340**:1-7. DOI: 10.1016/j.powtec.2018.09.007

[3] Franco P, Reverchon E, De Marco I. Zein/diclofenac sodium coprecipitation at micrometric and nanometric range by supercritical antisolvent processing. Journal of CO_2 Utilization. 2018;**27**:366-373. DOI: 10.1016/j.jcou.2018.08.015

[4] Yoon TJ, Son W-S, Park HJ, Seo B, Kim T, Lee Y-W. Tetracycline nanoparticles precipitation using supercritical and liquid CO_2 as antisolvents. Journal of Supercritical Fluids. 2016;**107**:51-60. DOI: 10.1016/j.supflu.2015.08.014

[5] Guamán-Balcázar MC, Montes A, Pereyra C, Martínez de la Ossa E. Production of submicron particles of the antioxidants of mango leaves/PVP by supercritical antisolvent extraction process. The Journal of Supercritical Fluids. 2019;**143**:294-304. DOI: 10.1016/j.supflu.2018.09.007

[6] Arango-Ruiz A, Martin A, Cocero MJ, Jiménez C, Londoño J. Encapsulation of curcumin using supercritical antisolvent (SAS) technology to improve its stability and solubility in water. Food Chemistry. 2018;**258**:156-163. DOI: 10.1016/j.foodchem.2018.02.088

[7] Kodama T, Honda M, Machmudah S, Diono W, Kanda H, Goto M. Crystallization of All trans–β–carotene by Supercritical Carbon Dioxide Antisolvent via Co–axial Nozzle. Engineering Journal. 2018;**22**(3):25-38. DOI: 10.4186/ej.2018.22.3.25

[8] Vorobei AM, Ustinovich KB, Pokrovskiy OI, Parenago OO, Lunin VV. Preparation of Hydroxypropylmethylcellulose Microparticles Using Supercritical Antisolvent Precipitation. Russian Journal of Physical Chemistry B. 2015;**9**:1103-1108. DOI: 10.1134/S1990793115080114

[9] Prosapio V, Reverchon E, De Marco I. Polymers' ultrafine particles for drug delivery systems precipitated by supercritical carbon dioxide + organic solvent mixtures. Powder Technology. 2016;**292**:140-148. DOI: 10.1016/j.powtec.2016.01.033

[10] Da Silva EP, Winkler MEG, Giufrida WM, Cardozo-Filho L, Alonso CG, Lopes JBO, et al. Effect of phase composition on the photocatalytic activity of titanium dioxide obtained from supercritical antisolvent. Journal of Colloid and Interface Science. 2019;**535**:245-254. DOI: 10.1016/j.jcis.2018.09.098

[11] Montes A, Williamson D, Hanke F, Garcia-Casas I, Pereyra C. Martínez de la Ossa E, Teipel U, New insights into the formation of submicron silica particles using CO_2 as anti-solvent. The Journal of Supercritical Fluids. 2018;**133**:218-224. DOI: 10.1016/j.supflu.2017.10.013

[12] Montes A, Wehner L, Pereyra C, Martínez de la Ossa EJ. Precipitation of submicron particles of rutin using supercritical antisolvent process. Journal of Supercritical Fluids. 2016;**2016**(118):1-10. DOI: 10.1016/j.supflu.2016.07.020

[13] Montes A, Wehner L, Pereyra C, Martínez de la Ossa EJ. Generation

of microparticles of ellagic acid by supercritical antisolvent process. Journal of Supercritical Fluids. 2016;**116**:101-110. DOI: 10.1016/j.supflu.2016.05.019

[14] Montes A, Wehner L, Pereyra C, Martínez De La Ossa EJ. Mangiferin nanoparticles precipitation by supercritical antisolvent process. Journal of Supercritical Fluids. 2016;**112**:44-50. DOI: 10.1016/j.supflu.2016.02.008

[15] Montes A, Williamson D, Hanke F, Garcia-Casas I, Pereya C, Martínez de la Ossa E, et al. New insights into the formation of submicron silica particles using CO_2 as anti-solvent. The Journal of Supercritical Fluids. 2018;**133**:218-224. DOI: 10.1016/j.supflu.2017.10.013

[16] Harrison JJ, Lee C, Lenzer T, Oum K. On-line in-situ characterization of CO RESS processes for benzoic acid, cholesterol and aspirin. Green Chemistry. 2007;**9**:351-356. DOI: 10.1039/b612055e

[17] Corazza ML, Filho CL, Dariva C. Modeling and simulation of rapid expansion of supercritical solutions. Brazilian Journal of Chemical Engineering. 2006;**23**(3):417-425. DOI: 10.1590/S0104-66322006000300015

[18] Huang J, Moriyoshi T. Fabrication of fine powders by RESS with a clearance nozzle. Journal of Supercritical Fluids. 2006;**37**:292-297. DOI: 10.1016/j.supflu.2005.11.024

[19] Türk M, Bolten D. Polymorphic properties of micronized mefenamic acid, nabumetone, paracetamol and tolbutamide produced by rapid expansion of supercritical solutions (RESS). Journal of Supercritical Fluids. 2016;**116**:239-250. DOI: 10.1016/j.supflu.2016.06.001

[20] Sodeifian G, Sajadian SA, Daneshyan S. Preparation of Aprepitant nanoparticles (efficient drug for coping with the effects of cancer treatment) by rapid expansion of supercritical solution with solid cosolvent (RESS-SC). The Journal of Supercritical Fluids. 2018;**140**:72-84. DOI: 10.1016/j.supflu.2018.06.009

[21] Uchida H, Nishijima M, Sano K, Demoto K, Sakabe J, Shimoyama Y. Production of theophylline nanoparticles using rapid expansion of supercritical solutions with a solid cosolvent (RESS-SC) technique. Journal of Supercritical Fluids. 2015;**105**:128-135. DOI: 10.1016/j.supflu.2015.05.005

[22] Montes A, Litwinowicz A, Gradl U, Gordillo MD, Pereyra C, Martínez de la Ossa EJ. Exploring high operating conditions in the ibuprofen precipitation by RESS process. Industrial and Engineering Chemistry Research. 2014;**53**:474-480. DOI: 10.1021/ie402408j

[23] Montes A, Bendel A, Kürti R, Gordillo MD, Pereyra C, Martínez de la Ossa EJ. Processing naproxen with supercritical CO_2. Journal of Supercritical Fluids. 2013;**75**:21-29. DOI: 10.1016/j.supflu.2012.12.016

[24] Huang Z, Sun GB, Chiew YC, Kawi S. Formation of ultrafine aspirin particles through rapid expansion of supercritical solutions (RESS). Powder Technology. 2005;**160**:127-134. DOI: 10.1016/j.powtec.2005.08.024

[25] Turk M, Bolten D. Formation of submicron poorly water-soluble drugs by rapid expansion of supercritical solution (RESS): Results for Naproxen. Journal of Supercritical Fluids. 2010;**55**:778-785. DOI: 10.1016/j.supflu.2010.09.023

[26] Wang J, Chen J, Yang Y. Micronization of titanocene dichloride by rapid expansion of supercritical solution and its ethylene polymerization. Journal of Supercritical Fluids. 2005;**33**:159-172. DOI: 10.1016/j.supflu.2004.05.006

[27] Yildiz N, Tuna S, Doker O, Calimli A. Micronization of salicylic acid and taxol (paclitaxel) by rapid expansion of supercritical fluids (RESS). Journal of Supercritical Fluids. 2007;**41**:440-451. DOI: 10.1016/j.supflu.2006.12.012

[28] Montes A, Merino R, De los Santos DM, Pereyra C, Martínez de la Ossa EJ. Micronization of vanillin by rapid expansion of supercritical solutions process. Journal of CO_2 Utilization. 2017;**21**:169-176. DOI: 10.1016/j.jcou.2017.07.009

[29] Werling JO, Debenedetti PG. Numerical modeling of mass transfer in the supercritical antisolvent process. Journal of Supercritical Fluids. 1999;**16**:167-181. DOI: 10.1016/S0896-8446(00)00054-1

[30] Montes A, Gordillo MD, Pereyra C, Martínez de la Ossa EJ. Supercritical CO_2 precipitation of poly (L-lactic acid) in a wide range of miscibility. Journal of Supercritical Fluids. 2013;**81**:236-244. DOI: 10.1016/j.supflu.2013.06.008

[31] Garay I, Pocheville A, Madariaga L. Polymeric microparticles prepared by supercritical antisolvent precipitation. Powder Technology. 2010;**197**:211-217. DOI: 10.1016/j.powtec.2009.09.015

[32] Teipel U, Kröber H, Krause H. Formation of energetic materials using supercritical fluids. Propellants, Explosives, Pyrotechnics. 2001;**26**: 168-173. DOI: 10.1002/1521-4087

[33] Reverchon E, De Marco I, Della Porta G. Tailoring of nano- and micro-particles of some superconductor precursors by supercritical antisolvent precipitation. Journal of Supercritical Fluids. 2002;**23**:81-87. DOI: 10.1016/S0896-8446(01)00129-2

[34] Lam UT, Mammucari R, Suzuki K, Foster NR. Processing of iron oxide nanoparticles by supercritical fluids. Industrial and Engineering Chemistry Research. 2008;**47**:599-614. DOI: 10.1021/ie070494+

[35] Reverchon E, Adami R, De Marco I, Laudani CG, Spada A. Pigment Red 60 micronization using supercritical fluids based techniques. Journal of Supercritical Fluids. 2005;**35**:76-82. DOI: 10.1016/j.supflu.2004.10.010

[36] Werling JO, Debenedetti PG. Numerical modeling of mass transfer in the supercritical antisolvent process: miscible conditions. Journal of Supercritical Fluids. 2000;**18**:11-24. DOI: 10.1016/S0896-8446(99)00027-3

[37] Dukhin SS, Zhu C, Pfeffer R, Luo JJ, Chavez F, Shen Y. Dynamic interfacial tension near critical point of a solvent–antisolvent mixture and laminar jet stabilization. Colloids and Surfaces A. 2003;**229**:181-199. DOI: 10.1016/S0927-7757(03)00326-1

[38] Heater KJ, Tomasko DL. Processing of epoxy resins using carbon dioxide as an antisolvent. Journal of Supercritical Fluids. 1998;**14**:55-65. DOI: 10.1016/S0896-8446(98)00106-5

[39] Randolph TW, Randolph AD, Mebes M, Yeung S. Sub-micrometer sized biodegradable particles of poly (L-lactic acid) via the gas antisolvent spray precipitation process. Biotechnology Progress. 1993;**9**:429-435. DOI: 10.1021/bp00022a010

[40] Fadli T, Erriguible A, Laugier S, Subra-Paternault P. Simulation of heat and mass transfer of CO_2–solvent mixtures in miscible conditions: Isothermal and non-isothermal mixing. Journal of Supercritical Fluids. 2010;**52**:193-202. DOI: 10.1016/j.supflu.2010.01.008

[41] Kwauk X, Debenedetti PG. Mathematical modeling of aerosol formation byrapid expansion of supercritical solutions in a converging nozzle. Journal of Aerosol

Science. 1993;**24**:445-469. DOI: 10.1016/0021-8502(93)90031-4

[42] Türk M. Influence of thermodynamic behaviour and solute properties onhomogeneous nucleation in supercritical solutions. Journal of Supercritical Fluids. 2000;**18**:169-184. DOI: 10.1016/S0896-8446(00)00080-2

[43] Helfgen B, Hils P, Holzknecht C, Türk M, Schaber K. Simulation of particle formation during the rapid expansion of supercritical solutions. Journal of Aerosol Science. 2001;**32**:295-319. DOI: 10.1016/S0021-8502(00)00080-X

[44] Helfgen B, Türk M, Schaber K. Hydrodynamic, Aerosol modeling of the rapid expansion of supercritical solution (RESS-Process). Journal of Supercritical Fluids. 2003;**26**:225-242. DOI: 10.1016/S0896-8446(02)00159-6

[45] Yamamoto S, Furusawa T. Thermophysical flow simulations of rapid expansion of supercritical solutions (RESS). Journal of Supercritical Fluids. 2015;**97**:192-201. DOI: 10.1016/j.supflu.2014.11.016

[46] Perez de Diego Y, Wubbolts FE, Jansen PJ. Modelling mass transfer in the PCA process using the Maxwell–Stephan approach. Journal of Supercritical Fluids. 2006;**37**:53-62. DOI: 10.1016/j.supflu.2005.07.002

[47] Campardelli R, Reverchon E, De Marco I. PVP microparticles precipitation from acetone-ethanol mixtures using SAS process: Effect of phase behavior. The Journal of Supercritical Fluids. 2019;**143**:321-329. DOI: 10.1016/j.supflu.2018.09.010

[48] Balabel A, Hegab AM, Nasr M, El-Behery SM. Assessment of turbulence modeling for gas flow in two-dimensional convergent–divergent rocket nozzle. Applied Mathematical Modelling. 2010;**35**:3408-3422

[49] Bouchard A, Jovanovic N, de Boer AH, Martín A, Jiskoot W, Crommelin DJA, et al. Effect of the spraying conditions and nozzle design on the shape and size distribution of particles obtained with supercritical fluid drying. European Journal of Pharmaceutics and Biopharmaceutics. 2008;**70**:389-401. DOI: 10.1016/j.ejpb.2008.03.020

[50] Martín A, Cocero MJ. Numerical modeling of jet hydrodynamics, mass transfer, and crystallization kinetics in the supercritical antisolvent (SAS) process. Journal of Supercritical Fluids. 2004;**32**:203-219. DOI: 10.1016/j.supflu.2004.02.009

[51] Bałdyga J, Kubicki D, Shekunov BY, Smithd KB. Mixing effects on particle formation in supercritical fluids. Chemical Engineering Research and Design. 2010;**88**:1131-1141. DOI: 10.1016/j.cherd.2010.02.016

[52] Wena Z, Liua B, Zhenga Z, Youa X, Pua Y, Li Q. Preparation of liposomes entrapping essential oil from Atractylodes macrocephala Koidz by modified RESS technique. Chemical Engineering Research and Design. 2010;**88**:1102-1107. DOI: 10.1016/j.cherd.2010.01.020

[53] Hezave AZ, Esmaeilzadeh F. Micronization of drug particles via RESS process. Journal of Supercritical Fluids. 2010;**52**:84-98. DOI: 10.1016/j.supflu.2008.01.019

[54] Türk M, Lietzow R. Formation and stabilization of submicron particles via rapid expansion processes. Journal of Supercritical Fluids. 2008;**45**:346-355. DOI: 10.1016/j.supflu.2008.01.019

[55] Sodeifian G, AliSajadian S. Utilization of ultrasonic-assisted RESOLV (US-RESOLV) with polymeric stabilizers for production of amiodarone hydrochloride nanoparticles: Optimization of the process parameters. Chemical Engineering Research

and Design. 2019;**142**:268-284. DOI: 10.1016/j.cherd.2018.12.020

[56] Moussa AB, Ksibi H, Baccar M. Simulation of particles transport and coagulation during the RESS process. European Physical Journal Applied Physics. 2008;**43**:253-261. DOI: 10.1051/epjap:2008117

[57] Reverchon E, Pallado P. Hydrodynamic modeling of the RESS process. The Journal of Supercritical Fluids. 1996;**9**:216-221. DOI: 10.1016/S0896-8446(96)90051-0

[58] A Program Package for Thermophysical Properties of Fluids, Ver. 12.1, PROPATH group

[59] Yamamoto S, Furusawa T, Matsuzawa R. Numerical simulation of supercritical carbon dioxide flows across critical point. International Journal of Heat and Mass Transfer. 2011;**54**:774-782. DOI: 10.1016/j.ijheatmasstransfer.2010.10.030

[60] Helfgen A, Türk M, Schaber K. Hydrodynamic and aerosol modelling of the rapid expansion of supercritical solutions (RESS-process). Journal of Supercritical Fluids. 2003;**26**:225-242. DOI: 10.1016/S0896-8446(02)00159-6

Chapter 18

Bioconvective Linear Stability of Gravitactic Microorganisms

Ildebrando Pérez-Reyes and Luis Antonio Dávalos-Orozco

Abstract

Interesting results on the linear bioconvective instability of a suspension of gravitactic microorganisms have been calculated. The hydrodynamic stability is characterized by dimensionless parameters such as the bioconvection Rayleigh number R, the gyrotaxis number G, the motility of microorganisms d, and the wavenumber k of the perturbation. Analytical and numerical solutions are calculated. The analytical one is an asymptotic solution for small wavenumbers (and for any motility number) which agrees very well with the numerical solutions. Two numerical methods are used for the sake of comparison. They are a shooting method and a Galerkin method. Marginal curves of R against k for fixed values of d and G are presented along with curves corresponding to the variation of the critical values of R_c and k_c. Moreover, those critical values are compared with the experimental data reported in the literature, where the gyrotactic algae *Chlamydomonas nivalis* is the suspended microorganism. It is shown that the agreement between the present theoretical results and the experiments is very good.

Keywords: bioconvection, hydrodynamic stability, Galerkin method

1. Introduction

Since many years ago, efforts in the experimental and theoretical investigation of the bioconvection phenomenon have been made. These efforts, which lead to the understanding of bioconvective instability, have produced novel and interesting applications. For example, Noever and Matsos [1] proposed a biosensor for monitoring the heavy metal Cadmium based in bioconvective patterns as redundant technique for analysis, a number of researchers [2–6] have been working on the control of bioconvection by applying electrical fields (as in galvanotaxis) to use it as a live micromechanical system to handle small objects immersing in suspensions, Itoh et al. [7, 8] use some ideas of bioconvection in a study for the motion control of microorganism groups like *Euglena gracilis* to manipulate objects by using its phototactic orientation (as in phototaxis), and more recently possibly bioconvection seeded the investigation of Kim et al. [9, 10] for using a feedback control strategy to manipulate the motions of *Tetrahymena pyriformis* as a microbiorobot, among others. Perhaps, further applications on biomimetics [11–13] at the nano- and microscale could be driven by this contribution.

The term bioconvection was first coined by Platt [14] as the spontaneous pattern formation in suspensions of swimming microorganisms. This phenomenon has some similarity with Rayleigh-Benard convection but originates solely from diffusion and the swimming of the organisms. Reviews about this topic have been published by Pedley and Kessler [15] and Hill and Pedley [16]. Ideas and theories on cellular motility can be found in the book of Murase [17], and the effect of gravity on the behavior of microorganisms is widely explained in the book of Hader et al. [18]. In 1975, Childress et al. [19] presented a model for bioconvection of purely gravitactic microorganisms and their results of a linear theoretical study, and later Harashima et al. [20] studied the nonlinear equations of this model. According to the model of Childress et al. [19], the critical wavenumber at the onset of the instability is always zero. In ordinary particles and colloidal suspensions, the internal degrees of freedom like the internal rotation or spin are important under some geometrical and physical conditions [21, 22]. The case of a suspension of microorganisms is not an exception. For this case, Pedley et al. [23] proposed a gyrotactic model for a suspension of infinite depth. Their model includes the displacement of the gravity from the geometric center in the organisms along their axis of symmetry. Hill et al. [24] performed an analysis of the linear instability of a suspension of gyrotactic microorganisms of finite depth using the model of Pedley et al. [23]. Hill et al. [24] found finite wavenumbers at the onset of the instability, but agreement with the experiment was not good. Later, Pedley and Kessler [25] reported a model for suspensions of gyrotactic microorganisms where account was taken of randomness in the swimming direction of the cells. In a study of the linear instability of the system based on the model of Pedley and Kessler [25], Bees and Hill [26] found disagreement between their theoretical results and the experimental data reported by Bees and Hill [27]. Several experimental investigations of bioconvection have been reported by Loeffer and Mefferd [28] and Fornshell [29], by Kessler [30] and Bees and Hill [27] who take into account the gyrotaxis, by Dombrowski et al. [31] and Tuval et al. [32] who take into account the oxitaxis, and more recently by Akiyama et al. [33] who observed a pattern alteration response characterized by a rapid decrease in the bioconvective patterns. Pattern formation has been observed in cultures of different microorganisms such as *Chlamydomonas nivalis*, *Chlamydomonas reinhardtii*, *Euglena gracilis*, *Bacillus subtilis*, *Paramecium tetraurelia*, and *Tetrahymena pyriformis*.

More recently, investigations have been reported for a semi-dilute suspension of swimming microorganisms where cell–cell interactions are considered [34–38]. On the other hand, Kitsunezaki et al. [39] investigated the effect of oxygen and depth on bioconvective patterns in suspensions with high concentrations of *Paramecium tetraurelia*. Bioconvection is also studied from other points of view in gravitational biology. Interesting results are also available in Refs. [40–42] about the pattern formation in suspensions of *Tetrahymena* and *Chlamydomonas* subject to different gravity conditions. Further results are due to Sawai et al. [43] who investigate the proliferation of *Paramecium* under simulated microgravity, to Mogami et al. [44] who report an investigation of the formed patterns by *Tetrahymena* and *Chlamydomonas* as well as a physiological comparison, to Takeda et al. [45] who give an explanation of the gravitactic behavior of single cells of *Paramecium* in terms of the swimming velocity and swimming direction, to Mogami et al. [46] who present theory and experiments of two mechanisms of gravitactic behavior for microorganisms, and to Itoh et al. [47] who investigate the modification of bioconvective patterns under strong gravitational fields.

This chapter presents interesting results about the bioconvective linear stability of a suspension of swimming microorganisms. Use is made of the equations presented by Ghorai and Hill [48, 49] some years ago. In their approach, Ghorai and Hill [48, 49] used a different dimensionalization scale for the concentration microorganisms which gives distinct meaning to the basic state for the concentration of microorganisms and a bioconvective Rayleigh number defined in terms of the mean cell concentration. To the authors best knowledge, those equations along with the change in the basic state and Rayleigh number definitions have not been used to determine the linear bioconvective instability in an infinite horizontal fluid layer and to compare the results with experiment. These results were obtained by means of both numerical and analytical techniques. The critical values of the Rayleigh number R_c and the wavenumber k_c, for fixed values of the gyrotaxis number G and the motility of microorganisms d, that characterize the hydrodynamic stability of the system are compared with the experimental data presented in Table I of Bees and Hill [27] and Table II of Bees and Hill [26] where the gyrotactic biflagellate alga *Chlamydomonas nivalis* is used as suspended microorganism. Below, it is shown for the first time that the numerical results have a very good agreement with the experimental data.

The chapter is organized as follows. The governing equations and boundary conditions [48, 49] as well as the basic state can be found in Section 2. Nondimensionalization and linearization of the system of equations is outlined in Section 3. In Section 4, use is made of an asymptotic expansion [50–53] method and a Galerkin method [54] to find limiting cases and predict critical values of R and k for the instability. The numerical calculations done by means of the shooting method along with the graphics corresponding to the marginal curves are given in Section 5. In Section 6, the experimental data [27] are compared with the numerical results. A discussion is given in the final section.

2. Equations of motion

We consider an infinite horizontal layer of a suspension of gyrotactic microorganisms. The fluid layer is bounded at $z^* = -H, 0$. The fluid where the cellular microorganisms swim is water with density ρ. Each cell has a volume υ and density $\rho + \Delta\rho$, where $\Delta\rho \ll \rho$. The suspension is considered dilute and incompressible. Density fluctuations in the suspension are small enough such that the Boussinesq approximation is valid and the corresponding governing equations are

$$\rho \frac{D\mathbf{u}^*}{Dt^*} = -\nabla p^* - n^* \upsilon g \Delta\rho \mathbf{k} + \mu \nabla^2 \mathbf{u}^* \tag{1}$$

$$\frac{\partial n^*}{\partial t^*} = -\nabla \cdot \mathbf{J}^* \tag{2}$$

$$\nabla \cdot \mathbf{u}^* = 0 \tag{3}$$

where t^* is the time, $\mathbf{u}^*$ is the suspension velocity, p^* is the pressure, $g\mathbf{k}$ is the acceleration due to gravity, $\mathbf{k}$ is the vertical unit vector, μ is the viscosity, n^* is the concentration of microorganisms, and $\mathbf{J}^*$ is the flux density of organisms through the fluid defined as

$$\mathbf{J}^* = n^* \mathbf{u}^* + n^* V_c \mathbf{p}^* - D_c \nabla n^* \tag{4}$$

where V_c is the cell swimming speed, $\mathbf{p}^*$ is a unit vector representing the average orientation of cells, and D_c is a scalar microorganism mass diffusion coefficient independent of the other parameters of the problem. Use is made of Cartesian coordinates with the z-axis in the vertical direction. The walls at $z^* = -H, 0$ are considered to be rigid. As pointed out by Hill et al. [24] although the top boundary is open to the air, algal cells tend to collect at the surface forming what appears to be a packed layer, and it is unlikely that the boundary is ever fully stress-free. Then the boundary conditions are

$$\mathbf{u}^* = 0 \, \text{at} \, z^* = -H, 0 \tag{5}$$

$$\mathbf{J}^* \cdot \mathbf{k} = 0 \, \text{at} \, z^* = -H, 0 \tag{6}$$

In the basic state, the fluid velocity is zero and $n^* = n_0^*(z)$ and $\mathbf{p}_0 = \mathbf{k}$. Thus for $n_0^*(z)$ from Eq. (2) with the boundary conditions (6), we have

$$n_0^*(z) = \frac{\overline{n} V_c H \exp\left[V_c z^* / D_c\right]}{D_c(1 - \exp\left[-V_c H / D_c\right])} \tag{7}$$

where $\overline{n}$ represents the average concentration of organisms. Eq. (7) is the same basic state as presented by Ghorai and Hill [48, 49] whose linear stability will be investigated. It differs from that of Childress et al. [19] and Hill et al. [24] by the coefficient

$$\frac{V_c H}{D_c(1 - \exp\left[-V_c H / D_c\right])}$$

3. Linear stability

We make the governing Eqs. (1–3) nondimensional by scaling all lengths with H, the time with H^2/D_c, the fluid velocity with D_c/H, the pressure with $\nu D_c \rho / H^2$, and the cell concentration with $\overline{n} H V_c / D_c$. Now the dimensionless variables are expressed without star. The boundaries are located at $z = -1, 0$ and the basic state is

$$\mathbf{u}_0 = 0, \;\; \mathbf{p}_0 = \mathbf{k}, \;\; n_0(z) = \frac{e^{dz}}{1 - e^{-d}}$$

where the nondimensional quantity $d = V_c H / D_c$ is the ratio of swimming speed of microorganisms and their representative mass diffusion velocity. Here, d is called the motility of the microorganisms. In order to investigate the linear stability of the system, small perturbations have to be considered. They are

$$\mathbf{u} = \mathbf{u}_0 + \delta \mathbf{u}_1, \;\; p = p_0 + \delta \mathbf{p}_1, \;\; \mathbf{p} = \mathbf{p}_0 + \delta \mathbf{p}_1, \;\; n = n_0 + \delta n_1$$

where $\delta \ll 1$. The components of $\mathbf{u}^1$ are (u_1, v_1, w_1). In this way, the nondimensional governing Eqs. (1–3) are linearized to order $O(\delta)$. Then, we have the following linear equations

$$Sc^{-1} \frac{\partial \mathbf{u}_1}{\partial t} = -\nabla p - R n_1 \mathbf{k} + \nabla^2 \mathbf{u}_1 \tag{8}$$

$$\frac{\partial n_1}{\partial t} = -w_1 \frac{d n_0}{dz} - d \frac{\partial n_1}{\partial z} - d n_0 \nabla \cdot \mathbf{p}_1 + \nabla^2 n_1 \tag{9}$$

$$\nabla \cdot \mathbf{u}_1 = 0 \tag{10}$$

where

$$S_c = \frac{\nu}{D_c}, \;\; R = \frac{\overline{n} \upsilon g \Delta \rho H^3 d}{D_c \nu \rho}$$

are the Schmidt and bioconvection Rayleigh numbers, respectively. Pedley and Kessler [55] give a definition of the vector $\mathbf{p}_1$ for swimming microorganisms with spheroidal shape. They determine $\mathbf{p}_1$ in terms of $\mathbf{u}_1$ that in nondimensional form is

$$\mathbf{p}_1 = G\left[(1+\alpha_0)\frac{\partial \mathbf{u}_{1\perp}}{\partial z} - (1-\alpha_0)\nabla_\perp w_1\right] \tag{11}$$

$$G = \frac{BD_c}{H^2} \tag{12}$$

where the subscript $\perp$ denotes the horizontal component, α_0 is the cell eccentricity, and G is the nondimensional form of the gyrotactic orientation parameter B. Finally after substitution of $\mathbf{p}_1$ and n_0, the governing equations become

$$\mathrm{Sc}^{-1}\frac{\partial \mathbf{u}}{\partial t} = -\nabla p - Rn\mathbf{k} + \nabla^2\mathbf{u} \tag{13}$$

$$\frac{\partial n}{\partial t} = \frac{de^{dz}}{1-e^{-d}}\left(-w + G\left[(1+\alpha_0)\frac{\partial^2 w}{\partial z^2} + (1-\alpha_0)\nabla_\perp^2 w\right]\right) - d\frac{\partial n}{\partial z} + \nabla^2 n \tag{14}$$

$$\nabla\cdot\mathbf{u} = 0 \tag{15}$$

with boundary conditions

$$\mathbf{u} = 0\,\mathrm{at}\,z = -1, 0 \quad \text{and} \quad \frac{\partial n}{\partial z} - dn = 0\,\mathrm{at}\,z = -1, 0 \tag{16}$$

where the superscripts have been deleted. Notice that the adimensionalization of the equations is different from that of Hill et al. [24]. Here, an application of a more general asymptotic analysis for any magnitude of d is used. An analytic Galerkin method and a shooting numerical method for the solution of the proper value problem allowed us to have an interesting perspective of the stability of the present problem under research. The results are used here to compare with the experimental data of the flagellated alga *Chlamydomonas nivalis*.

By elimination of the pressure from Eqs. (13–15), it is possible to obtain a coupled system of two equations, for w and n, to describe the instability of the system. The perturbations of the variables will be analyzed in terms of normal modes of the form

$$w = W(z)\exp\left[(k_x x + k_y y)i + \sigma t\right],$$
$$n = \Phi(z)\exp\left[(k_x x + k_y y)i + \sigma t\right]$$

where $k = \sqrt{k_x^2 + k_y^2}$ is the wavenumber of the disturbance and σ is the growth rate. The wavenumber is scaled as $k = k^*H$ corresponding to a nondimensional wavelength $\lambda = 2\pi/k$. Thus, the governing equations become

$$-Rk^2\Phi = \left(\frac{\sigma}{S_c} + k^2 - D^2\right)(k^2 - D^2)W \tag{17}$$

$$(\sigma + dD + k^2 - D^2)\Phi = \frac{de^{dz}}{1-e^{-d}}\left(-1 + G\left[(1+\alpha_0)D^2 - (1-\alpha_0)k^2\right]\right)W \tag{18}$$

subject to the boundary conditions

$$W = DW = D\Phi - d\Phi = 0 \ \text{ at } \ z = -1, 0 \tag{19}$$

where $D = d/dz$. The variables of the above problem can be changed in order to simplify the analysis. The change of dependent variable is

$$\Phi = F(z)e^{dz}$$

Then, Eqs. (17) and (18) and the boundary conditions Eq. (19) become

$$-Rk^2Fe^{dz} = \left(\frac{\sigma}{S_c} + k^2 - D^2\right)(k^2 - D^2)W \tag{20}$$

$$(\sigma - dD + k^2 - D^2)F = \frac{d}{1 - e^{-d}}\left(-1 + G\left[(1 + \alpha_0)D^2 - (1 - \alpha_0)k^2\right]\right)W \tag{21}$$

subject to the new boundary conditions

$$W = DW = DF = 0 \text{ at } z = -1, 0 \tag{22}$$

In this form, the equations are very similar to those of the well-known problem of thermal convection in an infinite horizontal fluid layer between nonconducting boundaries [50–53, 56]. The familiar fixed heat flux boundary condition is the main characteristic of those thermal convection problems and is analogous to that presented in Eq. (22). The equations derived by Childress et al. [19] can also be analyzed from the present view point of this change of variable. In the theory of thermal convection as in that of Childress et al. [19], a zero critical wavenumber is found as a result of the fixed flux boundary condition. In more recent models, which include the effects of gyrotaxis, the similarity with the thermal convection problem is not valid unless $G = 0$.

4. Asymptotic analysis

In this section, the eigenvalue problem stated in the system of Eqs. (13–15) with boundary condition Eq. (16) is investigated by means of two analytic methods. The magnitude of the marginal value of R is a function of all the other parameters. The way in which the solution of the stability problem is to be carried out is as follows. For a given value of d and G, we must determine the lowest value for R with respect to the wavenumber k. The values obtained are the critical Rayleigh numbers R_c at which instability will first occur.

4.1 Asymptotic analysis

We conducted a general asymptotic analysis in comparison with those used before [19, 24, 26] which included the restrictions of the limits $d \ll 1$ for shallow layers and $d \gg 1$ for deep layers along with different restrictions for G. In a similar way, as in other problems of convection, we follow the steps of Chapman and Proctor [51], Dávalos-Orozco [52], and Dávalos-Orozco and Manero [53]. Under the above conditions, the analysis is very complex, the reason why use has been made of the Maple algebra package. Thus, we look for a solution to Eqs. (20) and (21) using the following expansions:

$$W = W_0 + \varepsilon W_1 + \ldots, \tag{23}$$

$$\Phi = \Phi_0 + \varepsilon \Phi_1 + \ldots, \tag{24}$$

$$R = R_0 + \varepsilon R_1 + \ldots, \tag{25}$$

$$\sigma = \varepsilon \sigma_0 + \varepsilon^2 \sigma_1 + \ldots \tag{26}$$

where $\varepsilon \ll 1$. We also consider no restrictions for d and G and rescale the wavenumber as $k = \varepsilon^{1/2}\tilde{k}$. Thus, after substitution of expansions Eqs. (23–26) and the mentioned scalings in Eqs. (20) and (21) with boundary condition Eq. (22), we obtain the following systems of equations at different orders.

At order $O(1)$

$$D^4W_0 + \tilde{k}^2 R_0 e^{dz} F_0 = 0, \tag{27}$$

$$(D^2 + d)F_0 = 0 \tag{28}$$

subject to

$$W_0 = DW_0 = DF_0 = 0 \text{ at } z = -1, 0. \tag{29}$$

At order $O(\varepsilon)$

$$D^4W_1 - \left(2\tilde{k}^2 + \frac{\sigma_0}{Sc}\right)D^2W_0 + \tilde{k}^2 e^{dz}(R_0F_1 + R_1F_0) = 0, \tag{30}$$

$$\frac{d}{1-e^{-d}}\left[G(1+\alpha_0)D^2 - 1\right]W_0 - \left(\sigma_0 + \tilde{k}^2\right)F_0 + (D^2 + d)F_1 = 0 \tag{31}$$

subject to

$$W_1 = DW_1 = DF_1 = 0 \text{ at } z = -1, 0. \tag{32}$$

At order $O(\varepsilon^2)$

$$D^4W_2 - \left(2\tilde{k}^2 + \frac{\sigma_0}{Sc}\right)D^2W_1 + \left(\tilde{k}^4 + \frac{\sigma_0\tilde{k}^2}{Sc} - \frac{\sigma_1}{Sc}D^2\right)W_0 + \tilde{k}^2 e^{dz}(R_0F_2 + R_1F_1 + R_2F_0) = 0, \tag{33}$$

$$\frac{d}{1-e^{-d}}\left\{\left[G(1+\alpha_0)D^2 - 1\right]W_1 - G(1+\alpha_0)\tilde{k}^2 W_0\right\} - \left(\sigma_0 + \tilde{k}^2\right)F_1 - \sigma_1F_0 + (D^2 + d)F_2 = 0 \tag{34}$$

subject to

$$W_2 = DW_2 = DF_2 = 0 \text{ at } z = -1, 0. \tag{35}$$

The systems of equations at order $O(\varepsilon)$ and higher are inhomogeneous and must satisfy their corresponding solvability conditions allowing to compute the Rayleigh number R as an eigenvalue in terms of the other parameters of the problem. Solvability conditions are found as usual [57]: each inhomogeneous system is multiplied by the solution to the adjoint of the homogeneous system and integrated over the range of the independent variable. The resulting integral must vanish.

Thus, the solvability conditions at $O(\varepsilon)$ and $O(\varepsilon^2)$ are, respectively,

$$0 = \frac{d}{1-e^{-d}}\int_{-1}^{0} e^{dz}\left[G(1+\alpha_0)D^2 - 1\right]W_0 dz - \left(\sigma_0 + \tilde{k}^2\right)\int_{-1}^{0} e^{dz}F_0 dz \tag{36}$$

$$\begin{aligned} 0 = &\frac{d}{1-e^{-d}}\int_{-1}^{0} e^{dz}\left\{\left[G(1+\alpha_0)D^2 - 1\right]W_1, -G(1+\alpha_0)\tilde{k}^2 W_0\right\}dz \\ &-\left(\sigma_0 + \tilde{k}^2\right)\int_{-1}^{0} e^{dz}F_1 dz - \sigma_1\int_{-1}^{0} e^{dz}F_0 dz \end{aligned} \tag{37}$$

The solutions of the system of equations at leading order are

$$F_0 = 1, \quad W_0 = f_1(z,d)R_0\tilde{k}^2 \tag{38}$$

where the function $f_1(z,d)$ can be obtained from the authors upon request. For convenience, the solution F_0 has been normalized to 1. The next step is to evaluate the solvability condition Eq. (36) at $O(\varepsilon)$ and obtain an expression for σ_0

$$\sigma_0 = \tilde{k}^2\left\{\left[1 - d^2G(1+\alpha_0)\right]A_0R_0 - 1\right\} \tag{39}$$

The constant A_0 is large and can be obtained from the authors upon request. At order $O(\varepsilon)$ similar steps as those for solving the system of equations at $O(1)$ are followed to find F_1 and W_1. Then, algebraically F_1 is

$$F_1 = \left[G(1+\alpha_0)f_2(z,d) + f_3(z,d)\right]R_0\tilde{k}^2 \tag{40}$$

After substitution of F_0, W_0, F_1, and σ_0 into Eq. (30), the velocity W_1 can be calculated subject to its corresponding boundary condition Eq. (32). Because of the term e^{dz} appearing in the system of equations at $O(1)$, the expression of W_1 is very large and complicated and will not be given here. The evaluation of the solvability condition at order $O(\varepsilon^2)$ given in Eq. (37) yields

$$\begin{aligned}\sigma_1 = {} & \left[1 - d^2G(1+\alpha_0)\right]\tilde{k}^2A_0R_1 - \left\{\left[G^2(1+\alpha_0)^2A_1 + G(1+\alpha_0)A_2 + A_3\right]R_0^2\right. \\ & + G(1-\alpha_0)A_4R_0 + R_0\left[1 - d^2G(1+\alpha_0)\right]\left\{\left[G(1+\alpha_0)A_5 + A_6\right]R_0\right. \\ & \left. + A_7 + \left(\left[1 - d^2G(1+\alpha_0)\right]A_8R_0 + A_7/2\right)/Sc\right\}\tilde{k}^4\end{aligned} \tag{41}$$

The growth rate can now be obtained by substitution of σ_0 and σ_1 into the expansion for σ given in Eq. (26). However σ is omitted to save space but can be obtained from the authors upon request. Finally, use is made of the expansion Eq. (25) for R.

Now, the transition from stability to instability via a stationary state is investigated by setting $\sigma = 0$. Thus, the corresponding value of R for the marginal state is

$$\begin{aligned}R = {} & \frac{1}{\left[1 - d^2G(1+\alpha_0)\right]A_0} + \frac{k^2}{\left[1 - d^2G(1+\alpha_0)\right]A_0} \\ & \left\{\frac{G(1-\alpha_0)}{\left[1 - d^2G(1+\alpha_0)\right]} + \frac{G(1+\alpha_0)A_5 + A_6}{\left[1 - d^2G(1+\alpha_0)\right]A_0^2} + \frac{G^2(1+\alpha_0)^2A_1 + G(1+\alpha_0)A_2 + A_3}{\left[1 - d^2G(1+\alpha_0)\right]^2A_0^2} + \frac{A_7}{A_0}\right\}\end{aligned} \tag{42}$$

where some simplifications have been made with the use of R_0 obtained from Eq. (39). The functions $f_2(z,d)$ and $f_3(z,d)$ and the coefficients A_1 to A_8 appearing in the above expressions are functions of d and can be obtained from the authors upon request. The result for R_0 is

$$R_0 = \frac{1}{\left[1 - d^2G(1+\alpha_0)\right]A_0} \tag{43}$$

From the expression for the Rayleigh number given in Eq. (42), it is possible to calculate the limit for $d \ll 1$. In this case, we consider d and k to be of the same order in such a way that $k_d = k/d$ and $k_d \sim O(1)$. Then, under these assumptions, the approximation of $R(k,d,G,\alpha_0)$ is

$$R = 720\left\{1 + d^2\left[\frac{17}{420} + G(1+\alpha_0) + k_d^2\left(\frac{17}{462} - \frac{2}{7}G(5-2\alpha_0)\right)\right]\right\} + O(d^3) \quad (44)$$

Here we point out that in the present chapter, our definition of the Rayleigh number differs from that defined by Hill et al. [24]. If our approximation given in Eq. (44) is multiplied by $\frac{d}{1-e^{-d}} = (1 + d/2 + d^2/12 - d^4/720...)$, the same approximation by Hill et al. [24] is obtained. Moreover, if $G = 0$ this becomes that given by Childress et al. [19]. In the more general expression of R for a small wavenumber approximation, Eq. (42) has a special characteristic due to its dependence on the square of the wavenumber k. The first coefficient at zeroth order in k corresponds to R_0, and that at the second order in k is R_1. Even though in the experiments on bioconvection [27], only finite critical wavenumbers $k_c > 0$ have found these coefficients are very useful. For example, it can be shown from R_0 that if $A_0 > 0$ and

$$1 - d^2G(1+\alpha_0) > 0 \quad (45)$$

then $R_0 < 0$. This corresponds to a stable stratification, which is not the case here. The second coefficient R_1 in Eq. (42) is a very important result, because it provides information about the shape of the marginal curve with respect the critical wavenumber. That information can be obtained by making zero the coefficient R_1 and calculating the following critical value of the gyrotaxis number G_c

$$\begin{aligned} 0 = G^2(1+\alpha_0)^2\left[A_1 - d^2A_5 - d^2A_0^2\frac{1-\alpha_0}{1+\alpha_0}\right] + G(1+\alpha_0) \\ \left[A_2 + A_5 - d^2(A_6 + A_7A_0) + A_0^2\frac{1-\alpha_0}{1+\alpha_0}\right] + A_3 + A_6 + A_7A_0 \end{aligned} \quad (46)$$

From this equation, two admissible cases are possible when Eq. (45) is satisfied. First, for fixed values of d, α_0, and $G > G_c$, the marginal curve starts at $k = 0$ and then decreases monotonically. However, according to the numerical analysis presented below, the marginal curves in fact first decrease and then start to grow monotonically after a minimum is attained, at the critical wavenumber. In the second case, for fixed values of d, α_0, and $G < G_c$, the marginal curves start at $k = 0$ and then grow monotonically. Here, the critical wavenumber is always zero. The importance of these results is that the magnitude of G_c agrees very well with the results of the marginal curves found in the numerical analysis given below. This critical value determines the magnitude for which the curves have a finite critical wave number ($G > G_c$) or a zero critical wavenumber ($G < G_c$). It is of interest to note that some of the magnitudes of the gyrotaxis parameter G calculated from the data in the literature are very near but above of G_c. This is the reason why some of the curves found from numerical analysis are almost flat in a range of wavenumbers near to zero. Because the experimental critical wavenumbers found for gyrotactic bioconvection are always finite, we conclude that G_c is important to point out where to find the theoretical limitations of the model.

4.2 Analytic Galerkin method

Here use is made of the analytical Galerkin method to study the eigenvalue problem of Eqs. (17)–(18) with the boundary condition Eq. (19). This method has been used before by Pellew and Soutwell [58], Chandrasekhar [54], and Gershuni and Zhukovitskii [59]. Even though this is an approximate method, it has a very high precision. The advantage of the method is that it is possible to obtain an explicit expression of the Rayleigh number R. Here, it is supposed that $\sigma = 0$.

Briefly, the method consists in assuming a trial function which satisfies the boundary conditions for each of the dependent variables. Let that variable be Φ which after substitution in one of the equations of the problem allows for the exact solution of the other variables, let us say W. Both trial functions are now substituted into the other coupled equation. Then, use is made of the orthogonality properties of the solutions in this equation to obtain the proper value of the Rayleigh number as a function of the other parameters [60].

In this way, the proposed expansions of Φ and W are

$$\Phi = \sum_{n=0}^{\infty} B_n \exp(dz) \cos \pi nz \quad \text{and} \quad W = \sum_{n=0}^{\infty} B_n W_n \tag{47}$$

then, after substitution of Φ of Eq. (47) into Eq. (20), W_n is the solution of the following differential equation:

$$\left(D^2 - k^2\right)^2 W_n = -Rk^2 \exp(dz) \cos \pi nz \tag{48}$$

which is subjected to the conditions

$$W_n = DW_n = 0 \text{ at } z = -1.0. \tag{49}$$

The solution is

$$W_n = (a_1 \cos \pi nz + a_2 \sin \pi nz)e^{dz}k^2R + c_1e^{kz} + c_2e^{-kz} + c_3ze^{kz} + c_4ze^{-kz} \tag{50}$$

where c_1 to c_4 are constants of integration which can be found by evaluating in the boundary condition Eq. (49).

Next, Eq. (18) is multiplied by Φ and is integrated in the range $z = -1$ to $z = 0$, to get

$$\begin{aligned} &\frac{d}{1-e^{-d}}\int_{-1}^{0} \Phi\left(-1 + G\left[(1+\alpha_0)D^2 - (1-\alpha_0)k^2\right]\right)Wdz \\ &-\int_{-1}^{0} \Phi\left[D\left(e^{-dz}D\right) - k^2e^{-dz}\right]\Phi dz = 0 \end{aligned} \tag{51}$$

After substitution of Φ and W, given in Eqs. (47) and (50) and some simplifications, we obtain

$$\left| \begin{aligned} &\frac{d}{1-e^{-d}}\int_{-1}^{0} \Phi_m\left(-1 + G\left[(1+\alpha_0)D^2 - (1-\alpha_0)k^2\right]\right)W_n dz \\ &-\int_{-1}^{0} \Phi_m\left[D\left(e^{-dz}D\right) - k^2e^{-dz}\right]\Phi_n dz \end{aligned} \right| = 0 \tag{52}$$

This determinant, calculated with the help of the software Maple, is the solvability condition from which the eigenvalue R is calculated. The resulting algebraic expression of the integrals in this equation is very complex and will not be presented here. However, the first approximation of R, corresponding to the element (0,0) of the matrix in the determinant Eq. (52), is

$$R = \frac{\left(1-e^{-d}\right)A_{10}}{de^{-d}\left[1 + G(1-\alpha_0)k^2 - d^2G(1+\alpha_0)\right]A_9} \tag{53}$$

where A_9 and A_{10} are functions of the wavenumber k and d and can be obtained from authors upon request. This result is new because it includes, for the first time, all the parameters of the problem without any approximation. In the limit of $d, k, G \to 0$, R reduces to the well-known value of 720. Higher-order estimates of R can be obtained from Eq. (52), which provides a useful check on numerical calculations. The comparison of R given in Eq. (42), obtained from the asymptotic analysis, and that of Eq. (52) shows that in the limit $k \to 0$, the agreement was very good.

5. Numerical computations by a shooting method

Here, the shooting method [61] is used to solve the eigenvalue problem posed by the system of Eqs. (20) and (21) subjected to the boundary condition Eq. (22). Curves of marginal stability in the plane (k, R) were calculated for fixed values of the parameters d, G, and α_0. Notice that very good agreement was always found among the values of the R of the asymptotic analysis (in the limit $k \to 0$), of the Galerkin method, and of the numerical computations. Calculations were made in two ways. First, the parameters d, G, and α_0 were varied in order to obtain a representative set of marginal curves for the problem of bioconvection. Second, experimental data were also used to fix the values of d, G, and α_0 and used to find theoretical values of k_c and R_c that could be compared with their corresponding experimental values. Here, in particular, a selection is made of $\alpha_0 = 0.4$, which corresponds to the flagellated alga *Chlamydomonas nivalis*. **Figures 1–3** show marginal curves for different values of the gyrotaxis parameter G, while d remains fixed with magnitudes 0.1, 1, and 5, respectively. These figures clearly show the effect the gyrotaxis parameter G has on the critical wavenumber. When the magnitude of G is large enough, the critical wavenumber changes from zero to a finite value which increases with G, as shown by the squares located at the minimum value of R. Notice that it is found that the critical value G_c, which represents the magnitude at which the properties of the marginal curves change, from having $k_c = 0$ to $k_c > 0$, is very well approximated by Eq. (46). This critical value is important because it represents the magnitude of G below which the present theory ceases to predict the experimental results which always show critical wavenumbers $k_c > 0$.

In the curves shown in **Figure 1a–b**, the critical values of the gyrotaxis parameter are $G_c = 0.0306, 0.0266, 0.0060$, respectively. As mentioned above for $G > G_c$, the critical wavenumber is finite, and for $G < G_c$ the critical wavenumber is always zero. The combined effects of the velocity of the swimming of microorganisms, d,

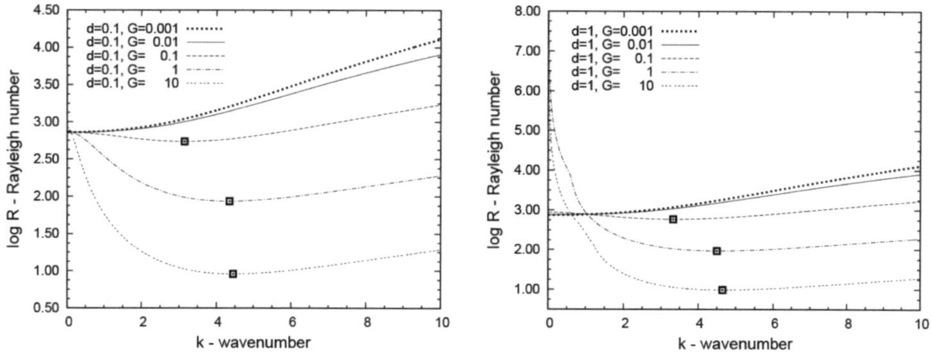

(a) (b)

Figure 1.
(a) Graphs of log R vs. k for fixed d = 0.1. (b) Graphs of log R vs. k for fixed d = 1. The black square markers indicate the position of the critical wavenumber and Rayleigh number.

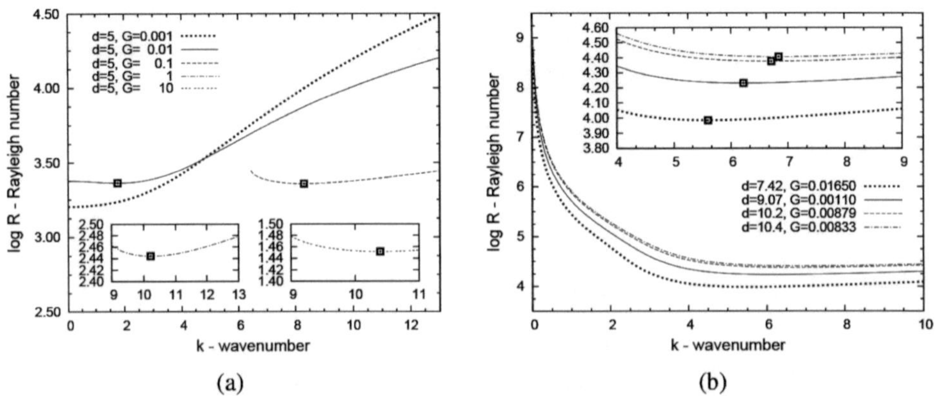

Figure 2.
(a) Graphs of log R vs. k for fixed d = 5. (b) Graphs of log R vs. k for experiments 35, 2, 4, and 9 with d increasing from the curve below to above. The black square markers indicate the position of the critical wavenumber and Rayleigh number.

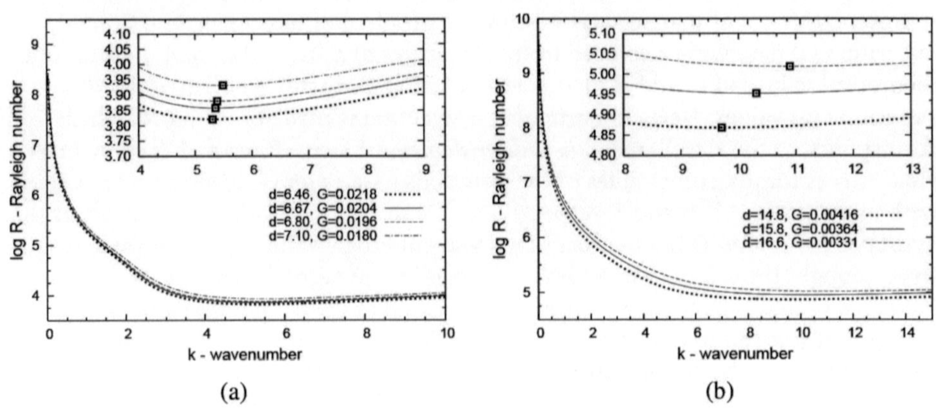

Figure 3.
(a) Graphs of log R vs. k for experiments 26–29, 24, 31, and 16 with d increasing from the curve below to above. (b) Graphs of log R vs. k for experiments 13, 10, and 20 with d increasing from the curve below to above. The black square markers indicate the position of the critical wavenumber and Rayleigh number.

and that of gyrotaxis, G, change the location of the critical wavenumber. Note also that for fixed d, when G increases, the system becomes more unstable. From the **Figure 1a** and **b**, it can be seen that the most unstable case corresponds to that for $d = 0.1$ and $G = 10$ where $k_c = 4.45$ and $R_c = 9.0618$. This may be understood by the fact that the accumulation of microorganisms near to the top of a shallow layer is faster than in a deeper one. This is due to the important role that the mass diffusion of microorganism D_c and the depth of suspension H play on the instability of the system. The value of G_c in the limit of $d, k \to 0$ can also be calculated from Eq. (46) by means of an asymptotic analysis. That is,

$$G_c = \frac{17}{132(5 - 2\alpha_0)} + O(d^2) \tag{54}$$

Here, some theoretical curves are presented of which some have a very good agreement and others a reasonable agreement with the experiments 2, 4, 9, 10, 13, 16, 20, 24, 26, 27, 28, 29, 31, and 35, performed by Bees and Hill [27].

The values for the motility d and the gyrotaxis parameter G used in **Figures 2** and **3** were calculated based on experimental data by Bees and Hill [26, 27], which

	Experimental results			Theoretical predictions					Error (%)	
EN	d_{BH}	k_c	R_{BHc} $(\times 10^6)$	d_{BH}	$d_{BH^2\eta}$	α_0	k_c	R_{BHc} $(\times 10^6)$	k	R_{BH}
2	44.7	5.11	3.25	40	16	0.2	51	5.0	898	53
				40	16	0.4	51	9.0	898	176
23	204	7.84	863	200	32	0.2	270	1700	3343	96

EN represents the experiment name. Subscript BH indicates that the definition of Bees and Hill, [26] for the parameters d and R is used. η is the gyrotactic parameter [26].

Table 1.
Experimental measurements of Bees and Hill [27] and their theoretical prediction [26].

in here are presented in **Tables 2** and **3** of the following section. In order to observe in detail the position of the critical point (k_c, R_c) in **Figures 2** and **3**, a local magnification is included.

Here, a comparison is done of our theoretical results of (k_c, R_c) with the theoretical ones presented by Bees and Hill [26] in their Table VI. According to Bees and Hill [26], experiments 2 and 23 in their Table V have (k_c, R_c) of comparable order with those in their Table VI. In our **Table 1**, we reproduce the comparison made by Bees and Hill [26] of their own theoretical and experimental results of their Table V, and we added the corresponding error in percent of the wavenumbers and Rayleigh numbers, respectively. Note that the value $\alpha_0 = 0.4$ corresponds to flagellated microorganisms such as *Chlamydomonas nivalis*, while $\alpha_0 = 0.2$ corresponds to nonflagellated. Notice that their experimental and theoretical values of d are not exactly the same.

For the sake of comparison of our theoretical results with those of the experiments, **Table 1** shows the percent of error calculated by taking the difference of the experimental and theoretical values and then dividing by the smallest one. In **Table 1**, the more realistic value $\alpha_0 = 0.4$ for *Chlamydomonas nivalis* is included, which corresponds to the second line of experiment 2 of Bees and Hill [26] predictions. It is clear from **Table 3** that our theoretical results show a very important improvement in the reduction of the percent error with respect to experiment 2.

6. Comparison with experiments

In this section a comparison is done of our theoretical results of R_c and k_c with the corresponding experimental values obtained by Bees and Hill [27]. Here use is made of the results of the 39 experiments shown in Table I of Bees and Hill [27]. Besides, the more realistic value of the parameter $\alpha_0 = 0.4$, corresponding to the flagellated algae *Chlamydomonas nivalis*, is also used to calculate d, G, and R.

In **Table 3**, the values of d, G, and R resulting from the experimental data are presented. Note in **Table 2** that the experimental results of the cell swimming speed V_s and of the cell diffusivity D_c are given inside a range of values. In this case, a particular value inside the range has to be selected. The swimming speed used here is $63{\times}10^{-4} cm/s$. The decision is based on the results obtained by Hill and Hader [62], Pedley and Kessler [25], and Bees and Hill [26]. The value of the cell diffusivity was decided to be that corresponding to an average over the range given in **Table 2**, that is, $D_c = 27.5 \times 10^{-5} cm^2/s$.

Very recent experimental measurements on the diffusivity for different microorganisms like the biflagellated alga *Chlamydomonas reinhardtii* have been reported by Polin et al. [63]. Bees and Hill [27] state that there is some evidence to suggest that cells of *Chlamydomonas nivalis* are not gyrotactic during the first week of subculturing; then if it is not the case for the cells of *Chlamydomonas reinhardtii*, more measurements for the parameters α_0, B, V_s, H, $\overline{n}$, and k_c would be needed in order to perform comparison between theoretical and experimental results. The definitions of d, G, and R are related with those of Bees and Hill [26] d_{BH}, η, and R_{BH}, respectively, as follows:

$$d = \frac{V_s^2 \tau K_2}{D_c K_1} d_{BH}, \quad G = \frac{D_c}{V_s^2 \tau} \eta, \quad R = \frac{K_2^2 \tau^3 V_s^5}{D_c^2 H K_1^2} \left(1 - \exp\left[-\frac{K_1 H}{K_2 V_s \tau}\right]\right) R_{BH} \tag{55}$$

where the constants $K_2 = 0.15$ and $K_1 = 0.57$ (see [26] for more details). τ is a direction correlation time which equals 1.3*s* in the nonflagellated case and 5*s* in the flagellated case. The data corresponding to the suspension depth H and the average cell concentration of microorganisms $\overline{n}$ of each experiment (see [27] for more details) have not been reported in **Table 3**. Only the parameters d, G, k, and R are presented in that table. It is also found that the value of the G of each experiment is greater (but sometimes near) than their corresponding critical value G_c of Eq. (46). Under these conditions, all the critical wavenumbers have to be $k_c > 0$.

By using the data of our **Table 2** and Table I of Bees and Hill [27], the experimental values for d, G, and R_E were calculated and listed in **Table 3**. The experimental value of the wavenumber k_E was also obtained from Table I of Bees and Hill [27] and was calculated as follows: the wavelength $\lambda_0(cm)$ is nondimensionalized with the corresponding suspension depth $H(cm)$ to get λ_E, and then the critical wavenumbers were calculated from $k_E = 2\pi/\lambda_E$. R_T and k_T are our theoretical wavenumber and Rayleigh number obtained by the shooting method. The curves of marginal stability corresponding to experimental results with good and very good agreement with theory are shown in **Figures 2** and **3**. As explained above, we have a substantial improvement in the agreement of the critical wavenumbers and Rayleigh numbers with respect to the experimental results (see **Table 3**). A great number of experimental data have been compared with the present theory in **Table 3**.

Name	Description	Value
ϑ	Cell volume	5×10^{-10} cm^3
g	Acceleration due to gravity	10^3 cms^{-2}
D_c	Cell diffusivity	5×10^{-5}–5×10^{-4} cm^{2}s^{-1}
ρ	Fluid density	1 gcm^{-3}
$\rho + \Delta\rho$	Cell density	1.05 gcm^{-3}
ν	Kinematic viscosity	10^{-2} cms^{-2}
V_c	Cell swimming speed	0–2×10^{-2} cms^{-1}
B	Dimensional gyrotaxis parameter	3.4 s
B	Including flagella	6.3 s
α_0	Cell eccentricity	0.20–0.31
α_0	Including flagella	0.4

Table 2.
Estimates and measurements of typical parameters for a suspension of the alga Chlamydomonas nivalis *[24, 25, 64, 65].*

NE	d	$G \times 10^{-2}$	RE	RT	kE	kT	Error k (%)	Error R (%)
1	7.63	1.56	7043.64	10384.23	5.67	5.65	0.353	47.6
2	9.07	1.10	10599.14	17075.28	5.12	6.22	21.5	61.1
3	8.36	1.30	23319.62	13531.06	8.59	5.95	44.4	72.3
4	10.2	0.879	24835.78	23783.25	5.96	6.71	12.6	4.42
5	11.9	0.636	12433.03	37984.11	6.81	7.69	12.9	205
6	16.7	0.326	59993.82	105787.08	6.61	10.9	64.9	76.3
7	9.14	1.09	4676.34	17544.33	6.01	6.31	4.99	275
8	8.73	1.19	7198.47	15309.02	7.25	6.09	19.0	112
9	10.4	0.833	20618.78	25437.60	8.33	6.84	21.8	23.4
10	15.8	0.364	88709.37	89546.43	8.34	10.3	23.5	0.943
11	6.46	2.18	3700.78	6613.53	5.24	5.31	1.33	78.7
12	12.1	0.621	39993.68	40261.72	5.66	7.94	40.3	0.670
13	14.8	0.416	77622.32	73779.00	7.85	9.71	23.7	5.20
14	8.80	1.17	8561.41	15655.83	6.93	6.11	13.4	82.9
15	7.28	1.71	4022.37	9146.74	5.46	5.54	1.46	127
16	7.10	1.80	6960.80	8540.98	6.43	5.49	17.1	22.7
17	10.7	0.788	18965.81	27708.92	4.16	7.02	68.7	46.1
18	10.7	0.788	18965.81	27708.92	8.32	7.02	18.1	46.1
19	10.7	0.788	18965.81	27708.92	4.89	7.02	43.5	46.1
20	16.6	0.331	107793.38	104192.94	8.7	10.9	25.3	3.45
21	8.80	1.17	8561.41	15655.83	7.01	6.11	14.7	82.9
22	8.13	1.37	6910.81	12466.94	6.11	5.84	4.62	80.4

NE	d	$G \times 10^{-2}$	RE	RT	kE	kT	Error k (%)	Error R (%)
23	10.7	0.791	41777.52	27799.28	7.84	7.06	11.0	50.3
24	6.67	2.04	6260.75	7202.62	6.07	5.36	13.2	15.0
25	4.26	5.01	1044.17	2301.46	6.22	4.82	29.0	120
26	6.46	2.18	5663.12	6613.53	6.71	5.31	26.4	16.8
27	6.46	2.18	5663.12	6613.53	6.05	5.31	13.9	16.8
28	6.46	2.18	5663.12	6613.53	5.37	5.31	1.13	16.8
29	6.46	2.18	5663.12	6613.53	5.94	5.31	11.9	16.8
30	7.83	1.48	33599.24	11220.81	7.46	5.74	30.0	199
31	6.80	1.96	6478.07	7585.38	6.00	5.39	11.3	17.0
32	4.47	4.56	4519.70	2583.99	6.66	4.86	37.0	74.9
33	2.70	12.4	475.88	2180.73	4.97	6.52	31.2	358
34	3.85	6.14	1958.29	4879.03	6.21	6.46	4.02	149
35	7.42	1.65	8259.40	9655.91	6.15	5.60	9.82	16.9
36	7.83	1.48	33599.24	11220.81	6.49	5.74	13.1	199
37	5.22	3.33	2418.53	3794.61	6.37	4.99	27.6	56.9
38	6.87	1.92	20572.53	22011.51	10.4	7.13	46.0	6.99
39	6.87	1.92	20572.53	22011.51	10.8	7.13	51.9	6.99

EN means experiment name, and subscripts E and T indicate experimental and theoretical data. Cell eccentricity $\alpha_0 = 0.4$ is used.

Table 3.
Experimental measurements of wavenumbers [27] and present theoretical predictions.

Some numerical results agree very well with experiments, as can be seen in the experiments 4, 10, 12, 13, 20, and 35 of **Table 3**. Others are good, such as the results of experiments 9, 16, 24, 26, 27, 28, 29, and 31. With respect to the other data in **Table 3**, it might be possible that if the experimental measurements are improved, the agreement with theory will be better. The results given here show that the approximate and numerical solutions of the system of governing equations presented in this paper may bring a light to the solution of many other problems of bioconvection.

7. Conclusions

The governing equations of bioconvection were used to investigate the problem of an infinite horizontal microorganism suspension fluid layer. The theoretical predictions of the critical wavenumber k_c and Rayleigh number R_c were compared with their experimental counterparts [27]. Very good, good, and fair agreements were found. But in general, we may say that our numerical results improve by far those obtained by Bees and Hill [26].

With the asymptotic analysis for $k < < 1$, it was possible to calculate a Rayleigh number not reported before without any restrictions on the magnitudes of d and G. This result is important because it was also possible to calculate a critical value of the gyrotaxis parameter G_c which indicates the boundary between the possibility of a marginal curve with $k_c = 0$ $(G < G_c)$ and another one with $k_c > 0$ $(G > G_c)$.

However, it is clear from the experimental results that the critical wavenumbers are finite and large and that the former case is not physical. Therefore, this G_c also defines the limit of validity of the theory. Note that it agrees very well with numerical analysis.

An analytic Galerkin method was also used to obtain a general expression of R without any restriction on the magnitudes of d, G, and $k \sim O(1)$. This gave us an explicit expression of R not reported before which proved to be very useful when checking with the numerical computations.

Numerical results have shown that the system becomes more unstable when the layers are shallow. The physical interpretation of such situation is that the accumulation of microorganisms near the top of the layer in the shallow case is faster than in the deeper case, due to the smaller depth of suspension H. A consequence of this is that the critical wavenumber is smaller for shallower layers. This can be explained by means of the boundary conditions of the microorganism concentration. If the parameter d tends to zero, the boundary conditions tend to those similar to the "fixed heat flux" boundary conditions of the problem of natural convection heated from below [50–53, 56]. Moreover, it has been shown above that by a change of variable, it is possible to transform the boundary conditions of the concentration into those similar to the "fixed heat flux" boundary conditions. In that problem it has been shown that the critical wavenumber tends to zero. However, due to the gyrotaxis, the critical wavenumber is not zero in the present problem if $G > G_c$, which, from the experimental results, is the case here. But notice in **Figures 1–3** that in fact, also in this case, the critical wavenumber decreases with a decrease of d. The change of the critical wavenumber with respect to G is also clear in the figures. The critical wavenumber decreases with a decrease of G.

Finally, we would like to point out that it is our hope that the results presented in this chapter may stimulate researchers to make more new and precise experiments on bioconvection in the near future.

Nomenclature

B	dimensional gyrotactic parameter, s
k	wavenumber
D_c	cell diffusivity, cm^2s^{-1}
d	motility of microorganisms
G	dimensionless gyrotactic parameter
g	acceleration due to gravity, cms^{-2}
H	layer depth, cm
$\mathbf{J}^*$	flux density of organisms
$\overline{n}$	average cell concentration
n^*	concentration of microorganisms
p	pressure
R	Rayleigh number
Sc	Schmidt number
t^*	time
V_c	cell swimming speed, cms^{-1}
$\mathbf{u}^*$	fluid velocity
$\mathbf{x}^*$	Cartesian coordinates
Greek symbols	
α_0	cell eccentricity
μ	viscosity, $gcm^{-1}s^{-1}$
ν	kinematic viscosity, cm^2s^{-1}
ρ	water density, gcm^{-3}
ϑ	cell volume, cm^{-3}
Subscripts	
BH	result obtained by Bees and Hill [26]
c	critical value
E	experimental result
T	theoretical result

Acknowledgements

The authors would like to thank Alberto López, Alejandro Pompa, Cain González, Raúl Reyes, Ma. Teresa Vázquez, and Oralia Jiménez for technical support. I. Pérez Reyes would like to thank the Programa de Mejoramiento del Profesorado (PROMEP).

Author details

Ildebrando Pérez-Reyes[1*] and Luis Antonio Dávalos-Orozco[2]

1 Facultad de Ciencias Químicas, Universidad Autónoma de Chihuahua, Chihuahua, Mexico

2 Departamento de Polímeros, Instituto de Investigaciones en Materiales, Universidad Nacional Autónoma de México, México, D.F., México

*Address all correspondence to: ildebrando3@gmail.com

References

[1] Noever DA, Matsos HC. A Biosensor for Cadmium Based on Bioconvective Patterns. Nasa tm–103523, NASA; 1990

[2] Itoh A, Toida H, Saotome Y. Control of bioconvection and its application for mechanical system: 2nd report, pitch optimization of electrodes array and driving of mechanical system by controlled bioconvection. Transactions of the Japan Society of Mechanical Engineers. Part B. 2006;**72**(716):972-978

[3] Kawaguchi T, Itoh A, Toyoda Y. Control of bioconvection and its mechanical application: 4th report: Generation of upward flow and its position control. Transactions of the Japan Society of Mechanical Engineers. Part B. 2002;**2002**(14):317-318

[4] Itoh A, Toida H, Saotome Y. Control of bioconvection and its application for mechanical system: 1st report basic effects of electrical field on bioconvection. Transactions of the Japan Society of Mechanical Engineers. Part B. 2001;**67**(662):2449-2454

[5] Itoh A, Toida H. Control of bioconvection and its mechanical application. In: 2001 IEEE/ASME International Conference on Advanced Intelligent Mechatronics. Proceedings; 2001. pp. 1220-1225

[6] Itoh A. Motion control of protozoa for bio mems. IEEE/ASME Transactions on Mechatronics. 2000;**5**(2):181-188

[7] Itoh A, Tamura W. Object manipulation by a formation-controlled Euglena group. In: Bio-mechanisms of Swimming and Flying. Kato Bunmeisha, Japan: Springer Japan; 2008. pp. 41-52

[8] Itoh A, Tamura W, Mishima T. Motion control of Euglena group by weak laser scanning system and object manipulation using Euglena group. In: 2005 IEEE/ASME International Conference on Advanced Intelligent Mechatronics. Proceedings. 2005. pp. 43-47

[9] Kim D, Brigandi S, Kim P, Kim MJ. Control of tetrahymena pyriformis as a microrobot. In: Microbiorobotics. Oxford, UK: Elsevier; 2012. pp. 277-287

[10] Kim DH, Brigandi S, Julius AA, Kim MJ. Real-time feedback control using artificial magnetotaxis with rapidly-exploring random tree (RRT) for *Tetrahymena pyriformis* as a microbiorobot. In: 2011 IEEE Int. Conf. Robotics Autom. IEEE; May 2011

[11] Ceylan H, Giltinan J, Kozielski K, Sitti M. Mobile microrobots for bioengineering applications. Lab on a Chip. 2017;**17**(10):1705-1724

[12] Magariyama Y, Kudo S, Goto T, Takano Y. An engineering perspective on swimming bacteria:high-speed flagellar motor, intelligent flagellar filaments, and skillful swimming in viscous environments. In: Bio-mechanisms of Swimming and Flying. Japan: Springer; 2004. pp. 1-12

[13] Itoh A. Euglena motion control by local illumination. In: Bio-mechanisms of Swimming and Flying. Japan: Springer; 2004. pp. 13-26

[14] Platt JR. Bioconvection patterns in cultures of free-swimming organisms. Science. 1961;**133**:1766-1767

[15] Pedley TJ, Kessler JO. Hydrodynamic phenomena in suspensions of swimming microorganisms. Annual Review of Fluid Mechanics. 1992;**24**:313-358

[16] Hill NA, Pedley TJ. Bioconvection. Fluid Dynamics Research. 2005;**37**:1-20

[17] Murase M. Dynamics of Cellular Motility. West Sussex, England: John Wiley and Sons; 1992

[18] Häder R, Hemmersbach D-P, Lebert M. Gravity and the Behavior of Unicellular Organisms. New York, USA: Cambridge University Press; 2005

[19] Childress M, Levandowsky S, Spiegel EA. Pattern formation in a suspension of swimming microorganisms: Equations and stability theory. Journal of Fluid Mechanics. 1975;**63**:591-613

[20] Harashima A, Watanabe M, Fujishiro I. Evolution of bioconvection patterns in a culture of motile flagellates. Physics of Fluids. 1988;**31**(4): 764-775

[21] Dávalos-Orozco LA, del Castillo LF. Hydrodynamic behavior of suspensions of polar particles. In: Encyclopedia of Surface and Colloid Science. New York, USA: Taylor and Francis; 2006. pp. 2798-2820

[22] Dávalo-Orozco LA. Natural convection in a polar suspension with internal rotation. Revista Mexicana de Física. 2002;**48**(S1):31-37

[23] Pedley TJ, Hill NA, Kessler JO. The growth of bioconvection patterns in a uniform suspension of gyrotactic micro–organisms. Journal of Fluid Mechanics. 1988;**195**:223-237

[24] Hill NA, Pedley TJ, Kessler JO. Growth of bioconvection patterns in a suspension of gyrotactic micro-organisms in a layer of finite depth. Journal of Fluid Mechanics. 1989;**208**: 509-543

[25] Pedley TJ, Kessler JO. A new continuum model for suspensions of gyrotactic micro-organisms. Journal of Fluid Mechanics. 1990;**212**(1):155-182

[26] Bees MA, Hill NA. Linear bioconvection in a suspension of randomly swimming, gyrotactic micro-organisms. Physics of Fluids. 1998; **10**(8):1864-1881

[27] Bees MA, Hill NA. Wavelength of bioconvection patterns. The Journal of Experimental Biology. 1997;**200**: 1515-1526

[28] Loeffer JB, Mefferd RB. Concerning pattern formation by free-swimming microorganisms. The American Naturalist. 1952;**86**:325-329

[29] Fornshell JA. An experimental investigation of bioconvection in three species of microorganisms. The Journal of Protozoology. 1978;**25**(1):125-133

[30] Kessler JO. Co-operative and concentrative phenomena of swimming microorganisms. Contemporary Physics. 1985;**26**(2):147-166

[31] Dombrowski C, Cisneros L, Chatkaew S, Goldstein RE, Kessler JO. Self-concentration and large-scale coherence in bacterial dynamics. Physical Review Letters. 2004;**93**(9): 2277-2282

[32] Tuval I, Cisneros L, Dombrowski C, Wolgemuth CW, Kessler JO, Goldstein RE. Bacterial swimming and oxygen transport near contact lines. Proceedings of the National Academy of Sciences of the United States of America. 2005;**102**:098103

[33] Akiyama A, Ookida A, Mogami Y, Baba SA. Spontaneous alteration of the pattern formation in the bioconvection of Chlamydomonas reinhardtii. Journal of the Japan Society of Microgravity Application. 2005;**22**(4):210-215

[34] Ishikawa T, Simmonds MP, Pedley TJ. Hydrodynamic interaction of two swimming model micro-organisms. Journal of Fluid Mechanics. 2006;**568**: 119-160

[35] Ishikawa T, Pedley TJ. Diffusion of swimming model micro-organisms in a semi-dilute suspension. Journal of Fluid Mechanics. 2007;**588**:437-462

[36] Ishikawa T, Pedley TJ. The rheology of a semi-dilute suspension of swimming model micro-organisms. Journal of Fluid Mechanics. 2007;**588**:399-435

[37] Ishikawa T, Locsei JT, Pedley TJ. Development of coherent structures in concentrated suspensions of swimming model micro-organisms. Journal of Fluid Mechanics. 2008;**615**:401-431

[38] Ishikawa T, Yamaguchi T, Pedley TJ. Bio-Mechanisms of Swimming and Flying, chapter Properties of a semi-dilute suspension of swimming microorganisms, pages 17–28. Springer Japan, Kato Bunmeisha, Japan, 2008

[39] Kitsunezaki S, Komori R, Harumoto T. Bioconvection and front formation of *Paramecium tetraurelia*. Physical Review E. 2007;**76**:046301

[40] Mogami Y, Baba SA. Amplified expression of the gravity effect on the spatio-temporal formation of bioconvection pattern. In: Proceedings of the Twenty-Fourth Space Utilization Symposium. Vol. 24; 2008. pp. 264-266

[41] Mogami Y, Ookida A, Baba SA. Bioconvection as a research model of gravitational biology. In: Proceedings of the Twenty-First Space Utilization Symposium. Vol. 21; January 2005. pp. 213-215

[42] Mogami Y, Yamane A, Gino A, Baba SA. Bioconvective pattern formation of tetrahymena under altered gravity. The Journal of Experimental Biology. 2004; **207**:3349-3359

[43] Sawai S, Mogami Y, Baba SA. Cell proliferation of *Paramecium tetraurelia* on a slow rotating clinostat. Advances in Space Research. 2007;**39**:1166-1170

[44] Mogami Y, Akiyama A, Baba SA. Comparative physiology of bioconvection. In: Proceedings of the Twenty-Second Space Utilization Symposium. Vol. 22; January 2006. pp. 222-223

[45] Takeda A, Mogami Y, Baba SA. Gravikinesis in Paramecium: Approach from the analysis on the swimming behavior of single cells. Biological Sciences in Space. 2006;**20**(2):44-47

[46] Mogami Y, Ishii J, Baba SA. Theoretical and experimental dissection of gravity-dependent mechanical orientation in gravitactic microorganisms. The Biological Bulletin. 2001;**201**:26-33

[47] Itoh A, Amagai K, Arai M, Mifune H. The effect of rotational gravitational field on bioconvection pattern. Transactions of the Japan Society of Mechanical Engineers. Series B. 1999; **65**(630):698-705

[48] Ghorai S, Hill NA. Development and stability of gyrotactic plumes in bioconvection. Journal of Fluid Mechanics. 1999;**400**:1-31

[49] Ghorai S, Hill NA. Periodic arrays of gyrotactic plumes in bioconvection. Physics of Fluids. 2000;**12**(1):5-22

[50] Pismen LM. Inertial effects in long-scale thermal convection. Physics Letters A. 1986;**116**(5):241-244

[51] Chapman CJ, Proctor MRE. Nonlinear Rayleigh–Bénard convection between poorly conducting boundaries. Journal of Fluid Mechanics. 1980; **101**(4):759-782

[52] Dávalos-Orozco LA. Magnetoconvection in a rotating fluid between walls of very low thermal conductivity. Journal of the Physical Society of Japan. 1984;**53**(7):2173-2176

[53] Dávalos-Orozco LA, Manero O. Thermoconvective instability of a second-order fluid. Journal of the Physical Society of Japan. 1986;**55**(2): 442-445

[54] Chandrasekhar S. Hydrodynamic and Hydromagnetic Stability. New

York, USA: Dover Publications, Inc.; 1981

[55] Pedley TJ, Kessler JO. The orientation of spheroidal microorganisms swimming in a fluid flow field. Proceedings of the Royal Society of London. Series B. 1987;**231**(1262): 47-70

[56] Gertsberg VL, Sivashinsky GI. Large cells in nonlinear Rayleigh–Bénard convection. Progress in Theoretical Physics. 1981;**66**(4):1219-1229

[57] Nayfeh AH. Introduction to Perturbation Techniques. New York, USA: John Wiley and Sons, Inc.; 1981

[58] Pellew A, Soutwell RV. On maintained convective motion in a fluid heated from below. Proceedings of the Royal Society of London. Series A. 1940; **176**(966):313-343

[59] Gershuni GZ, Zhukovitskii EM. Convective Instability of Incompressible Fluids. Jerusalem, Israel: Keter Publications; 1976

[60] Finlayson AB. The Method of Weighted Residuals and Variational Principles. Mathematics in Science and Engineering. New York, USA: Academic Press; 1972

[61] Press WH, Teukolsky SA, Vetteling WT, Flannery BP. Numerical Recipes. The Art of Scientific Computing. 3rd ed. Cambridge, USA: Cambridge University Press; 2007

[62] Hill NA, Häder D-P. A biased random walk model for the trajectories of swimming micro-organisms. Journal of Theoretical Biology. 1997;**186**:503-526

[63] Polin M, Tuval I, Drescher K, Gollub JP, Goldstein RE. Chlamydomonas swims with two gears in a eukaryotic version of run-and-tumble locomotion. Science. 2009;**325**(24):487-490

[64] Kessler JO. Individual and collective fluid dynamics of swimming cells. Journal of Fluid Mechanics. 1986;**173**: 191-205

[65] Jones MS, Le Baron L, Pedley TJ. Biflagellate gyrotaxis in a shear flow. Journal of Fluid Mechanics. 1994;**281**: 137-158

Section 8

Theoretical Approaches for Heat Transfer Problems

Chapter 19

Feedback Control of Rayleigh Convection in Viscoelastic Maxwell Fluids

Ildebrando Pérez-Reyes, Cornelio Alvarez-Herrera and Jonathan Rodríguez-Campos

Abstract

Control of Rayleigh convection in a viscoelastic Maxwell fluid is addressed here by considering a feedback from shadowgraphic visualizations. Here, a theoretical approach is made to the problem of the onset of convective motion through a source term in the lower thermal boundary condition. A numerical Galerkin technique is then used to study the linear hydrodynamic stability. Small relaxation times are considered for Prandtl numbers 1 and 10. Interesting results for the Rayleigh, the wavenumber, and the frequency of oscillations are presented along with discussion on the physical mechanism. In short, the linear hydrodynamic stability analysis states that suppression of convection may be favored.

Keywords: viscoelastic fluid, feedback control, Rayleigh convection

1. Introduction

The idea of control in physics and engineering, and other areas of applications, is not new. Several examples and interesting advances can be found in chemical processes where multivariable problems may appear in the way to obtain desired product quality or to guarantee safety of operators [1, 2]. At the same time, technology has been developed to meet the needs of the industry such as proportional integral derivative controllers, sensors, etc. On the other hand, new technology and lab developments needing control of variables may use different hardware. Onset of thermal convection and thermoconvective pattern formation are examples of phenomena aimed to be controlled.

Control of Rayleigh convection is a subject that has called the attention of researchers because of its connection to applications in electronics [3] and genetics [4], for example. In these cases understanding and modulation of convective motions are important so that the problem of Rayleigh convection may be coupled to a feedback control scheme. It is straight that heat transfer along the fluid layer should be monitored and controlled in order to modulate the convective motions.

The present manuscript is devoted to give an approach to the control of the Rayleigh convection phenomena in a viscoelastic fluid layer by a feedback scheme based on shadowgraphic visualization. This is a theoretical approach to the problem which actually introduces a modification in the lower thermal boundary conditions to modulate the heat transfer so that convection may be changed [5]. Since

IntechOpen

convective motions are undoubtedly connected to the hydrodynamic stability of the fluid layer, this theoretical framework is combined with the idea of a control feedback provided by shadowgraph technique measurements. The results presented in this work contribute to further developments in applications involving viscoelastic fluids.

Several investigations have been conducted in previous decades by authors interested in the selection of convective patterns or in avoiding convective motions, for example. However, to the best knowledge of the authors, perhaps the works of Singer et al. [6] and Singer and Bau [7] triggered interesting studies on the control of thermal convection. Singer et al. [6, 7] worked on the laminarization of chaotic convective motions by using a feedback control signal based on temperature measurements both theoretically and experimentally. Petrov et al. [8] reported results on the nonlinear control of convective motion in a liquid bridge which is based on temperature measurements that feed a nonlinear control algorithm controlling a thermoelectric element. Other works were devoted to the suppression of Rayleigh convection as motivated by the results (Singer et al.) [6, 7]. Later, Howle reported interesting results for the control of Rayleigh convection in Newtonian fluids both experimentally [9] and theoretically [5, 10]. In these works, Howle [5, 9, 10] has studied the coupling of linear hydrodynamic stability in a Newtonian fluid with the problem of controlling the supplied heat transfer to the fluid by using the familiar shadowgraph visualization technique. Howle [9] used an experimental arrangement for the visualization of convective motions to get information useful in the feedback of a set of heaters installed at the bottom of the lower boundary.

The mentioned investigations along with the possible improvements to applications [3, 4] have motivated the present study related to viscoelastic fluids. The main idea is to extend the results of Howle [5, 9, 10] about the control of thermal convection for viscoelastic Maxwell fluids. The thermal hydrodynamic stability of a viscoelastic fluid layer is investigated, while heat flux at the lower boundary is modulated by a feedback signal obtained from shadowgraphic visualization. Then, results of a numerical analysis are presented for the linear hydrodynamic stability.

The manuscript has been organized as follows. A brief introduction to the Rayleigh convection in viscoelastic fluids is given in Section 1. In Section 3 the governing equations of the problem in hand are presented along with an explanation of the physical nature of the system. In Section 3.1 the corresponding boundary conditions are introduced. Next, the linear hydrodynamic stability is shown in Section 4. General comments based on experimental tests on the shadowgraph visualization of the Rayleigh convection are presented in Section 5. Section 6 is devoted to expose results and discussion of the findings. Finally, the main conclusions are given in Section 7.

2. On the Rayleigh convection in viscoelastic fluids

The problem of Rayleigh convection in viscoelastic fluids has been a subject of interest for researches [11, 12] since several decades ago. The early works on this matter came up from the concern on the usage of viscoelastic fluid models by taking advantage of already known results and theory about Rayleigh convection in Newtonian fluids. In this way, the problem of Rayleigh convection in viscoelastic fluids can be represented as shown in **Figure 1**.

Although there are several models for the representation of viscoelastic fluids a few have been widely considered for thermal convection. These are the Maxwell

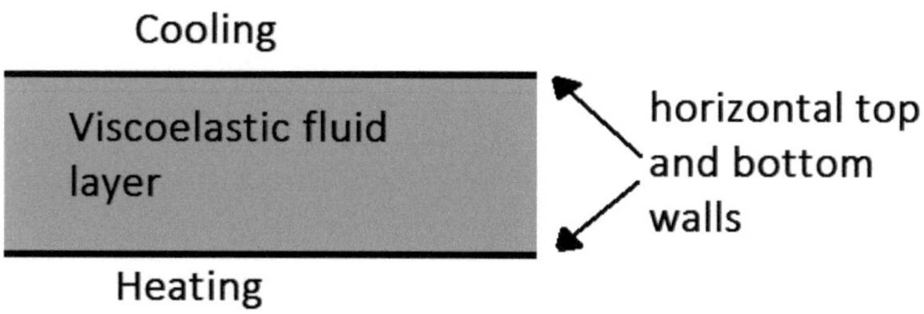

Figure 1.
The problem of Rayleigh convection in a viscoelastic fluid layer. As the critical conditions for the Rayleigh convection are achieved, convective fluid motions set in.

and the Jeffreys viscoelastic models which have linear and nonlinear constitutive equation representations. The following linear constitutive equation,

$$\left(1+F\frac{\partial}{\partial t}\right)\tau=2\eta_0\left(1+EF\frac{\partial}{\partial t}\right)\mathbf{e}$$

where τ is the stress tensor and $\mathbf{e}$ is the share rate tensor [13], corresponds to the Jeffreys viscoelastic fluid model and reduces to the Maxwell viscoelastic fluid model when $E=0$. These two viscoelastic fluid models predict time-dependent thermal convection also known as oscillatory convection [12, 15], which is of interest due to changes in the instability of the fluid. In other words, convective motions in the fluid may set in at lower critical conditions [12, 15] than for the Newtonian case. Finally, the mentioned findings represent a start point for developments on convective motion control.

3. Mathematical formulation

Here, the onset of thermal convection in a horizontal infinite viscoelastic Maxwell fluid layer heated from below is considered. The schematic of the system is shown in **Figure 2** where the optical setup along with proper data processing could give insight on the modulation of the heat supplied at the lower wall. This is in fact a feedback control scheme.

From the mathematical point of view, the linear hydrodynamic stability problem can be represented by the equation for the balance of momentum, the continuity

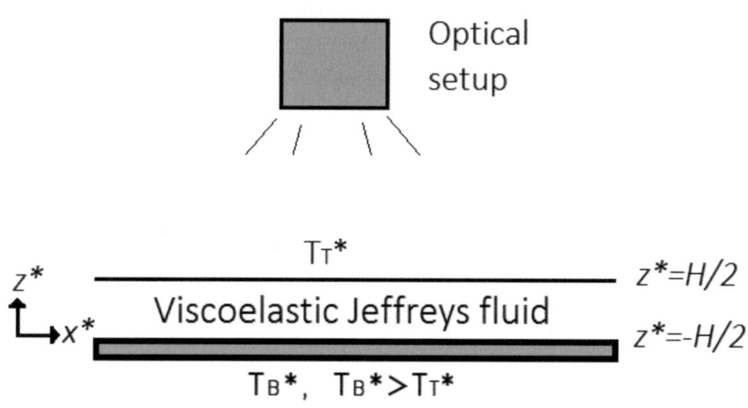

Figure 2.
*General schematics of the problem. The box for optical setup represents the setup for fluid visualization by the shadowgraph technique. Both walls are rigid but made of different materials since the one in the top should be a transparent media. * indicates dimensional variables.*

equation, the heat diffusion equation, and the constitutive equation for the viscoelastic Maxwell fluid. For short some familiar steps are avoided (see the publications of [14–16] for details). Thus, the governing equations are

$$(1+F\sigma)\left[\frac{\sigma}{Pr}\left(\frac{d^2}{dz^2}-k^2\right)W-Rk^2\Theta\right]=\left(\frac{d^2}{dz^2}-k^2\right)^2W \quad (1)$$

$$\left[\sigma-\left(\frac{d^2}{dz^2}-k^2\right)W\right]\Theta=W \quad (2)$$

where Eqs. (1) and (2) were obtained after a process of perturbation and nondimensionalization of the governing equations. The nondimensionalization of the variables was made with H for lengths, κ/H for velocity, $\kappa\mu/H^2$ for pressure, ΔT^* for temperature, and H^2/κ for time. Also, in Eqs. (1) and (2) $Pr=\nu/\kappa$, $F=\lambda\kappa/H^2$, and $R=\beta g H^3\Delta T^*/\nu\kappa$. Notice that pressure and velocity fields were separated by operating twice the rotational operator on the momentum balance equation. Also, the constitutive equation for the viscoelastic Maxwell fluid has been combined with the momentum balance equation. Next, normal modes $\exp\left[i(k_x x+k_y y)+\sigma t\right]$ were introduced so that the problem in hand could be reduced to a system of ordinary differential equations as shown in Eqs. (1) and (2). Then, Eq. (1) corresponds to the balance of momentum, and Eq. (2) corresponds to heat.

3.1 Boundary conditions

Eqs. (1) and (2) should be subjected to a proper set of boundary conditions. For the system in hand, the boundaries of the fluid layer are both rigid and solid walls so that the following mechanical conditions shall be used:

$$W=\frac{dW}{dz}=0 \quad \text{at} \quad z=-\frac{1}{2},\frac{1}{2}$$

Thermal conditions are adapted from those used in the study in the Rayleigh convection of a fluid layer bounded by a good thermal conducting wall at the top and by an insulating wall at the bottom [17]. This configuration of the thermal boundary conditions is preferred since control of convection can be made, in an experimental setup, by changing the heat flux at the bottom. Thus, the ideas of [10] are embraced here. For short, adjustments made in the supplied heat flux can be introduced in the model as an inhomogeneity in the bottom thermal boundary condition through the controller gain or magnitude of the made adjustments.

The boundary conditions for θ are

$$\Theta=0 \quad \text{at} \quad z=\frac{1}{2} \quad (4)$$

$$\frac{d\Theta}{dz}=\gamma k^2\int_{-1/2}^{1/2}\Theta dz \quad \text{at} \quad z=-\frac{1}{2} \quad (5)$$

4. Linear stability analysis

Eqs. (1)–(5) represent an eigenvalue problem for the Rayleigh number. The solution to this problem is then made by the Galerkin method which allows the calculation of the eigenvalue without completely solving for W and Θ. Then, this technique is numerically implemented to study the hydrodynamic stability of the viscoelastic fluid layer as follows (see the book of [18] or the monography of [19], for example).

First, a trial function for W, satisfying boundary condition (Eq. (3)), is proposed. This is

$$W_n = \sum_{n=1}^{N} (2z-1)^n (2z+1)^n \tag{6}$$

Next, W_n is substituted into the equation for Θ. The result of that substitution allows to solve for Θ from Eq. (2) subject to the corresponding boundary conditions (Eqs. (4–5)). The previous step gives the advantage of carrying more information on the solution of the problem and improving accuracy. Therefore, back in with Eq. (1) for W, the residual is calculated multiplying that equation by W_n as given in Eq. (6) and then integrating across the fluid layer. This orthogonalization process outputs a solvability condition from which R can be obtained. This is

$$\left| (1+F\sigma)\frac{\sigma}{\Pr}\left\langle W_n\left(\frac{d^2}{dz^2}-k^2\right)W_m\right\rangle - \left\langle W_n\left(\frac{d^2}{dz^2}-k^2\right)^2 W_m\right\rangle - (1+F\sigma)Rk^2\langle W_n\Theta_m\rangle \right| = 0 \tag{7}$$

where the angle brackets indicate the integral from $z=-1/2$ to $z=1/2$. Eq. (7) is a matrix of size (N, M) with $N=M=1$, the first approximation. As the size of the matrix is increased, the accuracy is improved and so is the complexity of the calculations. The condition given in Eq. (7) demands that the determinant of that matrix should be zero. From the resulting mathematical expression, the Rayleigh number R can be determined.

4.1 Numerical analysis

The hydrodynamic stability of the viscoelastic layer as modified by a varying heat flux supply at the bottom is studied numerically. First, a comparison between the results obtained from condition (Eq. (7)) and those reported by previous authors was carried out showing very good agreement. The critical Rayleigh, wavenumber, and frequency of oscillation were obtained by first fixing γ and ω and then minimizing the Rayleigh number with respect to the wavenumber. Critical R_c and k_c give critical ω_c. The process was repeated for a range of values of γ, and since Eq. (7) is valid for different viscoelastic fluids, the properties E, F, and Pr were mapped for some representative cases.

The validation of the present results was made by comparison of the results obtained from Eq. (7) with those reported by previous authors [10, 17]. Besides, these comparisons help to establish the order of approximation to be used in Eq. (7). For the case of Rayleigh convection in a Newtonian fluid layer the agreement with the results of reference [17], when the lower boundary is kept at constant heat flux and the top boundary is kept at constant temperature, is very good. In the present work, the critical Rayleigh and wavenumber are 1303.44 and 2.56, respectively, with a maximum error of 3.7%, for this case.

For the same conditions used by [10], from the condition (Eq. (7)), the critical Rayleigh and wavenumber are 3976.59 and 3.97, respectively, for $\gamma = 100$. The maximum error for this case is 2%. Here, the results of [10] were slightly extended. Curves of criticality for the Rayleigh and wavenumber are shown in **Figures 4–5** for a wide range of γ. These curves show that the Rayleigh number of very strong changes can be obtained at magnitudes of γ between 0 and 10. These are important results and have triggered further investigation in viscoelastic Maxwell fluids.

Comparison of the data for the hydrodynamic stability of viscoelastic Maxwell fluids was made with the results reported by [20]. In their work, Sekar and Jayalatha [20] made calculations for the hydrodynamic stability of a viscoelastic fluid heated from below subjected to different mechanical and thermal boundary conditions. Very good agreement for the corresponding case was obtained. However, an unexpected result is that the critical Rayleigh number is larger for the case of bottom rigid insulation wall and top rigid isothermal wall than for the case of two rigid isothermal walls [20]. Despite the large amount of results, published by [20], the physical mechanism of the mentioned finding is not explained. This change in the hydrodynamics of the fluid can be attributed solely to the viscoelastic nature of the fluid since the Newtonian fluid does not show this behavior.

For the numerical computations of this work, third-order, $n = 3$, approximations were used since a very good convergence was found at this order. Investigated higher-order approximations only give improvements smaller than 10^{-2}. Perhaps, the very good convergence was due to the contributions given by the solution of the heat equation instead of using trial functions for Θ.

The effect of heat flux modulation at the bottom wall on the hydrodynamic stability for the viscoelastic fluid was investigated as follows. Viscoelastic Maxwell fluids with $F = 0.1$ were investigated, while the Prandtl number was fixed at 1 and 10. Then, the critical Rayleigh number, the wavenumber, and the frequency of oscillation were determined.

5. An experimental possibility

In this section, some general comments are given based on preliminary results on an experimental work made on this problem. In view of the previous theoretical analysis to the problem of controlling the thermal convection in a viscoelastic Maxwell fluid layer, an experimental setup may be proposed. Since a feedback control strategy has been considered for this purpose, the scheme shown in **Figure 2** can be extended to include other features. **Figure 3** shows a very general experimental setup where some optomechanical and electromechanical parts are missing, but it still shows the experimental counterpart of the theoretical approximation.

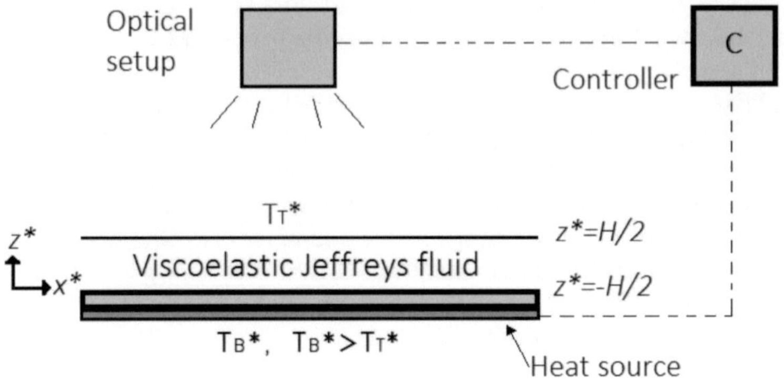

Figure 3.
Extension to the Rayleigh convection visualization schematics shown in ***Figure*** *2. Dashed lines indicate communication between the given hardware.*

Two possible goals may be set for the experiment: detection of the beginning of the Rayleigh convection and tuning for favoring the formation of a given convective pattern. For the present case, some experiments have been performed in an open loop for the setup shown in **Figure 3** where the detection of the beginning of fluid motion is visualized in the screen of laptop computer attached to a shadowgraph optical setup and further adjustments are manually introduced in a PID temperature controller.

Although convective motion was detected on the range of experimental conditions reported by previous authors, more work is needed to have a closed loop. In other words, the PID temperature controller should be changed by a PID shadowgraph controller able to feedback the heat supplied by the source (see **Figure 3**). This idea is also based on that early mentioned by Howle [5] who used a proportional control algorithm based on shadowgraph images of the experiment.

On the light of the experimental test, some comments can be made on the connection between theoretical predictions and experimental data. First, the controller output is related to a precise amount of Watts used by the heat source so that dynamical behavior is expected and has been found with the open loop tests, and as a consequence, the value of the parameter γ would change in time as well. Second, a thermal fluid may be used to supply the needed heat to the layer instead of an arrangement of heaters [5] which could give a delayed change in the Rayleigh number but more uniform behavior in the fluid layer horizontal extent.

6. Results and discussion

Very interesting results were found. **Figures 5a–5c** and **6a–6c** present the main findings on the linear hydrodynamic stability of the Maxwell viscoelastic fluid layer. Weak viscoelastic fluids, with F = 0.1, are investigated for understanding of the role played by the controller gain γ.

For the case of Pr = 1, the critical Rayleigh, the wave number, and the frequency of oscillation show unexpected behavior since at small values of γ the magnitude of

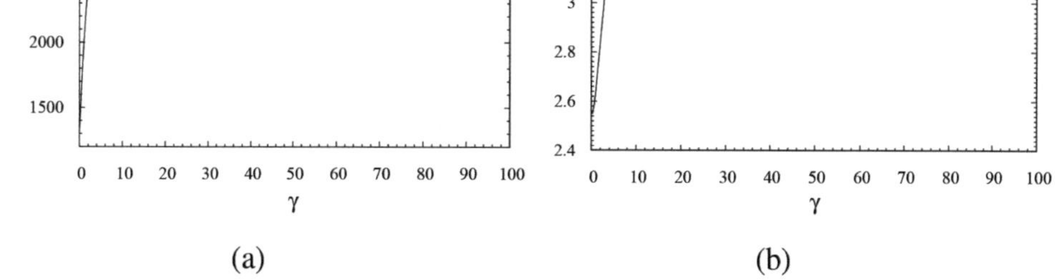

Figure 4.
Curve of criticality for the Rayleigh convection in a Newtonian fluid corresponding to R_c against γ (a) and to k_c against γ. This curve of criticality extends the results of [10]. At γ = 100, R_c = 3976.59, and k_c = 3.97.

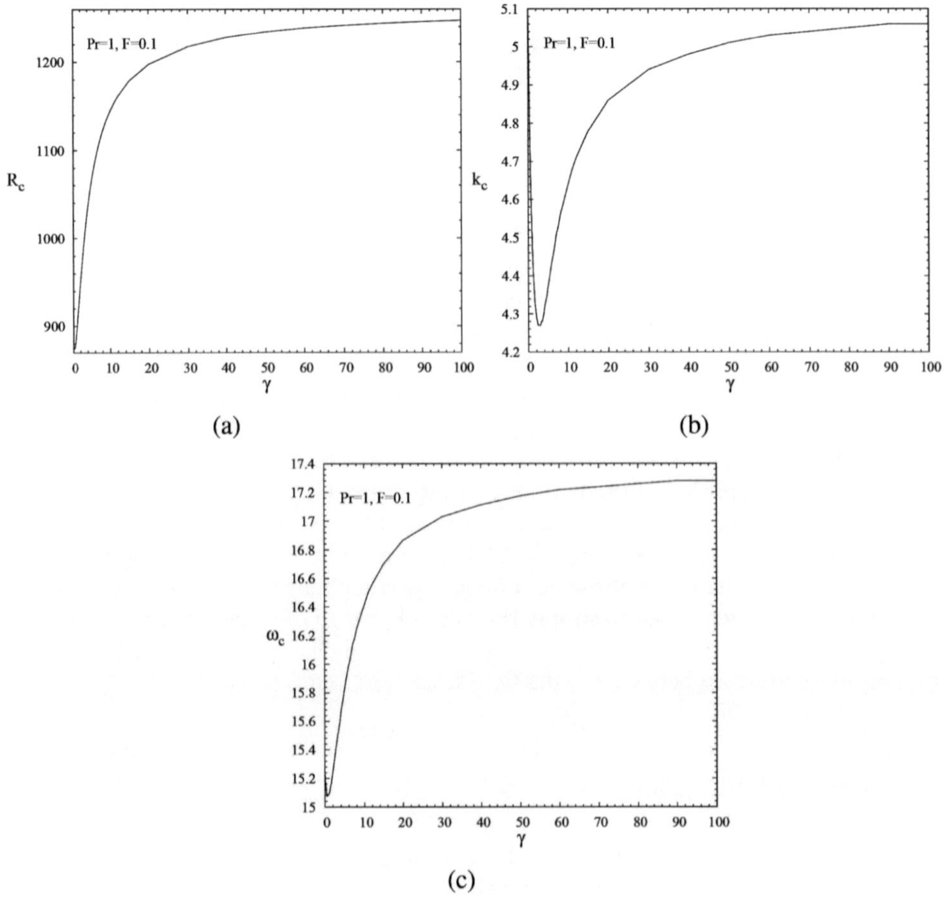

Figure 5.
Curve of criticality for the Rayleigh convection in a Maxwell viscoelastic fluid with Pr = 1 and F = 0.1. These curves correspond to R_c, k_c, and ω_c against γ.

R_c, k_c, and ω_c decreases. However, at certain values of γ, the same parameters start growing monotonically. On the other hand, for the case of Pr = 10, only the critical Rayleigh number behaves as for Pr = 1. The critical wavenumber and frequency of oscillation decrease monotonically with γ for Pr = 10.

The results on the hydrodynamics are unexpected and can be attributed to a coupling of the viscoelastic property *F* and to the nonzero heat flux bottom boundary condition. For the two values of the Prandtl number investigated, the fluid layer always stabilizes after certain critical value of γ. From the comparison with the curves for the Newtonian case, it can be said that fluid viscoelasticity triggers stronger nonlinear behavior of R_c, k_c, and ω_c (**Figures 4–6**).

The physical interpretation of the present results is as follows. Increasing γ means that temperature at the bottom is increased too. As the heat flux is increased, viscoelasticity helps to destabilize the system. At the same time, there must be a limit for the effect of small γ since the Rayleigh number *R* depends on the temperature difference which cannot be indefinitely increased. If temperature is increased with no limit, along with larger values of γ, the thermal energy should be released or converted into fluid motions, for example. Then, the oscillations in the fluid would help to diffuse the heat very quickly, while the layer becomes more stable with γ. This behavior is found in both systems (**Figures 5** and **6**).

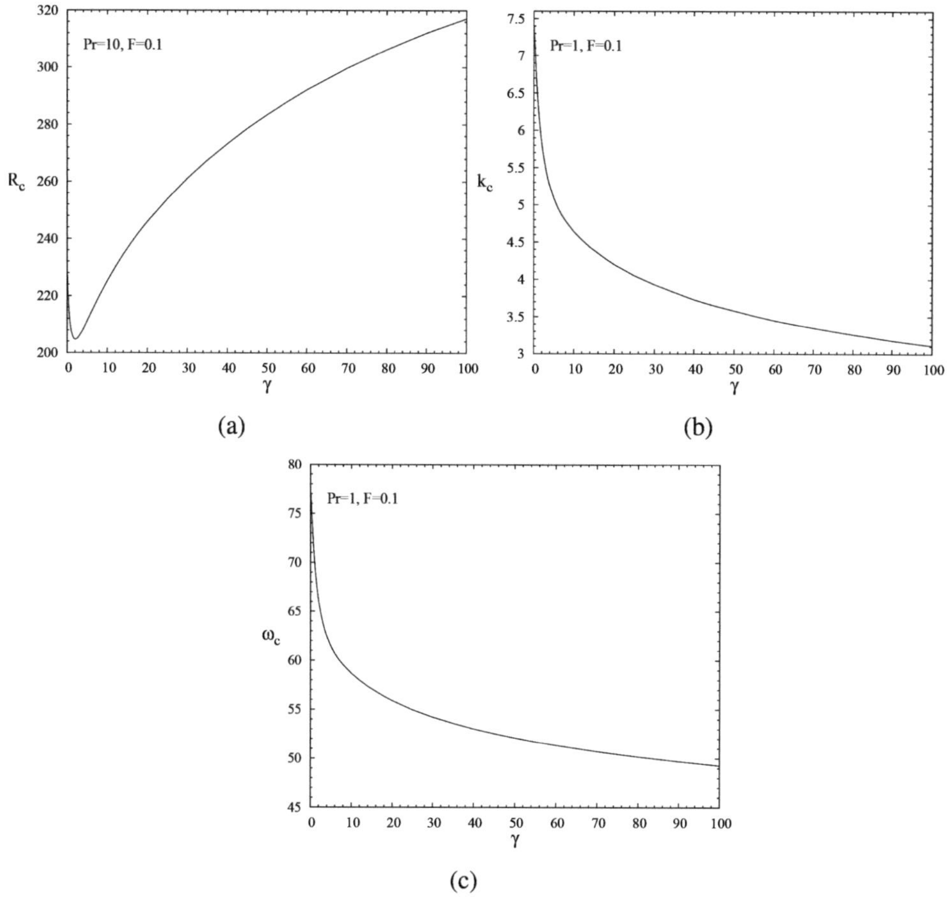

Figure 6.
Curve of criticality for the Rayleigh convection in a Maxwell viscoelastic fluid with Pr=10 and F=0.1. These curves correspond to R_c, k_c, and ω_c against γ.

7. Conclusions

In the present work, the effect of controller gain in the linear hydrodynamic stability of a viscoelastic Maxwell fluid was studied.

The main conclusion of this work is that convection in the fluid layer can be controlled, or at least it can be suppressed. This is a direct conclusion since the curves of criticality state that the hydrodynamic stability of the fluid layer is increased with γ. The coupling of the stability parameters gives unexpected behaviors at small γ, but to the best knowledge of the authors, it could happen experimentally.

Acknowledgements

I. Pérez-Reyes and J. Rodríguez-Campos would like to thank the financial support from CONACyT through the project Ciencia Básica-255839. Numerical computations were performed using Maple™.

Nomenclature

Greek letters	
F	dimensionless relaxation time
g	acceleration due to gravity
H	fluid layer depth
k	perturbation wavenumber
Pr	Prandtl number
R	Rayleigh number
T_B^*	bottom wall temperature
T_T^*	top wall temperature
W	vertical velocity perturbation
β	thermal expansion coefficient
γ	controller gain
Θ	temperature perturbation
κ	thermal diffusivity
λ	stress relaxation time
ν	fluid kinematic viscosity
μ	fluid dynamic viscosity
σ	complex parameter
σ_R	perturbation growth rate
ω	frequency of oscillation

Author details

Ildebrando Pérez-Reyes*, Cornelio Alvarez-Herrera
and Jonathan Rodríguez-Campos
Facultad de Ciencias Químicas, Universidad Autónoma de Chihuahua, Chihuahua, Chih., Mexico

*Address all correspondence to: ildebrando3@gmail.com

References

[1] Anderson NA. Instrumentation for Process Measurement and Control. 3rd ed. Taylor & Francis; 1997

[2] Luyben ML, Luyben WL. Essentials of process control. In: McGraw-Hill Chemical Engineering Series, McGraw-Hill. 1997

[3] Li M, Xu S, Kumacheva E. Convection in polymeric fluids subjected to vertical temperature gradients. Macromolecules. 2000;**33**: 4972-4978

[4] Braun D. PCR by thermal convection. Modern Physics Letters B. 2004;**18**(16):775-784

[5] Howle LE. Active control of Rayleigh–Bénard convection. Physics of Fluids. 1997;**9**(7):1861-1863

[6] Singer J, Wang YZ, Bau HH. Controlling a chaotic system. Physical Review Letters. 1991;**66**(9):1123-1125

[7] Singer J, Bau HH. Active control of convection. Physics of Fluids A. 1991; **3**(12):2859-2865

[8] Petrov V, Schatz MF, Muehlner KA, VanHook SJ, McCormick WD, Swift JB, et al. Nonlinear control of remote unstable states in a liquid bridge convection experiment. Physical Review Letters. 1996;**77**(18):3779-3782

[9] Howle LE. Control of Rayleigh-Bénard convection in a small aspect ratio container. International Journal of Heat and Mass Transfer. 1997;**40**(4): 817-822

[10] Howle LE. Linear stability analysis of controlled Rayleigh-Bénard convection using shadowgraphic measurements. Physics of Fluids. 1997; **9**:3111

[11] Sokolov M, Tanner RI. Convective instability of a general viscoelastic fluid heated from below. Physics of Fluids. 1972;**15**(4):534-539

[12] Vest CM, Arpaci VS. Overstability of a viscoelastic fluid layer heated from below. Journal of Fluid Mechanics. 1969; **36**:613-623

[13] Bird RB, Hassager O. Dynamics of Polymeric Liquids: Fluid Mechanics. Dynamics of Polymeric Liquids. Wiley; 1987

[14] Pérez-Reyes I, Dávalos-Orozco LA. Effect of thermal conductivity and thickness of the walls on the natural convection in a horizontal viscoelastic Jeffreys fluid layer. Journal of Heat Transfer. 2018;**140**(12):122501

[15] Pérez-Reyes I, Dávalos-Orozco LA. Effect of thermal conductivity and thickness of the walls in the convection of a viscoelastic Maxwell fluid layer. International Journal of Heat and Mass Transfer. 2011;**54**:5020-5029

[16] Takashima M. Thermal instability in a viscoelastic fluid layer. I. Journal of the Physical Society of Japan. 1972;**33**(2): 511-518

[17] Sparrow EM, Goldstein RJ, Jonsson VK. Thermal instability in a horizontal fluid layer: Effect of boundary conditions and non-linear temperature profile. Journal of Fluid Mechanics. 1964;**18**(04):513

[18] Finlayson AB. The method of weighted residuals and variational principles. In: Mathematics in Science and Engineering. New York, New York: Acedemic Press; 1972. p. 10003

[19] Chandrasekhar S. Hydrodynamic and Hydromagnetic Stability. New

York, New York: Dover Publications, Inc.; 1981. p. 10014

[20] Sekhar GN, Jayalatha G. Elastic effects on Rayleigh-Bénard convection in liquids with temperature-dependent viscosity. International Journal of Thermal Sciences. 2010;**49**(1):67-75